Know your food

Food ethics and innovation

AF771259

Know your food

Food ethics and innovation

EurSafe 2015
Cluj-Napoca, Romania
28-30 May 2015

edited by:
Diana Elena Dumitras
Ionel Mugurel Jitea
Stef Aerts

Wageningen Academic
Publishers

EAN: 9789086862641
e-EAN: 9789086868131
ISBN: 978-90-8686-264-1
e-ISBN: 978-90-8686-813-1
DOI: 10.3921/978-90-8686-813-1

Cover photo: Courtesy of Food-Stills
Christian Schuster, Duesseldorf, Germany
http://www.food-stills.de

First published, 2015

© Wageningen Academic Publishers
The Netherlands, 2015

This work is subject to copyright. All rights are reserved, whether the whole or part of the material is concerned. Nothing from this publication may be translated, reproduced, stored in a computerised system or published in any form or in any manner, including electronic, mechanical, reprographic or photographic, without prior written permission from the publisher, Wageningen Academic Publishers, P.O. Box 220, 6700 AE Wageningen, the Netherlands, www.WageningenAcademic.com copyright@WageningenAcademic.com

The individual contributions in this publication and any liabilities arising from them remain the responsibility of the authors.

The publisher is not responsible for possible damages, which could be a result of content derived from this publication.

Acknowledgements

The editors and the members of the organizing committee wish to express their gratitude towards the reviewers for their time, effort and guidance. Their help reviewing all papers in a relatively short timescale and their valuable contribution to ensure that a high standard is maintained are highly appreciated. A special thank you is addressed to the EurSafe board members who have shared their experience and offered their support in planning the conference. Special acknowledgements are addressed to the University of Agricultural Sciences and Veterinary Medicine Cluj-Napoca, Romania for offering their support and assistance in planning the 12th EurSafe Congress.

EurSafe 2015 Committees

Organising Committee

University of Agricultural Sciences and Veterinary Medicine Cluj-Napoca, Romania

Carmen Socaciu
Felix Arion
Diana Elena Dumitras
Rezhen Harun
Ionel Mugurel Jitea
Valentin Mihai
Iulia Muresan
Cristina Pocol
Alexandru Simu
Ramona Suharoschi
Dan Vodnar

Scientific Committee

Coordinator

Dumitras, Diana Elena, Associate Professor; Department of Economic Sciences, University of Agricultural Sciences and Veterinary Medicine Cluj-Napoca, Romania

Members (alphabetically ordered)

Aerts, Stef, Programme Director; Department of Agricultural and Biotechnological Sciences, Odisee University College, Belgium

Anjum, Rani Lill, Research Fellow; School of Economics and Business, Norwegian University of Life Sciences, Norway

Arion, Felix, Professor and Director; Department of Economic Sciences, University of Agricultural Sciences and Veterinary Medicine Cluj-Napoca, Romania

Bovenkerk, Bernice, Assistant Professor; Philosophy Group, Wageningen University, the Netherlands

Escajedo San Epifanio, Leire, Professor; Department of Constitutional Law and History of Political Thought University of the Basque Country – EHU, Spain

Gjerris, Mickey, Associate Professor; Department of Resource and Food Economics, University of Copenhagen, Denmark

Grimm, Herwig, Professor; Messerli Research Institute, Veterinary University Vienna, Medical University Vienna and University Vienna, Austria

Hagen, Kristin, Senior Researcher; EA European Academy of Technology and Innovation Assessment, Swedish University of Agricultural Sciences, Germany

Howard, Dawn, Chief Executive Officer, National Office of Animal Health, United Kingdom

Jitea, Ionel Mugurel, Associate Professor; Department of Economic Sciences, University of Agricultural Sciences and Veterinary Medicine Cluj-Napoca, Romania

Kaiser, Matthias, Professor; Centre for the Study of the Sciences and Humanities, University of Bergen, Norway

Lassen, Jesper, Professor; Department of Food and Resource Economics, University of Copenhagen, Denmark

Meijboom, Franck L.B., Associate Professor; Ethics Institute & Faculty of Veterinary Medicine, Utrecht University, the Netherlands

Meisch, Simon, Head of Junior Research Group 'Ethics of Science in the Research for Sustainable Development'; International Centre for Ethics in the Sciences and Humanities, University of Tübingen, Germany

Millar, Kate, Director, Centre for Applied Bioethics; Schools of Biosciences and Veterinary Medicine and Science, University of Nottingham, United Kingdom

Muresan, Iulia, Lecturer; Department of Economic Sciences, University of Agricultural Sciences and Veterinary Medicine Cluj-Napoca, Romania

Neeteson, Anne-Marie, Global Vice President Welfare and Compliance; Aviagen Group, United Kingdom

Olsson, Anna, Researcher; Laboratory Animal Science, IBMC-Instituto de Biologia Molecular e Celular, Universidade do Porto, Portugal

Pocol, Cristina Bianca, Associate Professor; Department of Economic Sciences, University of Agricultural Sciences and Veterinary Medicine Cluj-Napoca, Romania

Potthast, Thomas, Director and Professor; International Centre for Ethics in the Science and Humanities, University of Tübingen, Germany

Röcklinsberg, Helena, Associate Professor in ethics; Department of Animal Environment and Health, Swedish University of Agricultural Sciences, Sweden

Sandin, Per, Senior Lecturer in Bioethics and Environmental Ethics; Department of Crop Production Ecology, Swedish University of Agricultural Sciences, Sweden

Simu, Alexandru Vasile, Lecturer; Department of Economic Sciences, University of Agricultural Sciences and Veterinary Medicine Cluj-Napoca, Romania

Suharoschi, Ramona, Associate Professor; Department of Food Science, University of Agricultural Sciences and Veterinary Medicine Cluj-Napoca, Romania

Thompson, Paul B., Professor; Department of Philosophy, Michigan State University, USA

Van der Sluijs, Jeroen, Associate Professor, Centre for the Study of the Sciences and the Humanities, University of Bergen, Norway

Vodnar, Dan, Lecturer; Department of Food Science, University of Agricultural Sciences and Veterinary Medicine Cluj-Napoca, Romania

Voget-Kleschin, Lieske, Interim Professor for Environmental Ethics; Institute of Botany and Landscape Ecology, Greifswald University, Germany

Table of contents

Section 9. Consumer and citizen concerns

Section 10. Food cultures in comparison

Section 11. Food for the future

Section 12. Limits of food knowledge

Section 13. Poster session

Preface

Journalists typically do not want long-winded answers, they want short and crisp formulations, phrases that have the potential to result in a headline. The President of EurSafe and the Members of the Society are accustomed to being interviewed by journalists on some salient issues of food ethics, with questions like what makes food 'ethical' and so on. As soon as one goes into any detail about issues, things can become very complex and opaque. Often the journalists then come up with one crucial final question: what can I as a consumer do to promote ethical food? Luckily we have a very simple and rewarding answer to this: 'Know your food!' Just ask a number of simple questions about your food: What is this? Where does it come from? When was it harvested/slaughtered? Is it processed, and if so, how? How was it stored? Is it certified by any of the usual certifiers? Has it been approved as an ethical/fair trade/ sustainable/organic/animal welfare product, etc.? Do we know if it is healthy, or is it perhaps unhealthy? Many of our supermarkets, and even most of our restaurants would have a problem answering all these questions. Yet if we want ethical food, food that is good in more ways than just taste, we really should pressure individuals and organisations for good answers. No single answer, not even very detailed answers, will provide us with all we need to make the final judgement on whether the food is ethical or not. But without our questions we may never get 'ethics' in our food stuff. This may not be much to answer the journalist's query, but it is a necessary first step, a 'sine qua non!'

The food in our European countries is to a large extent the result of a long, and often complex and global value chain. Many actors come in between the producer and the consumer. Some people see this fact alone as a sign of diminished quality, they believe that proximity to the 'natural' source will assure the highest quality, and thus they erroneously believe the farmer and the fishermen enjoy the best and most healthy food. Fresh and natural food is regarded as a guarantee of optimal quality, in harmony with nature. Studies show this was not always true and for all we know, this was never true.

The food historian Rachel Laudan calls this the view of the Culinary Luddites[1], those who will always be defenders of a romanticist world-view and believers in a fall from grace, a lost paradise. But according to Laudan they have history 'back to front'. In order to both achieve a better taste and to reduce toxin levels our forebears ground, bred, baked, curdled, marinated, fermented and cooked their food stuff. It was usually the privilege of the rich urban classes to enjoy the safe and well-tasting processed food, the rural people who produced the food lived on more problematic diets. Thus it is basically an achievement of the 20th century to have industrialised the global food production to such an extent that in principle sufficient, good and healthy food could be available to and affordable by everyone. The fact that as many as a billion people are still chronically under-nourished relativizes recent achievements to a significant extent, and makes world hunger an ethical scandal.

Food markets are changing rapidly in Europe these days, and this is due to both regular food scares, and to the shifting tastes and fashions of the European middle class. While it is admittedly a market with various specialist pockets and niches, it is also an expression of a common search for a new food culture. We are no longer content with our regional specialities, we look for the seductive spices and exotic tastes from faraway places, the various culinary feasts of our globe. The processing of food, the trade of various food stuffs, and the industrialized ready-made or half-prepared meals are an essential ingredient of a life-style that many of us have become dependent on. Yes, there is a place for fresh products, for organic food, for locally produced food, for the kitchen garden, for 'slow food' and the traditional dishes of our ancestors, but all in all it looks like these will only add colour to a background which by sheer necessity

[1] cf.: Rachel Laudan (2001), 'A plea for culinary modernism: Why we should love new, fast, processed food', Jan 1, Gastronomica: The Journal of Food and Culture.

will continue to be dominated by industrialized food production of some kind. Given climate change, population growth, and the health issues related to our current diet, we will have to continue to innovate and look for improved crops and more effective agri- and aquaculture. Thus the complexity of the market will not be reduced, but rather is likely to increase. And this is why we have to learn to ask the intelligent questions: do you know you food?

We need to become conscious food citizens, and – as in other areas of our democracies – we have to find our way through the value diversity of the food sector. We have to make choices, and these choices need to be intelligent choices; else we risk not only to damage our own health, or the health and sustainability of our planet, or to show a lack of respect for living beings, but also to lose the positive value, in terms of human life quality, of our food cultures. So a food ethic needs to guide us on our journey to a better and more sustainable food culture, and the beginning of our travel guide consists of questions we all need to ask when we want to know our food.

This conference is the gathering of engaged scholars who want to help clarify some of these pressing issues. On many occasions we find that science does not hold all the answers, that there is considerable uncertainty about some of the crucial questions. But this is the trademark of ethics: it calls for our judgement and considered reflexive opinions. In the 'European Society for Agricultural and Food Ethics' (EurSafe) we have defined a forum to raise these questions, look for answers together, discuss pros and cons of various positions, and direct our future research interests. We need the trans- and inter-disciplinary frameworks of understanding, which this Society has now fostered for more than 15 years.

This forum meets in Cluj-Napoca this year, a city with significant cultural and scientific heritage in Romania, situated in the east of Europe, a venue which is glad to host the EurSafe conference in 2015, in the same year as the city is crowned as 'The Youth Capital of Europe'. The younger generation is obviously the most relevant addressee of the conference slogan: 'Know your food!' The enthusiasm of the young people, especially our students that exceed 100,000 in number, and their 'hunger' for new perspectives on food, will undoubtedly be a positive inspiration for the scholarly exchanges during the conference. 'Unity in Diversity', the official motto of EU, is also a very accurate description of our intention to bring together various scholars from different disciplines, united under an ethical umbrella, highlighting the culturally variable but also unifying role of values in our perceptions of food, and discussing low- and high-tech modes of food-production, based on a multitude of experiences, opinions and facts gathered all over the world.

EurSafe, both the Board and our Members, alongside the Organizers of the conference have no doubt that the peer-reviewed quality of the written contributions gathered in this volume will be matched by the participants' personal experiences in the scholarly discussions and friendly social environment of this conference.

Matthias Kaiser, President of EurSafe
Felix Arion, USAMV Cluj-Napoca

Keynote contributions

The role of Origin Consortia in fostering market access

N. Machado Rios, F. Russo and A. Siles*
United Nations Industrial Development Organisation (UNIDO), Clusters and Business Linkages Unit, Vienna International Centre. Wagramer Str 5m, 1400, Vienna, Austria; n.machado-rios@unido.org; f.russo@unido.org

Abstract

Since 2010, United Nations Industrial Development Organization (UNIDO) has been working in projects linking products and its origin through the Origin Consortia projects. An Origin Consortium is a tool for fostering market access of agrofood and handcrafts value chains in developing countries. These projects have been taking place in countries as Ecuador, Peru, Tunisia and Morocco. This article presents the UNIDO approach for Origin Consortia and partial results from the impact evaluation held in Peru among the producers of the Collective Mark for native potatoes Sumaq Sonqo.

Keywords: origin linked products, rural development, Sumaq Sonqo, Peru

Origin Consortia

One of the challenges faced nowadays is to preserve traditional knowledge and culture, but at the same time adjust to meet global market demands and combat poverty (Aylwin and Combe, 2014). Food production and consumption have a direct impact in Economic performance, livelihoods, territory dynamics, and environmental issues. Geographical indications and Collective Marks can be a tool for enhancing rural development and achieving those objectives (REDD, 2013).

The use of Origin Linked Products (OLP) is increasing in importance as a marketing strategy for branding and labelling products in a global market, adding value to products and establishing a reputation and reliability among consumers, both in developed and developing countries (Sautier and Van der Kop, 2006). This represents an opportunity on job creation and increased access for small and medium enterprises (SMEs) in markets, considering that the growth of smallholders' farms is a leading point for reducing rural poverty (Memedovic and Shepherd, 2009).

Origin Consortia are a tool for enhancing livelihoods and promoting rural development by the valorisation of traditional products of regional origin. Based on a system of cooperation and share of a collective reputation, SMEs in the agro-industrial sector can benefit of this new trend of consumers preference for authentic, typical and genuine products. Origin Consortia are developed by a jointly strategy of: (1) setting product specifications and assisting members to meet the commonly agreed requirements; (2) registering and promoting the group's collective brand; and (3) developing and implementing a joint territorial marketing strategy (UNIDO, 2010).

The main goal is to create a collective value-adding strategy increasing market access, but also contributing with other positive side effects from the collective organization as collective inputs purchase, share of costs, among others. An Origin Consortium may be established using different juridical forms, as associations, consortiums, or even a cooperative. Choosing the most adequate label to be registered is often part of the methodology of establishing an Origin Consortia. More than obtaining a geographical indication, a traditional product of regional origin should use the most appropriate legal card to increase market share. An Origin Consortium may be established using different juridical forms, as associations, consortiums, or even a cooperative.

The collective brand for an Origin-Linked Product (OLP) can be a Geographical Indication (GI), a Collective Mark (CM), a Certification Mark or another quality label developed and adapted for each specific situation. Choosing the most adequate label to be registered is often part of the methodology of establishing an Origin Consortia.

The Collective Marks Alliance in Peru

One of the first initiatives regarding the execution of Origin Consortia was the project held in Peru, 'Industrial clusters and consortia: promoting typical products to fight poverty in Peru'. Financed by the Italian Cooperation and implemented within a partnership with the Peruvian government represented by the National Institute for the Defence of Competition and Intellectual Property (INDECOPI) and Agro Rural (Program for Agro-Rural Productive Development of the Ministry of Agriculture and Irrigation), with a main objective of fostering rural development through the valorisation of typical food products in 8 of the poorest regions of Peru (UNIDO, 2013).

Peru owns an important agrobiodiversity patrimony, and farmers and peasant families are the ones conserving breeds and developing diversity through cultural ancestral practices (Ruiz, 2009).

Within the project activities, 4 products were registered as a Collective Mark in INDECOPI and other 5 products are in process of register (Table 1). The geographical indications 'Maíz Blanco Gigante Cusco' (Giant white corn from Cusco) and 'Huadquiña Machu Picchu Coffee' (Cusco) also received some technical assistance. Moreover, 350 professionals from public and private sectors were trained for the promotion and register of Collective Marks in Abancay, Lambayeque, Huancayo, Arequipa and Lima.

The Collective Marks Alliance continues to operate in the country after the end of the program, putting together different governmental institutions to support producers into the register of these marks.

The Sumaq Sonqo native potato

The highest concentration of different species of potatoes (*Solanum* sp.) is located in the Andean region. There are 8 different species domesticated and cultivated, but the main commercial species is *Solanum lycopersicum*, which has Peru as its centre of origin and diversification (INIA, 2009). Around 2000 varieties of potatoes have been recognized in the whole country, and in the Huancavelica region, 202 potato varieties are found.

COOPAPA (Cooperativas de Servicios Agrarios de Producción Agrícola 'Sumaq Sonqo') is a cooperative founded in 2012, assembling 500 producers of 24 associations of Huancavelica region in the province of Acobamba. COOPAPA producers come from 4 different districts: Acobamba, Rosario, Andabamba and Paucará. This region is known for its mountainous relief, and difficult road access, and also the highest poverty level in the country, 77.2%.

Table 1. Collective Marks developed by the project.

Collective Mark	Type of product	City/region	Status
Tallarín de Casa Abanquino	Special kind of spaghetti	Apurímac	Registered
Sumaq Sonqo	Native potato	Huancavelica	
Utku Muchik	Native cotton handcrafts	Lambayeque	
La Reina Olmosa	Honey	Lambayeque	

The Sumaq Sonqo potatoes are produced in a zone located in altitudes from 3,500 to 4,200 meters. Currently, 7 varieties of native potato are authorized in the Code of Practice of CM: *Amarilla Runtus, Peruanita, Huamantanga Morada, Queqorani, Camotillo, Huayro Rojo* and *Chacro*. These native varieties have been chosen because of their adaptation to local conditions (resistance to pests and other diseases, extreme weather conditions as low temperatures, frost and drought), durability (post-harvest dehydration resistance) and also because they possess the characteristics accepted and required by consumers.

Sumaq Sonqo means 'beautiful heart' in quechua language. The logo has been designed during the creation of the Collective Mark, and is composed by a potato using the typical hat from the 'Chopqa' Peruvian ethnicity. Even though this product is not a GI, the logo is combined with the commercial slogan 'Huancavelica, cradle of the native potatoes' in allusion to its geographical origin.

Sumaq Sonqo study case

During the first semester of 2014, an evaluation has been conducted among the producers of Sumaq Sonqo in a partnership with the field experts of CARE Peru. Structured interviews have been conducted among 72 members of COOPAPA, from the 250 authorized to use the Collective mark. The other producers can be admitted in the CM when the standards established in the Code of Practice are reached. A diachronic analysis has been made comparing the time before intervention (2010) and after intervention (2014).

One of the risks taken when developing an OLP marketing strategy is that the premiums will not be paid, or that that only large business captures the benefits, and the reward value paid on these products might not accrue to small producers (Bramley *et al.*, 2009) or the additional production costs can affect final profitability (Bramley, 2011).

Concerning these possible socioeconomic risks, the results from the evaluation have demonstrated positive impacts from the creation of the mark,

In 2013 Sumaq Sonqo potatoes had a higher price per Ton ($\bar{x}$=1,152; σ=105) than potatoes from other producers ($\bar{x}$=807; σ=64). In 2010, the price was almost the same for the two categories of producers ($\bar{x}$=732; σ=60 for producers outside of the CM and $\bar{x}$=737; σ=75 for Sumaq Sonqo producers). Important to stress that the CM was not still created in 2010 (Figure 1).

The increase on price and on market demand has also led to an increased need of labour. This better offer of jobs has decreased migration of rural workers to other regions of Peru outside of harvest period, as the agricultural practices established on the Code of Practice ask for more labour during other steps of production. Figure 2 shows the share of work in the native potato production.

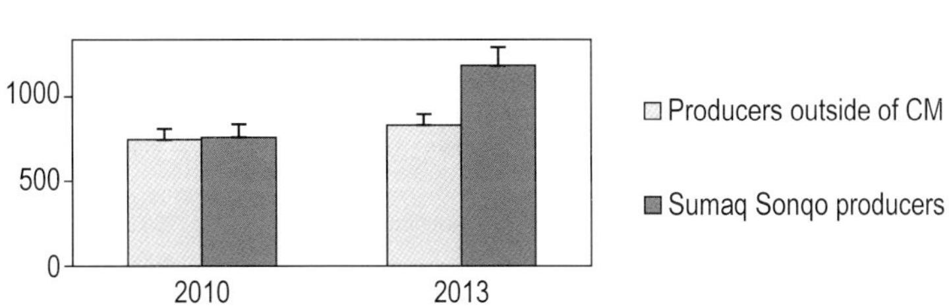

Figure 1. Prices paid to producers in Soles per ton of potato.

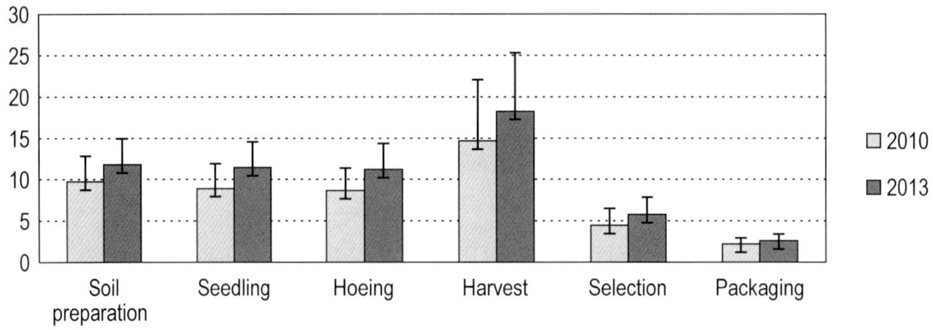

Figure 2. Average number of work days per farm in the Sumaq Sonqo production.

Salaries, paid on a daily basis, have also increased for both men and women were paid less than the national minimum wage, and now they have reached the basis salary (Table 2). Men are still better paid than women, probably because of the less availability of labour, or for the kind of activities executed, or still for the performance. One hypothesis is that men can harvest bigger quantities or work in bigger areas in a day than women. These qualitative issues, unfortunately, have not been investigated. Positions for women have also risen because of the selection that must be done before packaging.

Conclusions

Origin Consortium can be a very effective tool for fighting poverty and promoting rural development in developing countries by its strategy of fostering market access in local and global markets. The collective mechanism where an Origin Consortium is based brings benefits not only from the shared promotion and reputation of a selected product, but also for the associative activity related to that, from the share of costs, certification scheme, technicians, costs of transportation, among other benefits.

The use of voluntary standards should be evaluated case by case and answer consumers' demand. Geographical Indication, Certification Marks, and simple strategies as Collectives Marks can be used for promoting rural territories and enhance market access to producers that have long been excluded from the national economies. The choice of the most suitable one will vary depending on the type of product, market target, cultural aspects, and different strategies must be considered and evaluated. Innovation and creativity, including adapted strategies to producers and consumers' realities is the core of the success for developing resilient market channels. More than obtaining a geographical indication, an OLP should use the most appropriate legal card to increase market share. A Collective Mark can be a first step for organizing producing, and maybe leading producers for the later register of a GI.

Table 2. Salaries paid to rural labours in the Sumaq Sonqo production.

	Men	Women	Official minimum wage
2010	S/20 (deflated=22.03)	S/15 (deflated=16.52)	S/20 (deflated=22.03)
2013	S/30	S/25	S/25
Increase	36%	51%	13%

The potato produced by Sumaq Sonqo is now accessible for consumers in the biggest supermarkets of Lima. Without their structuration within a collective mark, middleman would still be buying the whole of the production and sell it to the distribution channels in the cities. This direct link from producer in the mountains and consumers in the capital by the coast are a link that was made possible thanks to the collective organisation and use of a common mark.

References

Aylwin, N. and Coombe, R.K. (2014). Marks indicating conditions of origin in rights-based sustainable development, Davis Law Review 47: 75-786.

Bramley, C., Biénabe, E. and Kirsten, J. (2009). The economics of geographical indications: towards a conceptual framework for geographical indication research in developing countries. In: The economics of intellectual property, WIPO, pp. 109-141, January 2009. Available at: http://tinyurl.com/oje7phc.

Bramley, C. (2011). A review of the socio-economic impact of geographical indications: considerations for the developing world, Worldwide Symposium on Geographical Indications, Lima, June 22-24, 2011. WIPO. Available at: http://tinyurl.com/ntz4k8g.

INIA (Instituto Nacional de Innovaccion Agraria) (2009). Peru: segundo informe sobre el estado de los recursos fitogeneticos para la alimentacion y la agricultura. Available at: http://tinyurl.com/pc8c7rr.

Memedovic, O. and Shepherd, A. (2009). Agri-food value chains and poverty reduction: overview of main issues, trends and experiences. Research and statistics branch working paper, 12/2008. UNIDO. Vienna.

MINAM (2009). Incidencia de la pobreza total segun departamentos 2009. Available at: http://tinyurl.com/lxp5oet.

REDD (2013). Study on the potential for marketing agricultural products of the ACP countries using geographical indications and origin branding (ACP-AGGI): Executive summary. Lausanne. Switzerland. Available at: http://tinyurl.com/o5bjl7v.

Ruiz, M. (2009). Agrobiodiversity zones and the register of native crops in Peru: learning from ourselves. Peruvian Society for Environmental Law – SPDA & Biodiversity International, pp 87.

Sautier, D. and Van de Kop, P. (2006). Origin-based marketing: a rural development tool? Conclusions and agenda for action and research. In: Van de Kop, P., Sautier, D. and Gerz, A. (ed.). Origin-based products: lessons for pro-poor market development. Bulletin 372. Kit & CIRAD, pp. 17-20 and 89-96.

UNIDO (2010). Adding value to traditional products of regional origin: a guide to creating a quality consortium. Business, Investment and Technology Services Branch. Technical paper series. UNIDO. Vienna.

UNIDO (2013). Enhancing the contribution of the private sector to sustainable industrial development and poverty reduction. Available at: http://tinyurl.com/nkevkjn.

Development of personalised food for the nutrition of elderly consumers

S. Forstner and A. Rusu*
Biozoon food innovations GmbH, Fischkai 1, 27572 Bremerhaven, Germany; forstner@biozoon.de

Abstract

The rising number of elderly within Europe and other developed countries increases the impact of several age related impairments such as chewing and swallowing difficulties that can decline the quality of life of the affected persons enormously. In order to increase their vital energy but especially to bring back their personal dignity when it comes to daily meal times, the EU funded project Performance is facing a new and holistic approach. Performance is a project with an innovative nutritional concept for frail and elderly persons enabling them to eat their daily meals in a form they are familiar with. The project is considering, a nutritional concept that focuses besides the general problems of the frail and elderly (e.g. decreased sensory perception, mastication and swallowing difficulties, etc.), on their special requirements such as nutrient demand and allows the industrial manufacturing and supply of these personalised meals. The supplied personalised food is modified in its texture and nutrient content, while keeping its traditional shape. This provides the frail and elderly with meals of a consistency that they can eat safely while having good taste and appearance – which is not the case with mash-type food. Due to the way of manufacturing, frail and elderly people will retain or even gain more appetite which can be directly linked to increased food consumption. As a result an improvement of the nutrition and health status and most important quality of life is likely to be expected from Performance personalised food products.

Keywords: performance, 3D food printing, dysphagia, elderly

Introduction

A major socio-economic problem that developed countries face is the aging population due to declining fertility and increasing life expectancy. The percentage of people over 65 years of age was below 15% until the beginning of the millennium; however, this group is expected to rise to 26% by 2050 (United Nations, 2011). This change implicates a rising number of health problems including impairments such as chewing and swallowing (dysphagia) problems.

Dysphagia is a condition describing swallowing difficulties of which a rising number of persons and especially elderly are suffering from (Alagiakrishnan *et al.*, 2013; Kayser-Jones and Pengilly, 1999, 2014; Park *et al.*, 2013; Van der Maarel-Wierink *et al.*, 2014). Degenerative diseases such as stroke, dementia, Parkinson's disease, and Alzheimer's disease but also certain stages of cancer can lead to these symptoms, whereby the larynx fails to close properly during swallowing, so that in the worst case, food or liquid can end up in the lungs and could lead to aspiration or even death (Kocdor *et al.*, 2014).

Frail and older people who suffer from dysphagia but also chewing difficulties are often being served with unappealing 'porridge-like' food. Beside the already existing pain during chewing and swallowing, this type of food often implies declining self-esteem and love of life, loss of appetite, and thus as a critical nutritional consequence nutrient deficiency, dehydration or even malnutrition (undernutrition). The daily meal time is for most of the elderly an important social event, whereas physical but also psychological reasons due to unappealing food served, is strengthening a circle which is difficult for the affected person to break (Figure 1).

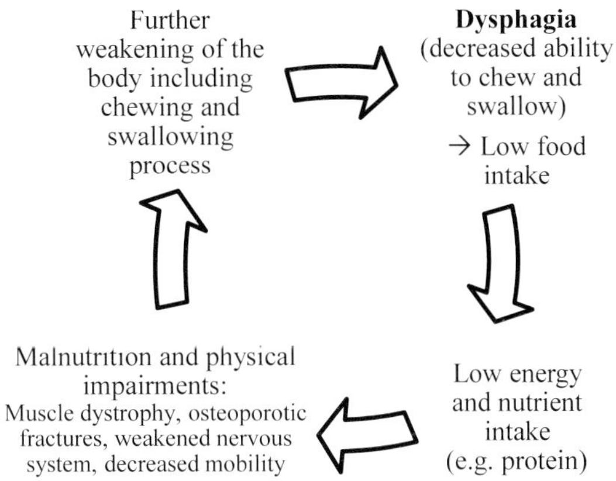

Figure 1. Circle of dysphagia patients inducing malnutrition.

In Germany, the smoothfood concept, is already widely applied in nursing homes and private care and is based on freshly handmade meals. Therefore, the components are prepared in the nursing home kitchens as usual, followed by the necessary pureeing step and the individual transformation into visual appealing food items. This allows appropriate and safe texture combined with palatable dishes and joy of eating.

Nonetheless, the ageing society and the correlated difficulty of malnutrition is not a problem which is limited to any borders but rather a global challenge. Aspects such as centralised kitchens, increased home care and especially the individual and personalised nutritional needs of elderlies complicate solutions like the smoothfood concept and therefore need to be addressed in Europe and other developed countries. Therefore, the Performance project, a European approach, is taking the development of a new concept to contribute towards a better life quality for the elderly population into consideration.

The Performance project

Concept

The main idea of the Performance project is to develop and validate a holistic, personalized food supply chain for frail and elderly in nursing homes, ambient assisted living facilities or at home (served by nursing or catering services). Performance is based on the smoothfood concept, but it aims at the simplification, personalisation and industrialisation of the visual appealing smoothfood meals in order to meet preferences and especially nutritional requirements of the individual person more specifically. The supply 'chain' in this case will resemble a loop with the elderly consumers forming the initial link (personal preference and needs) and the final link (consumption), demonstrated in Figure 2.

As a result of the Performance project, an overall concept will be available allowing automated manufacturing and supply of personalized, specially textured food for frail and elderly. The concept will cover the whole supply chain from the food producer using innovative 3D printing technology to the consumer's place, both in nursing facilities and at home. As nutrition is strongly correlated with aspects of health, physiological and psychosocial status, Performance is focusing not only on the personalised food processing innovations but especially in social innovation by improving quality of life for consumers. From ethical point of view, the serving of visual appealing meals should go as a matter

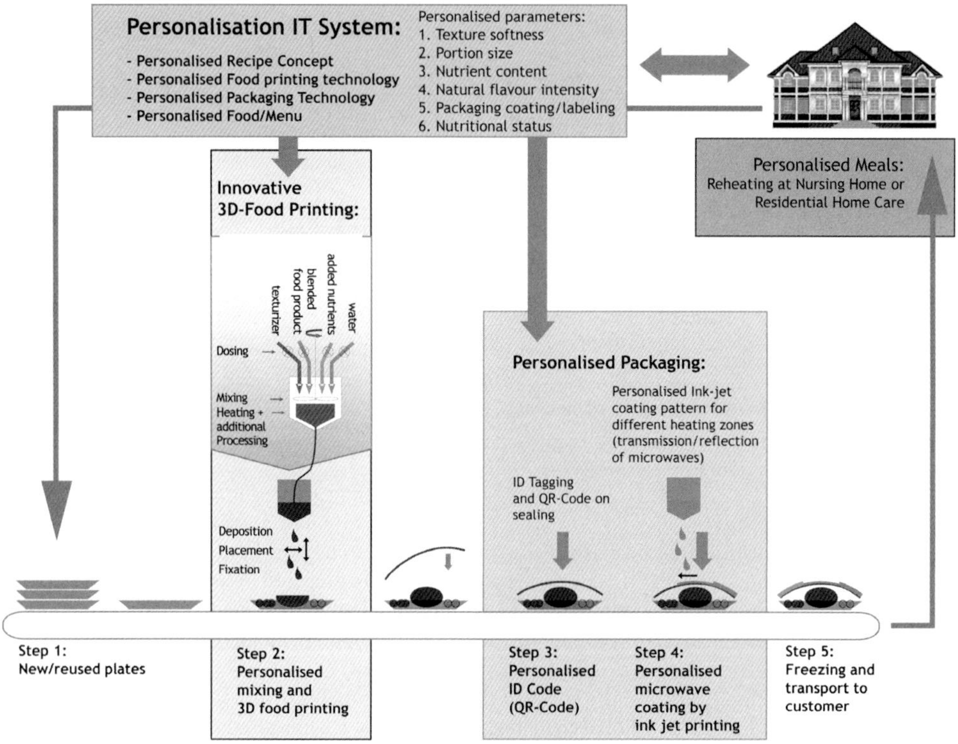

Figure 2. Performance personalised food supply chain (© 2012 www.eyecatcher-design.info).

of course not only in fancy restaurants, but also in nursing and care homes. Well-being, self-esteem and autonomy are of high importance in a person's life, inducing also a positive effect on the physical status.

Pre-results towards an improved life quality of elderly

One of the main challenges in the project is the interconnection of the printing material and the 3D printing technology in order to achieve stable and visually appealing food items. Through the excessive screening of various texturizing systems, this task was already achieved, forming at the same time the basis of the personalised Performance meals. The follow-up recipe development is realised in collaboration with the participating nursing homes to achieve the most palatable and visual appealing meals. 3D printing technology including correlated system expansion such as additional mixing devices were defined and optimised to operate in concordance with the printing material and its requirements.

For the scope of rapid manufacturing, a packaging design suitable for individual heating of the meals in a microwave oven was developed. Strong cooperation between project partners also enabled the pre-development of a specific software platform for the overall management of consumer requirements, food production and packaging processes and monitoring of real food consumption. The final validation of the prototype developments shall be performed through consumer tests.

Acknowledgement

This project received funding by the 7th Framework Programme of the EU, Collaborative Project. KBBE.2012.2.3-04 – Personalised approaches to food production and distribution.

References

Alagiakrishnan, H., Bhanji, R.A. and Kurian, M. (2013). Evaluation and management of oropharyngeal dysphagia in different types of dementia: a systematic review. Archives of Gerontology and Geriatrics 56:1-9.

Kayser Jones, J. and Pengilly, K. (1999). Dysphagia Among Nursing Home Residents. Geriatric Nursing 20:77-84.

Kocdor, P., Siegel, E.R., Giese, R. and Tulunay-Ugur, O.E. (2014). Characteristics of dysphagia in older patients evaluated at a tertiary center. The Laryngoscope 125(2):400-5. DOI: http://dx.doi.org/10.1002/lary.24917.

Park, Y.H., Han, H.R., Oh, B.M., Lee, J., Park, J.-a., Yu, S.J. and Chang, H. (2013). Prevalence and associated factors of dysphagia in nursing home residents, Geriatric Nursing 34: 212-17.

United Nations – Department of Economic and Social Affairs, Population Division (2011). World Population Prospects: The 2010 Revision.

Van der Maarel-Wierink, C.D., Meijers, J.M.M., De Visschere, L.M.J., de Baat, C., Halfens, R.J.G. and Schols, J.M.G.A. (2014). Subjective dysphagia in older care home residents: a cross-sectional, multi-centre point prevalence measurement, International Journal of Nursing Studies 51:875-81.

Section 1. Legal framework and good practices

'One animal is no animal' – consequences of measuring animal welfare at herd level

F. Lundmark[1], C. Berg[1], B. Wahlberg[2] and H. Röcklinsberg[1]*
[1]Department of Animal Environment and Health, Swedish University of Agricultural Sciences, Box 234, 53223 Skara, Sweden; [2]Department of Law, Åbo Akademi University, Fabriksgatan 2, 20500 Åbo, Finland; frida.lundmark@slu.se

Abstract

Animal welfare is by definition a characteristic of an individual animal. The aim of animal welfare legislation in Europe is usually to promote animal welfare and protect individual animals from suffering. However, several approaches have been made to create systems for measuring animal welfare at herd level. The aim of this study was to discuss how animals may be affected if a herd approach is used and how this relates to the current animal welfare legislation. Proportions of animals affected, sampling, thresholds and benchmarking are discussed. We conclude that systems based exclusively on group assessment would, if defined narrowly, not be in line with the intentions of a legislation that requires acceptable conditions and outcomes for every individual animal. However, the non-individual approach can be a valuable complement. As some standards use group measures it is important that both producers and consumers are aware of the limitations.

Keywords: animal welfare assessment, group level, individuals, intentions, legislation

Introduction

Animal welfare is by definition a characteristic of an individual animal (Butterworth, 2009). Animal welfare legislation usually has an individual approach to promote the welfare of animals and to protect them from suffering, for example the legislation in the UK (Radford, 2005), Sweden (Jordbruksverket, 2013), Finland (Wahlberg, 2011) and Norway (Stenevik and Mejdell, 2011). The legislation provides a minimum level of animal welfare, mainly through requirements aiming at preventing unnecessary suffering. This minimum level, which is to be achieved in a given country or other geographic region, does not aim at an ideal state of animal welfare. European consumers of meat or other animal-derived products are interested in animal welfare (European Commission, 2007) and hence this aspect of food ethics has attracted much attention. Furthermore, the European Court of Justice (ECJ) has interpreted that the objective and requirements of the EU animal welfare legislation cover all animals concerned, not only those who have suffered (C-455/06, C-491/06). In the control guidelines to the Swedish animal welfare legislation it is stated that 'Even if only one animal is affected by poor welfare this requires appropriate measures to be taken, since the animal welfare legislation is written with the perspective of an individual animal' (Jordbruksverket, 2013).

Traditionally, animal welfare legislation mainly focuses on resource and management input. Recently alternative approaches have been developed, focusing instead on the outcome state of the animal itself. As a result of the often large numbers of animals in modern farming it may, however, be a challenge both to stock-keepers, inspectors and auditors to control each and every individual animal. Hence, several attempts have been made to create systems for measuring animal welfare at herd level. This has developed into a specific research area and related conference events, such as the International Conference on the Assessment of Animal Welfare at farm and Group Level (WAFL).

The aim of this study was to discuss how an animal may be affected if a herd approach is used when animal welfare is assessed and measured, and how such an approach relates to the current legislation. Text analyses were made of the requirements and measurements in the Swedish animal welfare legislation and in three private standards in Sweden; Arlagården, Seal of Quality and KRAV (organic) concerning dairy cows. Also the assessment system developed within the Welfare Quality® project has been included in the discussion.

Different ways of measuring at herd level

There are different ways of performing control at herd level, some of which we will present here. The first is to consider the proportion of animals being affected by a certain condition. For example, according to Seal of Quality and Arlagården a case of non-compliance should normally not be reported unless a certain proportion of the animals are affected (for example a certain percentage of dirty or soiled animals). These standards require that every animal must be cared for by the stock-keeper and checked upon during a control visit, but still several animals have to be in poor condition before relevant action is actually requested. As expressed in the Seal of Quality standard; 'If only a few animals are affected, this should not be reported as a case of non-compliance' (Swedish Seal of Quality, 2012). The proportion of animals affected can also be used as a threshold. Seal of Quality, for example, used thresholds as a way of detecting systematic non-compliances at herd level. If mortality, dirtiness, wounds, poor general condition or lameness exceed certain levels then relevant measures must be taken and veterinary advice sought in order to solve the problems.

Benchmarking is a second way of using measures at herd level to compare the animal welfare level between farms. The Swedish industry-initiated programme 'Ask the Cow' provides a picture of a farm's animal welfare level compared to others with regard to body condition, hoof health, dirtiness, etc. Another industry-run system is the 'Animal Welfare Signals', which presents the results of mortality, diseases, etc. at a farm by comparing it to other farms, with information on if the farm is among the 50% top farms or among the 10% worst farms; i.e. using relative values. KRAV and to some extent Seal of Quality and Arlagården refer to these systems in their standards (Arla, 2013; KRAV, 2013; Swedish Seal of Quality, 2012).

The selection of a sample, i.e. only evaluating a subset of animals, is a third method used in relation to group assessment. None of the three Swedish private standards used sampling. This method builds on picking out sub-samples and is occasionally used for control of compliance with the legislation. According to the council Regulation (EC) N° 1099/2009 on the protection of animals at the time of killing (OJ 18.11.2009/L 303/1-30) regular checks should be carried out on a representative sample of animals to ensure that animals do not present any sign of consciousness after stunning. Another example of sampling is the Swedish Poultry Meat Association's animal welfare programme (Swedish Poultry Meat Association, 2014), which is authorized through the Swedish official animal welfare legislation. At the slaughterhouse a sample of 100 feet from each broiler flock is assessed and classified with respect to the occurrence of foot pad dermatitis. Depending on the result, measures may need to be taken by the farmer and the maximum permitted stocking density is set.

Potential risks related to the herd level approach

To an individual animal in poor condition it does not matter if all other animals in the group are in good shape and the overall assessment at farm level indicates a generally high animal welfare level. If a certain proportion of the animals are to be affected before non-compliance is recorded there is a risk that measures will not been taken in order to alleviate the situation for an individual animal in poor condition. There is also a risk that group assessment would shift the focus away from animals that need attention. The Welfare Quality® assessment model includes an adjustment for this phenomenon, aiming at reflecting a balance between the animals in poorest condition and the overall welfare at herd level (Viessier *et al.*, 2009). However, De Vries *et al.* (2013) reported that such evaluation of herds may lead to an increased focus on some welfare measures while diverting attention from improving others. If the assessment is made on only a sample of animals there is a risk that an animal in poor condition will not be assessed at all in the first place. Mullan *et al.* (2009) saw that even a large sample size was unable to mirror animal welfare in an accurate way with respect to conditions with a low prevalence.

The use of relative values to benchmark a farm, e.g. claiming that a farm is among the 20% best, does not necessarily mirror a high animal welfare level since it all depends on the status of the population observed (Viessier *et al.*, 2009). Benchmarking is used in the Welfare Quality® assessment protocols, as each farm get an overall assessment (excellent, enhanced, acceptable or not classified) based on the aggregated scores, on absolute aspiration values; i.e. not using relative values. However, also the use of aggregated scores has its limitations. For example, for a farm to be classified as 'excellent' does not necessary require it to reach the excellent level for all welfare principles. 'Excellent' does not always reflect the animal's point of view, but may be related to the normal farm standard, and the levels may hence have been adjusted so not all farms will end up in the lowest classes (Viessier *et al.*, 2009).

We saw that if similar requirements are used in different regulations (i.e. legislation and standards) there is a risk that there will be different opinions as to when these requirements should be considered being complied with or not, depending on if the assessments were made at individual or herd level. This can be highly confusing to the farmers, who are often expected to comply with several different animal welfare regulations at the same time.

Possibilities related to a herd level approach

Being healthy is an important aspect of animal welfare. It is generally acknowledged that in spite of good housing and proper management animals will now and then fall ill, and this does not in itself constitute a violation of the animal welfare legislation or standards. All three private standards in our study required that existing health problems at herd level must be analysed and prevented, not only that individual animals must be treated. The Swedish legislation requires any sick or injured animal to be given necessary care and, when relevant, veterinary treatment (Section 9, The Swedish Animal Welfare Act (1988:534)), but does not include specific herd level requirements. However, the guidelines for official animal welfare control state that it is important to differentiate between temporary and systematic problems. Measuring the prevalence of welfare problems at herd level could be a good complement to the individual approach as the herd data could indicate problems with the housing system or management routines. If this is observed and measures are taken, welfare problems for future animals could be prevented. A system with 'alarm thresholds' is used in other initiatives, for example by Welfare Quality®. There are also advantages related to benchmarking, as different farms can be compared to each other, and thereby giving incentives to farmers with lower animal welfare levels to make improvements and investments, and a motivation for the best farmers to stay on top (von Keyserlingk *et al.*, 2012).

For social, group-living animals it can be argued that the welfare definition in itself should include a 'good life of the group' aspect (Harfeld, 2013). Since several farm animal species are certainly social living, there are reasons for measuring the welfare of a group with reference to how well a group is functioning (Ohl and Putman, 2014).

Ethical considerations

Is it possible to talk about 'animal welfare' for a group of animals when the concept of animal welfare is a characteristic of an individual animal? Even when talking about measuring animal welfare at an individual level there will be different opinions on how this should be carried out, what aspects of welfare that are most important, e.g. health or behaviour, and how these different aspects should be summarized to describe the welfare of one individual (Fraser, 1995; Rushen and de Passillé, 2009). This will vary in relation to different farming species, circumstances, farmer's choices, etc. When measuring animal welfare at the herd level a range of ethical decisions have to be made, for example when it comes to how different criteria should be aggregated and what levels should be considered acceptable or not (Viessier *et al.*, 2009), e.g. how large a proportion of the animals can be allowed to be in poor condition at the same time as the overall welfare level is assessed as adequate. It is a challenge to communicate these decisions and limits to consumers, who are often concerned about individual animal's welfare and want to make informed choices based on this aspect.

In a recent study we saw that the species involved and the purpose behind the keeping of an animal has an impact on what welfare level the legislation finds acceptable (Lundmark *et al.*, 2013). Would it be acceptable to use group measures for pet and competition animals, like dogs and horses, which we tend to handle as individuals on an everyday basis, or is it mainly acceptable to use these kinds of measures for farm animals?

Group assessment in relation to animal welfare legislation and private standards

According to Yeates (2013), both individualistic and non-individualistic approaches can be useful when developing policies, but it is important to reflect on when to use which one. Both sub-set sampling and requiring a certain proportion of the animals to be negatively affected for non-compliance to be identified, are thoughts not in line with the present objectives of several pieces of animal welfare legislation. However, sampling is occasionally used when controlling for compliance with animal welfare legislation. If the intention is to safeguard the welfare of an individual animal the use of group measures only would create a goal conflict between the requirements in the regulation and the actual measurements. For example, if the requirement laid down in a regulation is that animals shall be 'clean enough' it appears more adequate to have control guidelines about what level is acceptable for an individual animal (proportion of dirty sections of the body) and not for a herd of animals (proportion of dirty animals). It is important that those who write the regulations ensure that the methods of measuring are in line with the aim of the regulation. Organisations behind private standards often have more room for manoeuvring when choosing which approach to apply, albeit having to use the legislation as a baseline. Private standards could therefore have a more progressive approach in order to improve animal welfare among the members (Main *et al.*, 2014). If the scope of a standard is limited to considering animal welfare at herd level only, there is a risk that a private standard claiming to have higher intentions than the legislation does in fact offer a protection level that is below the legislative level with respect to the individual animal.

Stenevik and Mejdell (2011) argues, from a legislative perspective, that when animal welfare is to be assessed for a group of animals the focus should be on the animals in poorest condition and the overall assessment should not be identical with an average welfare level. However, since there are also benefits with group assessments at herd and farm level these kinds of assessments can nevertheless be used as a complement to the individual approach.

Conclusions

Generally, current animal welfare legislation in Sweden and Europe is designed to protect the individual animal. An overall assessment of animal welfare at herd level can therefore never be allowed to distort the focus away from individual animals with poor welfare. Hence, measuring animal welfare exclusively at the group level is not entirely in line with the intentions of the legislation. However, measuring welfare at group level may be a good complement to identify systematic problems, make improvements at a farm for the future and benchmarking farms.

References

Arla (2013). Kvalitetsprogrammet Arlagården® version 4.0. Avaliable at: www.arla.se.

Butterworth, A. (2009). Animal welfare indicators and their use in society, In: Smulders, F.J.M. and Algers, B. (eds.) Welfare of production animals: assessment and management of risks. Wageningen Academic Publishers, Wageningen, the Netherlands, pp. 371-389.

De Vries, M., Bokkers, F.A.M., Van Schaik, G., Botreau, R., Engel, B., Dijkstra, T. and de Boer, I.J.M. (2013). Evaluating results of the welfare quality multi-criteria evaluation model for classification of dairy cattle welfare at the herd level. Journal of Dairy Science 96: 6264-6273.

European Commission (2007). Attitudes of citizens towards animal welfare. Brussels: Eurobarameter.

Fraser, D. (1995). Science, values and animal welfare: exploring the 'inextricable connection'. Animal Welfare 4: 103-117.

Harfeld, J.L. (2013). Telos and the ethics of animal farming. Journal of Agricultural & Environmental Ethics 26: 691-709.

Jordbruksverket (2013). Vägledning för kontrollmyndigheter m.fl. Bilaga Nötkreatur. Fastställd 4 Juni 2013.

KRAV (2013). Regler för KRAV-certifierad produktion, utgåva 2013. Available at: www.krav.se.

Lundmark, F., Berg, C., and Röcklinsberg, H. (2013). 'Unnecessary suffering' as a concept in animal welfare legislation and standards. In: Röcklinsberg, H. and Sandin, P. (eds.) The ethics of consumption – the citizen, the market and the law. Wageningen Academic Publishers, the Netherlands, pp. 114-119.

Main, D.C.J., Mullan, S., Atkinson, C., Cooper, M., Wrathall, J.H.M. and Blokhuis, H.J. (2014). Best practice framework for animal welfare certification schemes. Trends in Food Science & Technology 37: 127-136.

Mullan, S., Browne, W.J., Edwards, S.A., Butterworth, A., Whay, H.R. and Main, D.C.J. (2009). The effect of sampling strategy on the estimated prevalence of welfare outcome measures on finishing pig farms. Applied Animal Behaviour Science 119: 39-48

Ohl, F. and Putman, R.J. (2014). Animal Welfare at the Group Level: More than the sum of individual welfare? Acta Biotheoretica 62: 35-45.

Radford, M. (2005). Animal welfare law in Britain. Regulation and responsibility. Oxford University Press, Oxford, UK, 441 pp.

Rushen, J. and de Passillé, A.M. (2009). The scientific basis of animal welfare indicators. In: Smulders, F.J.M. and Algers, B. (eds.) Welfare of production animals: assessment and management of risks. Wageningen Academic Publishers, the Netherlands, pp. 391-415.

Stenevik, I.H. and Mejdell, C.M. (2011). Dyrevelferdsloven. Kommentarutgave. Universitetsforlaget, Oslo, Norway, 472 pp.

Swedish Poultry Meat Association (2014). Animal welfare programme. Available at: www.svenskfagel.se.

Swedish Seal of Quality (2012). IP Sigill Mjölk 2012:1. Available at: www.svensktsigill.com.

Viessier, I., Boutreau, R. and Perny, P. (2009). Scoring animal welfare: Difficulties and Welfare Quality® solutions. In: Keeling, L. (ed.). An overview of the development of the Welfare Quality® Project assessment systems. Welfare Quality® Reports No. 12. Uppsala, Sweden.

Von Keyserlingk, M.A.G., Barrientos, A., Ito, K., Galo, E. and Weary, D.M. (2012). Benchmarking cow comfort on North American freestall dairies: lameness, leg injuries, lying time, facility design, and management for high-producing Holstein dairy cows. Journal of Dairy Science 95:7399-7408.

Wahlberg, B. (2011). Reglering och förvaltning av produktions-och slaktdjurs välbefinnande – en offentligrättslig undersökning. Department of Law, Åbo Akademi University, Finland.

Yeates, J.W. (2013). Individualism and nonindividualism in the application of nonhuman animal welfare to policy. Journal of Applied Animal Welfare Science 16: 254-271.

Regulating food nanotechnologies: ethical and political challenges

V. Sodano
Department of Agriculture, University of Naples Federico II, via Università 96, 80055 Portici, Italy; vsodano@unina.it

Abstract

Whilst recognizing the need to tackle food nanotechnology risks, the EU has so far laid down scant regulatory provisions, allowing new nano products to enter the food market freely, thus depriving consumers of their right to be informed and hence to choose. After a brief presentation of the potential risks of food nanotechnologies, the paper explores the connections between neoliberalism, food nanotechnology innovation and the regulatory process, unveiling the power struggles and the ethical stances that lie beyond the EU regulatory choices. The main conclusion is that, in order for the benefits of new technologies to outweigh the costs, it is necessary to acknowledge the political issues which are at stake and allow the regulatory process to be dictated by deontological rather than utilitarian ethical principles. Moreover, direct forms of regulation should be put in place, such as pre-market authorization, mandatory labelling and the establishment of a public register of products and producers.

Keywords: neoliberalism, innovation, food, policy

Introduction

Nanotechnology generically refers to the manipulation of matter at the nanometer scale (a nanometer is 10^{-9} meters). At this scale, substances may present new properties, defined as quantum effects, which sharply differ from those of their bulk counterparts. With regard to the agri-food sector, there is a large assortment of possible applications already in existence, ranging from agricultural pest control to food processing and distribution (Cushen *et al.*, 2012). Many nanoproducts are already on the market (http://www.nanotechproject.org/cpi), such as: nanoemulsions and nanocapsules of pesticides, nanostructured food ingredients and additives, nanostructured delivery systems of food supplements, new food packaging, new food contact materials for food processing and storage. Besides the benefits they provide, which for the current applications are limited to shelf life, texture and nutritional improvement, nanofood technologies potentially pose many social, health and environmental risks.

Whilst recognizing the need to tackle food nano risks, the EU has so far laid down scant regulatory provisions, allowing new nano products to enter the market freely, thus depriving consumers of their right to be informed and hence to choose. Contrary to the conventional viewpoint, which accounts for the regulatory delay purely on the grounds of the difficulties involved in carrying out a sound risk assessment (Marrani, 2013; Brosset, 2013), this paper suggests that the delay mainly depends on the political and ethical choices made by the EU in its pursuit of a strict neoliberal policy agenda. The first part of the paper summarizes the potential risks of food nanotechnologies and shows how weak and ineffective the current EU regulatory framework for tackling them is. The second part explores the connections between neoliberalism, food nanotechnology innovation and the regulatory process, unveiling the power struggles and the ethical stances that lie beyond the supposedly democratic and value-free EU nanotechnology appraisal attitudes.

Nano food risks and the European regulatory framework

Nanotechnology application in the agro-food sector may produce many negative effects thus giving rise to various kinds of risks, such as health, environmental, economic, social and political risks. Health risks mainly depend on the ability of nanoparticles (NPs) to bypass cellular membranes, to pass through biological barriers (as, for instance, the blood-brain barrier) and to bio-accumulate with severe toxicological effects (Elsaesser and Howard, 2012; Hubbs *et al.*, 2013). Environmental risks are associated with the limited biodegradation of NPs and their interaction with living organisms, soil and aquatic ecosystems. So far the following negative impacts have been reported: phytotoxicity, damage to soil structure and fertility, reduction of microbial biomass and diversity, toxicity for algae and daphnidis (Mueller and Nowac, 2008). Economic risks arise from (Scrinis and Lyons, 2007): (1) the possible disruption of markets (for those traditional products which are replaced by new nanoproducts); (2) the displacement of workers due to a more capital-intensive mode of production; (3) the further consolidation of food systems, with the largest corporations more able to exploit the profit streams from the patents on the new technologies. As regards socio-political risks, it has been argued that the proliferation of nanotechnologies in the food systems might exacerbate social injustice, deepen the north south divide, and threaten the food sovereignty of local communities (Lyons *et al.*, 2012). Finally, because nanotechnology is a dual-use technology (being developed for military as well as civilian purposes) there may be military and security risks, as in the possible case of nanofood being used to spread disease in scenarios such as war and terrorist attacks.

During the first period of introduction (in the last decade of the last century) nanotechnologies entered the market without any specific forms of regulation. The first choice of public bodies all over the world was to consider nanomaterials equivalent to their bulk form. Following the concerns raised by independent agencies and non-governmental organizations (NGOs) due to the alarming results of toxicological research, in the early 2000s the debate surrounding nano-regulation intensified. In the EU, existing regulations have been deemed suitable for the new nano products and only a few provisions have been set out specifically for nanotechnologies, mainly those concerned with medical applications and workers' safety. As regards the food sector, there are only a limited number of regulations which provide specific provisions for nanomaterials, namely: Reg. 1333/2008 on food additives; Reg. 1332/2008 on food enzymes, Reg. 450/2009 on active and intelligent materials and articles intended to come into contact with food; Reg. 10/2011 on food contact plastic materials; Reg. 1169/2011 on novel foods.

The European approach has been to resort to soft regulation in the form of a plethora of documents, guidelines, recommendations and codes of conduct. There is a growing consensus as to the ineffectiveness and tardiness of the regulatory processes in the EU (Brosset, 2013; Marrani, 2013), which are generally ascribed to the technical problems affecting the risk assessment processes, such as the physicochemical characterization of nanomaterials, the identification of the toxicologically relevant dose and the estimation of human exposure (Shaktin, 2013). In order to overcome the difficulties faced when carrying out sound cost-benefit analysis, many scholars have suggested that EU policy makers ought to shift from the consequentialist perspective, endorsed when risk assessment is used a sole basis for policy choices, to the deontological perspective inherent in the precautionary principle, human rights doctrine and the informed consent principle (Kermisch, 2012; Ruggiu, 2013; Shrader-Frechette, 2007). To date, such a suggestion has not been welcomed by the European Commission which instead seems to dismiss any form of intervention, as proven by the cancellation of Article 18 of Regulation 1168/2011 which stipulated mandatory labelling for food containing manufactured nanomaterials.

The current difficulties and delay associated with EU legislation as far as delivering a clear and effective regulatory framework in the field of food nanotechnology is concerned can only partially be explained by the lack of toxicological knowledge and the technical obstacles posed by risk assessment. More accurately, they depend instead on the decision taken by the EU government to pursue a neoliberal policy agenda, which subordinates important social issues (health, environment and social justice) to strictly economic goals, in compliance with a utilitarian perspective.

Neoliberalism, food nanotechnology and EU regulatory process

Here I refer to the definition and conceptualization of neoliberalism set out by Harvey: 'neoliberalism is in the first instance a theory of political economic practices that proposes that human well-being can best be advanced by liberating individual entrepreneurial freedoms and skills within an institutional framework characterized by strong private property rights, free markets, and free trade' (Harvey, 2005: 2). I also accept Harvey's view that the worldwide neoliberalization process started in the 1970s has been from the very beginning a project aimed at achieving the restoration of class power, with the capitalist's class willing to regain economic and political power from the middle and working classes. Neoliberalism entails a theory of the role of the state anchored in the political tradition of contractarianism and in neoclassical economics. According to contractarianism, associated with Nozick's libertarian approach, a free society is one in which the state should have no power and duty other than that of securing private property rights and guaranteeing the proper functioning of markets. Markets, as defined by neoclassical economics, turn out to be the best means of resource allocation, provided that the basic hypotheses of the standard model are fulfilled, i.e.: rational economic actors; perfect competition; absence of externalities and public goods; perfect information. It is worth noticing that, because both contractarianism and neoclassical economics endorse the rational choice model, they propose an economization of society, with political, moral and social spheres merging into the economic sphere. Moreover, utilitarianism is the sole allowed ethical perspective and power relations are deemed not to exist in a world populated by free, independent, rational actors.

When tackling nanofood risks from such a neoliberal perspective, it turns out that the scope for public intervention shrinks dramatically, because it is restricted to the sole objective of enhancing economic efficiency. Only health, environmental and monopoly risks are taken into account, since they are deemed to be associated with market failures such as externalities, public goods and imperfect competition; accordingly, interventions are aimed at restoring market efficiency and not at enforcing human and nature rights or at levelling power imbalances in society. Moreover, only indirect interventions are undertaken, in the form of market incentives and soft regulation, ruling out command and control policy instruments such as standards, bans and mandatory product specification, testing and compliance. As regards the other risks – namely distributive, Northsouth divide, food sovereignty and security – since they involve social, moral and political spheres, they do not fall under the scope of intervention by the neoliberal state.

When analysed through the lens of the neoliberal approach, the causes and consequences of the stalling EU nano regulatory process can be better understood and some critical issues may be raised. First, it is clear that the approach adopted so far by the EU fully mirrors the 'theoretical' underpinnings of the neoliberal doctrine. The regulatory debate has been centred mainly on health risks, with scant reference to environmental risk and totally overlooking the remaining kinds of risks. The only provisions have been in the form of soft regulation, and the proposal of mandatory labelling for nano ingredients (which is a command and control intervention) has been suddenly withdrawn by the Commission, notwithstanding strong parliamentary opposition. The precautionary principle and the consequent possible moratoria on nanofood, has been cast aside so as not to penalize economic efficiency and the economic competitiveness of the European firms, consistent with the primacy of the economic sphere endorsed by neoliberalism. Moreover, regulatory decisions have been made which are dependent on the advice of scientists, considering the techno scientific sphere to be autonomous and value-free, in accordance with the neoclassical assumption of exogenous technological change.

Second, when looking at the effects of the regulatory choices, it emerges that they are also consistent with the political driver of neoliberalism, that is the restoration of economic power to the upper classes. In the case of the agri-food sector, the neoliberal turn has witnessed the growing power of big corporations and the expansion of the agribusiness model at the expense of traditional local food systems. New technologies have been identified as the backbone of the current corporate food regime since they permit the enactment of the neoliberal project in ways which were not previously possible (Bush, 2010; McMichael, 2009). Like biotechnologies, nanotechnologies, when backed by strict intellectual property laws, facilitate the privatization of living matter and natural resources, leading to a further commodification and consolidation within the global food system. Moreover, many applications (such as functional and personalized foods) are used for market differentiation strategies, boosting oligopolistic profits and eroding the niche markets previously occupied by traditional products and small businesses. In this framework it is clear that the EU choice of keeping nanotechnologies unregulated fosters corporate power and power relations within the food system, hence exacerbating class inequalities in society.

Having described the neoliberal agenda behind the EU nano regulation delay, it is worth questioning what is wrong with it and what might be an alternative modus operandi. Criticism of the EU neoliberal stances may be advanced on two grounds. The first encompasses all those arguments used by the supporters of political and economic approaches at variance with neoliberalism. A close examination of these positions is beyond the scope of this paper. It is only worth bearing in mind that both the more conservative alternatives (along the lines of the welfare approach endorsed by embedded liberalism) and the more radical alternatives (along the lines of those suggested by the anticapitalistic and food sovereignty movements) highlight the theoretically flawed premises of the rational choice model and the way in which the economization of the social and political spheres stressed by neoliberalism leads to a lack of human rights compliance. The second ground for criticism, on which I wish to focus, refers to the contradictory ethical stances framing the EU policy choices and the lack of transparency in the decision processes. Table 1 summarises the ethical stances that the state should endorse in order to deal with the various forms of nano-risks. It is clear that when exclusively subscribing to utilitarianism, as in the case of the neoliberal state, the only risks that can be addressed are the health and environmental risks when classified as efficiency losses due to market failures. Any other effective form of intervention to address the other risks is ruled out, since it would require the acceptance of some form of deontological ethics. The current policies adopted by the EU, as reported in the second row of Figure 1, bear witness to the EU's utilitarian stances; nevertheless, in the case of soft regulation, there emerges both a lack of transparency and a theoretical contradiction. Soft regulation includes measures such as communications, resolutions, recommendations, opinions and guidelines, which, in contrast to Directives, Regulations and decisions (hard law) are not binding on those to whom they are addressed. Soft regulation ultimately relies on the compliance of socially responsible firms, i.e. firms who take into account the possible effects of their behaviours on stakeholders and society at large. The neoliberal perspective stresses that corporations do not have any social responsibility other than to maximize shareholder value, as clearly stated in a famous article by the father of the neoliberal revolution (Friedman, 1970): 'the social responsibility of business is to increase its profits'. If that is the case, it is very unlikely that the guidelines and the recommendations set out by the EU will make firms take any action capable of reducing health and environmental risks beyond that which is required by the existing mandatory regulations. This reveals the lack of transparency on the part of the EU, which appears to be willing to manifest publicly a level of safety and warranty higher than that actually provided. The theoretical contradiction emerges if we assume that the EU refers instead to different concepts of corporate social responsibility such as the broader goal perspective and ethical theories (Sodano and Hingley, 2013), which look upon corporations as intrinsically moral institutions whose activities, beyond profit maximization, contribute to the achievement of goals of general interest, consistent with a deontological ethical approach. In this case, resorting to soft regulation would entail a theoretical contradiction because the deontological stances, refused as the basis of state intervention, are accepted de facto as the basis of firms' behaviour. Moreover, soft regulation can be considered also as

Table 1. The neoliberal approach to nano food regulation: ethical assumptions and EU choices.

Nano food risks	Ethical approach underlying state intervention	EU actual intervention
Health and environmental risks associated with market failures	Consequentialist approach: • utilitarianism	Indirect interventions (market incentives)
Health and environmental risks in terms of human and nature right	Deontological ethics: • human and nature right doctrine; • precautionary principle	Soft regulation
Market disruption economic inequality	Deontological ethics: • Rawls' theory of justice	No intervention
Northsouth divide	Deontological ethics: • Rawls' theory of justice; • human right doctrine	No intervention
Food sovereignty	Deontological ethics: • communities' autonomy and self determination	No intervention
Security	Deontological ethics: • constitutional principles, the state shall protect its citizens	No intervention

a questionable policy choice because the same neoliberal arguments against state intervention aimed at enforcing positive rights, as in the case of direct forms of regulation, fully apply to private organizations: paternalism and value judgments; disablement of the principle of autonomy; possible misconduct by government officials; inefficiency due to lack of knowledge and information; regulatory capture (cf. Stigler, 1971). As a consequence, resorting to hard regulation in favor of soft regulation turns out to be nothing but a way of shifting power from the public to the private sector, resulting in the loss of democratic control over choices and actions that deeply affect society.

Conclusion

Currently in the EU food nano products continue to enter the market in a regulatory vacuum, with consumers unaware of the new technology and its risks. The paper has suggested that the lack of appropriate regulations may depend on the neoliberal policy choices made by the EU. Such choices can be criticized on the grounds of three arguments. First, the contractarian and utilitarian stances, endorsed by the neoliberal worldview as basis for politics, prevent the implementation of effective policies to uphold human rights, protect the environment and promote social justice. Second, the lack of recognition of the political motivation of regulatory choices, concealed behind the efficiency argument, is to be interpreted as proof of poor transparency and fairness on the part of the EU government towards its citizens. Third, the resorting to soft regulation in order to overcome the negative social effects of a weak state, entails a power shift towards the private sector and consequent losses in terms of state sovereignty and democracy. The main conclusion is that, in order for the benefits of new technologies to outweigh the costs and be shared by society at large, it is necessary to acknowledge the ethical and political issues lying behind the regulatory choices and engage in a more transparent and fairer public discussion, which would bring technology innovation back under true democratic control. Meanwhile, relying on the precautionary principle, the following measures should be immediately put in place: stringent pre-market authorization requirements for the new nano products; mandatory labelling; the establishment of a public register of products and producers.

References

Brosset, E. (2013). The law of the European Union on nanotechnologies: comments on a paradox. Review of European, Comparative & International Environmental Law 22:155-162.

Bush, L. (2010). Can fairy tales come true? The surprising story of neoliberalism and world agriculture. Sociologia Ruralis 50:331-351.

Cushen, M., Kerry, J., Morris, M., Cruz-Romero, M. and Cummins, E. (2012). Nanotechnologies in the food industry. Trends in Food Science & Technology 24:30-46.

Elsaesser, A. and Howard, C.V. (2012). Toxicology of nanoparticles. Advanced Drug Delivery Review 64:129-137.

Friedman, M. (1970). The social responsibility of business is to increase its profits. New York Times Magazine 32-33:122-6.

Harvey, D. (2005). A brief history of neoliberalism. Oxford University Press, Oxford, UK, 256 pp.

Hubbs, A.F., Sargent, L. M., Porter, D.W., Sager, T.M., Chen, B.T., Frazer, D.G., Castranova, V., Sriram, K., Nurkiewicz, T.R., Reynolds, S.H., Battelli, L.A., Schwegler-Berry, D., McKinney, W., Fluharty, K.L., Mercer, R.R. (2013). Nanotechnology: toxicologic pathology. Toxicologic Pathology 41:395-409.

Kermisch, C. (2012). Do new ethical issues arise at each stage of nanotechnological development?. Nanoethics 6:29-37.

Lyons, K., Scrinis, G. and Whelan, J. (2012). Nanotechnology, agriculture, and food. In: Maclurcan D. and Radywyl N. (eds.) Nanotechnology and Global Sustainability. CRC Press, Taylor & Francis Group. Boca Raton, Florida, USA, pp. 117-139.

Marrani, D. (2013). Nanotechnologies and novel foods in European law. Nanoethics 7:177-188.

McMichael, P. (2009). A food regime genealogy. The Journal of Peasant Studies 36:139-169.

Mueller, N.C., Nowack, B. (2008). Exposure modeling of engineered nanoparticles in the environment. Environmental Science & Technology 42:4447-4453.

Ruggiu, D. (2013). Temporal perspectives of the nanotechnological challenge to regulation: how human rights can contribute to the present and future nanotechnologies. Nanoethics 7:201-215.

Scrinis, G. and Lyons, K. (2007). The emerging nano-corporate paradigm: nanotechnology and the transformation of nature, food and agri-food systems. International Journal for the Sociology of Food and Agriculture 15:22-44.

Shatkin, J.A. (2013). Nanotechnology, health and environmental risks. CRC Press, Taylor & Francis, London, UK, 265 pp.

Shrader-Frechette, K. (2007). Nanotoxicology and ethical conditions for informed consent. Nanoethics 1:47-56

Sodano, V. and Hingley, M. (2013). The food system, climate change and CSR: from business to government case. British Food Journal 115: 75-91.

Stigler, G. J. (1971). The theory of economic regulation. Bell Journal of Economics and Management Science 2: 3-21.

Ethics within legal limits: harm-benefit analysis according to the Directive 2010/63/EU

H. Grimm
Messerli Research Institute, Veterinary University Vienna, Medical University Vienna, University Vienna, Veterinaerplatz 1, 1210 Vienna, Austria; herwig.grimm@vetmeduni.ac.at

Abstract

According to the EU Directive 2010/63 on the protection of animals used for scientific purposes, the harm-benefit analysis is part of the project evaluation that has to be carried out by the competent national authorities. Ethical considerations have to be taken into account during the evaluation procedure. Therefore, all member states face the question how to understand the term 'ethical considerations' and how to implement these considerations in the evaluation procedure. A research project is currently carried out at the Messerli Research Institute Vienna to accomplish this task. An applicable methodology for the harm-benefit analysis is aimed for. The question arises which ethical aspects can be integrated into the project evaluation against the background of existing legal requirements. On the one hand, a great number of ethical requirements associated with animal protection are already reflected in the relevant law and regulations. On the other hand, the important question arises whether ethical considerations can exceed the legal framework. This second aspect is a vital issue since the harm-benefit analysis that integrates ethical considerations is a legally binding part of the project evaluation. In other words: If a project is turned down because of an imbalance of harms and benefits, this has to be done on a transparent, legal basis. Since ethical considerations are part of the harm-benefit analysis, a concise understanding in line with the legal basis is necessary. If projects were evaluated applying criteria that exceed the legal requirements, submitters would be treated in a way that is incompatible with the idea of a constitutional state and the principle of legality in particular. This leaves us with the question what is meant by 'taking ethical considerations into account' and whether its content can/should either be identified with 'legal requirements' or not. Further, even if this question is answered, it remains open how to integrate ethical considerations into the harm-benefit analysis and whether they can play a decisive role. I will address the first question on the basis of three possible understandings of the term 'taking ethical considerations into account'. Each implies different consequences for the harm-benefit analysis in animal research. However, only one understanding can be applied: It will be argued that ethical considerations as a part of the harm-benefit analysis can only be understood within the limits of the legal framework.

Keywords: animal experimentation, regulatory ethics, principle of legality, applied ethics

Introduction

The Directive 2010/63/EU on the use of live animals for scientific purposes had to be transposed into national law. According to the Directive, a harm-benefit analysis has to be part of the project evaluation. The harm-benefit analysis has to take ethical considerations into account (Article 38): '(2) The project evaluation shall consist in particular of the following: (...) a harm-benefit analysis of the project, to assess whether the harm to the animals in terms of suffering, pain and distress is justified by the expected outcome taking into account ethical considerations, and may ultimately benefit human beings, animals or the environment; (...)' (emphasis mine)

The main question of this passage is of course how to weigh incommensurable goods against each other in a harm-benefit analysis. Applied ethics can help to establish a modus vivendi to deal with this challenge as the relevant literature indicates (cf. Alzmann, 2010; Bass, 2012; Borchers and Luy, 2009; Borchers, 2009; Grimm and Binder, 2013). Since the Directive does not provide any explicit specifications on standards and criteria for 'ethical considerations', the passage triggers speculations such as the following: 'Now we have a legal basis to reduce animal research with ethical arguments'. Or: 'Ethics means respect for the 3Rs, so this is settled anyhow'. 'Ethics' is a synonym of 'animal protection'. Or: 'Ethics, in the sense of a moral philosophy e.g. Kant's, Bentham's or some contemporary philosopher's thinking, has to be integrated into the procedure of project evaluation'. At first sight, it seems as if someone put that term into the Directive in order to find out what people would make out of it. The Directive does not give explicit hints. However, I will argue that it does not leave the debate entirely open and a clarification is possible.

Ethics within legal limits and the principle of legality

For all member states the question arises how to implement 'ethical considerations' within the harm-benefit analysis as a part of the project evaluation. Therefore, it is crucial to apply an adequate understanding of the term. This section will clarify why 'ethics' can neither be identified with 'moral theory' nor 'public concern'. Despite the fact that 'ethics' or 'ethical' is often identified with what is considered as good by the public or moral philosophers, a quite different understanding is relevant in this context. 'Ethics' has to be understood within the limits of existing national law.

Ethics as a philosophical critique of law

From a moral philosophical perspective, one can approach the term 'ethical considerations' with regard to legally regulated fields from two perspectives: (1) a fundamental approach that puts existing law into doubt; and (2) regulatory ethics that reflects moral responsibility within the legal framework. Prominent ethicists like Peter Singer, Tom Regan, and Bernhard Rollins saw their responsibility and role in influencing, changing and developing existing animal law (cf. Grimm, 2012). Accordingly, the fundamental approach is highly influential in animal ethics. Here, ethics – as moral philosophy – is identified with a critique of existing legal frameworks. This can be done e.g. by addressing new issues of moral relevance like in the case of genetically modified organisms that existing law was ignorant of before. Further, inconsistencies can be addressed and ruled out. In short, ethicists identify shortcomings, new ideas that are not included in existing law, etc. This strategy is an important aspect of applied ethicists' responsibilities.

If we brought this idea back to the term 'taking ethical considerations into account' it would have to be understood as a moral philosophical endeavour addressing moral issues from a philosophical perspective without taking into account the normative force of existing law. Existing law would be turned from a normative framework into the object of criticism. To make the point even clearer: this understanding would make it possible to identify the content of 'ethical considerations' in the Directive with basically any ethical theory. For instance, an 'animal rights theory' which is – undoubtedly – an ethical position (cf. Francione, 2000; Regan, 2004 (1983)) could be used. In this theory, any use of sentient animals to gain knowledge would be morally evil. A consequentialist harm-benefit analysis is not in the scope of this theory and considered a moral wrong. But one could also understand 'ethics' in terms of a contractarian view, holding that animals have no moral status at all (cf. Cohen, 2008 (2001)). Since animals do not count morally in this theory, harms would not weigh heavily if they did at all. Also in this case, the harm-benefit analysis would not make much sense.

Besides the open question of what theory to use, this approach cannot provide the basis for the semantic clarification of 'ethical considerations'. The principles of constitutional states and the principle of legality in particular rule this option out (cf. Grimm and Binder, 2013). The latter principle aims at preventing arbitrariness of the state. Therefore, as a legal ideal, it requires all laws to be clear, ascertainable and non-retrospective. It requires authorities and decision makers to resolve disputes by applying legal rules and decide on a legal basis that has been declared beforehand. It prevents alterations of the legal situation retrospectively by discretionary deviation from established law. Since the national authorities have to approve or disapprove projects on a legal basis, the relevant aspects for the harm-benefit analysis have to be addressed implicitly or explicitly already in relevant national law. In other words, introducing new normative ideas from a philosophical point of view after the Directive and corresponding laws were brought into force is not in accordance with the principle of legality since the competent national authorities have to work and evaluate projects on a legitimate, legal basis. Otherwise they would act against one of the main principles and a cornerstone of constitutional states. As already mentioned, moral philosophy can assume the important responsibility and reflect existing law, but in this context this cannot serve as an appropriate understanding of 'taking ethical considerations into account'. The main reason is that authorities can only take normative aspects into consideration that are implicitly or explicitly included in the applicable law.

Ethics as public concern

A similar argument applies to the idea of identifying 'taking ethical considerations into account' with taking 'public concerns' into account. What the public thinks about animal research can lead to changes of existing law by the legislature on the basis of legitimate procedures. But existing law cannot be changed without a legally defined procedure in a constitutional state. Imagine a survey indicating that the majority of inhabitants of European member states were against animal research. This does not and must not change a thing in the project evaluation. Again, authorities have to work on the basis of applicable law. Even if it is true that public concerns are reflected in Directive 2010/63/EU and motivated its development, it's not public concerns that are binding for authorities but the law. Whether applicable law reflects public concerns is a different matter. However, public concerns may lead to political pressure and future reconsiderations of existing law. Surveys can give reason to change the law, but they cannot provide a basis for a legal treatment of citizens (submitters of research proposals) in constitutional states.

Ethics within legal limits

The relevant approach in this field can be plausibly addressed as 'ethics within legal limits'. Ethics has to be understood as a means to clarify moral responsibility within the framework of legal regulations. This understanding of ethics can – of course – not be identified with moral philosophy but only with one specific function of ethics in a particular context. This specific function is relevant in order to clarify the understanding of 'taking ethical considerations into account' in a meaningful way. The principle of legality provides a strong reason why regulations limit the bandwidth of 'ethical considerations' in this context. Since authorities have to make their decisions on the basis of this (limited) understanding, this approach can be labelled as 'regulatory ethics'. Despite the fact that regulatory ethics works 'within' the framework of existing law, there is work to do for ethicists. Ethics can and should envisage the moral position and consequences of relevant law. In this sense, ethics as a philosophical discipline explicates and uses the normative infrastructure of law, identifies normative principles from a moral philosophical point of view and uses them to e.g. argue for aspects that should be considered in the harm-benefit analysis. Coherence with the legally binding corner posts of the pertinent acts is of course a prerequisite.

From this point of view it is quite obvious how to elaborate the content of 'taking into account ethical considerations'. It can only be understood in terms of making the normative ideas of existing law explicit and applying them in order to get a clear picture of its content. If the term was interpreted incoherently with the legal regulations, it could not be part of the project evaluation by the authority since this would contradict the principle of legality. Therefore, as long as the principles of constitutional states are to be respected, the interpretation of 'taking ethical considerations into account' is restricted to the limits given in relevant acts and regulations.

Normative infrastructure of the Directive

Reflecting the moral nature of law that deals with the protection of animals used for scientific purposes has some tradition (cf. Binder, 2013; Grimm, 2013b; Grimm and Binder, 2013; Olsson *et al.*, 2003; Sandøe *et al.*, 2008). Some elements and related inconsistencies shall be sketched in the following.

Consequentialism

Whether an act is to be considered morally right or wrong depends on its consequences. The most obvious demonstration of this aspect in context of Directive 2010/63/EU is the harm-benefit analysis that is an obligatory element of the project evaluation. Only few non-consequentialist (deontological) elements are integrated into the Directive. Weighing harms and benefits is allowed in nearly every case, but not in all cases. Some exemptions are foreseen. For instance, inflicting severe or long lasting moderate pain, suffering, or distress that cannot be alleviated is (generally) prohibited, as well as the use of great apes for scientific purposes.

Pathocentricm

In principle, negative subjective experiences are the normative focus and measure for harms done to animals according to Directive 2010/63/EU. Animals are to be protected from suffering, pain and distress. If expected benefits can justify the expected harms, protection of animals is not granted entirely but only up to the possible extent. But it mainly depends on the most severe averse effect (severity classification) if the benefits can justify harms. The absolute number of animals used, death and killing seem to be of secondary importance. However, influential documents suggest that the number of animals is to be considered in the harm-benefit analysis, such as the 'Working document on project evaluation and retrospective assessment' which is in line with Recital 12 of the Directive in which the intrinsic value of animals is mentioned.

Non-speciesist anthropocentrism

Despite the fact that animals are considered morally relevant, the hierarchy of the animal kingdom is reproduced in the Directive. Although the species is not an explicit criterion to distinguish between animals and humans and within the group of non-human animals, a hierarchical logic can be found in the Directive. First, human interests are ranked higher than animal interests. Otherwise it would not be understandable why it is allowed to conduct research on animals but not on humans. Second, animals that are 'close' to humans are valued higher than others. This phenomenon is called 'non-speciesist anthropocentrism' (Grimm, 2013a) and comprises three defining elements:

- Moral consideration is not based on the criterion of belonging to a species (non-speciesist), but on the capacity ies of the individual.
- Full moral significance is only attributed to full-grown, rational human beings (anthropocentrism).
- Non-human animals are considered 'higher' and 'lower' in their moral standing, relative to the degree to which they resemble human persons in the biological or even social sense.

This can be illustrated by the protection of primates in the Directive 2010/63/EU. Primates – considered evolutionary the closest relatives of humans – are particularly protected because of their resemblance with humans (recital 33). But not only biological facts count. Cats and dogs are to be treated with special care because of the emotional bonds they can establish with humans (e.g. recital 26). This is of course not a consistent or logical ranking according to the idea of the least possible pain, suffering and distress. In general, the closer the animals are to humans (in an evolutionary or social respect), the more they are protected. From a political point of view this is understandable (cf. Eurobarometer, 2010; Ormandy and Schuppli, 2014), but if pain, suffering and distress were the criteria, closeness to the human species should not matter.

Moral individualism

This element is closely related to the above mentioned non-speciesist anthropocentrism. The core claim of moral individualism is to consider beings on the basis of their morally relevant capacities (cf. Crary, 2010; May, 2014; McMahan, 2005). Not being a member of a species or group is important, but the individuals' capacities. This is in line with the Directive. Recital 8 of the Directive makes this very clear: 'In addition to vertebrate animals including cyclostomes, cephalopods should also be included in the scope of this Directive, as there is scientific evidence of their ability to experience pain, suffering, distress and lasting harm'. The trumping criterion is not membership of a group (being a vertebrate) but the ability to experience pain, suffering, distress and lasting harm. Obviously, this is not consistent with non-speciesist anthropocentrism.

Considering these and maybe other elements of the normative infrastructure of the Directive, it seems possible to gain a better understanding of the term 'taking ethical considerations into account'. Developing this understanding is currently the aim of the mentioned Austrian project in which a methodology to carry out the harm-benefit analysis will be elaborated.

Conclusions

To summarise the results of this paper: Against the background of Directive 2010/63/EU that had to be transposed into national law in the EU member states, 'taking into account ethical considerations' has to be understood in the perspective of 'ethics within legal limits'. Ethics can neither be understood as fundamental, moral philosophical critique, nor as public opinion. I have argued that an appropriate understanding of ethics in this context has to be developed against the background of the principle of legality. Hence, ethics is to be understood as regulatory ethics where the principle of legality trumps ethical theory and public concern due to the foundations of constitutional states. Despite the fact that moral philosophy should question existing law in order to help to further develop it, this cannot be the understanding of ethics in this context. For an appropriate understanding of the term 'taking ethical considerations into account' in context of the harm-benefit analysis, the normative corner posts of the Directive and corresponding national law have to be made explicit. On this basis, the idea of a harm-benefit analysis that takes ethical considerations into account can be clarified.

References

Alzmann, N. (2010). Zur Beurteilung der ethischen Vertretbarkeit von Tierversuchen. Dissertation. Eberhard Karls Universität Tübingen. Tübingen, Germany.

Bass, R. (2012). Lives in the balance: utilitarianism and animal research. In: Garrett, J.R. (ed.) The ethics of animal research. Exploring the controversy. The MIT Press, Cambridge, Massachusetts, London, USA, pp. 81-105.

Binder, R. (2013). Rechtliche Grundlagen des Tierversuchs. In: Binder, R., Alzmann, N., Grimm, H. (eds.) Wissenschaftliche Verantwortung im Tierversuch. Ein Handbuch für die Praxis. Baden Baden, Germany, pp. 68-139.

Borchers, D. and Luy, J. (eds.) (2009). Der ethisch vertretbare Tierversuch. Kriterien und Grenzen. Paderborn.

Borchers, D. (2009). Ethiktools für Güterabwägung oder: wie pragmatisch dürfen Ethiker sein. In: Borchers, D. and Luy, J. (eds.) Der ethisch vertretbare Tierversuch. Kriterien und Grenzen. Paderborn, Germany, pp. 15-52.

Cohen, C. (2008 (2001)). Warum Tiere keine Rechte haben. In: Wolf, U. (ed.) Texte zur Tierethik. Stuttgart: Reclam, Ditzingen, Germany, pp. 51-55.

Crary, A. (2010). Minding what already matters: a critique of moral individualism. Philosophical Topics 38(1): 17-49.

Eurobarometer (2010). Science and technology. Eurobarometer Special 340/Wave 73.1. European Commission.

Francione, G.L. (2000). Introduction to animal rights: your child or the dog? Temple University Press, Philadelphia, USA

Grimm, H. (2012). Benthams Erben und ihre Probleme – Zur Selbstreflexion einer Ethik der Mensch-Tier-Beziehung. In: Zichy, M., Ostheimer, J. and Grimm, H. (eds.) Was ist ein moralisches Problem? Zur Frage nach dem Gegenstand der Angewandten Ethik. Freiburg im Breisgau/München. Alber, Germany, pp. 436-475.

Grimm, H. (2013a). Das Tier an sich? Auf der Suche nach dem Menschen in der Tierethik. In: Rippe, K.P. and Thurnherr, U. (eds.) Tierisch menschlich. Beiträge zur Tierphilosophie und Tierethik. Harald Fischer Verlag, Erlangen, Germany, pp. 51-95.

Grimm, H. (2013b). Ethik im Kontext des Tierversuchs. In: Binder, R., Alzmann, N. and Grimm, H. (eds.) Wissenschaftliche Verantwortung im Tierversuch. Ein Handbuch für die Praxis. Nomos, Baden-Baden, Germany, pp. 23-54.

Grimm, H. and Binder, R. (2013). Die Schaden-Nutzen-Analyse. In: Binder, R., Alzmann, N. and Grimm, H. (eds.) Wissenschaftliche Verantwortung im Tierversuch. Ein Handbuch für die Praxis. Nomos, Baden-Baden, Germany, pp. 55-67.

May, T. (2014). Moral Individualism, Moral Relationalism, and Obligations to Non-human Animals. Journal of Applied Philosophy 31(2): 155-168.

McMahan, J. (2005). Our fellow creatures. Journal of Ethics 9: 353-280.

Olsson, A.S., Robinson, P., Pritchett, K. and Sandøe, P. (2003). Animal research ethics. In: Hau, J. and Van Hoosier, G.L. Jr. (eds.) Handbook of laboratory animal science. Second edition. Vol. 1, Essential principles and practices, CRC Press, Boca Raton, Florida, USA, pp. 13-30.

Ormandy, E.H. and Schuppli, C.A. (2014). Public attitudes toward animal research: a review. Animals 2014 4: 391-408.

Sandøe, P., Christiansen, S.B., Hansen, A.K. and Olsson, A.S. (2008). The use of animals in experiments. In: Sandøe, P. and Christiansen, S.B. Ethics of animal use. Oxford University Press, Oxford, UK, pp. 103-117.

Regan, T. (2004 (1983)). The case for animal rights (1983). Second edition. Berkeley, California, USA.

Aquaculture and the precautionary principle in the New Zealand Supreme Court

M. Kaiser
The Centre for the Study of the Sciences and the Humanities (SVT), University of Bergen, Allegaten 34, 5020 Bergen, Norway; matthias.kaiser@svt.uib.no

Abstract

In 1997 the author published an article called 'Fish-farming and the precautionary principle: context and values in environmental science for policy' (Kaiser 1997), in which he discussed various precautionary strategies in view of unwelcome fish escapes from the sea cages. There is no unique strategy given precaution, he claimed, but instead there are choices of strategy depending on basic value assumptions. 15 years later many things have changed, both concerning the knowledge and uncertainties about aquaculture and concerning the legal background, including the precautionary principle. Some people claim that the precautionary principle has lost both status and some of its appeal in national and international law. In contrast to this, the Supreme Court of New Zealand has recently (2014) passed a judgement where the role of the precautionary principle in a complaint about aquaculture development was discussed in great detail. In particular its implications in view of adaptive management strategies were highlighted. The paper will provide information of the Court's reasoning and assess its findings in relation to earlier views on precaution and aquaculture.

Keywords: environmental law, uncertainty management, risk, adaptive management, coastal zones

Introducing the debate about precaution: four contested points

Does adopting the precautionary principle actually make any difference in our life? In other words, after many years discussing what the precautionary principle could mean (e.g. Sandin, 1999), and negotiating whether or not to adopt it in legally binding texts, we can now ask whether its explicit adoption has led to decision-making which would have been different without the precautionary principle? If it makes a difference, then how? In many ways, this would not simply be asking about an isolated principle, but it could even be seen as a significant test case how science, society, politics and law interact (Kaiser, 1997). The precautionary principle has always been at the interface between these different regimes and cultures.

There are certain points that have been hotly debated about the precautionary principle in particular, and in this article we shall endeavour to comment on them in the light of the NZ case discussed here.

(1) One point which often got confused by different authors was the difference between the precautionary principle managing uncertainty on the hand, and the problems of managing risks on the other. Uncertainty and risk got blended in many commentaries (Sunstein, 2005), not realizing how risk management presupposes some quantitative estimate of the risk to be managed. Yet, because not all future possible harm can be meaningfully assessed with quantitative methods of risk assessment, one has understood that there is a significant space of uncertainty with which we also have to deal in decision making about future development (De Sadeleer, 2003). And uncertainty comes from many possible sources, not necessarily restricted to not having done enough research yet. It may come from the complexity and unpredictability of reality, as for instance in stochastic or truly non-linear processes.

Another point (2) which was widely discussed (Van der Sluijs *et al.*, 2007) was whether applying the precautionary principle in practice meant to totally abstain from technological change, and stick to known practices. This was often implicit in criticisms which claimed that precautionary principle essentially is totally risk-aversive, and thus hampers all progress. If our forefathers had acted according to the precautionary principle we would never have gotten the wheel, the fireplace, the baking oven, or – in our days – the automobile, so claim people who oppose precautionary principle (e.g. Goklany, 2001). Yet proponents of precautionary principle have been claiming that endorsing the precautionary principle will not only promote the careful development and eventual improvement of new technological ideas, but also stimulate practices and encourage alternative technologies, as one has seen in environmental engineering. This debate is, of course, as much about progress and what the true nature of progress is, as it is about the precautionary principle. Yet, proponents of the precautionary principle were sometimes portrayed as anti-progressivists.

A further point (3) of discussion was if the precautionary principle allowed for gradual responses at all, or whether it is about a yes-or-no decision (cf. Kaiser, 2003). This confusion was to a large extent also caused by enthusiastic supporters of the precautionary principle, as they were to be found among several environmental NGOs. They often argued with a simple equation, namely that applying the precautionary principle implied that one cannot go ahead with the technology: if precautionary principle, then no risky new technology. A prominent example was the debate about gm food. Yet, other people argued that the precautionary principle could indeed in certain special cases imply a temporary moratorium, or perhaps even absolute abstention from a technology, but that in principle a wider range of precautionary responses would be available to the decision maker. Thus precautionary principle would allow for gradual responses, and that the proportionality of the response would matter.

Finally, a fourth point (4) had to do with the nature of the threat, which was at stake. Some people feared the possible harm had to be such that it could be given an economic (monetary) value. This fear was based on the assumption that somewhere at the end of the assessment it all would boil down to a classical form of risk-cost-benefit-analysis. Potentially this could lead to very counter-intuitive estimates, for instance when dealing with some local environmental consequence which in practice might be irreversible and thus the loss might be calculated as infinite, against a finite benefit. Yet other people claimed that the precautionary principle goes beyond a mere cost-benefit analysis. While it does consider various costs and various benefits, including the cost of foregone benefits (e.g. due to inaction), and while the precautionary principle does seek to achieve a sensible relation between costs and benefits, it does not necessarily rest on classical cost-benefit analyses (cf. Beder, 2006). So called intangible values or values which are perceived as important by many people without being measurable in importance, will play a part on the choice of precautionary strategies.

There are more points which were discussed in relation to the precautionary principle. The whole debate was in the end highly politicized, and even sociologically the debate seemed to reflect cultural divides and different legal systems (Cooney, 2005). I have addressed some of the more political aspects during the last EurSafe conference in Uppsala (Kaiser, 2013), and thus I shall leave it with this, and rather now turn to the New Zealand case.

Precaution about aquaculture development at the New Zealand Supreme Court

Aquaculture is not an entirely new activity in New Zealand. It has gradually developed since the 1980's, and by now a large portion of its products is exported to other countries. So far it is mainly mussels, oysters, and salmon, which are grown in aquaculture facilities. The government supports the growth of this sector, especially since 2007. Accordingly the industry is rather well regulated. Any aquaculture facility needs to apply for licenses to utilize the coastal waters they need. These applications are an integral part of the New Zealand Coastal Policy Statement (NZCPS). This together with the Resource Management Act (RMA, 1991, amended 2011) forms the basis for spatial and marine planning decisions through the local authorities. It is probably fair to say that the content of these legal frameworks resemble similar regulations and law of many other highly developed countries. Therefore it is also natural for the New Zealand court to look at – among others – Australia and Canada when dealing with concrete cases.

What is interesting in this case is that a more principled discussion was taken all the way to the Supreme Court of New Zealand, and that a significant part of the Court's findings were related to a detailed and specific interpretation of the precautionary principle. At the heart of the conflict which was brought to Court, was the plan of a company, the NZ King Salmon Company Ltd, to install new salmon cages in eight locations. The problem was that they wanted to develop this in an area which so far had been excluded from aquaculture development, and it would make planning changes necessary to transform the relevant area-plans from aquaculture prohibiting to discretionary activity areas, i.e. areas where licenses were handed out upon application given they do not infringe other interests or harm the environment. The government appointed a five-member ad hoc committee with various experts (hereafter: the Board) to decide the case. November 2011 the Board decided to grant plan changes in relation to four of the proposed sites. This decision was then appealed by two independent NGOs, the Environmental Defence Society (EDS) and the Sustain Our Sounds Inc (SOS). However, these two appellants addressed different issues. The EDS addressed their complaint only against one of the chosen sites and argued that this site should continue to be excluded from aquaculture development. The given reason was that this was a site of outstanding natural character and beauty, an outstanding natural landscape in the coastal environment. The SOS challenged all four plans on the grounds of the supposed danger to the water quality in the Sounds. The reason for this was the pressure on water quality from salmon feed and salmon faeces which might lead to eutrophication and increased algal blooms through increased nutrient concentration, especially nitrogen. The Supreme Court heard both cases together, while they were dealt with in separate judgements (SC 82/2013 (2014) NZSC 38 – hereafter simply referred to as: EDS-case – and SC 84/2013 (2014) NZSC 40 – hereafter referred to as SOS-case). The judgments are rather comprehensive in their reasoning, and only selected aspects of them shall be discussed here.

In the following I shall present and discuss some of the interesting points in the Court's findings using the four contested points about the precautionary principle mentioned above.

Risk versus uncertainty

In the SOS-case the expected effect of increased nutrient concentration played a central role. The factual basis for such an assessment proved problematic. First one lacked a reliable baseline. It was not quite clear what the current status of the sounds were. It was acknowledged that the trophic state of the water is indicative of their biological productivity (8; references are to the numberings of the chapters in the judgements). Second the effect of existing mussel farming was unknown, since such farms acted as nutrient sinks and counteracted eutrophication. The Court accepted that the current state might be oligotrophic/mesotrophic, and that the change into an eutrophic state would be biologically disastrous (10). An immediate exploration of current developments, baseline surveys, was thus called for (42).

Now, how is one to assess the possible changes in water quality when there is paucity of data? King Salmon presented an expert to the Board who offered three different models related to water quality, a mass balance model, a flushed aspatial model and a spatially explicit model, the SELFE model (43). Modifications of the models followed a peer-review process initiated by the Board. The Board accepted two of the models as particularly useful for the case at hand. However, strong criticism was raised against the company since it only provided model estimates for the initial feedings and biological loads, not for the maximum number applied for. This effectively excluded a worst case scenario.

Yet, in spite of impressive risk modelling, the Court stresses uncertainty:

> However, it (the Board) accepted evidence that such modelling would not necessarily provide any more certainty when attempting to quantify those effects (48).

> However, the Board considered that there remained considerable uncertainty 'as to the nature of the receiving environment, including the trends in their nutrient sources' and consequently in the ability of the Sounds to assimilate a significant increase in nutrients adequately (56).

In sum, the Board and the Court expressed a conviction about the utility of risk modelling, which could 'assist in the overall management of the Sounds' (48) but recognized that major uncertainties were inherent in the process. Such uncertainties would also make the employment of the precautionary principle a necessary management tool, as it is indicated in the NZCPS and RMA. I find this a remarkably clear recognition of the difference between risks and uncertainty, and the appreciation of the inherent uncertainties involved in nature management.

Not zero risk

It is also remarkable that the legal framework of NZ and the interpretation it received in these court cases at no point underestimates the principal benefits the commercial development of aquaculture could have. This is in line with their Coastal Policy statement on Aquaculture, where it reads: 'Recognise the significant existing and potential contribution of aquaculture to the social, economic and cultural well-being of people and communities' (NZCPS, 2010a). Already when King Salmon's proposals were judged to involve matters of national significance, one can see that any such decision was not easily referred to lower regional decision making levels. The initial decision to open these new sites for discretional aquaculture activity was based on an argument that they were particularly suited for salmon farming. 'The remoteness of the site and its location close to the Cook Strait made it attractive from a biosecurity perspective' (20, in EDS-case).

The relevant background is what is referred to as the 'single purpose of the RMA – the sustainable management of natural and physical resources' ((1227) in (35) EDS-case). This 'also allows for the balancing of conflicting considerations in terms of their relative significance or proportion in the final outcome' ((80) in (34) in EDS-case). This aspect of the balancing of conflicting interests goes through both judgements.

The precautionary principle is a central tool in the sustainable management of the NZ environment: 'Adopt a precautionary approach towards proposed activities whose effects on the coastal environment are uncertain, unknown, or little understood, but potentially significantly adverse' (NZCPS, 2010b). But it is by no means a dominating principle that would automatically sideline industrial interests and utilisation of natural resources. Precautionary principle can only be seen as a sub-ordinate tool to assess relevant options proportional to alternatives. As such it is certainly not stopping development and the criticism of 'progressivists' seem unfounded, at least when looking at the NZ practice.

Precautionary responses in gradual measures

One of the most remarkable features of the New Zealand Supreme Court deliberations is how adaptive management practices is treated as an adequate response to precaution (cf. also NZCPS, 2010b). The SOS-case was against all sites, based on fear for the quality of the water. I have also already noted that the available models were deemed good but also beset with major uncertainties. Thus the call for adopting the precautionary principle was evident. The question was which strategy this would imply.

This was the point where adaptive management entered the deliberations:

> The issue for the Court is when an adaptive management approach can legitimately be considered a part of a precautionary approach. This involves the consideration of the following: what must be present before an adaptive management approach can even be considered and what an adaptive management regime must contain in any particular case before it is legitimate to use such an approach rather than prohibiting the development until further information becomes available (124).

In this case the Court concluded that the available information though essentially inconclusive, still provided a basis for adaptive management to regulate production effectively so that critical thresholds were not passed:

> As to the threshold question of whether an adaptive management regime can even be considered, there must be an adequate evidential foundation to have reasonable assurance that the adaptive management approach will achieve its goals of sufficiently reducing uncertainty and adequately managing any remaining risk. The threshold question is an important step and must always be considered (125).

The discussion then entered some interesting principal points:

> The secondary question of whether the precautionary approach requires an activity to be prohibited until further information is available, rather than an adaptive management or other approach, will depend on an assessment of a combination of factors: (1) the extent of the environmental risk (including the gravity of the consequences if the risk is realised); (2) the importance of the activity (which could in some circumstances be an activity it is hoped will protect the environment); (3) the degree of uncertainty; and (4) the extent to which an adaptive management approach will sufficiently diminish the risk and the uncertainty (129).

What one sees here is a highly developed sensitivity to questions of uncertainty and precautionary strategies to manage the risks. The UNESCO/COMEST report on the precautionary principle (2005) called for precautionary interventions that would seek to avoid or diminish the envisaged possible harm, and it called for proportionality and continued monitoring. It appears that the NZ Supreme Court had these goals in mind when it opted for considering adaptive management as a precautionary strategy. The COMEST report is among those that were referred to as being consulted. Thus it does not come as a surprise when the final conclusion of the judgment states:

> The Board was entitled to consider that the adaptive management regime, reflected in both the plan and the consent conditions, was consistent with a proper precautionary approach (158).

The value of a scenic view

The final point I want to comment on may not be characteristic of only the precautionary principle, but may be of importance to sustainability in general, though it is clear that it is a crucial point for precautionary principle as well. I am talking about the consideration of values other than those, which easily can be transformed into monetary values. This is a point about ethics, and what kind of ethical approach we build upon. As long as things can be given a value which has a monetary equivalent, we can be reasonably sure that such a value can be handled through fairly traditional administrative procedures. Cost-benefit analyses demands this quantification into a unitary currency, and this is widely used in public administration. But what happens when things get more complicated, and certain values escape this quantification, or any ascription of such a value is totally ad hoc and the result of subjective whims?

Many people who engage in ethics will reply that other ethical traditions will accommodate these values better than utilitarian cost-benefit approaches. Deontological tradition based on respect and dignity could do this, but also some virtue ethics, or even rights-ethics approaches. But conservationist NGOs sometimes criticize public administration precisely because they cannot see that such values are readily admitted into their deliberations (cf. also Beder 2006: 129 ff).

Not so in the NZ case. In the EDS-case the question was raised whether one particular siting was of such a nature that it had an exceptional environmental value, in spite of not having particular use values or similar. It referred to the RMA and the occurrence of 'amenity values' in it. The Court states:

> The term 'amenity values' ... of this definition (in RMA) is itself widely defined to mean 'those natural or physical qualities and characteristics of an area that contribute to people's appreciation of its pleasantness, aesthetic coherence, and cultural and recreational attributes'. Accordingly, aesthetic considerations constitute an element of the environment (23).

Thus the Court finds it legitimate to exclude the area under consideration from aquaculture activities.

Conclusion

This paper examined four points of a general discussion around the precautionary principle in the context of the recent findings of the Supreme Court of NZ. Contrary to many precautionary principle skeptics I found that the precautionary principle did play a crucial role in the judgements of the Court, that it contributed to resolving a given conflict, and that this was done with considerable insight what precautionary approaches imply and do not imply. The court deliberations were without the rhetoric that often follow debates about precautionary principle in academia or in politics, but with an acute sense of the ethical nature of the choices involved.

Acknowledgement

I gratefully acknowledge the useful comments by an anonymous reviewer.

References

Beder, S. (2006). Environmental principles and policies – an interdisciplinary introduction, Earthscan, London, Sterling VA, USA.

Cooney, R. (2005). From promise to practicalities: the precautionary principle in biodiversity conservation and sustainable use. In: Cooney, R. and Dickson, B. (eds.) Biodiversity and the precautionary principle – risk and uncertainty in conservation and sustainable use. Earthscan, London, Sterling, VA, USA, pp. 3-17.

De Sadeleer, N. (2003). Environmental principles: from political slogans to legal rules. Oxford University Press, Oxford, UK.

Goklany, I.M. (2001). The precautionary principle – a critical appraisal of environmental risk assessment, Washington D.C., USA.

Kaiser, M. (1997). Fish-farming and the precautionary principle: context and values in environmental science for policy. Foundations of Science 2 (2): 307-341.

Kaiser, M. (2003). Ethics, science, and precaution – a viewpoint from Norway. In: Tickner, J. (ed.) Precaution, environmental science, and preventive public policy, Island Press, Washington D.C., USA.

Kaiser, M. (2013). Precaution or prudent vigilance as guiding the path to global food security?. In: Röcklinsberg, H. and Sandin, P (eds.), The ethics of consumption: the citizen, the market and the law, Wageningen Academic Publishers, Wageningen, the Netherlands.

New Zealand Coastal Policy Statement (NZCPS) (2010a). Aquaculture. Available at: http://tinyurl.com/qzx5zf4.

New Zealand Coastal Policy Statement (NZCPS) (2010b). Guidance note policy 3: Precautionary approach. Available at: http://tinyurl.com/qzrj2o2.

Resource Management Act (RMA) (1991). Available at: http://tinyurl.com/ncvsnzf.

Sandin, P. (1999). Dimensions of the precautionary principle. Human and Ecological Risk assessment 5: 889-907.

SC 82/2013 (2014) NZSC 38 and SC 84/2013 (2014) NZSC 40. Available at: http://tinyurl.com/osakxnd.

Sunstein, C.R. (2005). Laws of fear – beyond the precautionary principle, Cambridge University Press, Cambridge, UK.

UNESCO/COMEST (2005). The precautionary principle. A report of an ad hoc working group, endorsed by COMEST/UNESCO April 2005 (members of the expert group: M. Kaiser (chair), J.v.d. Sluijs (rapporteur), S. Beder, V. Hösle, A. Kemelmajer de Calucci, A. Kinzig; UNESCO secretariat: H. t. Have, S. Scholze, E. Kuok), UNESCO: Paris, France. Available at: http://tinyurl.com/k7883v9.

Van der Sluijs, J.P., Kaiser, M., Maxim, L. and Douguet, J. (2007). Le principe de precaution, un principe pour agir dans des situations d'incertitude. Risques, 72:75-81.

Stakeholders' perception on sustainability of genetically modified potato

F. Gillund[1], A. Myhr[1] and A. Hilbeck[2]*
[1]GenØk, Centre for Biosafety, Forskningsparken i Breivika, Postboks 6418, 9294 Tromsø, Norway; [2]Swiss Federal Institute of Technology, Institute of Integrative Biology IBZ, Universitätstrasse 16, 8092 Zurich, Switzerland; froydis.gillund@genok.no

Abstract

The importance of, but also challenges related to, assessing socio-economic impacts of genetically modified (GM) crops is receiving increasing attention among policymakers both within the EU and internationally. The Norwegian Gene Technology Act (1993) was one of the first legal frameworks for GMOs in the world with specific criteria for the assessment of contributions to sustainable development, benefits to the community and ethical justifiability. Experience has shown that practical implementation of these criteria can be challenging and that the issues to consider may differ depending on the GM traits introduced and the intended use of the GMO in question. During summer and fall of 2014 we organized two stakeholder workshops with the aim to gather perspectives on the sustainability of cultivating late blight resistant GM potato in Norway. 16 stakeholders involved in potato production in Norway took part in the workshops. The participants identified a number of conditions that GM potato with late blight resistance ought to satisfy in order to contribute to sustainable development. Among them, ethical considerations such as corporate control over genes and seed markets, transparency and openness during product development and risk assessment processes, opportunities for independent research and consumers' rights were recognized as issues that are important to consider in a sustainability assessment. The outcomes of the workshops are highly relevant for policy making and contribute to the operationalization of the assessment criteria regarding sustainable development in the Norwegian Gene Technology Act.

Keywords: agricultural biotechnologies, participation, sustainability assessment

Introduction

The requirements for risk assessment of genetically modified organisms (GMOs) in regulatory frameworks are typically restricted to documenting safety for the environment and human health. Many argue that the use of agricultural biotechnologies raises questions that go beyond these narrow and scientifically defined risk aspects (De Melo-Martin and Meghani, 2008). The need to consider broader concerns, such as socio-economic issues, in GMO regulations is increasingly recognized (UNEP, 2012; European Commission, 2011). Norway, with its Gene Technology Act (1993), was a pioneer in this respect. The Act requires an assessment of whether the deliberate release of a GMO contributes to sustainable development, benefits the community and is ethically justifiable. The Norwegian Biotechnology Advisory Board (NBAB), which has a special responsibility for assessing these criteria, has led several projects intending to contribute to the operationalization of these criteria. This work has shown that issues to consider in a sustainability assessment may differ depending on the type of GM crop in question (NBAB, 2014).

Funtowicz and Ravetz (1990) argue that involving an 'extended peer community', meaning all those that express a desire to participate in the resolution of an issue, will strengthen the quality of risk assessment processes. They particularly highlight that it is crucial to involve the public in assessments of innovations in science and technology when searching for sustainable solutions (Funtowicz *et al.*, 1999). Along these lines, Voß *et al.* (2006) emphasize that sustainability is an ambiguous and moving target that can only be ascertained and followed through processes of iterative, participatory goal formulation. These ideas are also reflected in the call for 'upstream public engagement' (e.g. involvement of the public at an early stage of development process (Wynne, 1992), the 'negotiated science' approach presented by Carr and Levidow (1999) and in the notion of 'opening up' the process of knowledge generation and decision making (Stirling, 2009). All these approaches resonate with the notion of responsible governance of research and innovation (Stigloe *et al.*, 2013).

We organized two workshops with the aim to identify relevant and important issues for a sustainability assessment of GMOs through stakeholder deliberation. The workshops concerned sustainability of cultivating GM potato with late blight resistance in Norway and took place in June and October 2014. 16 Norwegian stakeholders involved in potato production or agriculture in general participated. An important aim of the workshops was to contribute to the operationalization of the assessment criteria regarding sustainable development in the Norwegian Gene Technology Act. We will present some outcomes of the workshop discussions, focusing on the participants' understanding of and reflections on sustainability in this context and our experiences of facilitating such discussions.

The case: late blight and the development of GM potato with late blight resistance

Potato late blight, which is caused by a fungus-like organism called *Phytophthora infestans*, is the most devastating potato disease worldwide. Throughout the last century efforts have been made to develop commercial potato varieties with increased resistance to late blight using a variety of approaches. Due to the remarkable ability of *P. infestans* to overcome resistance, conventional potato breeders have only had limited success in developing commercial potato varieties with resistance that is lasting. One approach is to use genetic modification to introduce several genes coding for resistance against different strains of *P. infestans* and, thus, create 'stacked' GM potato varieties with a broad spectrum resistance expected to be more durable. Hence, the idea is to simulate a multi-gene quantitative trait by adding up several single-gene resistance traits (Vleeshouwers *et al.*, 2011). Several European companies and research institutes are involved in this research and field trials have taken place in several locations in Europe since 2006.

Cultivation of late blight resistant GM potatoes is a highly interesting case for a discussion about sustainable agriculture and food production for several reasons. Firstly, late blight is, as already mentioned, the most devastating potato disease worldwide, estimated to cause an annual global loss of 3.72 billion € (Haverkort *et al.*, 2008). Research has shown that the disease has become more aggressive in Europe during the last decade (Cooke *et al.*, 2011), and the problem might increase further due to climate change which is expected to cause more rainfall during the growing season. Secondly, current control measures depend heavily on fungicide applications. For instance, about half of the total fungicide consumption in Norwegian agriculture is currently used to control late blight on potatoes. Hence, late blight results in huge economic losses for producers and current control measures come with environmental costs. Thirdly, many of the late blight resistant GM potato lines currently undergoing field trials in Europe are cisgenic plants, i.e. all the introduced genes are derived from sexually compatible species (Haverkort *et al.*, 2008). Development of cisgenic plants is a new field of research within GM breeding, and consumer surveys have shown more acceptability towards this type of GM crops (Gaskell *et al.*, 2011).

Methodology: world café and problem formulation and options assessment

We organized two workshops with Norwegian stakeholders, including potato breeders, seed potato retailers, seed potato and potato producers (practicing both conventional, organic and biodynamic production forms), representatives from the potato processing industry, researchers in potato and plant pathogens, agricultural advisors, regulators and representatives of environmental and consumer interests. 16 stakeholders participated in the first workshop which lasted two days and took place in June 2014. All participants were invited to a one-day follow-up workshop in October 2014, in which 11 were able to participate.

We started the first workshop with a world café-style activity (Brown, 2005). Four questions were presented (one at a time) which the participants were asked to reflect upon and provide answers to in writing. The answers were shared and briefly discussed in small groups and summarized in plenary. The questions concerned the main challenges in potato production, understanding of sustainable agriculture and food production, expectations about the future of Norwegian potato production and understanding of the relationship between technology, humans and nature. The answers were collected on different posters and placed on the wall so that the participants could easily refer to them in later discussions during the workshop.

The main part of the discussions during the workshops was based on a framework called Problem Formulation and Options Assessment (PFOA). This is a framework for environmental risk assessment of GM plants developed by a group of international trans disciplinary researchers. It intends to integrate societal perspectives throughout the assessment process. The PFOA process emphasizes engagement with stakeholders in an iterative series of stages, from identification of the problem(s) through comparison of multiple technology solutions that could be used in the future with their relative benefits, harms, and risks. The method has been used and refined in many countries including assessment of Bt corn in Kenya, and Bt cotton in Brazil and Vietnam (Nelson *et al.*, 2009). We did not conduct a full PFOA during the workshops as we were primarily interested in identifying criteria for sustainability assessment of late blight resistant GM potato. Besides, as there has not yet been any applications for deliberative release of this type of GMO in the EU or elsewhere, the available information about the product is insufficient to conduct a full PFOA. The exercises during the workshops were however structured in four steps based on the framework: (1) characterization of the late blight problem in Norway; (2) identification of different strategies to control late blight; (3) comparing the benefits, disadvantages and knowledge gaps associated with conventionally bred potatoes and cisgenic GM potato with late blight resistance; and (4) identifying questions to assess whether cultivation of GM potato with late blight resistance contributes to sustainable development.

At the end of each workshop the participants were asked to fill in an evaluation form. These forms included questions for reflection on sustainability and to what extent their understanding of this concept had changed as a result of participating in the workshop.

Results: participants' perspectives on sustainability

Participants' reflections on sustainability during world café discussion

The participants used several keywords to describe sustainable agriculture and food production. Long term food production, recognizing the needs of future generations and responsible management of resources were the three most frequently mentioned issues. Several referred to the three pillars of sustainability and emphasized the maintenance of natural resources, economic profit for farmers and societal utility. A few mentioned that food production should be based on Norwegian resources, securing a populated countryside with a flourishing agriculture.

Participants' reflections on sustainability in the PFOA discussions

The participants suggested several conditions which a late blight resistant GM potato ought to satisfy in order to contribute to sustainable development:

- GM potato should contribute to reduce the use of fungicides to combat late blight.
- GM potato should be safe for human health and the environment in the long term.
- The resistance of the GM potato should be durable.
- GM potato should contribute to increased profitability for potato producers over time.
- GM potato should be accepted among consumers.
- Impacts on the diversity of commercial potato varieties and coexistence with conventional and organic agriculture should be assessed.
- Institutions developing GM potatoes should strive for transparency and good communication with the public and regulatory authorities.
- GM potato should be available for further breeding and independent research.
- Risk assessment and regulation should be based on the precautionary principle.

Many participants highlighted the need for further research to properly evaluate to what extent GM potato with late blight resistance satisfy these requirements. Many participants also stated that, based on what they had learned during the workshops, they were open and curious to this type of GM potato and had great faith in that cultivation of this GM potato would contribute to a more sustainable potato production. This positive attitude was mainly based on that: (1) cultivation of this type GM potato will reduce the need for fungicides; (2) many of the late blight resistance GM potato lines currently undergoing field trials are cisgenic plants; (3) there are few challenges related to coexistence and unintentional spread of GM potato in Norway; and (4) some of the institutions that develop this GM potato are publicly funded and strive for transparency and recognition of plant breeders' rights.

Feedback in evaluation forms regarding sustainability

The participants felt that the most relevant issues related to sustainability of cultivating GM potato with late blight resistance were discussed during the workshops. Some commented that the workshop discussions were a good starting point for an important debate, and everyone agreed that the findings are valuable contributions for a future impact assessment of this type of GM potato in Norway. Some commented, however, that they wanted a better clarification of what is meant by the concept of sustainability in this context, as well as a wish for more focused discussion. Most of the participants did not feel that their understanding of sustainability had changed after participating in the process.

Discussion

The intensity and engagement in the workshop discussions clearly showed that the participants found the topic interesting and important. Many ethical considerations were brought forward in the discussions. For instance, issues such as corporate control, patents, transparency and openness in the product development and risk assessment process, the need for independent risk research and consumers' right to know and choose what they eat were recognized as relevant issues to consider in a sustainability assessment. This underlines the importance of broadening the scope of risk assessments beyond an assessment of product quality and safety to human and environmental health. Some participants were however concerned that by identifying too many questions for assessing sustainability, the group could halt the development of a product that could potentially promote a more sustainable form of potato production than what is currently practiced in conventional potato production. This was based on an assumption that cultivating this type of GM potato would be more sustainable than chemical control, i.e. extensive use of fungicides, of the pathogen. This hesitancy to engage in the task among some of the participants clearly influenced the discussion.

While only a few expressed that their understanding of sustainability had changed, almost all felt they had learned more about different approaches to breeding for late blight resistant potatoes after participating in the workshops. When considering the role the workshops had played as a learning exercise for the participants, it may be concluded that they functioned more as an arena for exchange of technical knowledge, rather than an exercise that stimulated and broadened the base for reflection on sustainability aspects. The reflections shared in the world café activity and the arguments made in the remaining discussions do however indicate that the participants were quite familiar with the concept of sustainability. Many came with a clear understanding of the concept and its use in this context. Deep reflection on complex issues like sustainability requires time. As emphasized by some of the participants, time constraints did often limit the discussions to an exchange of opinions rather than thorough considerations and dialogue. Hence, with more time it is likely that the workshops could have reached more of its potential as an exercise for reflection.

The participants were never asked to state their opinion about this type of GM potato or its potential to contribute to a more sustainable potato production in Norway. Still, many were eager to do so, particularly those who held a positive view. While the stakeholders' urge to state opinions was welcomed, it also raised some challenges for the discussions. As this type of GM potato is not yet on the market, many of the arguments made by the participants were based on expectations, hopes and assumptions rather than on documented facts (for instance concerning the expected durability of the resistance in the GM potato plants or that transgenic plants are safer than cisgenic plants). The scientific information needed to verify many of the claims is not yet available, or characterized by scientific disagreements and uncertainties. This is common challenge for any technology assessment performed at an early stage which undoubtedly sets limits for a knowledge-based discussion.

Lack of, or disputed, knowledge can even be more problematic if there is some degree of imbalance in the power-relationships among the participants, meaning that some participants hold a higher academic degree or feel to hold more privileged expert knowledge in the field over that of others. In our case the participants represented a diverse range of actors with different interests, knowledge and experience. This allowed for a broad discussion based on both practical experience and knowledge with the subject. Participants with extensive technical expertise and advanced knowledge on the topics obviously helped to enrich the discussions. Still, in some instances, assumptions and expectations were presented as 'facts' by some of the more experienced participants, with the result that it became challenging for participants with less specific knowledge on the subject to express themselves or offer counter evidence and be heard for their views.

Conclusion

The stakeholders participating in the two workshops presented here contributed to the framing of a sustainability assessment of GM potato with late blight resistance by identifying relevant and important issues to assess. Several conditions which a late blight resistant GM potato ought to satisfy in order to contribute to sustainable development were suggested. This provides important insights to policymakers in Norway and elsewhere on issues that could be relevant to consider in a sustainability assessment. The workshop discussions did however reveal opposing views on what a sustainability assessment should entail. These views should be further elaborated through various forms of public involvement so that the full range of perspectives on sustainability is exposed and ideally taken into account in future decision-making processes. More research, intended to document whether this type of GM potato satisfies the conditions suggested by the stakeholders, as well as to investigate the assumptions underlying some of the claims put forward in the discussions, is needed.

References

Brown, J. (2005). The world café: shaping our futures through conversations that matter. Berrett Koehler, San Fransico, USA, pp 265.

Carr, S. and Levidow, L. (1999). Negotiated science the case of agricultural biotechnology regulation in Europe. In: Collier, U., Orhan, G. and Wissenburg, M. (eds.) European discourses on environmental policy, Ashgate Publishers, Aldershot, UK, pp. 159-72.

Cooke, L.R., Schepers, H.T.A.M., Hermansen, A., Bain, R.A., Bradshaw, N.J., Ritchie, F., Shaw, D.S., Evenhuis, A., Kessel, G.J.T., Wander, J.G.N., Andersson, B., Hansen, J.G., Hannukkala, A., Nærstad, R. and Nielsen, B.J. (2011). Epidemiology and integrated control of potato late blight in Europe. Potato Research 54: 183-222.

De Melo-Martin, I. and Meghani, Z. (2008). Beyond risk. Embo reports 9: 302-308.

European Commission (2011). Commission publishes report on socio economic aspects of GMO cultivation in Europe. Available at: http://tinyurl.com/klml7ha.

Funtowicz, S.O. and Ravetz, J. R. (1990). Uncertainty and quality in science for policy. Kluwer, Dordrecht, the Netherlands, 229 pp.

Funtowicz, S.O., Ravetz, J. and O'Connor, M. (1999). Challenges in the use of science for sustainable development. International journal for sustainable development 1: 99-107.

Haverkort, A.J., Boonekamp, P.M., Hutten, R., Jacobsen, E., Lotz, L.A.P., Kessel, G.J.T., Visser, R.G.F., Vossen, E.A.G. Societal costs of late blight in potato and prospects of durable resistance through cisgenic modification. Potato Research 51:47-57.

Gaskell, G., Allansdottir, A., Allum, N., Castro, O., Esmer, Y., Fischler, C., Jackson, J., Kronberger, N., Hampel, J., Mejlgaard, N., Quintanilha, A., Rammer, A., Revuelta, G., Stares, S., Torgersen, H. and Wager, W. (2011). The 2010 Eurobarometer on the life sciences. Nature Biotechnology 29: 113-114.

Nelson, K.C., Andow, D.A. and Banker, M. J. (2009). Problem formulation and option assessment (PFOA) linking governance and environmental risk assessment for technologies: a methodology for problem analysis of nanotechnologies and genetically engineered organisms. Journal of Law, Medicine and Ethics 37: 732-748.

Norwegian Biotechnology Advisory Board (NBAB) (2014). Herbicide-resistant genetically modified plants and sustainability. Available at: http://tinyurl.com/khwkl44.

Stilgoe, J., Owen, R. and P. Macnaghten. (2013). Developing a framework for responsible innovation. Research Policy 42: 1568-1580.

Stirling, A. (2009). Participation, precaution and reflexive governance for sustainable development. In: Agder, N. and Jordan, A. (eds.) Governing sustainability, Cambridge University Press, UK, pp. 193-225.

United Nations Environment Programme (UNEP) (2012) Report of the workshop on capacity building for research and information exchange on socio-economic impact of LMOs. Available at: http://tinyurl.com/m7tya6b.

Vleeshouwers, V.G.A.A, Raffaele, S., Vossen, J., Champouret, N., Oliva, R., Segretin, M.E., Rietman, H., Cano, L.M., Lokossou, A., Kessel, G., Pel, M.A. and Kamoun, S. (2011) Understanding and exploiting late blight resistance in the age of effectors. Annual Review of Phytopathology 49: 507-31.

Voß, J.P., Bauknecht, D. and Kemp, R. (eds.) (2006). Reflexive governance for sustainable development, Edward Elgar, Cheltenham, UK, 457 pp.

Wynne, B. (1992). Uncertainty and environmental learning: reconceiving science and policy in the preventive paradigm. Global Environmental Change 2: 111-127.

Socio-economic considerations in GMO Regulations: opportunities and challenges

R. Binimelis and A.I. Myhr*
Genøk-Centre of Biosafety, The Science Park, Pb 6418, 9294 Tromsø, Norway; rosa.binimelis@genok.no

Abstract

Socio-economic considerations (SECs) are included in many countries regulation frameworks on genetically modified organisms (GMO). At the same time, there are discussions about how to capture SEC and ethical impacts and how to implement assessment. With the intention to recognise the advances done so far, we present some important characteristics and issues of existing institutional frameworks, legislations and policies with provisions on SECs of GMOs. For doing so, information regarding the inclusion of SECs in biosafety decision-making was reviewed for 33 countries. We have analysed how SECs are defined in the different regulatory frameworks, taking into account which dimensions and aspects are assessed, and how the SEC are related to the environmental and health risk assessment of GMOs. Provisions for liability and redress in case of SECs harm are also briefly described, as well the role of public participation. Implementation of SEC provides both opportunities and challenges.

Keywords: sustainability, agriculture biotechnology, regulatory frameworks, impact assessments

Introduction

There is an increasing consensus internationally that decision-makers need to consider a broader range of issues and factors when assessing the use of genetically modified organisms (GMOs) in agriculture. They need to consider socio-economic considerations (SEC) and ethical issues and integrate these into broader 'impact assessments' that can take into account the context where the GMOs are meant to be introduced, as well as the systems they emerge from and support (Devos *et al.*, 2008; Pavone *et al.*, 2011). This inclusion of a broader range of issues is being called for in international frameworks (e.g. Article 26 of the Cartagena Protocol on Biosafety), European and African fora, as well as in an increasing number of national regulations. The Norwegian Gene Technology Act is the pioneering experience, emphasizing the need to consider social utility and contribution to sustainability, as well as direct and indirect impacts on agricultural practice and the socio-technological context.

With the increasing number of countries aiming for inclusion of a broader range of issues and factors, it has been recognised that there is a need for robust methodologies or frameworks able to capture ex-ante and ex-post SEC of GMO cultivation, for gaining grounded empirical knowledge, and for appropriate systems for data collection (Dalli, 2011; Eckerstorfer and Gaugitsch, 2013; Ludlow *et al.*, 2014). These gaps have driven, and are at the same time a result of, a rather contentious debate among GMO biosafety policy-makers and scholars on the desirability of including such considerations. Contentions have emerged around the scope, methods and disciplines involved, timing of consideration, baselines and comparators, criteria and indicators, 'endpoints' or targets, the role of public participation and the precautionary principle, the relationship with other fields of knowledge and with other the environmental and health risk assessment (Ludlow *et al.*, 2014; Spök, 2010).

The objective of this paper is to present some important characteristics and issues of existing institutional frameworks, legislations and policies with provisions on SEC of GMOs.

Methodology and information sources

The present study reviews information regarding the inclusion of SEC in biosafety decision-making in 33 countries (Figure 1).

The primary sources of information have been the laws, regulations and national policy frameworks, as well as other national official reports, if any. Additionally, the Convention of Biological Diversity Secretariat has compiled a series of summaries and reports (CBD, 2014a), as well as national surveys in order to assess how SEC on GMOs are taken into account (CBD, 2014a). These documents have also been reviewed for collecting information relative to the state of the art of the implementation of SEC in decision-making, as well as to the capacity-building needs. A literature review on the implementation of SEC in GMOs decision-making has also been conducted.

Characterisation of the legislative framework in countries that have included socio-economic considerations

Voluntary versus mandatory basis

There are 33 countries that have included socio-economic considerations in biosafety decision-making. The great majority of these countries have created new legislations for addressing biosafety, with the exception of Madagascar, Costa Rica and Venezuela. In 29 of the 33 analysed countries (81%), there is an obligation to take into account SEC, although the degree of implementation of this provision varies greatly among them. Although a considerable number of countries have legislation in place for assessing SEC in GMO-related decision-making, at least 40% of the analysed countries declare lacking implementation experience.

Objective of the legislations that have included socio-economic considerations

Most of the analysed regulations define the assessment endpoints deserving protection (e.g. protection goals) (21 of the total 33 countries), which would form the basis for the SE dimensions to be taken into account. Although most countries restrict the definition of SEC exclusively to social, economic aspects and, to a less extent, cultural/ethical aspects, many of the countries do also include health aspects (12 of the 33 analysed countries) and environment/ecosystem concerns (17 of the analysed countries) within the 'socio-economic' aspects of the regulatory framework (Figure 2).

Figure 1. Analysed countries (in black).

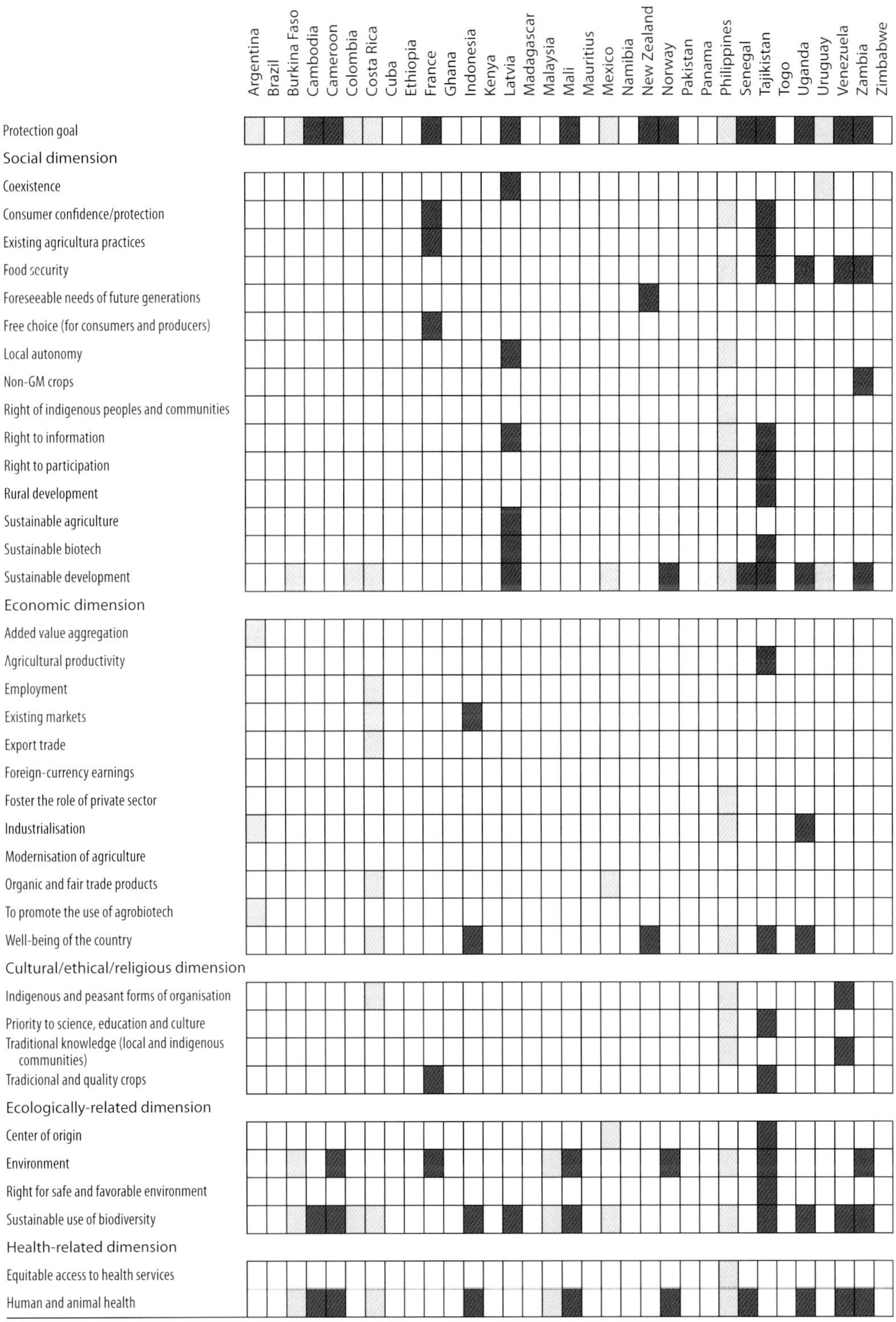

Figure 2. Socio-economic considerations taken into account in biosafety decision-making.

It seems that the countries that have approved GM crops (in grey) are more focused on the economic dimension policies. No difference is found for the rest of the dimensions (social, cultural, health-related and ecosystem-related), included as protection goals in the different regulatory frameworks. The protection goals that are considered as relevant may change as countries are gaining experience with implementation of SEC in decision-making.

Methodological approach and use of criteria and indicators for assessing socio-economic considerations

Many of the analysed regulatory frameworks lack suitable methodological approaches in order to assess SEC. In some cases, there is only an indication of the methods to use, e.g. cost-benefit principles. Many of the analysed frameworks do not have indicators or guiding questions, although some of the countries establish SEC as a condition for approval. These are mainly linked to the contribution of the GM crop to the benefit of the public interest and its contribution to the sustainable development of the country. In most of the cases, there is no indication on the appropriate methodologies for the assessment or on how to integrate divergent results, or on the baselines or the comparators that should be used for conducting SEC assessments, although the choice of appropriate baselines are crucial for the results of an assessment.

Options for SEC assessment in national frameworks: How does the SEC assessment relate to the environmental and health risk assessment

For the analysed countries, only two options were found for the relation between SEC assessment and the environmental and health risk assessment:
- SEC were assessed in the general risk assessment procedure;
- SEC were evaluated through an independent assessment (sometimes consecutive to the environmental and health risk assessment, others in a parallel but separated process).

Interestingly, in some countries SEC are defined as part of the environment, thus indirectly the SEC assessment is conducted together with the RA. For the great majority of cases, there is no specification for the level of analysis to be taken into account in the case of SEC assessment. Risk assessment is done as a response to market applications on a case-by-case basis. While the procedure for how to assess SEC does not differ that much between countries, there is a huge variation with regard to what aspects of SEC that need to be taken into account. In the cases in which more detailed descriptions of SEC aspects are given, usually the potential impacts address both micro (e.g. farm, households) and macro-levels (e.g. community or sector, country or region). In most of the procedures, however, the description of the considerations is so general that it is not possible to determine what level the legislation is referring to. In some of the analysed countries there are specific questions or criteria aiming at responding to concerns for (and/or from) more vulnerable sectors of the society.

Need to develop provisions for external and internal responsibility in GMO Regulation

Most of the countries that have been analysed do not have any provisions referring to external responsibility. In the case of Norway, there are guiding questions concerning the distribution of benefits, burdens and impacts from production and use among rich and poor countries, and also about global impacts on biodiversity and ecosystem functioning, focusing on differences between production and use.

Redress and liability systems in the assessed countries

All countries have implemented a redress and liability system, although not all of them have provisions related to redress and liability measures linked to socio-economic aspects in their frameworks. In a significant number of countries the SEC redress and liability system is linked to the admixture between GM and non-GM crops. In some of the cases, there is an explicit reference to the long-term character of SE impacts, having provisions that can cover potential delayed and indirect damage, as well as long-term damage after the introduction of GMOs (e.g. Togo, Zambia, Mali).

Participatory approaches in GMO Regulation

A participatory approach in GMO SE assessment may serve different purposes, ranging from informing and raising public awareness about GMOs, to the inclusion and representation of divergent societal opinions so as to increase transparency and legitimacy in the decision-making process. Public awareness and education is an important aspect raised by one third of the analysed biosafety regulatory frameworks. Most of the analysed countries have also regulated the access to information, by providing the public access to information on biosafety in general (for instance, by implementing National Biosafety Clearing Houses), as well as information on applications for GM crops approvals and on the authorised GM crops. In 30 of the 33 analysed countries, public participation is integrated in the decision-making on GMOs, although the role it plays in this process varies widely among them. In 14 of the countries public participation is binding or, at least, there is an obligation to take it into account when taking decisions on GMOs, while in the rest of the analysed countries, it plays an advisory role. However, important gaps exist on how these provisions are implemented, as well as to the weight that the outcome by public participation has in the final decision. Moreover, it was not possible to establish the role that public participation plays specifically in the assessment of SEC and ethical aspects, or if there are provisions to ensure that the public is engaged in a meaningful way (Fransen *et al.*, 2005), e.g. assuring that holistic arguments that fall outside the 'risk window' that only makes visible what has been predefined as a relevant risk (Jensen *et al.*, 2003) are not displaced by the purely technical issues that often dominate the debates on GMO risks (De Melo-Martin and Meghani, 2008).

Opportunities and challenges with the inclusion of socio-economic considerations

Methodological issues and approaches taken

The review points out that there are significant differences on how SEC are integrated and assessed into the studied national normative frameworks. Main differences are related to the status of the regulations (compulsory vs voluntary SE assessments) and approach. The research has made explicit that the inclusion of SEC responds to a diversity of national policy goals, which can partially explain the differences in framing, criteria and indicators. This shows the importance to acknowledge the value-choices in framing the socio-economic assessment in order to achieve and efficient and transparent assessment. Another interesting feature is the wide inclusion of health and environmental-related considerations as part of the SE assessment, blurring the separation between these dimensions that is demarcated by the traditional risk assessment.

Challenges, needs and recommendations

The review shows that there are inherent difficulties in the present methods and approaches that need to be employed to ensure that the implementation of SEC provisions becomes effective and can be performed in a systematic way. For instance, very few of the analysed regulations establish robust methodologies during the processes of framing, data gathering and assessment and decision-making related to SEC. The SE data to be assessed and that are identified may also be highly context-dependent. Transparent communication of uncertainties and knowledge-gaps due to lack of appropriate methods and data to address issues is therefore regarded as important (Eckerstorfer and Gaugitsch, 2013).

It must be expected that stakeholders use different conceptual frameworks, defined as a set of basic beliefs, values, attitudes and assumptions creating a frame through which they see themselves and the world, in their identification of which values are important to protect. In this sense, disagreement about the acceptability of risks is primarily a dispute over normative values (or policy goals) which cannot be resolved by more research but by making explicit the debate on values, and thus expanding the participation in the risk governance of GM crops (Devos *et al.* 2014). The difficulty of making normative values transparent can be illustrated by the example of the lack of use of appropriate comparators that could serve as baselines for assessing the impacts of by GMO crop introduction and cultivation. Further work needs also to be done on how to interweave expert and stakeholder input and other forms of knowledge.

Besides these challenges in problem-formulation, it is also necessary to make explicit the normative values during the assessment and decision-making phases. The implementation of SE assessments may result in contradictory or divergent results (not only between different SEC aspects but also between SEC and the environmental and health risk assessment), according to the framing used, the choice of assessment endpoints to be included, and the methods applied. This points out the importance for transparency, openness and accuracy in the communication of the assessment process. There is also a need to characterise the different roles played by stakeholders in the analysed regulatory frameworks at different phases of the assessment, e.g. at the beginning of the process so as to frame the issues, during the assessments so as to provide data, and at the end of the process for reviewing conclusions and providing opinions. Possible means for participation and/or consultation during SEC assessments need to be elaborated.

References

Convention of Biological Diversity (CBD) (2014a). Global overview of information on socioeconomic considerations arising from the impact of living modified organisms on the conservation and sustainable use of biological diversity. AdHoc Technical Expert Group on Socioeconomic Considerations.

Convention of Biological Diversity (CBD) (2014b). Results of the survey to gather information corresponding to indicators in the Strategic Plan. Available at: http://tinyurl.com/osgesog.

Dalli, J. (2011). Commissioner for health and consumer politics full picture of GMO cultivation is now shaping up. Hearing at the European parliament on the socio-economic dimensions of GMO cultivation Brussels, Belgium, 18 October. SPEECH/11/674.

De Melo-Martin, I. and Meghani, Z. (2008). Beyond risk. A more realistic risk-benefit analysis of agricultural biotechnologies. EMBO reports 9: 302-206.

Devos, Y., Maeseele, P., Reheul, D., Vanspeybroeck, L. and de Waele, D. (2008). Ethics in the societal debate on genetically modified organisms: a (re)quest for sense and sensibility. Journal of Agricultural and Environmental Ethics 21(1): 29-61.

Devos, Y., Sanvido, O., Tait, J. and Raybould, A. (2014). Towards a more open debate about values in decision-making on agricultural biotechnology. Transgenic Research 23(6): 933-943.

Eckerstorfer, M. and Gaugitsch, H. (2013). Summary report for the workshop 'framing socio-economic assessment in GMO & chemicals regulation', European Environment Agency, Copenhagen, 6-7 December 2012.

Fransen, L., La Vina, A., Dayrit, F., Gatlabayan, L., Santosa, D.A. and Adiwibowo, S. (2005). Integrationg socio-economic considerations into biosafety decision-making. The role of public participation. World Resource Institute White Paper, Washington, 47 pp.

Jensen, K.K., Gamborg, C., Madsen, K.H., Jørgensen R.B., Krayer von Krauss, M., Folker, A.P. and Sandøe P. (2003). Making the EU 'risk window' transparent: the normative foundation of risk assessment of GMOs. Environmental Biosafety Research 2: 161-171.

Ludlow, K., Smyth, S. and Falck-Zepeda, J. (eds.) (2014). Socio-economic considerations in biotechnology regulation. Natural resource management and policy series, Springer, New York, USA.

Pavone, V., Goven, J. and Guarino, R. (2011). From risk assessment to in-context trajectory evaluation – GMOs and their social implications. Environmental Sciences Europe 23(3): 1-13.
Spök, A. (2010). Assessing socio-economic impacts of GMOs. Issues to consider for policy development. Lebensministerium/Bundensministerium für Gesundheit, Vienna, Austria, pp 123.

The ethics of farmers to use land and equipment: case study of the rural area of Kurdistan Regional Government

R. Harun[1], I.C. Muresan[2], H.F. Arion[2] and D.E. Dumitras[2]*
[1]Department of Agricultural Extension, University of Sulaimani, Kurdistan Regional Government, P.O. Box 334 Sulaimani, Iraq; [2]Department of Economic Sciences, University of Agricultural Sciences and Veterinary Medicine Cluj-Napoca, Manastur St., No.3-5, 400372 Cluj-Napoca, Romania; rezhen25@yahoo.com

Abstract

Nowadays, the rural area of Kurdistan region is affected by many issues, between that the farmer ethics of using the land and equipment are one of the prominent issues in the region. The aim of this research is to find out the farmer ethics of using land and equipment properly for sustainable development of agriculture. The research area is represented by the Kurdistan Regional Government (KRG), located in the northeast part of Iraq. During June and September 2012, 236 farmers were interviewed. The field research indicates that the rural population of Kurdistan region is young and illiterate. The cultivated surfaces are between less than 2 ha until 100 ha. The farmers from KRG present non satisfaction with their lives and a weaker perception of community role in daily life. One of the main obstacles of the sustainable agricultural activity is represented by the lack of knowledge of the farmers regarding the ethical management and agricultural principles. A conceptual framework to sustain famers' ethics was developed starting from the principles of sustainable agriculture development and the influence of the farmers on these components. The proposed framework emphasizes the importance of paying a careful attention to the actions needed for a sustainable agriculture development in all decisions taken at the national level.

Keywords: sustainable development, no-till farming

Introduction

Ethics is related to social norms and moral principles (Falkenmark and Folke, 2002). Starting from this definition of ethics it is observed that it refers to the rightness or wrongness of certain actions. Thus, it may be asked what is right and what is wrong in the behaviour of farmers? What should the farmer do in order to act in accordance to the ethical standards or criteria?

Zimdahl (2006) provides a list of agriculture related questions that probably are addressed by many farmers, researchers, theoreticians, etc., that imply a choice between two different and even opposed situations which may result in other issues that are not necessary the optimal choice from an ethical point of view. For instance, is it better to increase the agricultural production to fulfill the entire demand or to pay more attention to the environmental harm? What would be the correct answer to this question?

On the other hand, it is noticed that one can talk about a mixture of ethics of modern agriculture, which comes from the fact that modern agriculture is considered to be superior to traditional agriculture when is related to the human resources and the environment of the farm, but the treatment of animals is inferior (Paarlberg, 2008).

It is clear that there are many unsolved issues when discussing agriculture and ethics. However efforts should be put into solving them step by step to assure that at least future generations will benefit of the fruitful and correct manners of acting. This can be achieved only by a proper use of the human resources and the environment, which are parts of the sustainable development. It is clear that the development of farming has to be based on sustainable development. However, it should not be neglected that in the end the purpose of any agricultural activity is to obtain a benefit, which in the case of an economic activity is the profit.

Farmers' action such as tillage, using pesticides and other 'unhealthy' activities might have negative impact on the environment. In the farmers' minds, the tillage seems to be right. It may be defined by the laws of the region or it may be just a cultural custom chosen because it helps to obtain more agricultural products. However, the law is not always a good guide to ethics (Burkhardt *et al.*, 2005) since some activities such as tillage is considered legal by the reality proves to be unethical, whereas other activities are illegal and not unethical (abuse chemicals or pesticides). Sullivan *et al.* (1996) suggests that to know the farmers' values and beliefs it would be useful to examine the competing philosophies in modern farmers. An understanding of the attitudes and motives of farmers definitely contributes to a better promotion of an environmentally and economically agriculture.

Aim of the study

This study has been designed with the primary goal to answer the question 'What is it about the farmer's ethics (meaning land and equipment users of the rural area of the Kurdistan region), that leads to the ethical linkage (primarily plowing, using machine, and the education of the farmer)?'. At the most basic level this question is broken down into seeking to understand the manner in which ethics is related to the farmer's activity.

On-farm investigations allowed the understanding of the extent of recent cropping practice change in the KRG and the ethical reasons behind the apparent unspreaded adoption of no-till. Moreover, a conceptual framework was built to sustain the farmers' ethics under a sustainable agricultural development.

The paper contributes not only to improved extension efforts around no-till in the Kurdistan region but to a broader understanding of the obstacles towards a sustainable agricultural development. The rural area of the Kurdistan is a large and important crop-growing region in Iraq.

Location of the study area

The Kurdistan Region is located southwest of Asia and northeast of Iraq and is composed of three governorates: Erbil (Hawler), Suleimani, and Duhok, located within the northern part of the Federal Republic of Iraq (Iraqi Constitution, Article 62). The area of the region is 42,812 km^2 (MoP, 2012), and represents 17% of the total area of Iraq.

Today, tillage is considered an old way connected with the development of agriculture (Sharma and Behera, 2008).

No-tillage was used around ancient times for the crops production (Derpsch, 2004). One of the birthplaces of the Neolithic Agricultural Revolution (7500-6500 BC) and discovery of agriculture was in the Kurdistan region. The horde of Shanidar and Jarmo were both ancient villages in the world, being located in the Kurdistan region. In this region, were found wild ancestors of two major cereals (wheat and barley) and of domesticated animals (goat, sheep, pig and cattle) (Braidwood *et al.*, 1983). For tillage, farmers used the stone axe (which was polished) and sickle for crops like, barley, rice, and wheat (Punia, 2006).

Demographic profile of farmers from Kurdistan Regional Government

In 2012, from June until September a survey was conducted on the farmers from KRG. The simple random sample, without replacement, continuous variable, with a relative error of 2.5% and 95% confidence interval was used to establish the size of the sample. The final sample size was 236 farmers. The questionnaire consisted in questions about: the socio-demografic profile of the farmers and data about the farm and farm activity.

The results of the analysis of survey show that people living in the KRG rural areas are young and less educated with an average age around 26 years. Although the average age of people from the rural space is around 26 years, the average age of the labor force of rural space is much higher (37.8 years), the average age of the head of household being around 50.25 years. More than 53% of the farmers are illiterate, while 28.81% have graduated only the primary school.

Agricultural surfaces land and agricultural equipment

The property size of the agricultural land owned by the farmers from Kurdistan ranges from less than two hectares until 100 hectares. Half of the farms have a surface between 2-10 hectares (26.7% from total range are between 2-5 hectares and 23.3% of the farms range are between 5-10 hectares). More than 40% of the famers own an agricultural surface higher than 10 hectares (18.6% from the total range have between 10-20 hectares and 22.03% have more than 20 hectares). The other 10% of the farmers own less than 2 hectares of agricultural land. The agricultural equipment of the farms is insufficient and is considered a stringent problem of the agriculture in KRG, being one of the major obstacles observed that impedes the efforts put by the farmers to increase the agricultural productivity. The most common agriculture equipment owned by the farmers are: tractors (37.9% of the respondents), plow stock (31.8%), disc harrow (26.3%) and seeder planting (10.2%). The use of disc harrow, hoe, and moldboard lead to soil inversion and is considered to be an ordinary farming practice (Pieri *et al.*, 2002).

Sustainable development of agriculture – from theory to the reality of Kurdistan Regional Government

The concept of sustainable development focuses on the improvement of the present conditions without affecting the satisfaction of the needs for the future generations (Giddings *et al.*, 2002; OECD, 2008). Generally, the meaning of sustainability for agriculture and rural development is often presented as being divided into three widely recognized dimensions: economy, environment, and society (Bowler, 2002; Hardi and Zdan, 1997, cited by Giddings *et al.*, 2002). The sustainable development evaluates all the economic, social and environmental impacts with the aim to establish the current situation the analyzed field and to develop optimal strategies that would lead to an improved situation.

- Economic. The economic dimension of sustainable rural development is prioritizing the assurance of agricultural supplies to farm and non-farm population, by maintaining satisfactory levels of the economic returns to the farms (Bowler, 2002). In the KRG, the agriculture activity is not profitable because of the low yield obtained by the farmers. This is the result of the lack of funds for agricultural inputs and the poor environment condition. In most of the cases, farmers practice a self-subsistence agriculture, which assures the daily necessary products for the household, thus, no-added value is registered for the obtained products. At the same time, being a family activity, the opportunity for developing a new business and creating new jobs in agriculture and upstream industries is limited. The oil exploitation is much more attractive from economic point of view, and people prefer to be involved in this activity instead of agriculture, without taking into consideration the long-term effects.

- Environmental. The use of natural resources, such as soil (land), water, has to be reproducible over succeeding generations (Bowler, 2002). The main concerns, as characteristics of the developing countries are the soil depletion, the water scarcity and the habitat loss (OECD, 2008). In the Kurdistan region, agriculture is damaging the natural resource base. This is happing because of the low-input farming and lack of knowledge found at the level of the farmers. The climate changes, the lack of precipitation, the low level of water in aquifers are problems with which the KRG farmers have to deal with. Because of the lack of water and proper agricultural technologies, farmers started to increase the cultivated surfaces, regardless the negative impact on the environment. In KRG, the negative impacts of agriculture on the environment are: habitat loss, beneficial nutrients and biomass loss, deployment of the natural structure of the soil. The choice of suitable instruments has been shifting towards good farming practices and environment enhancing actions by farmers. More direct instruments include various forms of cross compliance vigilances such as minimum tillage (OECD, 2008).
- Social. The social component of the sustainable development aims to enhance the life standard of the farmers and the society. This can be achieved by increasing the level of education of people from rural area and also by improving the managerial skills of the farmers (Hayati *et al.*, 2010). As mentioned before, most of the farmers from KRG are illiterate. They do not have the knowledge of sustainable development and they only 'look' at the short-run benefits, and neglect the potential benefits for the next generations. The farmers' behaviour could be explained by the tough economic and environment conditions in which they are living.

The farmers should be treated as key actors of the sustainable agricultural development because of the direct influence of their activity on all the three components of sustainable development. The agricultural activity has numerous functions: ecological, cultural, ecosystem, environment, social and landscape, as Fleskens *et al.* (2009) explained under the name of 'house of functions'. By analyzing these functions of agriculture one can notice that they cover all the principles of the sustainable agriculture. There are numerous examples of sustainable agriculture practices/approaches: crop rotation, conservation tillage, soil and water conservation measures (Kabat, 2013).

A conceptual framework to sustain the farmers' ethics

A conceptual framework of an ethical conceptual framework for sustainable agricultural was developed based on the arguments presented above regarding the ethics, the sustainable development of the agriculture, the functions of the agricultural activity and the role of the farmers and their influence on the sustainable development (Figure 1). The farmer as decision-maker had direct influence on the sustainability of the land usage. The used technology influences the depilation of the agricultural surfaces. It was observed that the farmers from KRG have and use agricultural equipment with high impact on the soil, affecting the environment component. Even that, because of the lack of funds, the possession of agricultural equipment by the farmers from KRG is low. Also the technology has direct impact on the production of agricultural goods, and in the end on the economic and social development of sustainable agriculture. The conceptual framework was bilt based on the study case of KRG, its design being possible to be adapted to others regions.

Traditionally the farmers were not informed and did not have acces to the principles of sustainable development of the agriculture. The lack of education of the farmers and the old average age of the farmers influence the way that the ethical principles are understood and applied by the farmers. Most of the famers complained about the lack of agricultural inputs, among which also agricultural equipment. They do not realize the impact that the tillage will have on the environment component, they take into consideration only the influence of tillage on the production and the economic component of the sustainable development. The extension services and the policy makers should pay attention to these

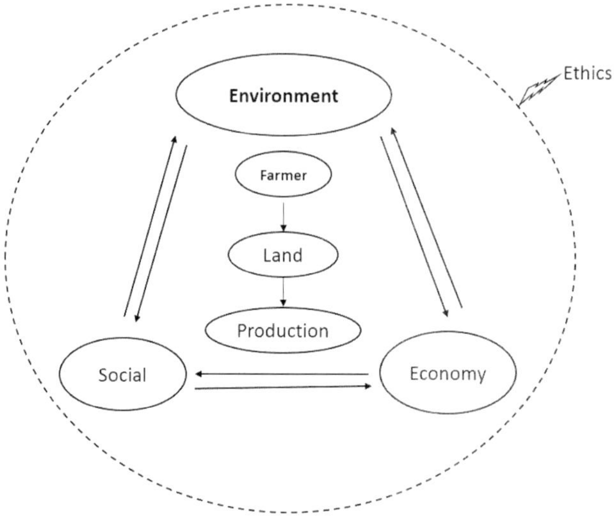

Figure 1. Conceptual framework to sustain farmers' ethics in Kurdistan Regional Government.

key aspects of the sustainable development of agriculture. The social component of the sustainable agricultural development should not be neglected. It is the main component on which the ethical principles should focuses first. Traditional agriculture with low impact on the environment can be an opportunity to develop and preserve the authenticity of local community, on one hand. On the other hand the local communities can process the agricultural produce and sell as local or traditional products. By this way new working places can appear, the life standard of local community can increase and arises the opportunity of international recognition.

Conclusions

Ethics remain an important and debated subject for the future development of the humanity.

The no-till farming in Kurdistan region has effect on the yield, which is quite below the one registered in the developed countries. Unfortunately there is no specific support for no-till farming and no legal framework, policies for changes in the behaviour of farmers. Thus, it may be asked: Would it be ethical to use a less intensive agriculture in order to decrease the negative impact on the environment, with assuring less working place, with lower production which does not assure the needed quantities of the agricultural suppliers?

It is hoped that the efforts put by researchers to enlighten the importance of paying a careful attention to the actions needed for a sustainable agriculture development will not be vanished and decision makers will act accordingly.

The development can be achieved by education of the young generation that is living the rural area of KGR in accordance with the ethical principles of agricultural development. By this way the people from rural area of KRG will be more aware about the effects of their actions and they will act in accordance with the sustainable development.

References

Bowler, I.R. (2002). Developing sustainable agriculture. Geography 87(3): 205-212.

Braidwood, L.S., Braidwood, R.J., Howe, B., Reed C.A. and Watson, P.J. (1983). Prehistoric archeology along the Zagros flanks. Oriental Institute Publication, Chicago, USA.

Burkhardt, J., Comstock, G., Hartel, P. G. and Thompson, P.B. (2005). Agricultural ethics. The mission of the council for agricultural science and technology. Council for Agricultural Science and Technology 29: 1-10.

Derpsch, R. (2004). History of crop production with and without tillage. The Journal of No-Till Agriculture 3(1): 150-154.

Falkenmark, M. and Folke C. (2002). The ethics of socio-eco hydro logical catchment management: towards hydro solidarity. Hydrology and Earth System Sciences 6(1): 1-9.

Fleskens, L., Duarte F. and Eicher I. (2009). A conceptual framework for the assessment of multiple functions of agro-ecosystems: a case study of Tra's-os-Montes olive groves. Journal of Rural Studies 25: 141-155.

Giddings, B., Hopwood B. and O'Brien, G. (2002). Environment, economy and society: fitting them together into Sustainable development. Sustainable development, 10: 187-196.

Hardi, P. and Zdan T. (1997). Assessing sustainable development: principles in practice. International Institute for Sustainable Development, Canada.

Hayati, D., Ranjbar Z. and Karami, E. (2010). Measuring agricultural sustainability, biodiversity, biofuels, agroforestry and conservation agriculture, Sustainable Agriculture Reviews 5:73-100.

Kabat, P. (2013). Water at a crossroads. Nature Climate Change 3: 11-12.

Ministry of Planning (MoP). (2012). regional development strategy for Kurdistan region, Kurdistan Regional Government (KRG).

Organisation for Economic Co-operation and Development (OECD) (2008). The organisation for economic co-operation and development (OECD) annual report.

Paarlberg, R. (2008). The ethics of modern agriculture. Society 46(1): 4-8.

Pieri, C., Evers, G., Landers, J., O'connell, P. and Terry, E. (2002). No-till farming for sustainable rural development. The international bank for reconstruction and development. Washington D.C., USA.

Punia, M.S. (2006). International agriculture research – initiatives and ethics. Department of Genetics and Plant Breeding, Hisar, India 1-5.

Sharma, A.R. and Behera, U.K. (2008). Modern concepts of agriculture. Conservation tillage. Available at: http://tinyurl.com/pguwzk9.

Sullivan, S., McCann, E., De Young, R. and Erickson, D. (1996). Farmers' attitudes about farming and the environment: a survey of conventional and organic farmers. Journal of Agricultural and Environmental Ethics 9(2): 123-143.

Zimdahl, R. (2006). Agriculture's ethical horizon. Academic Press. Inc. New York, USA.

The ethics of agricultural policies in the New Member States: a Romanian case-study

I.M. Jitea, D.E. Dumitras and A.V. Simu*
University of Agricultural Sciences and Veterinary Medicine Cluj-Napoca, Department of Economic Sciences, 3-5 Calea Mănăștur, 400372 Cluj-Napoca, Romania; mjitea@usamvcluj.ro

Abstract

To become a New Member of the EU means, among other criteria, to apply immediately after joining the entire common legislative framework known under the name of the Community Acquis. In the agricultural field, except some deregulations, the New Member States apply the common agricultural policy (CAP) immediately after joining. Thus, the farms become eligible for direct payments or for investment grants supported from the second CAP pillar. The paper questions whether there are any ethical issues induced by the agricultural financial distribution of the payments between farms and rural communities. The link between the environmental friendly techniques used by the small subsistence and semi-subsistence farms and their access to payments and investments was the first issue that is analysed. Also, the type of investments promoted in the rural areas is investigated in relation to the need for a sustainable rural development. The data show that the subsistence and semi-subsistence farms are the ones that have the most difficult access to any type of payments. This causes an important ethical consequence for the CAP's policy-makers. How can be a public policy be justified when only the large farms in terms of economical size have access to any form of financial support? On the other hand, the investments promoted by the rural communities tend to support the so called Western type architecture even if the New Member States villages are particular interesting for their traditional architecture. In Romania, for example, the last 10 years caused more damage on the rural architecture as compared with the Communist Regime because the wood architecture structure was replaced by coloured concrete ones.

Keywords: common agricultural policy, small farms, rural touristic facilities

Introduction

The agricultural policies are represented by the legislative framework designed to regulate the internal production, the local agro-food market and the international exchanges with agro-food products (Ellis, 1992). In the EU, the common agricultural policy (CAP) was created in the late 50's in order to eradicate the lack of food and the rural poverty. In time, it knew important reforms that were targeted to achieve different societal goals. Starting with the 92's MacSharry reform, its main instruments represented by the intervention prices, the export subsidies and the import protection have been reduced (Daugbjerg and Swinbank, 2007). The 2003 Fischler reform and the 2008 Health Check that followed changed the allocation philosophy of the direct income support (Regulation (EU) 1782/2003; 73/2009). It became decoupled from production with important cross-compliance rules.

According to Article 49 of the Treaty on EU, a State has to meet certain relevant criteria in order to become a New EU Member. In agriculture, this means to apply the CAP Acquis immediately after joining. For Romania this was a difficult task, because the farm structures were dominated by a huge number of small subsistence and semi-subsistence farms (around 3 million). For the first Pillar, the analyses already made by different researches showed that the financial supports went especially to the big farms (Swain, 2013) (12.05% from the first CAP pillar in Romania was devoted to only 0.03% of farms). So it is argued that the Western policy model developed in the first CAP pillar was not adapted to the Eastern agricultural realities (Swain, 2013). In the same time, the EU membership has not led to any form of farm consolidation. The semi-subsistence farms still remain the dominant characteristics of Romanian agriculture even after accession as proved by Hubbard *et al.* (2014). Moreover, these payments have not yet produced any positive effects on the Romanian overall agricultural productivity due to the low share of high value-added products in the total agricultural production (Jitea and Pocol, 2014).

For the Romanian rural area, the traditions inherited from generation to generation are a real treasure that can sustain the local rural development by promoting rural touristic destinations (Iorio and Corsale, 2010; Turnock, 1999). In the last years, due to a special measure of the Romanian National Rural Development Plan (RNRDP), investments in agro-tourism were sustained by the means of the non-refundable EU funds. In this context, the paper examines whether the investments made with this type of funding promoted the local specific architecture and the local traditions. Thus, the paper questions whether there are any ethical issues induced by the agricultural financial distribution of the payments between farms and rural communities. The link between the environmental friendly techniques used by the small subsistence and semi-subsistence farms and their access to payments and investments is the first issue analysed. Also, the type of investments promoted in the rural areas is investigated in relation to the need for a sustainable rural development.

Material and method

The results of the Romanian rural development program were analysed. Between 2007 and 2013 the program was divided in four axis: (1) 'competitiveness'; (2) 'environment and land management'; (3) 'economic diversification and quality of life' and (4) 'leader'. The paper investigates the outcomes from: measure 112 'The Installation of the young farmers', Measure 141 'Supporting the semi-subsistence agricultural holdings' (axis 1), and Measure 313 'Encouragement of tourism activities' (agro-tourism included) (axis 3). Information about these measures was gathered from several official institutes such as the Romanian Agency of Payments for Rural Development and Fishing, Romanian National Institute of Statistics and Eurostat. Data were analysed using statistical indicators: averages; ratios; weights (Horn, 1993). During on-farm visits, the touristic investments were observed and compared with the local traditional architecture. The observations were made in a recognised Romanian rural touristic destination, the Apuseni Mountains. The touristic facilities were identified using the on-line touristic sites that promote the local facilities (www.turistinfo.ro). 25 of them were visited between 2012 and 2014.

Ethical dimensions in funds allocation – the Romanian second common agricultural policy pillar case

In terms of number, the Romanian agriculture is still dominated by small subsistence and semi-subsistence farms (Table 1). They were created immediately after 1989 when the country experienced the political shift from the communist regime to the democratic organisation. The land was restituted to the former owners in small plots. It was a form of social protection in a period in which industry was hit by important economic failures.

These types of farms are low in financial capital and are usually managed by old farmers. In this context their access to different types of non-refundable funds was limitative or even prohibitive. The mandatory need to co-finance the projects in different axes and measures prevented them from applying for funds. Therefore, they had access to only 8 measures of the Romanian National Rural Development Plan (Table 2): those that do not need any co-financing funds.

The total public expenses in these 8 measures represents around 3.5 billion € that represents 40% from the entire budget. So, even from the initial design, 90% of Romanian farms could have access to less than half of the entire second CAP pillar budget.

Figure 1. The study region.

Table 1. The distribution of the Romanian farms in different size classes.

Size class	Number of farms in %			Utilised agricultural area in %		
	Romania		EU[a]	Romania		EU[a]
	1989[b]	2010[c]	2010[c]	1989[b]	2010[c]	2010[c]
0-2 ha	99.9	74.3	49	17.2	12.9	2
2-10 ha		23.5	31		25.8	9
10-50 ha		1.6	15		8.4	21
>50 ha	0.07	0.6	6	82.8	52.8	66
Total (million)	2.72	3.859	11.996	13.306	13.306	171.428
Average size (ha)				0.4[d] 2,376[e]	3.44	14.28

[a] EU with 27 members

[b] Sabates-Wheeler, 2005.

[c] Eurostat database, 2014.

[d] Individual farms.

[e] State farms.

The measures that provided the easier access to funding for the subsistence and semi-subsistence farms were:

- Measure 112 – 'The Installation of the young farmers' designed to motivate the young people to start a business in agriculture;
- Measure 141 – 'Supporting the semi – subsistence agricultural holdings' created to increase the competitiveness of agricultural exploitations which are in undergoing restructuring by encouraging the increase of the production and its diversification according to the market demands.

Table 2. The comparative analysis of the financial allocation in different axes and measures and the access to funds for the subsistence and semi-subsistence Romanian farms (million €) (RNRDP, 2013).

Axis 1			Axis 2			Axis 3			Axis 4		
Measure	Public expenses	Access to funds	Measure	Public expenses	Access to funds	Measure	Public expenses	Access to funds	Measure	Public expenses	Access to funds
111	78.9	Yes	211	673.5	Yes	312	456.2	No	4.1	296.1	No
112	302.2	Yes	212	471.9	Yes	313	304.1	No	4.21	7.6	No
114	0		214	1,270.7	Yes	322	1,610.9	No	4.31	63.1	No
121	1,028.5	No	215	372.9	Yes						
122	2.3	No	221	3.3	No						
123	963.2	No									
125	523.6	No									
141	357.6	Yes									
142	22.4	Yes									
143	12.3	Yes									

Results show (Table 3) that only 12,976 farms were financed through the Measure 112 and 60,870 farms by Measure 141. This represents only 2% from the entire number of the subsistence and semi-subsistence farms (Table 1). In this context, the positive effects produced by this small number of financed farms are rather limited.

Table 3. Comparative situation of projects under the Measure 112, 141 at 31 December 2013 (APDRP, 2013).

Year	Submitted projects		Selected projects		Contracted projects	
	No.	Public value[1]	No.	Public value[1]	No.	Public value[1]
Measure 112 – farm economical size: 6-40 European size units						
2008	661	13.78	512	11.42	500	10.05
2009	2,702	58.52	2,297	48.93	2,252	47.99
2010	3,209	64.42	2,899	57.74	2,845	56.71
2011	7,948	220.97	4,058	100.17	3,899	96.42
2012	7,974	272.39	3,680	120.85	3,480	114.57
Total	22,494	630.08	13,446	339.11	12,976	325.74
Measure 141 – farm economical size: 2-8 European size units						
2008	6,442	48.32	6,262	46.96	5,614	44.44
2010	29,974	224.81	28,250	211.88	27,270	116.88
2011	26,943	202.07	14,000	105.00	13,504	81.05
2012	25,487	191.15	15,031	112.73	14,482	65.17
Total	88,846	666.35	63,543	476.57	60,870	307.54

[1] In million €.

Investments were developed especially in the mixed sector (Table 4). This type of farming is particular interesting for biodiversity protection because they preserve a mosaic landscape. From this point of view the investments support the low input traditional farming techniques.

Innovation or kitsch in the investments made in touristic facilities

In the last years, the rural facilities designed to host the rural tourists increased. One source of funding was possible by accessing the Measure 313 'Encouragement of tourism activities' which was set to sustain the development of tourism activities in rural areas to increase employment and other alternative revenues. This measure also refers to the agro-tourism activity, the main condition being that the applicant proves that he is active in agriculture as well. In 2012, a total of 1292 projects were selected, among which 619 projects including agro-tourism activities. Unfortunately the investments did not respect the local traditional architecture. The new innovative architectural construction style transformed the Romanian villages in shiny coloured castles houses (Figure 2). Ilies (2008) pointed out that the presence of non-authentic and kitsch architecture elements was mainly caused by the lack of an optimal institutional framework protecting the authenticity of the rural areas. One common explanation of the choice towards the kitsch style was expressed by some owners as a wish to connect to the global world, characterized by modernity (Iancau, 2011). On the other hand, tourists expressed their wish to spend their vacations in a specific traditional environment and not only to participate to traditional activities, as most of the owners believe it is sufficient for the preservation of the rural area.

Conclusions

Due to important financial constraints, the Romanian subsistence farms have access to only a limited number of measures from the rural development program. On the other hand the new investments do not promote the local traditional architecture. The question that remains to be answered for the future is how the above revealed problems can be corrected in the next CAP financial allocation (2014-2020). Solutions for a better distribution of funds should be immediately found, such that even the semi-subsistence farms can have the opportunity to develop their agriculture activity, which at the end will

Table 4. Measure 112 by agriculture sectors and 141 by cultivated surface (APDRP, 2013).

Measure 112			Measure 141		
Agriculture sector	No.	Public value[1]	Cultivated surface	No.	Public value[1]
Field crops	3,176	84.45	<5 ha	45,108	290.51
Horticulture	1,713	45.33	5-10 ha	11,322	74.72
Viticulture	141	2.80	>10 ha	4,440	30.98
Permanent crops	423	8.29			
Milk and dairy products	427	14.52			
Ruminant (without dairy)	324	11.57			
Pigs	194	6.00			
Poultry	60	1.77			
Mixed production	6,517	150.96			
Other	1	0.04			
Total	12,976	325.73	Total	60,870	396.21

[1] In million €.

Figure 2. Comparison between tradition architectonic style (left) and modern rural touristic facilities (right). Personal gallery (Marisel commune, Cluj County, October 2014).

be in the benefit of the development of the area. This way, the migration of villagers may decrease as well, and people will be more likely to practice agriculture. On the other hand, even if there are cases where the architecture was 'damaged' by the inclusion of kitsch elements, the institutional framework should be improved in order to sustain the preservation of the authenticity.

References

Daugbjerg, C. and Swinbank, A. (2007). The politics of CAP reform: trade negotiations, institutional settings and blame avoidance. Journal of Common Market Studies 45(1): 1-22.

Ellis, F. (1992). Agricultural policies in developing countries. Cambridge University Press, Cambridge, UK, 359 pp.

Eurostat Database (2014). Availabe at: http://ec.europa.eu/eurostat.

Horn, R. (1993). Statistical indicators for the economic & social sciences. Cambridge University Press, Cambridge, UK: 229 pp.

Hubbard, C., Luca, L., Luca, M. and Alexandri, C. (2014). Romanian farm support: has European Union membership made a difference? Studies in Agricultural Economics 116: 100-106.

Iancau, M. (2011). Orange or pink? Colours, houses and modernity in rural Romania. International Review of Social Research 1(2): 85-98.

Ilieş, A., Ilieş, D., Josan, I., Grama, V. and Gozner, M. (2008). Romanian rural tourism between authentic/traditional and modern/contemporary. The case of Crisana and Maramures area (I). GeoJournal of Tourism and Geosites 2(2): 140-148.

Iorio, M. and Corsale, A. (2010). Rural tourism and livelihood strategies in Romania. Journal of Rural Studies 26 (2): 152-162.

Jitea, I.M. and Pocol, C.B. (2014). The common agricultural policy and productivity gains in Romanian agriculture: is there any evidence of convergence to the Western European realities? Studies in Agricultural Economics 116: 165-167.

Regulation (EU) 1782/2003 of the European parliament and of the Council establishing common rules for direct support schemes under the common agricultural policy and establishing certain support schemes for farmers. Official Journal of the European Union, 46 (L 270): 1-69.

Regulation (EU) 73/2009 of the European parliament and of the Council establishing common rules for direct support schemes for farmers under the common agricultural Policy and establishing certain support schemes for farmers. Official Journal of the European Union, 52 (L 30): 16-99.

Romanian Agency of Payments for Rural Development and Fishing (APDRP) (2013). Annual report for the implementation progress of the Romanian national rural development plan.

Romanian National Rural Development Plan (RNRDP). (2013). Romanian government. Bucharest, pp. 429-430.
Sabates-Wheeler, R. (2005). Cooperation in the Romanian countryside: an insight into post-soviet agriculture. Lexington Books, Oxford, UK, 53 pp.
Swain, N. (2013). Agriculture 'east of the elbe' and the common agricultural policy. Sociologia Ruralis 53(3): 369-389.
Turnock, D. (1999). Sustainable rural tourism in the Romanian carpathians. The Geographical Journal 165(2): 192-199.

Section 2. Dimensions of agricultural production

Sharing values or maximizing profits? Ethical discourses and sustainable food initiatives

E. de Bakker and H. Dagevos*
LEI Wageningen UR, Alexanderveld 5, 2585 DB The Hague, the Netherlands; erik.debakker@wur.nl

Abstract

In this paper we pose that in a social context dominated by neoliberal politics utilitarian and contractarian ethics tend to hinder and can even smother critical alternatives of our food system that have more holistic narratives on the consequences of our food choices. First of all, we will argue that the dominant (neo) liberal view on society and economy, featured by the principles of value pluralism and freedom of choice, has a strong affinity with utilitarian and contractarian ethics. We will then take a closer look at 11 sustainable food initiatives in the Netherlands, ranging from larger agricultural projects with an international orientation to small-scale initiatives of crowdbutching. Particularly initiatives operating at the fringes of the agri-food system showed affiliation with an ethical virtue approach questioning the ends of our agricultural and food system. Instead of focusing on the maximization of profits these initiatives were thinking in terms of shared values. To some extent this approach shows a resemblance with the strategy of creating shared values that has been announced by Porter and Kramer as an innovative route that could reinvent capitalism. However, when we probe a bit deeper it turns out that more profound issues are at stake here. Resuming the fundamental tension between 'oikonomia' and 'chremastike', as originally brought to the fore by Aristotle, we finally formulate the following hypothesis regarding the ethical discourses on our food system: in a situation where economy and society are dominated by neoliberal politics, more virtue ethics are needed as a counterweight in the ethical and political debate on our current system of food production and consumption.

Keywords: food system, social entrepreneurship, neoliberalism

Introduction

In the prologue of the third edition of 'After virtue', a classic work that offers a controversial perspective on our understanding of ethics, Alasdair MacIntyre states: 'On the dominant liberal view, government is to be neutral as between rival conceptions of the human good, yet in fact what liberalism promotes is a kind of institutional order that is inimical to the construction and sustaining of the types of communal relationship required for the best kind of human life' (2007: xv). This paper can be seen as a reflection on this statement in the light of the condition of neoliberalism that made its appearance on the world stage around 1980, focusing on our current food system and the ongoing debates on sustainability. In line with MacIntyre we also point to some empirical examples that illustrate how practice itself urges a more profound debate on the goods and ends that are internal to our food system.

Let us state in advance that we do not necessarily affiliate ourselves with the Aristotelian-Thomistic position of virtue ethics that MacIntyre defends, but we appreciate his historical-empirical approach to understand and value ethical positions. Though this paper can only offer a finger exercise in this respect we believe that such an approach can illuminate why certain ethics (or ethical ideas) tend to be beneficial for neoliberal politics in our current time. The irony is that this beneficial effect is often not intended (particularly a deontological ethic can even be very critical of neoliberalism), but it happens. Let us first take a closer look how liberalism links closely to neutrality and pluralism that are also vital aspects of utilitarian and contractarian ethics.

Liberalism and lifestyles: neutrality and pluralism

A lifestyle can be defined as 'a more or less integrated set of practices which an individual embraces, not only because such practices fulfil utilitarian needs, but because they give material form to a particular narrative of self-identity' (Giddens, 1991: 81). Such a particular narrative or identity, seen in a positive way, is closely connected to our fulfilment of a good life. According to liberalism, this endeavour for good life should be primarily seen as an individual endeavour, not a communal one. Therefore, a liberal state should be neutral with respect to citizens' conceptions of what it means to live a good life; citizens are conceived as 'individuals living out their vision of the good life' (Noll, 2014: 212).

Neoliberalism can be described as 'a theory of political economic practices that proposes that human well-being can best be advanced by liberating individual entrepreneurial freedoms and skills within an institutional framework characterized by strong private property rights, free markets, and free trade' (Harvey, 2009: 2). In this way it represents a restricted version of liberalism that promotes free markets as the most efficient means to distribute goods in a fair manner. The idea of a neutral state is combined with an 'economist' view in which the state should meddle as little as possible in trade relationships and free markets. Producers and consumers are depicted as sovereign agents 'driven by an individual utilitarian orientation and seek to maximize personal benefits at the lowest possible cost' (Kjaernes, 2012: 146) and it is believed that this will result in economic prosperity, fair distribution and efficient solutions for societal needs.

Although the more restricted 'economist' view makes a crucial difference, both from a liberal and neoliberal perspective lifestyle choices are regarded as private matter and, accordingly, political efforts to define the 'good life' more specifically are valued as futile or even as dangerous because they can lead to dictatorial politics repressing individual lifestyle choices and/or self-identities. The neutrality of the government is inextricably linked to the idea of a plurality of lifestyles that should be respected as long as they are within the boundaries of the law. 'Since it is not clear which lifestyle could apply for common affirmation in pluralistic societies, nor why consumers should identify with one lifestyle anyway, political liberalism holds it that this pluralism should be accepted as a general fact of life' (Beekman, 2000: 189). Not only politicians or decision-makers, but also citizens and opinion makers often cling to this idea of pluralism and individual freedom, which may result in a reproach of narratives on the good life as a form of moralism and paternalism (e.g. Haen, 2014).

Neoliberal politics and ethics

Whatever one thinks of neoliberalism, the strength of this political vision is that many feel attracted by its promise of individual freedom that is associated with a diversity of lifestyles. Also in Europe we have seen the rise of a neoliberal politics in the last decades. Entrepreneurial freedom is valued highly and the states should retire where possible so that free market operations can flourish. This development supports Harvey's claim that neoliberalism has become hegemonic as a mode of discourse and that the (ethical) belief that the social good will be realized in the domain of the market 'has become incorporated into the common-sense way many of us interpret, live in, and understand the world' (2009: 3). On EU level, the European Charter of Fundamental Rights and the Treaty on the EU both illustrate the dominance of neoliberal principles that are difficult to accommodate to with an Aristotelian concept of the good life, but show a more close relationship with either contractarian ethics (fair treatment of rights and duties) or utilitarian ethics (economic prosperity to provide happiness for all). One can discuss whether the constitutional rights of citizens are protected effectively, what claims on the social security system are legitimate or which policy is best suited to stimulate (sustainable) economic growth, but one should keep oneself from designing and implementing perfectionist ideals that guide people in their daily lives. In other words, the philosophical justification of current EU policy is largely built on utilitarian and deontological values and not on virtue ethics that would include a more specific account of what 'happiness' and 'fairness' should be and that would formulate more specific ends both for the individual and the community/society. The Treaty on the EU offers a broad account of what the EU should achieve (e.g. balanced economic growth and price stability, highly competitive social market economy, equality between men and women, protecting and improving the quality of the environment, economic and social cohesion and solidarity among Member States), but these goals are strongly linked to values that promote economic growth and do not entail a more specific or holistic conception of the 'good life'.

Practice and ethics: sustainable food initiatives in the Netherlands

The transition towards a more sustainable agricultural and food system represents an interplay between both small- and large-scale initiatives aiming at changes in production, distribution and consumption. One of the key issues that all these initiatives address in one way or another is the well-known gap between producers and consumers. In research financed by the Ministry of Economic Affairs we studied a selection of sustainable food initiatives to get more insight into best practices that could inspire reconnection between the food system and society. Starting from an inventory of several initiatives (about 50) we selected 15 cases to study in depth, making sure that this selection was a good cross-cut of different motives, strategies and ambitions (our selection also contained 4 cases outside the food domain, but in this paper we will concentrate on the 11 food domain cases). Our interest was not so much the impact these initiatives have on realizing a more sustainable food system but particularly to what extent they establish or (could) enable more connectedness. We clustered the cases in their relationship to the dominant regime of agricultural and food practices, or as one could call this in more popular terms: 'business as usual' (Table 1). This relationship turned out to be an insightful criterion for understanding the basic (socio-economic) value positions of the different initiatives. We clustered the cases in their relationship to the dominant regime of agricultural and food practices ('close', 'mixed' or 'challenging'). 'Close' meant that the initiative was strongly supported by 'regime players'. 'Mixed' meant that the initiators were trying to find partners within the common regime practices. 'Challenging' meant that the initiatives were shaking up common regime beliefs by new approaches or ideas starting from different principles.

Because of the set-up of this conference book it is not possible here to give a detailed account of the goals and achievements of these food initiatives and which means they deployed. Our focus is on some

Table 1. Eleven sustainable food initiatives.

Food initiatives	Goals in keywords	Position towards regime
Floriade 2022	international expo, creating a green city/neighbourhood	close relationship
Sugar Hothouse	bringing horticulture back to city and citizens	
Agri Meet Design	farmers and designers innovating together	
Biopark Terneuzen	business join in collaborative environmental strategy	
Network Farmer-Beer-Water	building local community, environmental strategy	mixed relationship
Feed Design Lab	sharing knowledge, building sustainable protein chain	
Network Peelhorst	re-building trust with and through local community	
Meat your own	connecting meat producers and consumers, short chain	challenging
Extraordinary pigs	direct selling to consumers, fair/transparent product	
Buy a cow	direct selling to consumers, fair/transparent product	
Oregional	co-operation farmers and citizens, fair prices/products	

main lessons and the ethical positions that can be distinguished in the different value positions of the initiatives; particularly the 'challenging' initiatives were very insightful for this.

Main lessons

A lesson we learned from these good practices is that the drivers to connect more strongly with other business or consumers are diverse: sustainable food initiatives can be based on: (1) 'higher', collective ideals, but also on (2) mixed motives that combine ideals with commercial thinking or on (3) plain motives of 'enlightened' self-interest. More strategies are possible and seem also necessary in a society characterized by pluralism (many different lifestyles) and different needs for connection. What works for one group in a specific setting, can be of little effect for another group in another setting. Entrepreneurs and (network or process) facilitators should follow the way that is closest to their heart and do what they are most good at. 'Be authentic, stick to your ideas and qualities'. Further, we noticed that all initiatives shared a willingness to have a true dialogue with others. Reconnection demands empathy and ears that can really listen. In case one is mainly occupied with sending messages and expects that other parties should change their opinions or behaviour, the efforts of connection will often stumble. These things may sound trivial but (re)connecting in a pluralist society is far from simple and it demands creativity and accuracy, open-mindedness and soul-searching.

However, the most important lesson for our purpose was the fact that the more far-reaching initiatives for sustainability, those that challenged the food system, were built on non-economic values that preceded purely economic cost-benefit thinking. These good practices also showed a lot of creativity in their search for financial resources. Here we also came across more holistic narratives on a sustainable food system going through a substantial transition: transparency and craftsmanship, more awareness concerning the origins of our food and food culture, fair prices for fair products and so on. Though respondents did not talk in ethical concepts or were in all cases very explicit about their (moral) values, their passions and practices were exemplifying alternative (communal) perceptions on the good life. Oregional, which paired regional identity with building a local community of farmers and citizen-consumers, was the most striking and visionary example of an alternative 'good food life'. It struck us that these best practices can be seen as part of a broader trend of entrepreneurship that takes creating shared value and not maximizing profits as the starting point of their business.

Creating shared value

Creating shared value is a business strategy concept of Porter and Kramer (2011) that rests on the assumption that the competitiveness of a firm is strengthened when societal needs are integrated into business activity by creating additional values. They even announced it as a new innovative approach that reinvents capitalism. The traditional divide between economic concerns and social ones must be opened up to have a broader, more effective working of Adam Smith's invisible hand. 'It opens the doors of the pin factory to a wider set of influences. It is not philanthropy but self-interested behaviour to create economic value by creating shared value. (...) Survival of the fittest would still prevail, but market competition would benefit society in ways we have lost' (Porter and Kramer, 2011: 17). The sustainable food initiatives of our research clearly show affinity with creating shared values, but closer scrutiny reveals that more profound issues are also at stake here. Porter and Kramer are pointing to innovative pathways but still adhere to the neoliberal thought that markets should do their work to produce the maximum of benefits for all. The role of governments and NGOs is to think along these lines and encourage companies to pursue shared value. 'Companies must take the lead in bringing business and society together' (Porter and Kramer, 2011: 4). It can be said that Porter and Kramer represent a utilitarian ethics: our moral task is creating a maximum of prosperity and happiness and markets are the best way for achieving this.

Economy and society: what comes first?

Should society serve economy or should economy serve society? Probably most people will agree with the latter, but the neoliberal answer to this question basically turns things upside down. Of course, it is argued, economy serves society, but to keep the economy competitive and guarantee a prosperous and happy society we must be willing to sacrifice social provisions of our welfare state so that we can stay ahead of other countries and will be able to generate further economic growth. Therefore, an economy that serves society requires a society willing to subject itself to the harsh laws of economic markets to avoid a lack of economic growth followed by a decline of prosperity and happiness. What does this mean for the transition debates in the light of the 'Grand Challenges' that also included sustainable agriculture? As we already argued, both current neoliberal principles of EU policy and utilitarian and deontological ethics play down discussions on the good life (virtue ethics) because they are not in line with value pluralism and individual freedom. This results in discussions on sustainable food systems staying on the safe side, looking mainly for reforms within the system as it is, averse to more fundamental debates about (the consequences of) our lifestyles and routines. The 'Grand Challenges' are to be met without 'Big Narratives', thereby narrowing down discussions and giving not much space to more 'revolutionary' perspectives. It could be said that the more challenging initiatives of our research address this transformation on a more fundamental level by setting up alternative practices that tell us in a more holistic (and narrative) manner how we could attain to an economy that serves the community (again) and not the other way around. In this sense they definitely transcend the economist view of Porter and show affiliation with Aristotle's premise that economic activities are always firmly integrated into an overall socio-political context and conceptions of the 'good life'. Although Aristotle's vision on economy has been dismissed by modern economists because this was seen as obsolete, in more recent years his thoughts received new attention (Dierksmeier, 2009; Dierksmeier and Pirson, 2009; Solomon, 2004). Particularly Aristotle's difference between oikonomia as 'the concern for morally adequate individual and public household management' (Dierksmeier and Pirson, 2009: 418) that should always be judged or evaluated in the light of communal, political endeavour for the good life and the practice of sheer money-making (chremastike) that will obfuscate communal goals as it becomes an end in itself. This distinction points to a fundamental issue that is largely neglected in the mindset of neoliberalism.

Discussions on food and sustainability need more virtue ethics

Kramer and Porter's concept of creating shared value is an interesting approach but their overall view on economy is troublesome from an Aristotelian perspective. Severing the realm of economic life from moral and political life, as they do, runs the risk of only stimulating chrematistic organizations and a corresponding cultural mind-set that interfere with (higher) ends and (communal) happiness. Particularly initiatives operating at the fringes of the agri-food system showed affiliation with such an ethical virtue approach questioning the prevailing economic yardstick of maximizing profits and by taking non-economic values as a starting point. If such an approach might seem strange and unrealistic for people, because economy – they argue – is all about shareholder value that is unchangeable, they should be reminded that the current prevalence of shareholder value is historically a quite recent event that came up with neoliberal politics (Heilbron *et al.*, 2013).

Utilitarian and deontological ethics are not to be identified with neoliberalism or supporters of the status quo, but within the neoliberal setting of today they can (unintentionally) contribute to blocking more fundamental oikonomic discussions and be advantageous for chrematistic practices. By marginalizing or declining narratives about the right way to live, work and consume, they make it easy for regime players and defenders of the status quo to smother debates on alternative lifestyles and holistic approaches that break more fundamentally with our current food choices and daily routines. However, these debates are vital and would be in the advantage of sustainable food initiatives striving for more fundamental system changes. Therefore, the following working hypothesis regarding the ethical discourses on our food system is posed: in a situation that economy and society are dominated by neoliberal politics more virtue ethics are needed as a counterweight in the ethical and political debate on our current system of food production and consumption, urging both policymakers and ethical 'rivals' to take a broader perspective on happiness and what we can know and hope for. Both politics and ethics could benefit from this.

References

Beekman, V. (2000). You are what you eat: meat, novel proteins foods, and consumptive freedom. Journal of Agricultural and Environmental Ethics 12(2): 185-196.

Dierksmeier, C. (2009). The freedom-responsibility nexus in management philosophy and business ethics. Journal of Business Ethics 101(2): 263-283.

Dierksmeier, C. and Pirson, M. (2009). Oikonomia versus chremastike: learning from Aristotle about the future orientation of business management. Journal of Business Ethics 88(3): 417-430.

Giddens, A. (1991). Modernity and self-identity. Polity Press, Cambridge, UK, 264 pp.

Haen, D. (2014). The paradox of E-numbers: ethical, aesthetic, and cultural concerns in the Dutch discourse on food additives. Journal of Agricultural and Environmental Ethics 275(1): 27-42.

Harvey, D. (2009). A brief history of neoliberalism. Oxford University Press, Oxford, UK, 247 pp.

Heilbron, J., Verheul, J. and Quak, S. (2013). The origins and early diffusion of 'shareholder value' in the United States. Theory and Society 43(1): 1-22

Kjaernes, U. (2012). Ethics and action: a relational perspective on consumer choice in the European politics of food. Journal of Agricultural and Environmental Ethics 25(2): 145-162.

MacIntyre, A. (2007). After virtue. Third edition. University of Notre Dame Press, Notre Dame, USA, 286 pp.

Noll, S. (2014). Liberalism and the two directions of the local food movement. Journal of Agricultural and Environmental Ethics 27(2): 211-224.

Porter, M.E. and Kramer, M.R. (2011) Creating shared value: how to reinvent capitalism – and unleash a wave of innovation and growth. Harvard Business Review, January-February 2011: 1-17.

Solomon, R.C. (2004). 'Aristotle, ethics and business organizations'. Organization Studies 25(6): 1021-1043.

Biogas production and society: evidence from Germany

K. Zander*, I. Christoph-Schulz and D. Bürgelt
Thünen Institute of Market Analysis, Bundesallee 50, 38116 Braunschweig, Germany; katrin.zander@ti.bund.de

Abstract

Biogas production has expanded tremendously in recent years and has become an important feature of modern agriculture. Within this development public subsidies have played a major role by making biogas production from raw plants highly competitive. Social acceptance should be important in contemporary political decision making: that is why we decided to ask people for their perceptions of and their attitudes towards biogas production in agriculture. We used a two-step approach, starting with the conduction of focus groups discussions in three German cities, followed by an online survey. The content of the online survey was based on the outcome of the focus group discussions. By means of factor and cluster analyses we classified the respondents into homogenous groups with regard to their attitudes towards biogas production. The results show that citizens' knowledge of biogas production and of its possible consequences is low. Generally, the production of biogas is well accepted by German citizens. They acknowledge biogas production to be a form of renewable energy, which contributes to overall sustainability goals. But, people are concerned about competition of biomass production for electricity and feed and food production, having in mind that hunger is an important issue worldwide. According to the test person's opinion organic waste should be preferred over fresh biomass. In contrast to what had been expected from earlier media reporting, neither changes in the landscape due to high shares of maize nor significant annoyances of people living close to biogas facilities caused by noise, smell or dirty roads were feared. The factor and cluster analyses resulted in the following groups of people: 'Opponents', 'Indifferents' and 'Supporters' of biogas production. The results of our research confirm earlier research outcomes on German citizens' attitudes in favour of fostering the use of renewable energies. Consistently, people also support the use of biogas. But, people are concerned about the likely competition of land used for the production energy plants with that used for food and feed. Considering that biogas production from energy plants causes quite high CO_2-abatement costs, resolute political action towards replacing energy plants in biogas production is urgently needed.

Keywords: renewable energy, social preferences, attitudes

Introduction

The biogas production has expanded tremendously in recent years in Germany. This is mainly due to extensive support schemes run by the German government which in turn are part of the sustainability strategy of the German government from 2002. It is estimated that about 80% of the substrate of biogas production in Germany is from raw plants, resulting in a share of the arable land used for maize production for biogas production of about 5 to 7% in 2010 (WBA, 2011).

There is a discussion ongoing with regard to the competition of plant production for food, feed and energy. Increasing plant production for energy purposes is supposed to reduce the world wide amount of food and feed and to increase food and feed prices (Leopoldina, 2012; The Guardian, 2012; WBGU, 2008). This debate also reflects in people's concerns (Gold, 2011). Due to the high subsidies granted, biogas production based on fresh plants has become highly competitive. Land prices increased markedly which result in reduced economic performance of particularly milk production and organic farming (Meyer and Priefer, 2012). Other effects of the biogas technology are impacts on the landscape due to high share of maize in crop rotations, the negative impact of these high maize shares on the environment, and the concern that genetically modified maize may be used (e.g. Ehrenstein *et al.*, 2012; WBA, 2011; WBGU, 2008). Biogas facilities may cause smell and noise disturbances, as well as inconveniences because of dirty roads (Devine-Wright, 2008; Gold, 2011).

Social acceptance is an important feature of contemporary political decision making. It is well known that the majority of the citizens are in favour of the use of biomass (Eurobarometer, 2007; Forsa, 2009). However, little is known about people's opinion on the use of biogas production. Thus, the aim of this research was to close this gap by investigating people's perception and attitudes towards biogas production in Germany. Groups of citizens with homogenous attitudes towards biogas production were identified.

Methods

A two-step research approach was used: first focus groups were carried in order to get an idea of people's knowledge and perception of biogas production. Focus groups are loosely-structured interactive discussions among a small group of respondents (6 to 12), accompanied by a trained facilitator. The strength of this technique lies in discovering the unexpected, which may result from a free-flowing discussion. Focus groups offer the opportunity to gain deeper insights into consumers' perceptions by letting them exchange thoughts and opinions in a dynamic manner (Halkier, 2010; Wilson, 1997). The discussions were audio- and video-typed and verbatim transcribed. The data analysis followed the approach of thematic coding: a technique which builds on the researchers' expertise in the extraction of the information relevant to the research question (Kuckartz, 2007). In this research in total 6 focus groups were conducted in three German cities (Leipzig, Münster and Stuttgart) with 6 to 8 participants each. Participants were recruited by an online panel run by a commercial market research agency. About half of the participants in each group were women and participants were of different age and education.

Second, a standardised online survey with about 1500 test persons was conducted based on the outcome of the focus groups. Participants were nearly distributed according to German averages regarding gender, region and income. Respondents older than 65 years were underrepresented. Socio-demographic variables and eight true/false questions regarding respondents' knowledge on agriculture in general were questioned. Additionally, respondents faced a seven-point Likert scale about their attitudes, perception and expectations towards biogas production ranging from 'totally agree' to 'do not agree at all'. Using the data of these 15 statements an exploratory factor analysis was conducted to define the underlying structure in the data matrix and to analyse respondents' attitude towards biogas production (Hair *et al.*, 2014). All statements were pretested (using MSA – Measurement of sampling adequacy, Kaiser-Meyer-Olkin (KMO) and Bartlett Test). The calculated measures ensure suitability for factor analysis. Based upon these findings, a cluster analysis was applied (Churchill and Nielsen, 1995).

Results

Within focus groups participants were first requested to discuss their associations with bioenergy and biogas production. Afterwards definitions of both terms were presented to ensure that all participants were on the same level of knowledge. Photos of biogas facilities and cultivation of plants for bioenergy were shown in order to discuss participants' perception. Finally, five statements taken from literature and media were presented and discussed in detail.

The associations with bioenergy in agriculture were manifold. Participants had associations such as 'photovoltaic' and 'wind wheel' which are forms of renewable energy but not of bioenergy. The discussion about biogas had a positive tone because biogas was seen as a reasonable alternative way of energy production within the general change in energy policy. The local production of biogas from liquid and solid manure and of organic waste was positively appraised. Some participants worried about the smell of biogas production or the possibility of an explosion. In general, participants' knowledge of bioenergy and biogas was low.

The first statement that was brought into the discussion claimed a competition between the cultivation of plants for biogas and for food. Participants already discussed this point intensively in the first section of the focus groups. They felt that it was not ethically correct to grow plants for energy production on fields that could be used for food, as long as people are suffering from hunger anywhere in the world. 'It is unpopular to burn plants or ordinary food and feed'(participant in Leipzig). Participants suggested replacing raw plant materials by solid and liquid manure, content from bio-waste containers or garden waste. At this point it became clear that some participants thought that biogas was already basically produced from waste and that some participants had never heard of a competition between the cultivation of plants for energy and food.

The second statement said that biogas constitutes a valuable contribution for the general change in energy policy. Most participants agreed with this statement but some were not sure whether the contribution of biogas is really valuable as its relevance might be rather low. 'The change in energy policy will only work with local energy production. And biomass is well suitable' (participant in Münster). Critical statements were related to possibly rising electricity prices and the carbon footprint of using energy plants.

The impact of the cultivation of energy plants such as maize and of biogas facilities on the landscape was discussed by means of the third statement. Only a few participants had realised a higher share of maize in recent years and only participants in Leipzig regretted the more monotonous landscape. 'I don't mind biogas facilities but the maize that I have to pass for hours (when driving)' (participant in Leipzig). Interestingly, participants in Münster, where a higher share of maize can be found in surrounding landscape, did not show stronger concern on landscape issues than participants in Stuttgart or Leipzig. Generally, participants appreciated a larger diversity of land use but also had a pragmatic perspective and accepted certain impairments to reach the larger goal of energy change. The biogas facilities itself were not really recognised by many participants. They just became aware of biogas facilities during the discussions and assessed these as only slightly disturbing especially compared to wind wheels or coal-fired power sites.

The fourth statement claimed that residents are negatively affected by the smell or traffic noise of biogas facilities. Most participants had difficulties to discuss this aspect as they had no own experience or any corresponding second-hand information. However, they assumed that residents might be disturbed by smell and noise. 'So, I can imagine that residents are affected by traffic noise' (participant in Münster). The few participants with experience told of small or no impairments through smell from biogas sites.

The last statement about biogas said that some are making a lot of money with biogas production at the expense of society. Participants had difficulties to follow this statement. Higher acceptance through participation in decision making financial contribution was not relevant in all focus groups. The only concern in this regard were raising prices for electricity.

A factor analysis was conducted in order to pool the 15 statements on respondents' attitudes towards biogas production from the online survey. (A table with estimation results of the factor analysis is available from the authors on request.) A principle component analysis with promax rotation on level 4 was carried out and three factors were identified which describe respondents attitude towards biogas production. Cronbach's Alpha, the measure of reliability, ranges from 0.615 to 0.816. This presents an acceptable to good range (Hair *et al.*, 2014).

The factors were titled as (numbers in brackets indicate the explained variance):
- support of biogas production (24.3%);
- claims towards biogas production (18.0%);
- concerns about biogas (8.7%).

The first factor 'Support' includes the positive effects of biogas production. It was assumed, e.g. that areas that could not be cultivated in another way were used, that there was a possible contribution to the energy change and that new jobs would be generated. The second factor 'Claims' covers requests as well as claims towards biogas production. The use of agricultural land for biogas production should be limited by the state, food prices should not increase and farms should be more diversified regarding the plants cultivated. Biogas should not be produced from plants that alternatively could be used for food production. This factor describes respondents' apprehension that biogas production might compete with food or feed production. The third factor 'Concerns' reflects possible negative consequences of biogas production: people in Africa might suffer from starvation due to European biogas production, prices for land might increase and local residents might be affected negatively by noise and smell.

Based upon the findings of the factor analysis a cluster analysis was carried out to assign respondents to groups with as much homogeneity within and heterogeneity between them (Churchill and Nielsen, 1995). Three clusters were identified:
- opponents of biogas production;
- indifferents of biogas production;
- supporters of biogas production.

Figure 1 shows the mean deviation of each cluster from the grand factor mean of the overall sample for all three factors. The first cluster, 'Opponents', consists of 434 respondents (29%). Compared to the average respondent they did not believe in the advantages of the biogas technology and therefore did not recommend its application. They valued concerns higher than members of the other clusters and depicted more requests and claims regarding biogas production. Compared to the other clusters respondents were significantly (using cross tables) more often male, were better educated, had higher incomes and had a better knowledge of agriculture (seven or eight right answers out of eight).

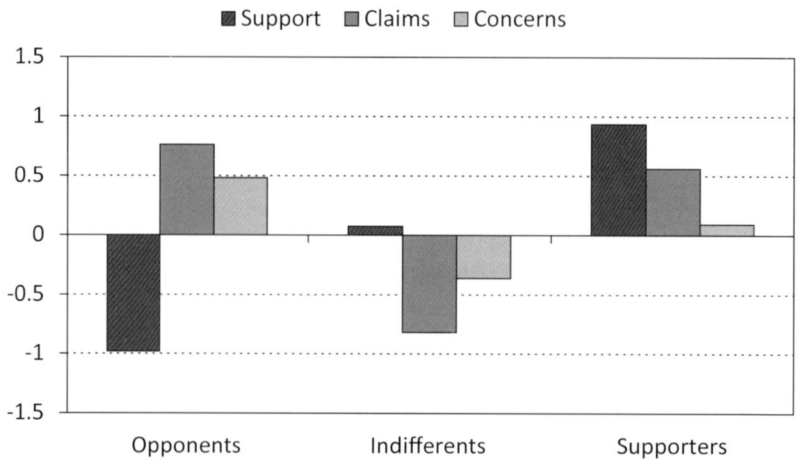

Figure 1. Standardised factor values of the three clusters.

With 679 participants (45%) the second cluster 'Indifferents' constituted the largest group. Their support of biogas production was only a little bit above average. Their claims towards and concerns about biogas production were below average. This can be interpreted as low involvement with regard to biogas production (a list of all statements is available from the author on request). These participants were more often female, were less educated, had lower incomes and only little knowledge about agriculture (zero to three correct answers).

'Supporters' of biogas production formed the third group with 406 participants (27%). They strongly recognised the advantages of biogas production and had average concerns towards this technology. At first sight it seems surprising that they also had above average values for the factor claims. People do not support this technology unrestrictedly, they want biogas production without competition for food and feed and therefore with limited land for the production of energy plants. Supporters are more often male, are less educated, have lower incomes and average knowledge about agriculture.

Discussion and conclusions

The subsidies for biogas production based on fresh plant material are highly discussed in Germany because of their impact on the environment, for their possible impact on worldwide food security and food prices. Another important issue in Germany are increasing land prices which reduce competitiveness of other agricultural activities such as milk production and organic farming. Our research shows that even though knowledge about biogas production is often limited, people care for the competition of biomass production for biogas with feed and food production which is in line with earlier research (Gold, 2011). They request to only use organic wastes instead of raw plants. Interestingly, they are only slightly disturbed by landscape changes due to high shares of maize in crop rotation or by the biogas facilities itself as mentioned by Ehrenstein *et al.* (2012), WBA (2011) or WBGU (2008). Respondents highly agreed to the statement that the government should become active to prohibit rising prices for food because of bioenergy production. Thus, people's opinions are well reflected in the ongoing discussions on biogas production and also in the adjustment of the support policy for biogas production (Leopoldina, 2012; The Guardian, 2012; WBA, 2011).

Policy makers are asked to care for efficiency in decision making. This means that policy in favour of renewable energies needs to consider the costs of different options: biogas production from maize production has very high relative CO_2-abatement costs (WBA, 2007). The long-running nature of support policies in the context of renewable energies make sound evaluation of the alternatives necessary considering the interest of all social groups.

References

Churchill, G.A. and Nielsen, A.C. (1995). Marketing research methodological foundations. The Dryden Press, Fort Worth, Texas, USA.

Devine-Wright, P. (2008). Reconsidering public acceptance of renewable energy technologies: a critical review. In: Grubb, M., Jamasb, T. and Pollitt, M. (eds.) Delivering a low carbon electricity system: technologies, economics and policy. Cambridge University Press, Cambridge, UK, pp. 443-461.

Ehrenstein, U., Strauch, S. and Hildebrandt, J. (2012). Akzeptanz von Biogasanlagen. Hintergrund, Analyse und Empfehlungen für die Praxis. Available at: http://tinyurl.com/mfnehvs.

Eurobarometer (2007). Energy technologies: knowledge, perceptions, measures. Available at: http://tinyurl.com/kaxbosp.

Forsa (2009). Mehrheit der Bundesbürger ist für Ausbau Erneuerbarer Energien bei unverminderter Förderung.

Gold, S. (2011). Bio-energy supply chains and stakeholder. Mitigation and Adaptation Strategies for Global Change 16: 439-462.

Hair, J., Anderson, R.E. and Tatham, R.L. (2014). Multivariate data analysis. Prentice Hall International Editions, 734 pp.

Halkier, B. (2010). Focus groups as social enactments: integrating interaction and content in the analysis of focus groups data. Qualitative Research 10(1):71-89.

Kuckartz, U. (2007). Einführung in die computergestützte Analyse qualitativer Daten. 2. Ed. VS Verlag für Sozialwissenschaften, Wiesbaden, Germany, 268 pp.

Leopoldina (ed.) (2012). Bioenergie: Möglichkeiten und Grenzen. Kurzfassung und Empfehlungen. Leopoldina, München. Available at: www.lwopoldina.org.

Meyer, R. and Priefer, C. (2012): Ökologischer Landbau und Bioenergieerzeugung – Zielkonflikte und Lösungsansätze. Arbeitsbericht Nr. 151. Büro für Technikfolgenabschätzung beim Deutschen Bundestag. Berlin, Germany.

The Guardian (2012). Secret report: biofuel caused food crisis. Internal World Bank study delivers blow to plant energy drive. Available at: http://tinyurl.com/q8ay4bu.

Wilson, V. (1997). Focus groups: a useful qualitative method for educational research. British Educational Research Journal 23(2): 209-224.

Wissenschaftlicher Beirat für Agrarpolitik beim BMELV (WBA) (2007). Nutzung von Biomasse für die Energiegewinnung – Empfehlungen an die Politik. Berichte über Landwirtschaft, Sonderheft 216.

Wissenschaftlicher Beirat für Agrarpolitik beim BMELV (WBA) (2011): Förderung der Biogaserzeugung durch das EEG. Stellungnahme zur geplanten Novellierung des Erneuerbare-Energien-Gesetzes. Berlin, Germany.

Wissenschaftlicher Beirat der Bundesregierung Globale Umweltveränderungen (WBGU) (2008). Zukunftsfähige Bioenergie und nachhaltige Landnutzung. Berlin, Germany.

Ethical distinctions between different kinds of plant breeding

B.K. Myskja[1*], *H.J. Schouten*[2] *and M. Gjerris*[3]
[1]*Department of Philosophy and Religious Studies, Norwegian University of Science and Technology, Dragvoll, 7491 Trondheim, Norway;* [2]*Wageningen University and Research Centre, Plant Breeding, Droevendaalsesteeg 1, 6708 PB Wageningen, the Netherlands;* [3]*Department of Food and Resource Economics, University of Copenhagen, Rolighedsvej 25, 1958 Frederiksberg, Copenhagen, Denmark; bjorn.myskja@ntnu.no*

Abstract

The article discusses whether there are ethically significant distinctions between different forms of plant breeding. We distinguish different forms of plant breeding according to the kind of technology and degree of human intervention compared to plant reproduction occurring in nature. According to the dominant scientific view, the main concerns are issues of biosafety that are dealt with through risk assessment. Thus, the techniques are ethically equivalent and only the resulting product is of interest. In parts of the societal and philosophical discourse, however, there are attempts to distinguish ethically between these approaches, often relying on ambiguous concepts such as 'naturalness'. We argue that a virtue-based approach can be used to explicate the assumptions behind such distinctions that are relevant for scientific and public discourse, and support a conclusion that there are ethical differences between plant breeding methods. The framework can contribute to an improved dialogue between the scientific community and the wider public by making the scepticism towards GM-technology more intelligible.

Keywords: virtue ethics, naturalness, new plant breeding techniques, GM plants

Introduction

Modern biotechnology has changed the art of plant breeding, albeit accompanied by controversy. The debate is split: the scientific community has discussed genetic modification as a method for improving the traits of food plants, focussing on biosafety, holding that if these risks are equivalent to conventionally bred counterparts, non-contained use of the GM plant is acceptable. A significant group within the European public has taken a wider normative approach to the ethics of plant biotechnology, holding that the acceptance of risk for GM depends on what benefits to expect (Gaskell *et al.*, 2004). In addition, many have objected to the technology based on intrinsic normative assessments, such as 'naturalness', generally not part of the scientific discourse (Gaskell *et al.*, 2011). Only seldom have any of these discourses dealt with the differentiation and refinement of plant biotechnology in different techniques as having normative significance. For the dominant scientific view the difference in methods are arguably irrelevant, as the matter is one of evaluating the product rather than the technology. For the public, the details in technical distinctions are not known or heeded. However, public surveys that take into account technological differences, find that significant proportions of the public judge them to be of moral relevance (Gaskell *et al.*, 2011); the most important criterion for acceptance being the perceived naturalness of different kinds of breeding techniques (Mielby *et al.*, 2013). Most scientists would not recognize the idea of naturalness as a relevant concept in this debate – especially not when the concept also gains a normative status.

In the following, we will describe different technological possibilities of intentional changes in the traits of plants and discuss to what extent these technologies differ ethically significantly. We will suggest a virtue-based approach that takes into account both the scientific risk assessment approach and a modified concept of naturalness as a way to make a more nuanced and integrated approach to the ethical evaluation of different kinds of plant breeding. Due to limited space, we will not discuss the politics and regulation of different techniques although the ethical evaluation clearly may be relevant for regulation or for guiding consumer choices. For the same reason, we will only briefly discuss important ecological and socio-economic issues related to new plant varieties.

Different breeding techniques

We distinguish different forms of plant breeding according to the kind of technology and degree of human intervention compared to plant reproduction occurring in nature:

- Conventional breeding by means of crossings and selection, based on phenotypic selection only.
- DNA marker assisted selection.
- Undirected mutation breeding using irradiation or chemical treatment.
- Directed mutation breeding, using novel genome editing techniques.
- Using transgenic plants as intermediates, but the resulting cultivar does not have genetic alterations compared to conventional breeding (so-called null-segregants).
- Cisgenesis, which refers to introducing natural genes from crossable relatives that can be used in conventional breeding too. However in case of cisgenesis this is performed in a targeted way, by only introducing the desired gene(s) only, using genetic modification techniques, rather than random mixing and recombination of genes (alleles) among whole genomes, as occurs in crossings.
- Intragenesis. Similar as cisgenesis, but now new combinations of genetic elements, such as coding region and promoter form the crossable plants, are made and introduced in a plant, possibly leading to new expression patterns of the gene. In case of cisgenesis, the gene has always its native promoter and native terminator (Schouten *et al.*, 2006).
- Transgenesis, which refers to introduction of foreign genes, using genetic modification techniques (Lusser *et al.*, 2012).

Narrow scientific and broad ethical assessments

According to the dominant scientific view, these methods are at the outset ethically equivalent as it is the end result, the GM-plant that is relevant. However, they differ in degree of interventions, speed of alterations, and whether the changes could have occurred without human intervention. This can be reflected in a biosafety-based differentiation as we see in the European Food Safety Authority (EFSA) evaluation (EFSA, 2012), where the biosafety of cisgenic plants for food, feed and environment is considered similar to that of conventionally bred plants, whereas intragenic plants can lead to additional risks. It is typical for this approach that the evaluation of risks is based on an implicit utilitarian approach, decoupled from two kinds of question: the first concerns the relation these possible risks have to potential benefits and the second the evaluation of the broader impacts of the technology use has on farming and society. One can argue that these questions are not parts of the scientific biosafety assessment but concern the broader assessment, which is a political task. But a full ethical assessment should include all relevant aspects of the implementation of the technology in modern society, and a narrow scientific risk assessment will therefore leave us with an incomplete evaluation. It will also leave us without resources for handling the question of the perceived naturalness of the technology.

A broader ethical assessment would ideally integrate the scientific risk assessment. The problem is that key concepts in the public discourse on plant biotechnology such as 'naturalness' are difficult to handle to scientists as they are not easily definable or measurable and apparently based on scientific illiteracy. A successful integration could perhaps be achieved if these concepts could be shown to contain more than just scientific misunderstandings and allegedly unfounded metaphysical assumptions. But that presupposes that the implicit narrow utilitarian framework for scientific assessment is adjusted and that 'naturalness' can be shown to carry meaningful content within such a broadened framework. We suggest a virtue ethical approach, based on Hursthouse (1991) and Sandler (2007), which is consequentialist in nature but avoids an instrumental conception of value, where objects or acts are merely evaluated according to how they affect happiness. A virtue theory is, from our perspective, well suited for framing the contested concepts and making the assumptions behind them relevant for scientific and public discourse.

Virtue ethics and plant breeding

The core idea of virtue ethics is that the rightness of actions is not expressed by rule-following or calculation of outcomes, but in the character of the acting person or group. In this sense, it is an agent-focused approach to morality (Baron *et al.*, 1997: 177). The good acts are those as the good person of practical wisdom will choose; the person who knows and can make sense of what is good in life. Having virtue is an integral part of living a good life, a life in *eudaimonia*, which means human flourishing. As we will return to later, this can be interpreted as including nature in the sphere of entities with intrinsic value – in the ethical community so to speak. When we evaluate a field such as plant breeding from the point of view of virtue ethics, we judge different courses of action according to what vices or virtues they express, and to what extent these are integral to a life of human flourishing. This approach is holistic in the sense that the acts are valued not on external criteria, but according to how they are an integrated part of a good human life, or in the case of a particular practice such as plant breeding, part of a good practice of this sort. Thus, we need to ask what the virtues of plant breeding are, or to be more precise: and in what way different choices within plant breeding can contribute to a flourishing human life and – as an integral part of that – the respectful treatment of nature. Virtue ethics is also relational in the sense that human flourishing presupposes that we are social beings, realizing ourselves in moral communities. This social sphere is not necessarily limited to humans but can be said to include animals, plants and nature as such.

This obviously opens up a discussion of when a human life can be said to be flourishing. In that sense any theory of virtues demand a 'thick' understanding of what is worthwhile as Rosalind Hursthouse calls it (Hursthouse, 1991). This is, however, no more disqualifying for this type of ethical theory than it is for more utilitarian or deontological approaches that also need to give 'thick' descriptions of happiness respectively rationality to be able to engage in substantial ethical discussions (Hursthouse, 1991). In the case of a virtue ethics approach focusing on questions of plant breeding, the virtues needed would be focusing not only on human relationships, but also on the relationship between humans and nature.

In order to identify the relevant virtues we have to analyse the broader context of plant breeding. The aim of industrial-scale breeding is typically to produce better plants for farming, meaning that good breeding practices are those that lead to good farming practices and to the kind of food products that are good. Good here in all instances meaning 'lead to a flourishing life'. It is not sufficient that the plants are acceptable from a pure biosafety perspective. For example, not only the effect of introduction a gene on the safety of food, feed and environment should assessed, but also the possibly changed use of chemicals due to that introduced gene, and the ecological footprint of these chemicals, socio-economic consequences for farmers, etc. The virtues of a good plant breeder includes taking seriously the needs and concerns of those who are going to grow and eat or use the plants, but also the environment where the plant will become an element. As Sandler (2007) points out, key virtues in assessing agricultural practices such as the use of GM plants include environmental sustainability and environmental stewardship.

Sandler (2007) argues that the current dominant GM crops are expressive of a domination and manipulation approach to agriculture that is contrary to these virtues, when a more comprehensive approach is the adequate response. The problems we face are connected to modern lifestyle choices and 'social, economic, and institutional factors that contribute to our environmental and agricultural challenges' (Sandler, 2007: 136). Thus, we have no reason to accept the environmental and socio-economic consequences posed by the current commercially available GM varieties. Note that these are not arguments against using the GM technology as such, but only against the current implementation of the technology. To assess whether GM plants are ethically acceptable, we need a more comprehensive approach where the issue is to what extent they contribute to solve significant challenges: 'alleviate the suffering of the impoverished, eliminate global inequalities, or reduce the negative ecological effects of modern agriculture' (Sandler, 2007: 137). Sandler mentions Golden Rice as a case of a GM variety that could be acceptable within a comprehensive virtue ethical evaluation. The essential question here is not the utilitarian whether the technology results in greater aggregated happiness, but the comprehensive question whether it is integrated in agricultural practices that are sound, environmentally friendly and productive.

Although Sandler's general conclusion regarding the framing of the GM technology appears warranted, his analysis can and should be expanded, to capture the 'richness and complexity of our relationship with the natural environment' (Sandler, 2007: 3) which is relevant in using biotechnology in plant breeding. The characteristics of the technology itself make a difference in the wider assessment. In this context, the making sense of a concept such as 'naturalness' within an environmental virtue ethics approach can provide the required sophistication to the discussion.

Naturalness

The idea that the concept of naturalness has a role to play in the ethical discussions related to plant breeding technologies is, to say the least, contested. This both in the sense that plants developed through GM-technologies would be more 'unnatural' than e.g. plants developed through conventional breeding technologies and in the sense that 'natural' should in some sense be normatively different from the 'unnatural' (Leyser, 2014; Reiss and Straughan, 2000; Weale, 2010). Nonetheless there is no doubt that in the public sphere the idea that GM-technology is to at least some extent more 'unnatural' than more conventional technologies is widespread and that this plays a role for the ethical evaluation of the technology (Gaskell *et al.*, 2011; Mielby, 2013; Verhoog, 2003). This is by proponents of the technology often explained as a 'misconception' of the technology, as it is no more or no less unnatural than conventional plant breeding that also meddles with the genes of the plants. The belief that the controversy regarding GM-plants can be made to disappear through enhanced information of the public on the nature of the technology is today often labelled the 'knowledge deficit' or 'knowledge gap' model (Cook *et al.*, 2004) and has in many instances been shown not to be true as acceptance of the technology does not seem to grow with the level of information (Lassen *et al.*, 2006).

As Sagoff (2001) points out, already J.S. Mill's (1806-1873) thoughts on 'nature' show that if we look for absolutes in our definition of 'nature', we end up with either a notion of nature as the totality of what is, which makes everything, including all human activities such as agriculture and GM-technology 'natural', or a notion where that which is natural is that which is untouched by humans. But as it will be very hard to find anything on the face of the planet in the Anthropocene age that has not already been touched by humans, this leaves us with no nature at all (Sagoff, 2001). The task from a virtue ethics perspective must be to see whether the notion of 'naturalness' that plays a role in the public objections to GM-technology, but from a narrow scientific perspective do not make much sense, can be translated into certain virtues and vices pertaining to the relationship between humans and nature thus balancing between the two Millean extremes. This will enable a discussion of the values underlying the critique of GM-technology from the 'naturalness' point of view between proponents and opponents of the technology that could move beyond the futile discussions based on the knowledge deficit model.

If we assume that the public views on naturalness are connected to the kind of wider ethical assessment suggested above, we can talk of degrees of naturalness. If we also accept that what the naturalness argument objects to is the lack of respect for the complexity of nature as something that we only partially can and should control (Myskja, 2006), then we can say that showing such respect is a central virtue in plant breeding. Then classical breeding is the paradigm of a practice that displays humility regarding human potential for control over nature, expressing respect for the value of nature independent of human interests. If we are going to accept more radical interventions, we must first ensure that potential negative consequences are acceptable, and that they do provide benefits that are distributed in accordance with the potential burdens. This can be understood as less morally problematic than to infringe on the values expressed in the 'naturalness' objection

Assessment of different breeding techniques

When assessing the different techniques according to this virtue approach, the significant point is not the technique as such, but how it affects the traits of the plant with regard to effect on the ecological system, food, feed, and agricultural practices. Conventional breeding and DNA marker assisted selection are both expressive of respect for our limited capacity for controlling the complexities of nature, and are based on crossings and uncontrolled reshuffling of DNA during the meiosis and fertilisation, followed by selection on phenotypic traits and/or DNA markers. Cisgenesis respects the natural crossing borders between plants, and cannot lead to novel phenotypic traits, as compared to conventional breeding. From this point of view cisgenesis is close to conventional breeding. Intragenesis uses DNA of the plant itself and from crossable relatives, but novel combinations can be made between genetic elements, leading to traits or gene expression patterns that can be novel or altered compared to nature or conventional breeding. Transgenesis introduces foreign DNA, and can lead to novel traits, and therewith novel biosafety issues. Transgenesis can be regarded as less natural, and leading to less natural products, compared to cross breeding and cisgenesis. Spontaneous mutations occur frequently in nature and conventional breeding. However, untargeted mutation breeding leads to a far higher mutation frequency. Directed mutation leads to very few mutations only, which also could occur spontaneously in nature. The use of transgenic intermediates leading to end products without genetic changes, does eventually not lead to novel traits. According to the principles we suggest, the end-product should be classified as similar to end products of DNA marker assisted breeding.

Conclusions

Based on a virtue ethics framework, we argue that there are ethical differences between plant breeding methods, with implications for policy and regulatory issues. The framework can contribute to an improved dialogue between the scientific community and the wider public as it makes the scepticism towards GM-technology more intelligible.

References

Baron, M.W., Petitt, P. and Slote, M.A. (1997). Three methods of ethics: a debate. Wiley, Hoboken, New Jersey, USA.

Cook, G., Pieri, E. and Robbins, P.T. (2004). 'The scientists think and the public feels': expert perceptions of the discourse of GM food. Discourse & Society 15 (4): 433-449.

European Food Safety Authority (EFSA) (2012). Scientific opinion addressing the safety assessment of plants developed through cisgenesis and intragenesis. EFSA panel on genetically modified organisms (GMO). EFSA Journal 10 (2): 2561.

Gaskell, G., Allansdottir, A., Allum, N., Castro, P., Esmer, Y., Fischler, C., Jackson, J., Kronberger, N., Hampel, J., Mejlgaard, N., Quintanilha, A., Rammer, A., Revuelta, G., Stares, S., Torgersen, H. and Wagner, W. (2011). The 2010 Eurobarometer on the life sciences. Nature Biotechnology 29 (2): 113-114.

Gaskell, G., Allum, N., Wagner, W., Kronberger, N., Torgersen, H., Hampel, J. and Bardes, J. (2004). GM foods and the misperception of risk perception. Risk Analysis 24(1): 185-194.

Hursthouse, R. (1991). Virtue theory and abortion. Philosophy & Public Affairs 20 (3): 223-46.

Lassen, J., Gjerris, M. and Sandøe, P. (2006). After dolly – ethical limits to the use of biotechnology on animals. Theriogenology 65: 992-1004.

Leyser, O. (2014). Moving beyond the GM Debate. PLoS Biology 12 (6). DOI: http://dx.doi.org/10.1371/journal.pbio.1001887.

Lusser, M., Parisi, C., Plan, D. and Rodríguez-Cerezo, E. (2012). Deployment of new biotechnologies in plant breeding. Nature Biotechnology 30: 231-239.

Mielby, H., Sandøe, P. and Lassen, J. (2013). Multiple aspects of unnaturalness: are cisgenic crops perceived as being more natural and more acceptable than transgenic crops? Agriculture and Human Values 30 (3): 471-480.

Myskja, B.K. (2006). The moral difference between intragenic and transgenic modification of plants. Journal of Agricultural and Environmental Ethics 19: 225-238.

Reiss, M.J. and Straughan, R. (2000). Improving nature? The Science and Ethics of Genetic Engineering. Cambridge University Press, Cambridge.

Sagoff, M. (2001). Genetic Engineering and the Concept of the Natural. Philosophy and Public Policy Quarterly 21 (2): 2-10.

Sandler, R. (2007). Character and Environment. New York: Colombia University Press.

Schouten, H.J., Krens, F.A. and Jacobsen, E. (2006). Do cisgenic plants warrant less stringent oversight? Nature Biotechnology 24:753. DOI: http://dx.doi.org/10.1038/nbt0706-753.

Verhoog, H. (2003). Naturalness and the genetic modification of animals. Trends in Biotechnology 21 (7): 294-297.

Weale, A. (2010). Ethical arguments relevant to the use of GM crops. New Biotechnology 27 (5): 582-587.

Environmental virtue ethics – big hype for nothing?

P. Moula
Division of Philosophy, KTH School of Architecture and the Built Environment, Brinellvägen 34, 100 44 Stockholm, Sweden; moula@kth.se

Abstract

Environmental virtue ethics has grown to be a vital part of environmental philosophy and with regard to agricultural and food ethics virtues are proposed to guide us on how to manage our agriculture and food production. In this paper I will critically examine one central question for environmental virtue ethics: Do we need new virtues in light of our environmental challenges? By investigating the virtue 'being rightly oriented toward nature' as proposed by Hursthouse I highlight: (1) that many of the new proposed virtues are different from the old virtues in the sense that they are not justified as virtues by the notion of eudaimonia; I argue (2) that the question of whether new virtues are needed demands a specification of what 'need' the new virtues are supposed to address.

Keywords: environment, eudaimonia, inherent worth, vice

Introduction

Virtue ethics has a long and diverse history going back to the writings of Aristotle and Plato. Since then virtue ethics has seen a modern revival which is often attributed to Anscombe (1997: 202). In her important 1958 essay, 'Modern Moral Philosophy', Anscombe argues that the concepts of moral duty and moral obligation are harmful and ought to be jettisoned because they are survivals of earlier conceptions of ethics which no longer survive. Hursthouse describes virtue ethics as following:

> 'Virtue ethics' is a term of art, initially introduced to distinguish an approach in normative ethics which emphasizes the virtues, or moral character, in contrast to an approach which emphasizes duties or rules (deontology) or one which emphasizes the consequences of actions (utilitarianism). (Hursthouse, 1999: 1)

In comparison to with utilitarian approaches or deontic frameworks virtue ethics can be seen to include more into what constitutes a moral life. Such things as emotions, desires and habits are essential for virtue ethics.

What then are virtues? At the most fundamental level virtues are excellent dispositions, regarding actions and emotions, manifested in habituation. Dispositions are not something one can easily get rid of or acquire overnight. It is important that the virtues are habituated and come from a firm character. For example, kindness is not possessed by someone who is kind on occasion or when it suits her. A person has the virtue of kindness when she has a habituated emotional and behavioural disposition to be kind and it is a part of her life. Swanton's conception of virtue, which she takes from Aristotle, maintains that virtues are dispositions in which both reason and emotion are well ordered (Swanton, 2003: 8).

More recently virtue ethics has been of growing interest within environmental ethics. Environmental ethics is a broad field, described in the Stanford Encyclopedia of Philosophy as 'the discipline in philosophy that studies the moral relationship of human beings to, and also the value and moral status of, the environment and its nonhuman contents'. Hursthouse explains environmental ethics differently but also the role of environmental virtue ethics:

> Environmental ethics is concerned with the articulation and defence of what I shall call 'the green belief'- the belief, namely, that a fairly radical change in the way we engage with nature is imperative. Environmental virtue ethics, then, is concerned with articulation and defending the green belief in virtue ethics terms, rather than in the terms of its two rivals, utilitarianism and deontology. (Hursthouse, 2007: 155)

Do we need new virtues?

Within the field of environmental virtue ethics several authors have proposed and discussed new virtue. The basic question I ask in this paper is this: Do we need a new set of virtues in light of our environmental challenges?

First however, we need to make some clarifications. If we ask ourselves whether we need new virtues we must ask what we might need them for. A distinction has to be made between needing new virtues on order to actually solve our environmental challenges or whether we need them in order for virtue ethics to, in theory, be able to meet our challenges. I assume that by adding new virtues people will not become virtuous, and as long as people are not it is difficult to see how our challenges will be met (see discussion on akrasia). However, let's assume that everyone were virtuous, or virtuous enough, in accordance to the old virtues. Would this be enough to deal with our problems or does our environmental situation make need for new virtues?

There is no straightforward demarcation of 'old' and 'new' virtues. As mentioned virtue ethics has a rich history, and so does the language of virtues. Throughout history some virtues have fallen out of popularity while others have gained and many have been given new meaning. In more recent times Louke van Wensveen (2000) lists in an appendix of her book 189 virtues she claims occur in post-1970 environmental literature. Whether or not all these are actual virtues it is clear the task of dividing them into old and new is not easy. With that being said some distinctions can be made. Hursthouse (2007) for example explicitly mention prudence, practical wisdom, compassion, and proper humility as old virtues and of course justice has to be mentioned as well.

Let us then try to answer the main question. If people were virtuous enough in accordance with the old virtues would our problems be solved? Hursthouse writes:

> It seems clear that much of what is wrong about our current practices with regard to nature springs from these familiar and ancient human vices (pride and vanity) played out, in environmental ethics, on an unfamiliar stage. And it may well be that if we could find a way of releasing many human beings from the grip of these familiar vices, the change in our current ways of going on would be so extraordinarily radical that it would indeed adequately set the scene for all the changes that environmentalists dream of'. (Hursthouse, 2007: 157)

My intuition is that if we got rid of these familiar and ancient human but personal vices and replace them with familiar personal virtues we would come a long way, but not all the way. The reason for my intuition at this stage is not because of some lacking some virtues but the nature of our environmental problems. Some of our environmental challenges are reasonably viewed as problems of collective actions, not of individual action. Ronald Sandler writes:

> Many environmental problems are longitudinal collective action problems. They arise from the cumulative unintended effects of a vast amount of seemingly insignificant

> decisions and actions by individuals who are unknown to each other and distant from each other. (Sandler, 2010: 167)

Brian Treanor has pointed out that much of what has been written about environmental virtue focuses on what he calls personal virtue – individual actions, characteristics and dispositions. However, he claims that personal virtues are not fully adequate for effecting social change and therefore addressing most environmental problems. He argues that if environmental virtue ethics is not to fail it need to focus on: 'virtue politics' – collective actions, characteristics, or dispositions – or in what might be called 'public virtues', actions, characteristics, or dispositions that benefit the community rather than the individual (Treanor, 2010: 9)'.

These virtues are well grounded in the Aristotelian virtue tradition and are in no way new. If we combine these virtues with the old personal virtues we have a case that a proper consideration of the old virtues will result in an answer to our question in this paper with a 'no'. One way of reasoning about whether they will be enough or not is by looking at the proposed new virtues to see if something is missing. Here is a short list with some proposed new environmental virtues: patriotism (Cafaro, 2010); radical hope (Thompson, 2010); simplicity (Gambrel and Cafaro, 2010); forgivingness (Norlock, 2010); wonder (Hursthouse, 2007); being rightly oriented towards nature (Hursthouse, 2007). Due to limitations in space I will focus on the new virtue of being rightly oriented towards nature which is popular and brings out the role many of these new virtues play.

Being rightly oriented towards nature

Hursthouse rightly points out that in the literature on environmental virtues a lot of people suggest a virtue of 'respect for nature'. Originally the term was used by Paul Taylor, not as a virtue but as an 'ultimate moral attitude'. For Taylor, the attitude consists in respecting the inherent worth of all living things. Taylor claims that all living things have a *telos*, which makes it the case, he writes, that they can be harmed or benefitted. *Telos* can be understood in slightly different ways. Hursthouse writes that things which have a *telos* have a 'good of their own' (Hursthouse, 2007). In my opinion using the word 'good' might to strong. The question whether all living things have a good is an open question even if they have a telos (which means they are biologically directed) but in this paper I will use Hursthouse's understanding.

The notion of all living things having inherent worth brings with it well known problems. Either the notion does not allow for degrees which make it practically impossible to use, or we allow for degrees which might allow for human beings, or persons, to have superior weight which for some would compromise the purpose of respecting nature. Whatever we think about Taylor's idea there are some challenges when wanting to adopt it into virtue ethics. I will focus on what I think will be the biggest problem, and also in the end what I think is characteristic for these proposed new virtues, namely that they break from the idea that virtues ought to be beneficial also for their processor.

With regard to what things ought to be virtues, and most importantly why, the strongest tradition in virtue ethics refer to the notion of eudaimonia, translated as human well-being or human flourishing. On this account the virtues are excellences in character or intellect which benefit their possessor; they help their possessors achieve eudaimonia which is the highest human good. If being rightly oriented towards nature is a virtue it should be explained either how it contributes to the well-being of its possessor or it has to be justified by something other than eudaimonia.

In his seminal book Character and Environment from 2007, Ronald Sandler chooses the second route. Sandler claims that we are able to acknowledge inherent worth in the world and that this worth makes a claim on us. 'The inherent worth of living things establishes the good of living organisms as a noneudaimonistic end to be incorporated into the pluralistic teleological account of what makes a character trait a virtue' (Sandler, 2007: 72). The thought is that there are entities, organisms and collectives, which have a *telos*, good of their own, that we have to take into moral account regardless of whether they promote our human flourishing not. Sandler proposes a pluralistic account of what makes a character trait a virtue and opens up for noneudaimonistic considerations.

Let us for the sake of argument accept this new virtue of being rightly oriented towards nature as a genuine virtue and go back to our first question. Does it improve the possibility for virtue ethics to be able to respond to our environmental challenges? Perhaps it does, but mainly it seems to respond to those who claim that we need an ethic that considers all living and perhaps even non-living things in nature as morally considerable. It seems that we cannot fully answer our first question without answering what we mean by environmental challenges. We can mean two things:

1. Our environmental challenges consist of meeting threats to the future existence and/or well-being of human beings (and sentient animals if not anthropocentric).
2. Our environmental challenges consist of meeting threats to the future existence and/or well- beings of human beings, sentient animals and the well-being of all living, and perhaps also non-living things.

With this distinction in mind we are better equipped to answer our first question. If we by environmental challenges mean (1) then there is a strong case to be made that the old virtues hold their ground. I agree with Hursthouse (2007) that the extreme cases, for example people who deny climate change or simply do not care about it, would disappear if people had the traditional virtues. How then about the people who are not extreme cases, would they change (radically) if they were more virtuous? This is somewhat unclear. Many philosophers who have written about personal environmental virtues tend to claim that if we were virtuous we would live radically different lives, but they also often include new virtues. Hursthouse seems to be of the opinion that it does not really matter whether we add the new virtue of 'being rightly oriented towards nature' or not. If we are to be virtuous we need to change radically. She says that 'anyone who had adequately reconfigured the old' virtues accordingly to the context of environment 'could be truly described as having acquired the new'. Thereafter she says that the virtue of 'being rightly oriented toward nature':

> ...is not a character trait we see manifested by any academic philosophers who, inevitably, lead lives of standard Western, materialistic comfort, driving to shop at their supermarkets, buying new clothes, listening to opera on their CD players, dining in restaurants, writing their books and articles on computers, jetting to international conferences to present their views on environmental ethics, and teaching them to their students in large, land-occupying, buildings. (Hursthouse, 2007: 168)

I believe the details in the quote of Hurthouse can be questioned but let's assume that people who have the personal virtues will make rather radical lifestyle changes. On top of that lets assume that the traditional public virtues are habituated in people as well which would lead to people having goodwill towards member of one's community, concern for one's local environment and having a political environment (Treanor, 2010). On top of that assume that our society is well structured in order to make possible and promote collective actions for the well-being of public good. We now have virtuous citizens who do not have personal vices such as greed, self-indulgence, short-sightedness, and lack of compassion as driving forces and neither do they have political vices such as apathy and passivity. They have habituated behaviour which promotes the public good as well as the individual good and they take part in collective action for causes such as global justice and the well-being of future generations. This is a strong argument for why the traditional virtues, if well understood and properly habituated in society, is enough for securing at least the well-being of human and future human beings.

How then about sentient beings other than humans? Would a well-ordered society with virtuous people secure the well-being of sentient and future sentient animals? This is a large question, only a very rough sketch can be provided here. Let us imagine a truly compassionate person, one who cares about the well-being of other human beings for their own sake. Now imagine this person having knowledge about the biological traits/structure of some other beings which make it very probable that these beings have the ability to feel pleasure and pain. Imagine that this virtuous person observes situations where these animals behave in ways which are reasonable to assume to be expressions of pain, like screaming and trying to flee. Is it likely that this person would be indifferent towards this animal? I think not. Remember that the virtuous human has a habituated disposition to feel compassion and try to prevent (undeserved) human suffering, this is not a reaction she can choose; it is there. It seems implausible that virtuous people who have strong emotional and behavioural dispositions towards human suffering would be completely indifferent to the animals. They would have a hard time justifying it, if our aim is to be good virtuous human beings, and to response to human suffering, why should we be indifferent towards animal suffering? But also they would have a psychological make-up which seems difficult, if not impossible to have. How can a person who is genuinely compassionate towards all forms of undeserved human suffering not feel any compassion towards undeserved suffering in the forms of vertebrate/mammal animal suffering?

How about if we have (2) in mind when we speak of environmental challenges? Then some new virtues might be necessary and the reason can perhaps best be seen in the discussion of 'being rightly oriented towards nature'. We must first ask ourselves weather the old virtues cover concerns for living but not sentient things for their own sake. Do the virtues of compassion, justice and wisdom include all living things in nature as part of their concern? Is the scope of the old virtues so broad to include concern for plants and trees for their own sake? It is hard to see how this could be the case. As mentioned above, earlier virtue ethics has a rich history and the just treatment or well-being of plants is not included in the traditional discussion and description of concern for the old virtues.

It should also be noted that Hursthouse criticizes Taylor's account of being limited only to living thing, or the animate. She says that the respect ought to be extended toward such things as spider webs, fossils, the Grand Canyon and so forth. If we are to treat non-living things in nature as morally considerable it is hard to see what this means in practice. If I see an empty spider web in my house, am I not to remove it? Or am I to remove it an I certain way in order to treat it with respect? Or am I supposed to have a special emotional disposition towards removing it; perhaps I should do it with a sense of unhappiness?

The basic question is this: If I am truly compassionate and virtuous with regard to the old virtues, would I have reason to treat trees and spider webs as morally considerable in themselves because I have the old virtues habituated? Recalling that being virtuous is caring about the right things for the right reasons I cannot see the answer being 'yes'. Perhaps someone would claim that the last man argument (Peterson and Sandin, 2013) shows why a virtuous person acts wrong when destroying non-sentient things. However, what the last man argument, a hypothetical scenario in which the last person on earth has the choice to ensure the lives of all living things, tries to do is to show that non-sentient things have value of their own, inherent value, and that we therefore should take these into moral consideration. The old virtues do not cover direct moral concern for these things, if one believes the last man argument to be strong then that gives us reason to discuss new virtues such as 'being rightly oriented toward nature'.

In the discussions on agriculture and food ethics some philosophers have proposed new virtues to help us move forward. I conclude this brief sketch trying to our answer the question of whether we need new virtues to meet our environmental challenges by noting that it cannot be answered without first spelling out our challenges. If we believe our challenges consists of securing the wellbeing of humans and sentient animals we have a good case that our traditional virtues with a mix of personal, public and political virtues are sufficient. If we believe our environmental challenges also consist of respecting and treating all living and perhaps non-living things in nature with moral status we have a good case that we do need new virtues.

References

Anscombe, G.E.M. (1958). Modern moral philosophy. Philosophy 33(124): 1-19.

Gambrel, J.C. and Cafaro, P. (2010). The virtue of simplicity. Journal of Agricultural and Environmental Ethics 23: 85-108.

Cafaro, P. (2010). Patriotism as an environmental virtue. Journal of Agricultural and Environmental Ethics 23: 185-206,

Hursthouse, R. (1999). On virtue ethics. Oxford University Press, Oxford, UK.

Hursthouse, R. (2007). Environmental virtue ethics. In: Walker, R.L. and Ivanhoe, P.J. (eds.) Working virtue. Oxford University Press, Oxford, UK, pp. 155-171.

Norlock, J.K. (2010). Forgivingness, pessimism, and environmental citizenship. Journal of Environmental and Agricultural Ethics 23: 29-42.

Peterson, M. and Sandin, P. (2013). The last man argument revisited. Journal of Value Inquiry 47: 121-133.

Sandler, R. (2007). Character and environment: a virtue oriented approach to environmental ethics. Columbia University Press, New York, USA.

Sandler, R. (2010). Ethical theory and the problem of inconsequentialism: why environmental ethicists should be virtue-oriented ethicists. Journal of Environmental and Agricultural Ethics 23: 167-183.

Swanton, C. (2003). Virtue ethics: a pluralistic view. Oxford University Press, Oxford, UK.

Thompson, A. (2010). Radical hope for living well in a warmer world. Journal of Agricultural and Environmental Ethics 23:43-59.

Treanor, B. (2010). Environmentalism and public virtue. Journal of Environmental and Agricultural ethics 23: 9-28.

Van Wensveen, L. (2000). Dirty virtues. Humanity Books, Amherst, New York, USA.

Necessary new approaches towards sustainable agriculture – innovations for organic agriculture

N. Hagemann[1,2*] *and T. Potthast*[2]
[1]*Geomicrobiology, Department of Geosciences, University of Tuebingen, Hoelderlinstraße 12, 72074 Tuebingen, Germany;* [2]*International Centre for Ethics in the Sciences and Humanities, University of Tuebingen, Wilhelmstraße 19, 72074 Tuebingen, Germany; nikolas.hagemann@uni-tuebingen.de*

Abstract

Food production is one of the most pressing issues for a sustainable development (SD). Instead of mainstream paths like the utilisation of insect protein, highly efficient soilless hydroponics, aquaculture or precision agriculture, this article focusses on various approaches of organic agriculture. Organic Agriculture is subject to both scientific and non-academic research as well as commercial application. In many case studies, it has been demonstrated that organic agriculture can be productive, economic and able to reduce the carbon footprint of food. However, its comprehensive applicability and suitability to feed the world is still subject to both scientific and political debates. This article does not intend to adjudicate on this general question, but wants to shift the perspective on organic agriculture and highlight challenges and what we regard as some of the next necessary innovations. We understand SD as a continuous process of seeking inter- and intra-generationally fair and justifiable solutions within the 'safe and just space for humanity'. The general suitability of organic agriculture as a tool towards SD is already proven. Yet the gradual improvement of organic agriculture and its mainstreaming by food labelling and law, we shall argue, are not sufficient in the endeavour towards SD in food production. New ideas need to be included into research and tested as additional and maybe alternative paths. This article introduces the concepts of: (1) stockfree/vegan-organic farming; (2) Terra Preta techniques for soil fertility; and (3) featured biodiversity approaches as a quality marker for agriculture. They are discussed with respect to their potential to combat the loss of biodiversity, unsustainable land use change and soil degradation. These issues are relevant challenges for humankind and intergenerational justice not only in the realm of agricultural ethics.

Keywords: sustainable development, stockfree-organic, vegan-organic, Terra Preta, featured biodiversity, agricultural ethics

Global challenges of food production and organic agriculture

Food production is among the most pressing issues for a sustainable development (SD). On the one hand, 870 million people suffered from undernourishment in the period of 2010-2012. However, there is a positive trend as this number decreased by 39% compared to 1990-1992 (FAO, 2013: 70). On the other hand, agriculture is one of the key activities contributing to the human environmental impact. This includes the responsibility for 30% of the anthropogenic greenhouse gas emissions, 70% of the global freshwater use and the use of 30% of the total land surface (FAO, 2013: 200). This situation results in a dual challenge, to continue and improve the positive trend in nutrition while limiting the environmental impact to a certain extent. This article ethically builds on the concept of sustainable development to highlight the need for new approaches in research to tackle this challenge.

Broadly speaking, to date there are two competing agricultural concepts: 'industrial'/'conventional' and 'organic'/'agro-ecological'. Industrial agriculture is based on the principles of specialisation, simplification and concentration (Kirschenmann, 2007) and is characterised by the use of inorganic, soluble fertilisers and synthetic pesticides. Using breeds of both animals and plants that are optimised specifically for high productivity, it aims at high yields of few types of standardised products. Organic agriculture uses non-soluble, natural inputs for soil fertility management and does not use synthetic inputs. Looking at the selection of breeds and the conditions of livestock, there is a wide spectrum within organic agriculture.

We will focus on organic agriculture (OA) because of the following three arguments. First, the 'myth of nitrogen fertilization for soil carbon sequestration' (Khan *et al.*, 2007) is confuted, i.e. the application of inorganic nitrogen fertilisation does not increase soil fertility and soil organic matter content in the long term. Second, organic agriculture decreases yields globally by no more than 8 to 25% compared to conventional methods on neighbouring fields. This was evaluated in several meta-analyses that also highlighted individual cases of yield increases by organic management (Badgley *et al.*, 2007; De Ponti *et al.*, 2012; Seufert *et al.*, 2012; Stanhill, 1990). Third, industrial agriculture contributes to the human impact on global 'planetary boundaries', especially with respect to the loss of biodiversity and the influence on the nitrogen cycle (Rockström *et al.*, 2009). However, the comprehensive applicability and suitability of organic agriculture to feed the world is still questioned and subject to both scientific and political debates. This article does not intend to adjudicate on this general question of agricultural and food ethics, but wants to shift the perspective on organic agriculture and highlight the next necessary innovations.

Organic agriculture: history, environmental benefits and limits

Bio-dynamic agriculture was the first OA movement and started already in the 1920s in Europe and Mexico (Demeter, 2014). Most popular organic farming associations were founded in the 1970s, while the EU started to promote organic farming in 1989 (BÖLW, 2014). The EU approved their first source of law-regulating criteria for OA in 1991 as the EU Regulation N° 2092/91. Today, 5.7% of the total utilised agricultural area of the EU is under certified organic management. This also includes the areas under conversion. The respective ratio for the member states ranges from less than 1% in Bulgaria and Malta to 18.6% in Austria (Eurostat, 2014).

As described above, OA typically results in lower yields per hectare compared to industrial management, but this yield is less dependent on external inputs and nutrients are used more efficiently (Mäder *et al.*, 2002). Cederberg and Mattsson (2000) in a life cycle assessment of organic and conventional milk production in Norway showed that OA reduces external nutrient input. It can serve local conservation goals, too, and thus contribute to aesthetic values. Biodiversity generally benefits from OA especially with regard to soil organisms (Bengtsson *et al.*, 2005; for a critique of the methodological approaches of comparison and a counter-examples cf. Hole *et al.*, 2005).

Sustainable development and the necessity for 'next steps' in organic agriculture

According to the Brundtland-report, the globally most widely accepted document, Sustainable Development (SD) is a 'development which meets the needs of current generations without compromising the ability of future generations to meet their own needs' (WCED, 1987). This includes a continuous process of seeking inter- and intra-generationally just solutions within the 'safe and just space for humanity' (Raworth, 2012). Undoubtedly, safeguarding food security, food safety and food sovereignty while at the same time maintaining biodiversity and the functions of ecological systems is a common denominator for all approaches of operationalising SD.

The general suitability of organic agriculture as a tool towards SD is already proven as it reduces the environmental impact of agriculture as described in the previous section and results in positive effects including the promotion of farmers' socio-ecological resilience (Milestad and Darnhofer, 2003). But with regard to the above-mentioned dual challenge, we conclude that the gradual improvement of OA and its further mainstreaming by food labelling and law (contested differences in standards and labelling notwithstanding) will not suffice for the endeavour towards SD in food production.

Thus, first, we need to set more ambitious goals towards our agriculture to keep pace with the major challenges of rising energy prices, climate change, ecological degradation and economic stress of farmers and food insecurity (Kirschenmann, 2007; Lal, 2010).

Second, given the notion of uncertainty of predicting the sustainability of agricultural practices and technologies in general (Rigby and Cáceres, 2001: 23), a broader variety of carefully selected OA approaches is more likely to serve as a pool of potential solutions to challenges of future food supply. Some of these approaches will be discussed in the following section. This selection of the examples does not intend to be exclusive. However, some alternative paths are not included on purpose: utilisation of insect protein, highly efficient soilless hydroponics, aquaculture or precision agriculture. These approaches are based on technology and thus will get more attention and research funds in the future anyway, as they are not affected by the structural discrimination of agro-ecological approaches (cf. Vanloqueren and Baret 2009).

New approaches beyond organic

New approach: stockfree-organic agriculture

In general, livestock is considered an essential part of agricultural systems under organic management as they provide manure as a fertiliser or essential raw material for quality compost. Vegan-organic or stockfree organic (both terms are understood synonymous in this text) agriculture works without any animal inputs. Although 25% of all organic farms in Germany do not have livestock, most of them are not stockfree as they buy animal inputs (Bonzheim, 2014). Visak (2007) argues that a stockfree agriculture should be the consequence of the moral rejection of animal production by most theories of animal ethics. According to Fairlie (2007), stockfree agriculture is the answer to the question 'Can Britain feed itself?' referring to the high demand of land by livestock farming. Stockfree agriculture is seen as a chance to create a productive, space-saving and climate-friendly (no animal and less compost-borne greenhouse gas emissions, nitrogen input exclusively by biological nitrogen fixation) agriculture. In Great Britain and North America a 'stockfree organic' certification is available (http://www.stockfreeorganic.net) and there are a few stockfree-organic farms e.g. in Germany (Bednarczyk, 2014).

A search on Web of Science™ (www.webofknowledge.com, basic search, 03.12.2014) using 'stockfree' does not result in a single article while 'vegan agriculture' or 'vegan farming' results in a few articles; however, none of them is evaluating stockfree agricultural practices. This was also found by Bonzheim (2014), while Schmidt (2004) found 12 relevant research projects in Europe.

However, this paper does not intend to promote stockfree agriculture as a global goal, since: (1) not everyone shares the strong ethical assumptions on the moral status of animals; (2) on a practical perspective of prudence, there is evidence that animals in agriculture are necessary in certain cases. Cows and sheep are needed to mimic large herbivore herds in the grasslands of Southern Africa to fight desertification (Savory and Parsons, 1980). Here, animals support both biodiversity and food security (maintaining grassland as a habitat and as arable land) as well as climate change mitigation (grasslands as carbon storage). In addition, two thirds of the world's arable land are not suitable for cropping, but for livestock. Nevertheless, stockfree agriculture should be developed as one future building block for OA and further ethical reflection on the human-husbandry-relation.

New approach: terra preta technique

'Terra Preta do Indio' or Amazonian Dark Earths (ADE) are soils with significantly higher agronomic potential than adjacent non-anthropogenic soils found in the Amazon basin (Lehmann *et al.*, 2003). They are of anthropogenic origin and contain charcoal. Thus, they store significant amounts of carbon that has a ^{14}C age of at least 1.000 years (Glaser, 1999). Motivated by the promising characteristics of ADE, the concept of biochar was defined as 'charred organic matter (that) is applied to soil in a deliberate manner, with the intent to improve soil properties' (Lehmann and Joseph, 2009). Biochar proved to create benefits when applied to soil in various climates, including a reduction of greenhouse gas emissions from agricultural soil (Cayuela *et al.*, 2013; Ippolito *et al.*, 2012).

However, Jeffery *et al.* (2011) found that biochar increases yields significantly, but with a grand mean of just 10%. For this purpose, they meta-analysed 16 studies, only four of them using organic fertilisers. Thus, research that is more specific is needed to integrate the biochar concept into OA. From the standpoint of agricultural ethics, Terra Preta approaches are well-suited candidates for improving OA, again as one future building block of SD.

New approach: featured biodiversity

Biodiversity is defined by the UN Convention on Biological Diversity as the variability among living organisms, including the 'diversity within species, between species and of ecosystems' (UN, 1992, Article 2). We define 'featured biodiversity' as agricultural approaches that actively promote biodiversity with the intent not only to increase economic yield per unit of area and time but also the resilience of the agro-ecosystem against pests and other disasters by supporting biodiversity. Traditional intercropping of e.g. carrots and onion to prevent both onion fly and carrot fly damage, or agro-forestry (trees can be seen as nuclei for 'featured biodiversity'; on the importance of biodiversity in agro-ecosystems cf. Altieri, 1999; Brussaard *et al.*, 2010; Kotschi, 2007). However, there are only few documented and researched case studies of explicit strategies to actively promote biodiversity on cultivated land. Schmidt (2010) considers biodiversity as a 'quality marker' in agriculture and suggests measures to 'biodiversify' vineyards. He stresses the importance of soil care including leguminous intercropping and mulching. Fruit trees introduce a new dimension of ecological niches into the vineyard that are complemented by designated 'biodiversity hot spots' that should be created on 0.5% of the cultivated land. These measures significantly reduced the impact of the vinegar fly *Drosophila suzukii* because ants feed on their larvae. This pest caused severe damages in neighbouring vineyards under conventional management. Kirschenmann (2007) gives the example of a 50% rice yield increase by the conversion of an industrial rice farm into a duck/fish/rice organic farming system.

However, the concept was already questioned. Phalan *et al.* (2011) found 'land sparing' (high yield farming coupled with protection of natural habitats) more beneficial to both biodiversity (represented by 167 bird and 220 tree species) and crop yields than 'land sharing' which integrates both agriculture and conservation on the same land. Yet the authors admit that their data just refers to forest-dominated tropical ecosystems in Ghana and India. There is a need for similar studies in other ecosystems. Additionally they did not comment on potential positive effects of land sharing towards a resilience of the agro-ecosystems. For integrating the goals of both agriculture and biodiversity protection under the overall goal of SD, an integrative approach is indispensable, not least with respect to a transdisciplinary participation and empowerment of different societal groups and the socio-economic factors influencing the pathways (not) taken. In any case, ethically speaking, the values of biodiversity are not at all at odds with SD (Potthast, 2014).

Conclusions and outlook

We have argued for the need to go 'beyond organic' to achieve a significant contribution of OA to sustainable development in the future as well. This was illustrated by introducing three examples of 'new' agricultural techniques that lack adequate attention in science so far. While stockfree agriculture and featured biodiversity have not received the appropriate attention by science, biochar and Terra Preta are already 'hot topics', but are not focussed on the needs of organic agriculture. These new approaches need to be researched in practice to verify their claims and to guide these approaches from niche to broad acceptance and application. This article focussed on the methods of cultivation of land in food production. However, the need for being progressive in aiming at sustainable agriculture is evident for all other aspects of agriculture including animal welfare and food distribution.

However, 'innovation' cannot be generated by science and technology alone, but it is a product of a societal process, supported also by academia. Thus, we close by addressing the need for transdisciplinary research. In order to continue a positive trend by employing OA and to meet the challenges of the future, it is necessary to go beyond today's understanding of organic farming and progress to the next steps by integrating new elements.

References

Altieri, M.A. (1999). The ecological role of biodiversity in agroecosystems. Agriculture, Ecosystems & Environment 74(1-3): 19-31.

Badgley, C., Moghtader, J., Quintero, E., Zakem, E., Chappell, M.J., Avilés-Vázquez, K., Samulon, A. and Perfecto, I. (2007). Organic agriculture and the global food supply. Renewable Agriculture and Food Systems 22(2): 86-108.

Bednarczyk, S. (2014). Biovegane Landwirtschaft – Gurke ohne Tier. die tageszeitung. Available at: http://www.taz.de/!143950/.

Bengtsson, J., Ahnström, J. and Weibull, A.C. (2005). The effects of organic agriculture on biodiversity and abundance: a meta-analysis. Journal of Applied Ecology 42(2): 261-269.

Bonzheim, A. (2014). Die bio-vegane Landwirtschaft in Deutschland: Definition, Motive und Beratungsbedarf. Hochschule für Nachhaltige Entwicklung. Bachelor thesis. Available at: http://tinyurl.com/n8kp352.

Brussaard, L., Caron, P., Campbell, B., Lipper, L., Mainka, S., Rabbinge, R., Babin, D. and Pulleman, M. (2010). Reconciling biodiversity conservation and food security: scientific challenges for a new agriculture. Current Opinion in Environmental Sustainability 2(1): 34-42.

Bund Ökologische Lebensmittelwirtschaft (BÖLW) (2014). Wie ist der ökologische Landbau entstanden? Bund Ökologische Lebensmittelwirtschaft (BÖLW), Berlin, Germany. Available at: http://www.boelw.de/biofrage_01.html.

Cayuela, M.L., Sánchez-Monedero, M.A., Roig, A., Hanley, K., Enders, A. and Lehmann, J. (2013). Biochar and denitrification in soils: when, how much and why does biochar reduce N2O emissions? Scientific Reports 3.

Cederberg, C. and Mattsson, B. (2000). Life cycle assessment of milk production – a comparison of conventional and organic farming. Journal of Cleaner Production 8(1): 49-60.

Demeter (2014). Die Demeter Historie. Demeter, Darmstadt, Germany. Available at: http://tinyurl.com/masjttd.

De Ponti, T., Rijk, B. and Van Ittersum, M.K. (2012). The crop yield gap between organic and conventional agriculture. Agricultural Systems 108: 1-9.

Eurostat (2014). Area under organic farming – EU28, European Commission.

Fairlie, S. (2007). Can Britain feed itself? The Land 4: 8.

Food and Agriculture Organization of the United Nations (FAO) (2013). Statistical yearbook: world food and agriculture. Food and Agriculture Organization of the United Nations, Rome, Italy. Availabe at: http://tinyurl.com/ons2x29.

Glaser, B. (1999). Eigenschaften und Stabilität des Humuskörpers der Indianerschwarzerden Amazoniens. Bayreuther Bodenkundliche Berichte 68.

Hole, D.G., Perkins, A.J., Wilson, J.D., Alexander, I.H., Grice, P.V. and Evans, A.D. (2005). Does organic farming benefit biodiversity? Biological Conservation 122(1): 113-130.

Ippolito, J.A., Laird, D.A. and Busscher, W.J. (2012). Environmental benefits of biochar. Journal of Environmental Quality 41(4): 967-972.

Jeffery, S., Verheijen, F.G.A., Van der Velde, M. and Bastos, A.C. (2011). A quantitative review of the effects of biochar application to soils on crop productivity using meta-analysis. Agriculture, Ecosystems & Environment 144(1): 175-187.

Khan, S.A., Mulvaney, R.L., Ellsworth, T.R. and Boast, C.W. (2007). The myth of nitrogen fertilization for soil carbon sequestration. Journal of Environmental Quality 36(6): 1821-1832.

Kirschenmann, F.L. (2007). Potential for a new generation of biodiversity in agroecosystems of the future. Agronomy Journal 99(2): 373-376.

Kotschi, J. (2007). Agricultural biodiversity is essential for adapting to climate change. GAIA-Ecological Perspectives for Science and Society 16(2): 98-101.

Lal, R. (2010). Managing soils for a warming earth in a food-insecure and energy-starved world. Journal of Plant Nutrition and Soil Science 173(1): 4-15.

Lehmann, J. and Joseph, S. (2009). Biochar for environmental management: an introduction. In: J. Lehmann and Joseph, S. (eds.) Biochar for environmental management – science and technology. Earthscan, London, UK, 416 pp.

Lehmann, J., Pereira da Silva, J., Steiner, C., Nehls, T., Zech, W. and Glaser, B. (2003). Nutrient availability and leaching in an archaeological Anthrosol and a Ferralsol of the Central Amazon basin: fertilizer, manure and charcoal amendments. Plant and Soil 249(2): 343-357.

Mäder, P., Fliessbach, A., Dubois, D., Gunst, L., Fried, P. and Niggli, U. (2002). Soil fertility and biodiversity in organic farming. Science 296(5573): 1694-1697.

Milestad, R. and Darnhofer, I. (2003). Building farm resilience: the prospects and challenges of organic farming. Journal of Sustainable Agriculture 22(3): 81-97.

Phalan, B., Onial, M., Balmford, A. and Green, R.E. (2011). Reconciling food production and biodiversity conservation: land sharing and land sparing compared. Science 333(6047): 1289-1291.

Potthast, T. (2014). The values of biodiversity. In: Lanzerath, D. and M. Friele (eds.) Concepts and values in biodiversity. Routledge, London, UK, pp. 131-146.

Raworth, K. (2012). A safe and just space for humanity: can we live within the doughnut. Oxfam Policy and Practice: Climate Change and Resilience 8(1): 1-26.

Rigby, D. and Cáceres, D. (2001). Organic farming and the sustainability of agricultural systems. Agricultural Systems 68(1): 21-40.

Rockström, J., Steffen, W., Noone, K., Persson, Å., Chapin, F.S., Lambin, E.F., Lenton, T.M., Scheffer, M., Folke, C. and Schellnhuber, H.J. (2009). A safe operating space for humanity. Nature 461(7263): 472-475.

Savory, A. and Parsons, S.D. (1980). The savory grazing method. Rangelands: 234-237.

Schmidt, H. (2004). Viehloser Ackerbau im ökologischen Landbau – Evaluierung des derzeitigen Erkenntnisstandes anhand von Betriebsbeispielen und Expertenbefragungen. University of Kassel. Availabe at: http://orgprints.org/5020.

Schmidt, H.P. (2010). Biodiversität ist die Grundlage für Terroirqualität. Journal für Terroirwein und Biodiversität. Available at: http://tinyurl.com/q8gtvtb.

Seufert, V., Ramankutty, N. and Foley, J.A. (2012). Comparing the yields of organic and conventional agriculture. Nature 485(7397): 229-232.

Stanhill, G. (1990). The comparative productivity of organic agriculture. Agriculture, Ecosystems & Environment 30(1-2): 1-26.

United Nations (UN) (1992). Convention of biological diversity. Montreal, Canada.

Vanloqueren, G. and Baret, P.V. (2009). How agricultural research systems shape a technological regime that develops genetic engineering but locks out agroecological innovations. Research Policy 38(6): 971-983.

Visak, T. (2007). Vegan agriculture: animal-friendly and sustainable. In: Zollitsch, W., Winckler, C., Waiblinger, S.,Haslberger, A., (eds.) Sustainable food production and ethics. Preprints of the 7th Congress of the European Society for Agricultural and Food Ethics (EurSAFE) 2007, Vienna, Austria, September 13-15, 2007, Wageningen Academic Publishers, Wageningen, the Netherlands, 550 pp.

World Commission on Environment and Development (WCED) (1987). Report of the world commission on environment and development: our common future. Available at: http://www.un-documents.net/our-common-future.pdf.

Towards a sustainable dairy chain in the Netherlands – the opinion of dairy farmers and their advisors

C.C. de Lauwere[1*], *A.C.G. Beldman*[1], *J.W. Reijs*[1], *G.J. Doornewaard*[1], *A. van den Ham*[1], *A.C. Hoes*[1] *and A.P. Philipsen*[2]
[1]*LEI Wageningen UR, P.O. Box 35, 6700 AA Wageningen, the Netherlands;* [2]*Livestock Research, Wageningen UR, P.O. Box 338, 6700 AH Wageningen, the Netherlands; carolien.delauwere@wur.nl*

Abstract

In the Sustainable Dairy Chain (SDC), the dairy industry and its representatives strive to make the Dutch dairy sector the world leader in sustainability. Focus is on climate and energy, animal welfare and health, maintenance of grazing and environment and biodiversity. A survey with closed questions and pre-programmed answers was finished by 150 conventional dairy farmers. Questions were asked about SDC, the measures farmers took to make their farm more sustainable, the feasibility of these measures, ethical stances of dairy farmers regarding SDC, the farmers' willingness to change, the role of advisors, etc. In addition, 25 in-depth interviews were carried out with veterinarians, feed suppliers, banks, accountants and independent advisors (five in each group). The results of the survey and the in-depth interviews showed that both dairy farmers and their advisors had a positive attitude towards the goals of SDC and their feasibility, and that measures are already taken (or are planned to be taken) to achieve those goals. However, dairy farmers also believe that certain sustainability measures do not suit their farm, or they are unsure about these measures because they do not have enough knowledge or concrete tools are missing. The farmers' advisors share this view. Dairy farmers often see concrete, animal-related goals as important, such as reducing the use of antibiotics, maintaining grazing and lengthening the lifespan of the cows. On the other hand, less concrete goals tend to be seen as less important, such as improving biodiversity, using sustainably-produced soy and palm kernel and reducing phosphate in manure. Approximately half of the conventional dairy farmers who completed the questionnaire indicated that they wished to keep their farm up-to-date and approximately one third said that they preferred to see how things developed before making changes. This should be taken into account in the communication about SDC and its goals. Advisors can play an important role in this because they influence the dairy farmers' decisions to implement sustainability measures. However, to be able to do that, they need to know SDC first.

Keywords: sustainable dairy farming, dairy industry, farmers, advisors

Introduction

Background – the sustainable dairy chain

The Sustainable Dairy Chain (SDC) is an initiative of the Dutch Dairy Organisation (NZO) and the Dutch Federation for Agriculture and Horticulture (LTO Nederland) in which the dairy industry and dairy farmers strive to make the Dutch dairy sector the world leader in sustainability. The parties involved dedicated themselves to generate future support from both the market and society (www.duurzamezuivelketen.nl/eng/home). They have formulated goals to realize a more sustainable dairy sector under four themes: climate and energy, animal welfare and health, maintenance of grazing and environment and biodiversity.

The role of dairy farmers

Dairy farmers play an important role in the realization of the SDC goals. They must be able and willing to take measures on their farm which contribute to the realization of the SDC goals. Therefore, it is important for the SDC to know how dairy farmers think about sustainable dairy farming and the goals strived for by the SDC. To get to know this, it is important to take into account that the choice of a dairy farmer to take certain measures depends on his attitude to this measures and the SDC goals (Jansen *et al.*, 2010), but also on the opinion about these measures and the SDC goals of other people who are important to the farmer (Fielding *et al.*, 2005, 2008), and the extent to which the farmer perceives that the measures are suitable for him and his farm (e.g. does the farmer belief that he has enough space on his farm to take the measure or that he has enough time or knowledge or money to take the measure?). This fits in the Theory of Planned Behaviour (Ajzen, 1991; Ajzen, 2002), which has proven to be useful to understand consumers' behaviour but also is applied in the agricultural domain to understand farmers' decisions (Breukers *et al.*, 2012; Colémont and Van den Broucke, 2008; De Lauwere *et al.*, 2012). The attitude of farmers to the SDC and the measures they are willing to take also depends on the type of farmer. After all, different types of dairy farmers exist who reason from different backgrounds, values and ethical stances (Bock and Van Huik, 2007; De Rooij *et al.*, 2010; Mepham, 2008). Therefore this also should be taken into account when trying to understand how dairy farmers think about the SDC and its goals.

The role of advisors

Dairy farmers don't take decisions regarding the SDC on their own. They are also affected by their family and by advisors, although they are not always aware of that (Nolan *et al.*, 2008). Therefore it is also important for the SDC to know how advisors think about sustainable dairy farming and the goals strived for by the SDC.

Material and methods

A written survey among 309 conventional dairy farmers was carried out in the autumn of 2012 to find out how dairy farmers think about the SDC and its goals. The farmers could finish the questionnaire via a link on the internet or they could send it back by e-mail or ordinary mail. Eventually, 150 farmers finished the questionnaire (response rate: 48.5%). Of them, 10.7% returned the questionnaire after the first e-mail, 10.0% returned it after the second e-mail and 79.3% returned it by ordinary mail without further e-mail reminder. The farmers were asked 112 structured questions with pre-programmed answers about their farm, their way of farming, their attitude to the importance and feasibility of the SDC goals, the SDC-related measures they already took or were planning to take, the extent to which family and advisors affected their decisions regarding these measures and their values and ethical stances. In addition, 25 in-depth interviews were carried out with veterinarians, feed suppliers, banks, accountants and independent advisors (five of each group). They were asked questions about the developments in the dairy sector, the goals of the SDC, the dilemma's they perceived among SDC goals and between SDC goals and the developments and the way they brought up sustainability issues in their discussions with dairy farmers.

Results

Dairy farmers about the sustainable dairy chain, its goals and their feasibility

The interviewed dairy farmers had a positive attitude towards the goals of SDC and their feasibility, and they already took measures or had plans to do so to achieve these goals. The most mentioned measures which the farmers already took or were planning to take, were reduction of the use of antibiotics, maintenance of pasturing, using less gas and electricity and producing roughage more efficiently (mentioned by 85, 68, 68 and 67% of the respondents respectively). However, dairy farmers also mentioned that certain sustainability measures did not suit their farm (for example maintenance of pasturing at big farms), or that they were unsure about certain measures because knowledge about these measures is lacking or concrete tools are missing (for example regarding the production of sustainable energy).

The interviewed dairy farmers were asked which three (out of eleven) SDC goals they found most important and which three goals they found least important. They mentioned reduction of the use of antibiotics, maintenance of pasturing and lengthening of the lifespan of the cows as the three most important goals, and improvement of biodiversity, sustainable concentrate (containing sustainable soy and palm kernel) and the reduction of phosphate in manure as the three least important goals (Table 1).

Almost half of the interviewed dairy farmers (48.9%) found it logical that SDC undertook activities to make the dairy chain more sustainable. However, 25.5% of the interviewed dairy farmers understood the activities of the SDC but did not like it that their dairy factory interfered with their operational management and 25.5% found the additional demands to make the dairy chain more sustainable nonsense. About the feasibility of the SDC goals, 62.7% of the interviewed dairy farmers said that these were feasible, 33.6% found the ambitions too high and 3.7% found the ambitions not high enough.

Differences between dairy farmers

The farmers' willingness to change was studied. Of the interviewed farmers, 52.4% said that they wanted to keep their farm up-to-date (and thus were willing to change), 33.6% preferred to wait and see how things developed at other farms before they made a change themselves, 5.6% called themselves a forerunner and 8.4% said that they did not want to change. The farmers in the latter group had significantly less cows and less ha of (owned and rented) land than the farmers in the other groups ($F=6.2$; $P<0.01$ and $F=3.6$; $P<0.05$ respectively). They also relatively often belonged to the group of dairy farmers that was planning to quit farming within 10 years ($\chi^2=29.5$; $P<0.01$) (Figure 1).

It was also studied whether farmers took their decisions about sustainability issues on the basis of the possible consequences which a decision may have (e.g. 'big dairy farms fit in the Netherlands because they are more competitive'), on the basis of virtues (e.g. 'big dairy farms do not fit in the Netherlands because small dairy farms cannot compete with them'), or on the basis of rights and duties (e.g. 'farmers should have the right to develop their farm the way they want'). It appeared that the interviewed dairy farmers were not consequent in their way of reasoning. Sometimes they reasoned on the basis of consequences, sometimes on the basis of virtues and sometimes on the basis of rights and duties. Questions were also asked to find out how the interviewed farmers thought about animal welfare, consumer related issues and farm size. It appeared that the most interviewed dairy farmers were more concerned about animal welfare than about consumer concerns. Of the interviewed dairy farmers, 70.6% preferred the

Table 1. Most important and least important farming goals according to the interviewed dairy farmers (% of farmers).

Most important farming goals	%	Least important farming goals	%
Reduction of use of antibiotics	73.7	Improvement of biodiversity	54.0
Maintenance of pasturing	63.7	Sustainable concentrate	45.6
Lengthening of the lifespan of the cows	59.6	Reduction of phosphate in manure	37.6

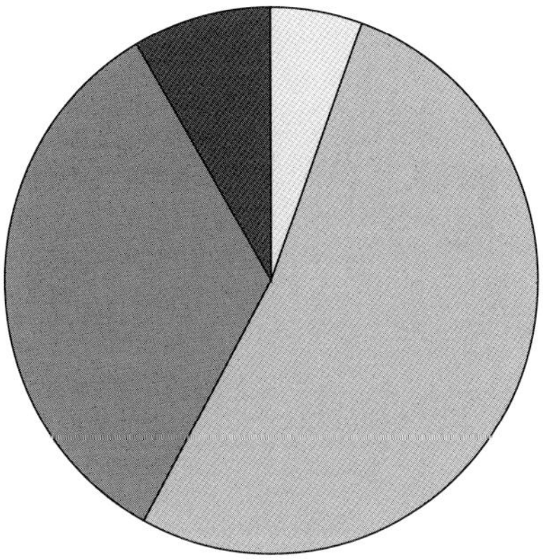

Figure 1. The dairy farmers' willingness to change (% of respondents).

preposition 'animal welfare is good if all natural needs of the animals are fulfilled' above the preposition 'animal welfare is good if the animals produce well', However, 70.1% of the interviewed dairy farmers preferred the preposition 'I am prepared to apply more sustainable production methods at my farm if the consumer is prepared to pay for that' above the preposition 'I think that it is important to produce sustainably, even if the consumers do not want to pay for it', and 69.5% of them preferred the preposition 'consumers who say that the welfare of dairy cows leaves much to be desired, do not know where they are talking about' above 'It makes me think if consumers say that the welfare of dairy cows leaves much to be desired'. The opinions of the dairy farmers concerning the maximum size of dairy farms were divided: 51.1% preferred the preposition 'my farm should not get that big that pasturing is no longer possible' and 48.9% preferred the preposition 'I accept that pasturing is no longer possible if my farm gets bigger'.

Dairy farmers about their advisors

It was asked in the survey to what extent advisors or family affected the decisions of the interviewed dairy farmers. More than 50% of the interviewed farmers called their family supporting in different decisions regarding sustainability such as pasturing, the reduction of the use of antibiotics, the building of a new cowshed, the production of roughage more efficiently and contributing to nature development. Although the interviewed dairy farmers said that most of their decisions were not affected by their advisors, more than 50% of them did call their advisors supporting in the areas in which they were specialized: the veterinarian for decisions regarding the use of antibiotics, the constructor, the feed supplier and the bank for decisions regarding the building of a new cowshed and the feed supplier for decisions regarding the production of roughage and the reduction of N and P in concentrate.

Advisors about the sustainable dairy chain goals

The in-depth interviews with advisors showed that there was a lot of similarity between the SDC goals and the work of the advisors. Advisors endorsed the fact that the SDC strives for more sustainability in the dairy sector. Like the interviewed dairy farmers, the interviewed advisors found the reduction of antibiotics, the lengthening of the lifespan of the cows and the maintenance of pasturing the most important goals, besides – and unlike the interviewed dairy farmers – the reduction of phosphate in manure and the improvement of the nitrogen efficiency. The interviewed advisors found the improvement of biodiversity, sustainable concentrate (containing sustainable soy and palm kernel) and the reduction of greenhouse emissions the three least important goals. The interviewed dairy farmers thought the same about biodiversity and sustainable concentrate. The interviewed advisors had objections against the concept of sustainability. They called it a catch-all term and rather spoke about more concrete issues with the farmers like the improvement of feed efficiency, the lengthening of the lifespan of the cows or animal health.

It was obvious that 19 of the 25 interviewed advisors hardly knew something about the SDC, if at all. According to the interviewed advisors, point of special interest was that the SDC goals are not always specified in detail on the level of the dairy farm. For example, it is not clear what the (positive or negative) consequences are for the dairy farmers if they do or do not take measures to contribute to the SDC goals. Besides, it is not always clear which concrete measures the dairy farmers actually can take do to contribute to the goals. Indicators or tools to be able to control the progress in the achievement of the goals are also missing. It would be useful if SDC can improve this. This would make it easier for the advisors to give more concrete advices to the dairy farmers and would be helpful to get the dairy farmers to start moving in the direction of a more sustainable dairy sector.

Discussion and conclusions

Dairy farmers and their advisors are positive about the SDC and its goals. Both dairy farmers and their advisors found animal related goals – such as the reductions of the use of antibiotics and the lengthening of the lifespan of the cows – more important than less concrete goals like the improvement of biodiversity and the use of sustainable concentrate.

The interviewed dairy farmers did not take their decisions regarding measures to improve sustainability on their farm on their own. Their family was involved in most decisions regarding sustainability and, although the interviewed dairy farmers said that most of their decisions were not affected by their advisors, more than 50% of them did call their advisors supporting in the areas in which they were specialized. For example, the veterinarian for decisions regarding the use of antibiotics, the constructor and the bank for decisions regarding the building of a new cowshed and the feed supplier for decisions regarding the production of roughage and the reduction of N and P in concentrate. This is an indication that 'important others' (like family and advisors) do affect the decisions of farmers. Family and advisors thus can play an important role in the realisation of the SDC goals at farm level. For the SDC this means that they should direct their communication about SDC and its goals not only at farmers but also at their family and advisors. The in-depth interviews with the advisors showed that they are willing to take their part. However, to be able to do so, they should get more familiar with the SDC and its goals. In that case, the SDC could make more use of their services and at the same time, it would give the advisors more legitimacy in their daily work because the SDC initiative is widely supported by the dairy sector. The SDC should take this into account and should act accordingly, for example by communicating more actively about SDC in farmers' and advisors' journals and other media these groups like to use. According to the interviewed advisors, the SDC should make more clear also which concrete measures farmers can take to improve sustainability on their farms and what the – positive or negative – consequences are if farmers do or do not take measures to contribute to the SDC goals.

With regard to the dairy farmers, it is also important for the SDC to take into account that farmers do not take decisions only on the basis of the consequences these may have (focusing on the outcome of the act) but also on the basis of virtues (focusing on the actor him/herself) or rights and duties (focusing on the intention to act) (Mepham, 2008). Moreover, different types of farmers can be distinguished on the basis of their opinions about animal welfare, consumer related issues and farm size. This has been found by others as well (Bock and Van Huik, 2007; De Rooij *et al.*, 2010). This means that SDC can use different types of arguments to convince the dairy farmers that they should take measures to contribute to the SDC goals.

Another important finding in the survey is that some dairy farmers are more willing to change than others. SDC should use different strategies to get these farmers started in the direction of more sustainability on their farms. Jansen *et al.* (2010) distinguished a central and a peripheral route of communication for farmers who were respectively willing or not willing to take measures to reduce mastitis on their farms. The first group was actively involved in the development of measures and tools to reduce mastitis on their farms, while the second group were offered ready-made solutions to do so. Analogously, farmers who are willing to take sustainability measures on their farms – the farmers who wanted to keep their farm up-to-date according to the survey – probably should be involved more actively in the development of tools and measures to improve sustainability at farm level, while farmers who preferred to wait and see how things developed at other farms before they made a change themselves should be offered ready-made solutions. However, at first, insight into these ready-made solutions is needed; both for the farmers and their advisors.

References

Ajzen, I. (1991). The theory of planned behaviour. Organizational Behavior and Human Decision Processes 50: 179-211.

Ajzen, I. (2002). Perceived behavioural control, self-efficacy, locus of control and the theory of planned behaviour. Journal of Applied Psychology 32 (4): 665-683.

Bock, B.B. and Van Huik, M.M. (2007). Animal welfare: the attitudes and behaviour of European pig farmers. British Food Journal 109 (11): 931-944.

Breukers, A., Van Asseldonk, M.A.P.M., Bremmer, J. and Beekman, V. (2012). Understanding growers' decisions to manage invasive pathogens at the farm level. Phytopathology 102 (6): 609-619.

Colémont, A. and Van den Broucke, S. (2008). Measuring determinants of occupational health related behaviour in Flemish farmers: an application of the theory of planned behaviour. Journal of Safety Research 39: 55-64.

De Lauwere, C.C., Van Asseldonk, M.A.P.M., Van 't Riet, J.P., De Hoop, J.G., Ten Pierick, E. (2012). Understanding farmers' decisions with regard to animal welfare: the case of changing to group housing for pregnant sows. Livestock Science 143(2-3): 151-161.

De Rooij, S.J.G., De Lauwere, C.C., Van der Ploeg, J.D. (2010). Entrapped in group solidarity? Animal welfare, the ethical positions of farmers and the difficult search for alternatives. Journal of Environmental Policy & Planning 12 (4): 341-361.

Fielding, K.S., Terry, D.J., Masser, B.M. and Hogg, M.A. (2008). Integrating social identity theory and the theory of planned behaviour to explain decisions to engage in sustainable agricultural practices. British Journal of Social Psychology 47, 23-48.

Fielding, K.S., Terry, D.J., Masser, B.M., Bordia, P. and Hogg, M.A. (2005). Explaining landholders' decisions about riparian zone management: the role of behavioural, normative and control beliefs. Journal of Environmental Management 77: 12-21.

Jansen, J., Steuten, C.D.M., Renes, R.J., Aarts, N. and Lam, T.J.G.M. (2010). Debunking the myth of the hard-to-reach farmer: effective communication on udder health. Journal of Diary Science 93: 1296-1306.

Mepham, B. (2008). Bioethics. Oxford University Press, Oxford, UK.

Nolan, J. M., Schultz, P. W., Cialdini, R. B., Goldstein, N. J., and Griskevicius, V. (2008). Normative social influence is underdetected. Personality and Social Psychology Bulletin, 34: 913-923.

Section 3. Animal husbandry

Is bigger better? Farm size and animal welfare

J.A. Robbins, M.A.G. von Keyserlingk, D. Fraser and D.M. Weary*
Animal Welfare Program, Faculty of Land and Food Systems, University of British Columbia, 2357 Main Mall, Vancouver, BC V6T 1Z4, Canada; jesserobbins.ubc@gmail.com

Abstract

The intensification of agriculture is well underway and one prominent feature of intensification has been the shift to fewer and larger farms. Critics suggest that increasing farm size is inimical to animal welfare because it erodes animal care values and makes it impossible to provide individual care and attention. We evaluated the empirical evidence for these claims using data from more than 100 studies, in a variety of farmed species. Research from the human organizational literature looking at relationships between organization size and various outcomes was also reviewed. Our analysis suggests that the relationship between farm size and animal welfare is far from straightforward. Although smaller farms are more likely to rear animals in more naturalistic conditions, larger farms are more likely to implement science-based, standard operating procedures, train their employees, utilize technology to track and monitor animals and implement costly changes to improve welfare. We found no evidence that farmers on large farms view animals differently, but we did find that cases of animal neglect and mistreatment are more likely to occur on smaller farms. We argue policy efforts focused on farm size are misguided; instead policy makers should try to generalize beneficial animal welfare practices independent of farm size.

Keywords: factory farms, intensification, stocking density, technology

Background

A central feature of agricultural intensification has been the shift towards fewer and larger farms. May have view this transition as largely detrimental, especially with respect to animal welfare, leading to widespread antipathy towards larger farms. A representative sample of USA citizens, found a majority (57%) agreed or strongly agreed that, 'farm animals raised on small farms have a better life than those raised on large farms' (Lusk *et al.*, 2007). Farmer's on the hand appear to view farm size as having much less impact on animal welfare than the general public (Vanhonacker *et al.*, 2008). In an attempt to arbitrate this disagreement, here we summarize some preliminary findings of a literature review examining the relationship between farm size and animal welfare.

Methods

We conducted a literature review of animal welfare studies that include some measure of farm size – defined as number of animals – as a covariate. Our preliminary review includes more than 100 studies and is largely qualitative in nature. Indicators of animal welfare were considered as anything impacting an animal's biological health and functioning, mental state or ability to perform 'natural' behaviours.

Preliminary findings

1. Health/disease. A variety of studies have found group-level prevalence of diseases increases with farm size. However, for animal welfare, the proportion of affected cows within a herd is more relevant, and this number tends to decrease with increasing herd size (Hill *et al.*, 2009), meaning that individual animals are at a lower risk when housed on larger farms. Multiple studies of dairy herds have observed decreased somatic cell counts, indicative of improved udder health, in larger dairy herds (e.g. Oleggini *et al.*, 2001).

Lameness in dairy cattle has been studied extensively and most studies have found increasing rates of lameness with increasing dairy herd size (Alban, 1995; Holzhauer, 2006), yet other work with larger herd sizes contradicts these findings. For example, recent studies of dairy herds in the USA (Chapinal *et al.*, 2013) and China (Chapinal *et al.*, 2014) have found lower rates of lameness in larger herds.

2. Mortality. There is little evidence of a simple consistent relationship between farm size and mortality of the animals. Although larger veal farms were found to have reduced mortality rates (Lensink *et al.*, 2000), the majority of studies have found higher cow mortality rates on larger dairy farms (Alvåsen *et al.*, 2012; Thomsen and Sørensen, 2009). A large-scale study in the USA, reported dairy calf mortality decreased as herd sized increased (NAHMS, 2002), whereas the majority of other studies have failed to find any effect of farm size (e.g. Walker *et al.*, 2012). Similarly, some studies of swine mortality have failed to find any effect of herd size (Wolter *et al.*, 2001), whereas others have found both positive (Christensen *et al.*, 1995) and negative associations (Sasaki and Koketsu, 2008).
3. Human animal interactions (HAI). Larger farms typically have higher worker to animal ratios, so in this sense may be at a disadvantage in terms of providing individualized care. Less frequent HAI on larger farms explains why indicators of fear such as avoidance distance and tonic immobility tend increase with farm size (e.g. Bilčík *et al.*, 1998). However, simply considering the frequency of human animal interactions fails to recognize that such interactions can be positive, negative or neutral (Zulkifli, 2013). Several studies have failed to detect differences in negative HAI (Lensink *et al.*, 2000; Waiblinger and Menke, 1999). There does however appear to be a clear reduction in the frequency of positive human interactions as herd size increases. For example, Waiblinger and Menke (1999) reported Austrian farmers with larger herds were less likely to brush their cows.
4. Management. Variability in animal welfare indicators across similar production systems suggests management has a strong effect on animal welfare and worker satisfaction have been identified as a key factor in ensuring animal welfare (Rushen and de Pasille, 2010). Worker satisfaction is associated with compensation and farm worker compensation is positively correlated with farm size – even after controlling for differences in worker skill level (Hurley *et al.*, 1999).
5. Multi-dimensional measures. One study, evaluating animal welfare on 194 Irish beef herds using 33 environmentally and resource-based indicators found a negative relationship between level of welfare and farm size (Mazurek *et al.*, 2010), but another study of 20 commercial sheep farms in Great Britain found no relationship between flock size and overall animal welfare scores (Stott *et al.*, 2012). Studies in Denmark failed to find any relationship between farm size and statutory animal welfare violations (Andrade and Anneberg, 2014; Czekaj, 2013). A recent study in Sweden found smaller, more diverse swine farms were more likely to have multiple statutory animal welfare violations (Hess *et al.*, 2014).

Preliminary conclusions

Contrary to popular opinion, these preliminary results do not support any simple relationship between farm size and animal welfare. The shift towards larger farms creates some welfare risks, but can also provide benefits for animal welfare. Most studies are limited by the inclusion of very few indicators of welfare and/or a small range of farm sizes. Determining the effects of farm size is complicated as the number of animals is frequently confounded with stocking density and other qualitative differences in production systems (intensive vs extensive). Statistical correlations between farm size and indicators of animal welfare – absent casual explanations – should be interpreted cautiously. Factors associated (and often confounded) with farm size (e.g. limited resources such as capital and time), professional management, worker training and specialization, long-term planning horizon, adoption of technology, awareness of – and sensitivity to – shifting consumer attitudes, etc.), are likely to be better predictors of farm animal welfare. Attempts to limit farm size to improve animal welfare are likely to fail or be ineffective.

Acknowledgments

The authors wish to thank the many farmers who have welcomed us onto their farms and discussed the ideas presented in this paper – as well as member of the UBC Animal Welfare Program members for their frequent discussions and feedback.

References

Alban, L. (1995). Lameness in Danish dairy cows: frequency and possible risk factors. Preventive Veterinary Medicine 22(3): 213-225.

Alvåsen, K., Jansson Mörk, M., Hallén Sandgren, C., Thomsen, P T. and Emanuelson, U. (2012). Herd-level risk factors associated with cow mortality in Swedish dairy herds. Journal of Dairy Science 95(8): 4352-4362.

Andrade, S.B. and Anneberg, I. (2014). Farmers Under Pressure. Analysis of the Social Conditions of Cases of Animal Neglect. Journal of Agricultural and Environmental Ethics 27(1): 103-126.

Bilčík, B., Keeling, L.J. and Newberry, R.C. (1998). Effect of group size on tonic immobility in laying hens. Behavioural Processes 43(1): 53-59.

Chapinal, N., Barrientos, A.K., von Keyserlingk, M.A.G., Galo, E. and Weary, D.M. (2013). Herd-level risk factors for lameness in freestall farms in the northeastern United States and California. Journal of Dairy Science 96(1): 318-328.

Chapinal, N., Liang, Y., Weary, D.M., Wang, Y. and von Keyserlingk, M.A.G. (2014). Risk factors for lameness and hock injuries in Holstein herds in China. Journal of Dairy Science 97(7): 4309-4316.

Christensen, G., Vraa-Andersen, L. and Mousing, J. (1995). Causes of mortality among sows in Danish pig herds. The Veterinary Record 137(16): 395-399.

Czekaj, T.G., Nielsen, A.S., Henningsen, A. and Forkman, B. (2013). The relationship between animal welfare and economic outcome at the farm level. IFRO Report 222. Department of Food and Resource Economics, University of Copenhagen, Denmark.

Hess, S., Bolos, L.A., Hoffmann, R., and Surry, Y. (2014). Is animal welfare better on small farms? Evidence from veterinary inspections on Swedish farms. In: Agri-food and rural innovations for healthier societies. 14th Congress of European Association of Agricultural Economists. Ljubbljana, Slovenia.

Hill, A.E., Green, A.L., Wagner, B.A. and Dargatz, D.A. (2009). Relationship between herd size and annual prevalence of and primary antimicrobial treatments for common diseases on dairy operations in the United States. Preventive veterinary medicine 88(4): 264-277.

Holzhauer, M., Hardenberg, C., Bartels, C.J.M., and Frankena, K. (2006). Herd-and cow-level prevalence of digital dermatitis in the Netherlands and associated risk factors. Journal of Dairy Science 89(2): 580-588.

Hurley, T.M., Kliebenstein, J. and Orazem, P.F. (1999). The structure of wages and benefits in the US pork industry. American Journal of Agricultural Economics 81(1): 144-163.

Lensink, J., Boissy, A. and Veissier, I., (2000). The relationship between farmers' attitude and behaviour towards calves, and productivity of veal units. Annales de. Zootechnie 49: 313-327.

Lusk, J.L., Norwood, F.B. and Pricket, R.W. (2007). Consumer Preferences for Farm Animal Welfare: Results of a Nationwide Telephone Survey. Working Paper, Department of Agricultural Economics, Oklahoma State University.

Mazurek, M., Prendiville, D.J., Crowe, M.A., Veissier, I., and Earley, B. (2010). An on-farm investigation of beef suckler herds using an animal welfare index (AWI). BMC veterinary research 6(1): 55.

National Animal Health Monitoring System (NAHMS) – Dairy (2002). Part I: Reference of Dairy Health and Management in the United States. USDA Animal and Plant Health Inspection, Fort Collins, CO.

Oleggini, G.H., Ely, L.O. and Smith, J.W. (2001). Effect of region and herd size on dairy herd performance parameters. Journal of Dairy Science 84: 1044-1050.

Rushen, J. and de Passille, A.M. (2010). The importance of good stockmanship and its benefits. In: Improving animal welfare: a practical approach. CABI, Wallingford, England, UK.

Sasaki, Y. and Koketsu, Y. (2008). Mortality, death interval, survivals, and herd factors for death in gilts and sows in commercial breeding herds. Journal of Animal Science 86: 3159-3165.

Stott, A.W., Vosough Ahmadi, B., Dwyer, C.M., Kupiec, B., Morgan-Davies, C., Milne, C.E., Ringrose, S., Goddard, P., Phillips, K. and Waterhouse, A. (2012). Interactions between profit and welfare on extensive sheep farms. Animal Welfare 21: 57-64.

Thomsen, P.T. and Sørensen, J.T. (2009). Factors affecting the risk of euthanasia for cows in Danish dairy herds. The Veterinary Record 165(2): 43-45.

Vanhonacker, F., Verbeke, W., Van Poucke, E., and Tuyttens, F.A. (2008). Do citizens and farmers interpret the concept of farm animal welfare differently? Livestock Science 116: 126-136.

Waiblinger, S., and Menke, C. (1999). Influence of herd size on human-cow relationships. Anthrozoos 12: 240-247.

Walker, W.L., Epperson, W.B., Wittum, T.E., Lord, L.K., Rajala-Schultz, P.J. and Lakritz, J. (2012). Characteristics of dairy calf ranches: morbidity, mortality, antibiotic use practices, and biosecurity and biocontainment practices. Journal of Dairy Science 95(4): 2204-2214.

Wolter, B.F., Ellis, M., Curtis, S.E., Augspurger, N.R., Hamilton, D.N., Parr, E.N. and Webel, D.M. (2001). Effect of group size on pig performance in a wean-to-finish production system. Journal of Animal Science 79: 1067-1073.

Zulkifli, I. (2013). Review of human-animal interactions and their impact on animal productivity and welfare. Journal of Animal Science and Biotechnology 4: 25.

Collective dimensions in animal ethics

B. Bovenkerk and M.F. Verweij*
Wageningen University, Hollandseweg 1, 6706 KN Wageningen, the Netherlands; bernice.bovenkerk@wur.nl

Abstract

Due to its emphasis on experiential interests, animal ethics tends to focus on individuals as the sole unit of moral concern. Many issues in animal ethics can be fruitfully analysed in terms of obligations towards individual animals, but some problems require reflection about collective dimensions of animal life in ways that individualist approaches can't offer. Criticism of the individualist focus in animal ethics is not new; it has been put forward in particular by environmental ethics approaches. However, the latter tend to be so far removed from the concerns of animal ethicists that both groups talk at cross purposes. We think the gap between environmental and animal ethics could be bridged by focusing more on the collective dimensions of our concerns with animals. In our paper we explore various (novel) ways of conceptualising the moral relevance of collectiveness in animal life. We draw on insights from public health ethics, as this field of inquiry has also developed – at least partly – in response to individualist approaches in human bioethics, creating more room for recognizing the value of population health, interpersonal relations, solidarity, and ways in which a collective is constituted.

Keywords: animal liberation, ecocentrism, groups, public health ethics

Introduction

Most animal ethicists take individual animals as the sole locus of moral concern. While they have forcefully argued that we need to take the interests of sentient animals into account in our moral decisions, some such decisions have an impact beyond the interests of individual animals. Such situations seem to call for reflection on collective dimensions regarding our dealings with animals. Environmental ethicists argue that collectives such as populations or ecosystems should be the locus of moral concern. However, with their primary focus on collectives, environmental ethicists seem to have a blind spot for the interests of individual animals. Animal and environmental ethicists appear to talk at cross purposes. We will argue that the gap between the two approaches may be bridged by focusing more on the collective dimensions in our concerns with animals. As we will explain, even when they hold that only individual animals could reasonably be attributed moral status, animal ethicists should take into account collective dimensions. It is fruitful to seek analogies to the various ways in which human bioethics – notably ethics of public health – is increasingly acknowledging the collective dimensions of ethical problems in relation to human health and welfare. We present several ways to conceptualise the moral relevance of collectivities in public health ethics and seek possible analogies in animal ethics. As we will show, some ways in which collectives could be conceptualised correspond better with the individualistic perspective than others.

Individualism in animal ethics

In animal ethics we encounter a strong focus on the question of what animals possess moral status. This is often framed as a question about what entities possess those capacities necessary to have interests. Interests, in turn, tend to be understood as experiential interests. For example, Peter Singer (1999) takes as a basic premise that equals should be treated equally and unequals unequally. Equality is here to be understood in terms of interests. If a specific action will cause an equal amount of suffering to two different animals (or to an animal and a human being), the animals have an equal interest in avoiding that action and therefore we should treat these animals equally. An entity that is not conscious cannot experience interests and in Singer's view therefore does not have interests. Tom Regan (2004) argues that all beings who can subjectively experience their lives (in other words, all subjects-of-a-life) have inherent value and should be treated with equal respect. Clare Palmer (2010) argues that our specific relationship with an animal influences the question of how we should treat this animal. For example, when we have domesticated or captured an animal we have specific duties of care towards this animal that we do not have towards wild animals. Yet, when it comes to the question of what class of animals we have duties towards in the first place, she takes the capacity to have experiential welfare as the determining criterion. In other words, according to all these authors, we only need to think about the question of how to treat an animal if that animal can subjectively experience its own interests. Due to this emphasis on experiential interests, animal ethics tends to focus on individuals as the sole unit of moral concern (Palmer, 2010). After all, groups or collectives do not have conscious experiences.

Many issues in animal ethics can be fruitfully analysed in terms of obligations towards individual animals, but some problems require reflection about collective dimensions of animal life in ways that individualist approaches can't offer. For example, the question of whether it is morally permissible to genetically modify or breed animals for specific characteristics addresses the level of animal species or populations, rather than individuals. When we selectively breed animals to exhibit particular characteristics we often unintentionally breed along unwanted characteristics, which may lead to health problems. Think of English Bulldogs that due to our preference for big heads and flat faces have breathing difficulties and cannot give birth naturally. As Palmer (2011) has pointed out, the intervention in the genome of these animals is carried out before the animals are born and therefore (as long as the animals still have a life worth living) we cannot say we have harmed these particular animals. After all, had we not bred them in this way they would not even exist. If we want to argue that we have harmed these animals, for example because we have violated their integrity (Bovenkerk *et al.*, 2002), we need to make a claim about the species or population that we have changed. Similarly, the concern that many have with conserving animal biodiversity is not directed at the preservation of the lives of individual animals, but of their species. Think of the initiative to establish gene banks – biodepositories of genetic material of animals and plants that face extinction; the idea behind this seems to be that even if all individual specimens of a particular species go extinct, the species itself can still survive by bringing it back into existence. Disregarding for the moment the question of whether this is a desirable way to go about biodiversity conservation, the underlying assumption seems to be that we have reasons to preserve a species that surpass the reasons that we have to protect all the currently living individuals of that species. However, only these individuals have experiential welfare, whereas the species that they belong to do not.

Animal versus environmental ethics

Criticism of the individualist focus in animal ethics is not new; it has been put forward in particular by environmental ethics approaches. However, the latter tend to be so far removed from the concerns of animal ethicists that both groups talk at cross purposes. Environmental ethics, in particular ecocentrists, aim to formulate duties that we may have towards collectives such as populations, species, and ecosystems. They also attribute interests to natural entities, but these are interests of a non-experiential kind, such as an interest in living or flourishing. They argue that collectives can have interests that surpass, or even run counter to, the composite interests of their individual members. Johnson (1992) gives the example of a population of prey animals. While it is in the interest of the population that old or weak animals are killed by predators, it is obviously not in the interest of the individual victims in question. However, this example does not prove that populations are morally relevant, simply that a group can have independent interests; an extra argument is needed to make the step from simply interests to morally relevant interests.

When we think of specific clashes between animal liberationists and ecocentrists, we cannot help but think that both have a valid point of view, but both emphasize different aspects of the same problem.

For example, in the Netherlands we have witnessed a heated debate about the introduction of large herbivores into a newly established nature area, the Oostvaardersplassen (cf. Klaver *et al.*, 2002). The horses and cattle in question are part of a program of dedomestication. Because the aim is to create a wild ecosystem and the animals are perceived to be part of such an ecosystem, nature managers have decided that the animals should become wild by leaving them alone as much as possible. This means also letting them starve in winter when insufficient food is available. The value of wildness that ecocentrists champion in this case clashes with the value of animal welfare that animal liberationists want to protect. How could we reconcile these different viewpoints?

Ways to conceptualise collective dimensions

We think the gap between environmental and animal ethics could be bridged by focusing more on the collective dimensions in our concerns with animals. A variety of considerations exists between the idea of animal ethics that every concern should be reducible to respect and care for individual animals on the one hand, and the ecocentric focus on care for only collectives such as ecosystems on the other hand. In this paper, we do not address the question of what should be the basis of moral status, but rather we ask what types of moral concerns may be linked to the collective dimensions of moral problems. We think that there are different ways to conceptualise collectivity that could possibly be morally relevant. Drawing on insights from public health ethics, we explore various (novel) ways of conceptualising the moral relevance of collectiveness in animal life. We do this because public health ethics has also developed partly in response to mostly individualist approaches in human bioethics, creating more room for recognizing the value of population health, interpersonal relations, solidarity, and ways in which a collective is constituted.

What do we mean when we are speaking of collective dimensions in our concerns with animals? In public health ethics, collective dimensions have been conceptualized in at least four ways, and most of these may have relevant analogies in animal ethics. Firstly, in public health ethics it is natural to be concerned with the health and welfare of the public, understood as just the aggregation of the health of all individuals in the group (Verweij and Dawson, 1997). In ethical concerns about human health, numbers make a difference. If we need to choose between an intervention that would make either a large number of people better off, or just a smaller number of people, then, other things being equal, we should choose the former. This will be obvious for utilitarians, but numbers are also relevant in many non-consequentialist approaches, such as contractualism (Hirose, 2001). A similar line of reasoning holds for our treatment of animals; an animal experiment, for example, is regarded as more problematic the more animals are used. So one way to be morally concerned with collectives is through the aggregation of welfare or claims of individuals.

Secondly, we can think of group welfare as something that goes beyond the mere aggregate of individual members' welfare. In public health as well as for animal health we can see the value of realizing and maintaining herd immunity (Fine, 1993). Collectives with herd immunity are, in some senses, better off, than groups without herd immunity. Each individual benefits from being protected within the herd, even those who are not immune, or those who are ill and weak, and therefore more vulnerable to certain infections. The value of herd immunity is not necessarily reducible to aggregate welfare, because, at least in theory, the aggregate health of the group might be better if the weaker individuals were not protected and only the healthier part of the population would survive. This is most clear if aggregate health is understood as average health. Precisely because we are not only concerned with numbers or averages, but also with how 'strong' the group is as a group this concern cannot be merely reduced to concern for the group as consisting of the aggregate of many individuals. In the animal domain, the idea of herd immunity is equally applicable. But there are more reasons for being concerned with group health in distinction to the (aggregation of) health of individuals. For example, it is possible that certain risk factors only exist because of the genetic composition of the group. A population of animals can be healthy on average, but because its gene pool is not very diverse, the group as a whole could be considered unhealthy, or at least vulnerable.

Thirdly, in the realm of public health it is often considered ethically problematic if a population shows structural inequalities in health, notably where such inequalities are linked to societal and economic factors (Marmott, 2012). The question which inequalities are indeed to be considered as inequities is a complex matter, but the focus on distribution of health and quality of life within a population is clearly another ethically relevant concern about that population – at least so in relation to human health and quality of life. Could a parallel be drawn with animal ethics? In other words, does it make sense to see unequal distributions of health, welfare or freedom within a group of animals as unjust? Arguably, if we have an obligation to care for the wellbeing of those animals, we have good reason to give priority to the condition of animals that are worst off: those that are ill or in pain, that are suffering, or otherwise lack in quality of life. Whether this also means that we should take away inequalities within a group of animals is a matter of debate. Are such inequalities within the group to be considered morally problematic in relation to the group as such, or, in relation to the question of how the group (e.g. a group of cattle) is treated by relevant care-givers (e.g. farmers)? In other words, is distributive justice among animals a meaningful concept?

Fourthly, a number of authors in bioethics reject a mere individualistic approach to public health and emphasise more collectivist or communitarian values, such as solidarity. Solidarity can be considered in various ways, but in its strongest form it refers to a value that is intrinsically linked to the identity of the group as a whole, where individuals within the group also, in an important sense, see themselves as part of a group rather than as solitary individuals (Dawson and Verweij, 2012; Prainsack and Buyx, 2011; Verweij, 2015). Personal identity is then at least partly experienced as being constituted by one's role in, or membership of, the group. In public health and many other areas of societal life we have good moral reasons to strengthen and honour such social bonds, as they are a necessary basis for flourishing of the group as well as individuals within the group. Similar forms of community can be found in the animal kingdom. For example, there are many accounts of groups of animals that protect their weaker individuals. A threat to one individual can evoke a counter-attack of the entire group. In some cases, individuals can even be identified by their role in the group in a much more extreme sense. In a beehive, individual worker bees sacrifice themselves for the other bees, in particular the queen bee, in order to safeguard the continued existence of the whole group. A second way in which we can speak about group identity is when we are dealing with capacities that can only be understood in collective terms. For example, ants or bees appear not to possess a high degree of intelligence on an individual level, but collectively they have a type of intelligence with which they can pursue highly advanced goals. The collective could in this case be regarded as an organism in its own right.

Wildness and integrity

We have explored possible analogies between moral concerns in public health- and animal ethics, seeking various ways of how collectives or dimensions can be morally relevant. Of course, in environmental ethics it is not uncommon to look at collective dimensions, but in this exploration we are particularly interested in approaches that can be related to the moral status of individual animals. This applies to the first three, and somewhat 'more or less' to the fourth. In the realm of animal ethics we can also find other considerations that we can understand as collective dimensions, namely those that have to do with the species concept. For example the 'integrity argument' which states that interfering with an animal's 'essence' violates that animal's integrity (Bovenkerk *et al.*, 2002). It is not obvious that violating animal integrity is violating that particular individual's integrity. Integrity appears to be a notion that refers to 'how we think a type of animal should be'; it refers to the 'cowness of a cow'. This is a view about what animals are like as a group. By referring to some kind of shared essence of a species or population or breed, we are making claims on a collective level. Nevertheless, seeing that argument as fully 'collective' or essentialist and ignoring the link with particular individuals could be an oversimplification. Similarly, the value of wildness pertains to a group of animals rather than to individuals. While arguments based on wildness or integrity seem to be difficult to sustain within an individualist-focused animal ethics, analysing them in terms of collective values might offer possibilities to reinterpret them and give them a stronger basis.

Conclusion

We have presented a diversity of ways of thinking about collectives. Between the two extremes of on the one hand a sole focus on individuals and on the other hand a sole focus on collectives a variety of positions exist. We hope to have shown that thinking about collectives can co-exist with an individual perspectives on ethics. The individual and collective dimensions are not necessarily mutually exclusive. The next step will be to investigate the ways in which an exploration such as this will help to clarify practical dilemma's where animal and environmental ethics clash.

References

Bovenkerk, B., Brom, F.W.A. and Van den Bergh, B.J. (2002). Brave new birds: the use of 'animal integrity' in animal ethics. Hastings Center Report 32(1): 16-22.

Dawson A.J. and Verweij M.F. (2012). Solidarity: a moral concept in need of clarification. Public Health Ethics 5(1): 1-5.

Fine, P.E.M. (1993). Herd immunity: history, theory, practice. Epidemiologic reviews 15(2): 265-298.

Hirose, I. (2001). Saving the greater number without combining claims. Analysis 61(4): 341-342.

Johnson, L.E. (1992). Toward the moral considerability of species and ecosystems. Environmental Ethics 14 (2): 145-57.

Klaver, I., Keulartz, J. and Van den Belt, H. (2002). Born to be wild. Environmental Ethics 24(1): 3-21.

Marmott, M. and Bell, R. (2012). Fair society, healthy lives. Public Health 126(1): S4-S10.

Palmer, C. (2010). Animal ethics in context. Columbia University Press, New York, USA, 215 pp.

Palmer, C. (2011). Is something wrong with my bulldog? Breeding, ethics and harm to animals. In: Drenthen, M., Keulartz, J. (eds.) Old world and new world perspectives on environmental philosophy. Springer International Publishing, Cham, Switzerland.

Prainsack, B and Buyx, A. (2011). Solidarity: reflections on an emerging concept in bioethics. Nuffield Council on Bioethics, London, UK.

Regan, T. (2004). The case for animal rights. University of California Press, Berkeley, USA, 425 pp.

Singer, P. (1999). Practical ethics. Second edition, reprinted. Cambridge University Press, Cambridge, USA, 352 pp.

Verweij, M.F. (2015) How (not) to argue for the rule of rescue. Claims of individuals vs group solidarity. In: Cohen, I.G., Daniels, N. and Eyal, N. (eds.) Identified versus statistical victims. an interdisciplinary perspective. Oxford University Press, New York, USA, pp. 137-149.

Verweij M.F. and Dawson, A.J. (2007). The meaning of 'public' in 'public health'. In: Dawson A.J. and M.F. Verweij (eds.) Ethics, prevention, and public health. Oxford University Press, Oxford, UK, pp. 13-29.

One health as a normative concept: implications for food safety at the wildlife interface

J. Nieuwland[1] and F.L.B. Meijboom[2]*
[1]Institute for philosophy, Leiden University, Reuvensplaats 3-4, 2311 BV, Leiden, the Netherlands; [2]Ethics Institute, Utrecht University, Janskerkhof 13a, 3512 BL Utrecht, the Netherlands; j.nieuwland@hum.leidenuniv.nl

Abstract

Starting from the recognition of the interdependence of human, animal, and ecosystem health, the concept of One Health promotes multispecies and integrative approaches to health. While zoonoses (diseases that can be transmitted from non-human animals to human beings) have been a major impetus for the development of One Health, it is the human-animal-ecosystem interface in general that is emphasised. One Health has received little attention from (applied) ethics, despite its clear normative dimension. The aims of the paper are twofold, by showing: (1) the importance of explicating the normative dimension of One Health, and following from this; (2) the need for integrated ethics at the human-animal-ecosystem interface. A holistic understanding of health is presented, together with the acknowledgment of the moral status of animals, to discuss the normative implications of One Health. The discussion will address, in particular, the implications for food safety with respect to wildlife health.

Keywords: multispecies health, badgers, bovine tuberculosis, animal ethics

Health at the wildlife interface

The One Health concept emphasizes the interdependence of human, animal, and ecosystem health (e.g. Cook *et al.*, 2004; Zinsstag and Schelling, 2011). Based on this recognition, health is promoted within all these three domains. This has brought attention to the wildlife interface of human and animal health. The issue of bush meat consumption in Africa, and it's devastating consequences for both public health and wildlife, have illustrated the significance of wildlife interactions understood in terms of food safety.

In the UK, European badgers (*Meles meles)* are believed to contribute to the increase of bovine tuberculosis (bTB), which is caused by *Mycobacterium bovis*. There is however some uncertainty about the extent that badgers can be blamed for this (Hancox, 2002). Due to its zoonotic nature, bTB has been a target for public health interventions such as pasteurization of milk and control programs, which have been successful in minimizing the threat to public health (Pfeiffer, 2013). Despite prevalent disagreement about both its efficacy and justification, culling of badgers has been implemented with the aim to reduce bovine tuberculosis infections (Mathews, 2009). This case about food safety provides an opportunity to investigate the normative implications of a One Health perspective, especially because it explicitly addresses the health of human beings, domesticated animals, and wild animals in the context of a shared natural environment.

One Health as a normative concept

Multispecies and integrative approaches to health go beyond the domain of public health. The latter is primarily concerned with human health, and only indirectly interested in animal health. Both in principle and in effect, however, public health cannot be isolated from other disciplines such as veterinary medicine. This notion has been taken up by 'one medicine', an integration of human and veterinary medicine based on the premise that 'both sciences share, as a general medicine, a common body of knowledge in anatomy, physiology, pathology, and the origin of disease in all species' (Zinsstag and Schelling, 2011: 149). Zoonotic diseases such as bTB are paradigm examples for this line of thinking. But still, this can leave unaddressed the environmental factors supporting health and underlying disease emergence. In addition to being interdependent, the health of humans and animals is inextricably connected to ecosystems. This recognition lies behind the formulation of the Manhattan Principles (Cook *et al.*, 2004), which have been key reference points for One Health ever since.

In its aim for a integrative and multispecies approach to health, One Health lacks a systematic evaluation of its normative dimension (cf. Capps and Lederman, 2014; Rock and Degeling, 2014); i.e. in what way are values and ethical considerations involved? There are, in particular, three reasons in support of bringing out the normative dimension. First, health tells us how organisms should function (e.g. Nordenfelt, 2006). There is, however, no agreement on what this exactly entails for humans, animals, or ecosystems. If the concept of health is to guide policy in a multispecies context, its normative implications should be made explicit. The inclusion of animals is a second reason for philosophical scrutiny. If animals have moral status, this involves specific normative implications for the understanding of One Health. The third reason pertains to the integrative nature of One Health. There are multiple disciplines involved with their own area of interest and methodology. Furthermore, as One Health is adopted at an international level, it will be rooted in different worldviews. This will affect the two reasons just discussed, but also include other aspects such as environmental values and framing assumptions. Leach and Scoones (2013: 16) recognize the normative import when they state that 'the challenge is therefore not just combining disciplines, data and models, but more importantly, deliberation around framing assumptions, cultural understandings, policy narratives, politics and values'. This underlines the importance of reflecting on such aspects and how these influence the design and outcome of One Health approaches.

The conceptualisation of health in a multispecies approach

An overarching concept of multispecies health invites one to consider the meaning of health (cf. Lerner and Berzell, 2014). Arguments have been put forward to understand human health in bio-statistical terms as the absence of disease (Boorse, 1977). This is opposed to the holistic 1946 WHO definition of health as 'a state of complete physical, mental and social well-being and not merely the absence of disease or infirmity'. The utopian aim at completeness and its static understanding of health, however, have been points of major criticism with regard to this definition (e.g. Forrest, 2014). Recent contributions to the debate (Nordenfelt, 2006) are holistic in the sense of understanding health in relation to subjective experience, as instrumental to attaining vital goals.

Animal health is discussed far less than its human counterpart, although there is a large body of literature on animal welfare (Nordenfelt, 2006). How health of animals is conceptualised, can depend on discipline, context, and species (Lerner and Hofmann, 2011). Within veterinary medicine it is often understood as normality or the absence of disease, or associated with context-specific factors, such as productivity and reproduction (Gunnarsson, 2006). With regard to the badger case, it has been suggested that high productivity of cattle could have negatively impacted their immune status, making them more susceptible to bTB infection (Mathews, 2009). This could raise doubt regarding the adequacy of productivity as a criterion for health.

Comparing health concepts across different domains can help to reflect on the assumptions involved in context and species-specific understandings of health (Vieweger and Döring, 2014). If animal health is different from human health, what are the arguments supporting such discrepancy? Can the debate on human health be extrapolated to animal health? Nordenfelt (2006) has argued that to a large extent, there is nothing in the way of doing exactly that. According to him, health also enables animals to attain vital goals, albeit understood in terms of animal welfare. This is different from health as absence of disease, which does not take animal welfare as a goal. Following from this, the question arises whether health promotion is important in itself, or because it serves other goals. Furthermore, who are the beneficiaries of these goals? It is due time to look at the value of health in a multispecies approach.

Cross-species comparison of health

The health of animals can hold instrumental value with respect to human interests. From an epidemiological perspective, wild animals in particular are considered important. Animals or animal populations can function as sentinels for early detection and early warning of zoonotic disease outbreaks, as animal models for human disease, or be reservoir hosts for zoonotic pathogens (Rabinowitz *et al.*, 2008). Badgers are identified as maintenance hosts for *M. bovis*, which means that they constitute an independent source for tuberculosis infection in cattle. Interestingly, this is believed to be the result of spillover from cattle herds a little more than 100 years ago (Palmer and Thacker, 2012). Anyhow, there is more to badger health rather than its epidemiological value. It is – especially in a holistic understanding – valuable for badgers themselves. These two ways of valuing animal health can overlap and/or conflict.

A multispecies approach to health involves both questions about health as an interspecific concept and questions about distribution. Any One Health approach has to grapple with distributing the costs and benefits across the entities involved, perhaps without ever resolving conflict of interest (Rock and Degeling, 2014). Then, to what extent can agricultural practices be justified when the spatial configuration of landscapes is associated with risks to wildlife health or the use of culling strategies? In addition to the need for humans to justify their treatment of domesticated animals, these human-animal interactions (e.g. agriculture) should also be put into the perspective of wild animals, in this case, badgers.

Assumptions regarding the moral status of animals assign the boundaries for such comparison across species. A purely anthropocentric account (attributing moral status to humans exclusively) will be interested in bTB and the risk imposed to public health. From a narrow public health perspective, what counts is reducing or preferably eliminating the risk to human health. One could argue that this largely applies to the current situation in the UK (Torgerson and Torgerson, 2010). For 'one medicine', even when public health objectives are met, this is unsatisfactory, as the underlying issue remain unsolved. As already discussed, a One Health approach would also look at the environmental factors involved.

However, this broadening of scope can also have other normative implications. By going beyond mere public health, animals have become part of health policy. This brings along new questions about their position from a normative point of view. Despite the holistic aims of One Health at improving human and non-human health, the interests of the former appear to remain at the forefront (Hanrahan, 2014). A zoocentric account (attributing moral status to sentient animals) would argue that public health is too narrow, especially if it generally discounts animal health and welfare for the benefit of human health (cf. Akhtar, 2013). Animal health should be promoted, also irrespective of human interests. This is also supported by Nordenfelt's theory of animal health. Framed within a 'one medicine' perspective, this could entail sharing risks between sentient beings rather than fully eliminating risk for one species only (cf. Rabinowitz *et al.*, 2008). If one takes seriously the environmental factors that support human and animals health, as acknowledged by One Health, this will affect how ecosystems are managed and protected. This inevitably puts health of both humans and animals into its shared environmental context.

Health in context

For wildlife, health is evidently not a mere individual affair. First of all, animal health is embedded in and dependent on environments. Considerations that go 'beyond the skin' of the individual contribute to health in significant ways (Forest, 2014). This is not limited to physical environments, but includes the broader spectrum of socioecological interactions as the background for health (Hanisch *et al.*, 2012; Stephen, 2014; Zinsstag and Schelling, 2011). Second, wildlife health is generally assessed at the level of the population, rather than at the individual level. Going beyond absence of disease, a holistic understanding of wildlife population health can be formulated in terms of sustainability and resilience (Hanisch *et al.*, 2012). This appears to outweigh considerations at the individual level, but one could also argue that individual animals living in the wild are dependent on their populations. The vital goals of wild animals are then considered inseparable from considerations regarding the health of the population. How does this correspond with duties towards wild animals?

A principle of avoiding unnecessary harm to individual animals in the wild swiftly expands the normative scope of One Health, especially if one considers the influence that human beings exercise on nature. Anthropogenic activity such as habitat encroachment and agriculture can be important drivers for the emergence of tuberculosis infection in wild animals (Palmer and Thacker, 2012). However, human responsibility does not have to be limited to negative duties only (Swart, 2005). Swart argues for a principle of 'non-specific care', which entails taking care of the habitat of wild animals. The relation between animals and the environment is mediated by both dependency and adaptation. For Swart, this relation elicits non-specific care, which by definition is directed at the population rather than the individual level. Wild animals such as badgers are part of social groups and embedded in a particular environment, both of which are important to protect according to Swart. This contextual approach thereby matches the holistic articulation of wildlife health, both regarding the relevancy of socio-ecological factors, and the level of health assessment.

If one applies the principle of non-specific care, there is a need to integrate the health of animals in the way ecosystems are managed and protected. Furthermore, this involves considerations regarding future animal populations. Protecting habitats is in the interest of animals currently living in the wild, as it is for future animal populations. Vice versa, considerations about the sustainability and resilience of future animal populations affect how current populations are managed. From a One Health perspective, such considerations are in significant ways connected to public health.

Concluding remarks

The concept of One Health provides an integrative perspective on health that involves new normative questions. Different understanding of health and assumptions regarding moral status of animals influence One Health approaches in important ways.

Regarding the fate of the badgers, the debate appears primarily centred on the implementation of culling strategies and considerations regarding the effectiveness of such a measure. This may obscure the fact that even if culling strategies are effective, which is disputed, these measures still have to be justified from an ethical point of view. The aforementioned Manhattan Principles (Cook *et al.*, 2004) recommend to 'limit the mass cull of free-ranging wildlife species for disease control to situations where there is scientific evidence that a wildlife population poses an urgent, significant threat to public health and safety'. The culling of badgers does not appear to be compatible with this One Health principle. At the very least, it is an invitation to further examine the meaning of urgent and significant threats.

Stephen (2014) has expressed his concern about One Health reinforcing a disease-focused outlook at the expense of wild animals. Although this concern might be appropriate, it need not follow from One Health if health is conceptualised in a more holistic way. Throughout the badger case, similar to discussions on food-borne zoonotic diseases (cf. Romero Barrios *et al.*, 2013), it appears that the predominant understanding of health amounts to absence of disease. There is, on the one hand, a mainly anthropocentric and, on the other, a more zoocentric perspective on moral status. Both sides largely appear to share a similar understanding of health. This discussion entails the risk of reducing badgers to their role as maintenance hosts in the disease ecology of bTB. A holistic understanding of health would take into consideration the socio-ecological dynamics of these badger populations in relation to agro-environmental factors. Mathews *et al.* (2006) have illustrated this by relating the presence of ungrazed grassland and plentiful hedgerows to decreased interaction at the wildlife/livestock interface. Taking into account these environmental needs of badgers resulted in a lower incidence of bTB. Furthermore, a holistic understanding puts health into the perspective of animal welfare. Especially if one takes a zoocentric perspective, questions arise about the ways badgers are affected by bTB and whether there is a duty to reduce bTB infections in these populations, for example, by vaccination strategies.

Diseases that jump species, such as bTB, connect human and animal health in important ways. Although targeting diseases is of vital importance, more terrain has to be covered to adequately promote multispecies health. Holistic understandings, as discussed, are especially insightful in providing additional coverage. Such a perspective would recognize environmental drivers of infectious diseases, but furthermore be interested in determinants of health understood in terms of sustainability and resilience (Hanisch *et al.*, 2012; Stephen, 2014). In line with Nordenfelt's holistic approach, this entails that 'healthy individuals or populations need to have a minimal set of resources, functions, and capabilities that operate within an environment that enables them to cope with changes and challenges to meet expected end points' (Stephen, 2014: 429). The promotion of multispecies health, as envisioned by One Health, will benefit from integrating health objectives of wildlife populations understood in these terms.

Acknowledgments

We would like to thank Bruno Verbeek and Raymond Corbey for their valuable comments. Arcus Foundation is gratefully acknowledged for funding.

References

Akhtar, A. (2013). The need to include animal protection in public health policies. Journal of Public Health Policy 34:549-559.

Boorse, C. (1977). Health as a theoretical concept. Philosophy of Science 44:542-573.

Capps, B. and Lederman, Z. (2014). One health and paradigms of public biobanking. Journal of Medical Ethics, 41(3): 258-262. DOI: http://dx.doi.org/10.1136/medethics-2013-101828

Cook, R.A., Karesh, W.B. and Osofsky, S.A. (2004). The Manhattan principles on 'one world one health', Wildlife Conservation Society, New York, USA.

Forrest, C. B. (2014). A living systems perspective on health. Medical Hypotheses, 82(2): 209-214. DOI: http://dx.doi.org/10.1016/j.mehy.2013.11.040.

Gunnarsson, S. (2006). The conceptualisation of health and disease in veterinary medicine. Acta Veterinaria Scandinavica, 48(20). DOI: http://dx.doi.org/10.1186/1751-0147-48-20.

Hanisch, S., Riley, S. and Nelson, M. (2012). Promoting wildlife health or fighting wildlife disease: insights from history, philosophy, and science. Wildlife Society Bulletin 36(3): 477-482.

Hancox, M. (2002). The great badgers and bovine TB debate. Journal of Agricultural Science 139: 223-226.

Hanrahan, C. (2014). Integrative health thinking and the one health concept: is social work all for 'one' or 'one' for all? In: Ryan T. (ed.) Animals in social work: why and how they matter. Palgrave Macmillan, Basingstoke, UK, pp. 32-47.

Leach, M. and Scoones, I. (2013). The social and political lives of zoonotic disease models: narratives, science and policy. Social Science & Medicine 88: 10-17.

Lerner, H. and Berzell, M. (2014). Reference values and the problem of health as normality: a veterinary attempt in the light of a one health approach. Infection Ecology and Epidemiology 10(4). DOI: http://dx.doi.org/10.3402/iee.v4.24270.

Lerner, H. and Hofmann, B. (2011). Normality and naturalness: a comparison of the meanings of concepts used within veterinary medicine and human medicine. Theoretical Medicine and Bioethics 32: 403-412.

Mathews, F. (2009). Zoonoses in wildlife: integrating ecology into management. Advances in Parasitology 68: 185-209.

Mathews, F., Lovett, L., Rushton, S. and Macdonald, D. (2006). Bovine tuberculosis in cattle: reduced risk on wildlife-friendly farms. Biology Letters 2: 271-274.

Nordenfelt, L. (2006). Animal and human health and welfare: a comparative philosophical analysis. CAB International, Wallingford, UK.

Palmer, M. and Thacker, T. (2012). *Mycobacterium bovis*: a model pathogen at the interface of livestock, wildlife, and humans. Veterinary Medicine International. DOI: http://dx.doi.org/10.1155/2012/236205.

Pfeiffer, D. (2013). Epidemiology caught in the causal web of bovine tuberculosis. Transboundary and Emerging Diseases, 60(S1): 104-110.

Rabinowitz, P., Odofin, L. and Dein, F. (2008). From 'us vs them' to 'shared risk': can animals help link environmental factors to human health? EcoHealth 5: 224-229.

Rock, M. and Degeling, C. (2014). Public health ethics and more-than-human solidarity. Social Science & Medicine. DOI: http://dx.doi.org/10.1016/j.socscimed.2014.05.050.

Romero Barrios, P., Hempen, M. and Messens, W. (2013). Quantitative microbiological risk assessment (QMRA) of food-borne zoonoses at the European level. Food Control 29(2): 343-349.

Stephen, C. (2014). Towards a modernized definition of wildlife health. Journal of Wildlife Diseases 50(3): 427-430.

Swart, J. (2005). Care for the wild: an integrative view on wild and domesticated animals. Environmental Values 14(2): 251-263.

Torgerson, P. R. and Torgerson, D. J. (2010). Public health and bovine tuberculosis: what's all the fuss about? Trends in Microbiology 18(2): 67-72.

Vieweger, A. and Döring, T. (2014). Assessing health in agriculture – towards a common research framework for soils, plants, animals, humans and ecosystems. Journal of the Science of Food and Agriculture. DOI: http://dx.doi.org/10.1002/jsfa.6708.

Zinsstag, J. and Schelling, E. (2011). From 'one medicine' to 'one health' and systemic approaches to health and well-being. Preventive Veterinary Medicine 101: 148-156.

Best broiler husbandry system and perceived importance of production aspects by Dutch citizens, poultry farmers and veterinarians

M. van Asselt[1,2*], *E.D. Ekkel*[1], *B. Kemp*[2] *and E.N. Stassen*[2]
[1]*CAH Vilentum, University of Applied Sciences, Applied Research, Drieslag 4, 8251 JZ Dronten, the Netherlands;* [2]*Wageningen University, Animal Sciences, P.O. Box 338, 6700 AH Wageningen, the Netherlands; m.van.asselt@cahvilentum.nl*

Abstract

The objectives of this study were to explore the views of Dutch citizens, poultry farmers and poultry veterinarians regarding: (1) best broiler husbandry systems; (2) the importance of production aspects of broiler production and (3) the relation between best husbandry system and the importance of production aspects. Therefore, we surveyed by an online questionnaire the perceptions of Dutch citizens, poultry farmers and poultry veterinarians. According to our findings citizens perceived the free-range system with outdoor access as the best husbandry system and farmers and veterinarians perceived the conventional system as the best system. Citizens, farmers and veterinarians scored broiler health and food safety as the two most important production aspects and they scored animal welfare as equally important. Higher perceived importance of (indicators of) animal welfare and environment, together with a lower perceived importance of economic aspects, such as cheap broiler meat, broiler farmer income and broiler growth rate, was related to the choice for a more extensive system. The aspects food safety, risk broiler farm to human health and taste of broiler meat were not related to the choice for the best system. This study is a first step to analyse based on which moral values people weigh dilemmas such as animal welfare versus risk broiler farm to public health. These insights can contribute to the development of poultry production systems that are more socially accepted.

Keywords: perception, stakeholders, animal welfare, food safety, survey

Introduction

During last 60 years several broiler husbandry systems have been developed ranging from intensive to more extensive systems. Recent years public concerns regarding production aspects like animal welfare, food safety, public health and environmental impact have been increased (Harper and Henson, 2001). The intensive (indoor) systems were developed to guarantee food security and are characterized by efficiency and little space per animal, but animal welfare is criticized. More extensive (outdoor) systems are characterized by more space per animal, slow growing breeds, and outdoor access, and can contribute to better animal welfare (Tuyttens *et al.*, 2008). The extensive systems seem therefore to comply better with public concerns about animal welfare.

In systems with outdoor access broilers are more in contact with the environment which may result in increased food safety and public health risks as for instance *Campylobacter* and Avian influenza (Kijlstra *et al.*, 2009). Moreover, the environmental impact of extensive systems is debated to be higher than the impact of intensive systems (Bokkers and De Boer, 2009). So, each system has strengths and weaknesses regarding various aspects like animal welfare, food safety, and environment causing dilemma's among and within the systems.

To accomplish public support for the broiler production sector, the public opinion should be taken into consideration when adapting husbandry systems (Swanson *et al.*, 2011). However, previous research has mainly focused on the quantification of economic, ecological and animal welfare aspects (Bokkers and De Boer, 2009). The public opinion has mainly been studied for single broiler production aspects, like broiler welfare (Tuyttens *et al.*, 2014). Moreover, it has been shown that animal welfare (Fraser *et al.*, 1997) and public health risks are in general perceived differently by citizens and experts (Jensen *et al.*, 2005). However, the perception of multiple broiler production aspects and the relation with the perceived best broiler husbandry system in different stakeholder groups has not been studied. Insight in these perceptions may help to set priorities in broiler production to accomplish a more socially supported production sector. In general perceptions are influenced by interests, knowledge and moral values (Te Velde *et al.*, 2002) and can be explained by socio-demographic characteristics such as gender, age, educational level and animal related experiences (Kendall *et al.*, 2006). Consequently, citizens, poultry famers and poultry veterinarians may have different views and these stakeholder groups are therefore included in this research.

The objectives of this study are: (1) to explore the views of citizens, farmers and veterinarians about best broiler husbandry systems; (2) to investigate the perceived importance of production aspects of various broiler production by citizens, farmers and veterinarian; (3) to explore the relation between perceived importance of production aspects and perceived best husbandry system.

Material and methods

Survey

Through March and April 2014 a quantitative survey was done by an online questionnaire. The questionnaire was developed through literature review and stakeholder input. Based on literature we selected 10 production aspects, relevant for broiler production (see results) (e.g. Bokkers and De Boer, 2009). Stakeholder input was gathered from a consulting group consisting of experts and representatives of citizens, poultry farmers and poultry veterinarians. Finally the questionnaire was pilot tested for comprehensibility and clearness of the questions.

Participants were asked which husbandry system they thought was best for broiler production. As answer option, the EU recognized broiler husbandry systems (Directive 2007/43/EC) present in the Netherlands were included (see results). Because no description of the husbandry systems was given, an answer option 'I don't know' was included. For the 10 aspect of broiler production all participants were asked 'How important do you think the aspects are for broiler production?'. The respondents could rate the importance of the 10 aspects on a five-point Likert scale ranging from very unimportant to very important.

The survey was done among citizens of the CentERpanel (www.centerdata.nl), which is a representative selection of Dutch citizens. Of these citizens 2,259 (67.6%) completed the questionnaire. Veterinarians registered with the Section Poultry Health (VGP) of the Royal Veterinary Association of the Netherlands (n=144) were invited to fill in the questionnaire. The questionnaire was completed by 51 veterinarians (35.4%). Of these 51 veterinarians, 40 met our definition of poultry veterinarian (working minimal 30% of time as a vet in the poultry production sector). Poultry farmers were invited by both a digital newsletter of the Dutch organization of poultry farmers (NOP) and an article on the website of Poultry production (a professional journal for the poultry production sector) and 100 farmers completed the questionnaire.

Statistical analyses

Data were analysed using SPSS 19.0. For the relation between perceived best system and citizens, farmers and veterinarians, Pearson's Chi-square was calculated. To analyse the association between importance scores and (1) stakeholder group and (2) perceived best system, one-way ANOVA was applied. When variables were significant according to Levene's test, the results of the Welch test were reported. Post-hoc Games-Howell test for multiple comparisons was done to compare mean importance scores among the stakeholder groups and perceived best husbandry system.

Results

Perceived best broiler husbandry system

Almost half of citizens viewed the free-range system with outdoor access as best husbandry system for broiler production (Table 1). More than half of the poultry famers and poultry veterinarians (63,4%) viewed the conventional system as the best husbandry system. Perceived best system and being citizen, farmer or veterinarian were significantly associated (χ^2 (8)=505.3, P<0.001).

Within the group citizens, the best husbandry system was associated (P<0.001) with socio-demographic characteristics. Citizens that viewed the organic system as best system, more often were female, between 25 and 34 years old, higher educated, lived during childhood in a town, have pets, eat meat once or less a week, and donate to an animal welfare or nature organization. On the contrary, citizens that viewed an indoor system as best system more often were male, between 45 and 54 years old, lower educated, lived during childhood on a farm, have no pets, eat meat more than 5 time a week, and do not donate to an animal welfare or nature organization.

Perceived importance of production aspects

Citizens, farmers and veterinarians scored the importance of broiler production aspects on a 5-point Likert scale (Table 2). The mean importance of the aspects was, except for the aspect cheap meat, valued above 3.0, which implicates that the respondents attached some importance to the aspects. The mean importance of cheap meat was 3.0 or lower, indicating it is not perceived important. All stakeholder groups perceived broiler health and food safety, with mean scores above 4.4, as very important. Citizens, farmers and veterinarians scored broiler welfare, food safety and cheap meat equally. The scores of farmers and veterinarians differed for the aspects taste and broiler famer income. Citizens perceived broiler growth rate, broiler health, system meets natural needs of broilers, environmental friendly and

Table 1. Best broiler husbandry system according to citizens, poultry farmers and poultry veterinarians.

Best broiler husbandry system	Citizen		Farmer		Veterinarian	
	Count	%	Count	%	Count	%
Conventional	156	6.9	60	60.0	26	63.4
Free-range	265	11.7	28	28.0	13	31.7
Free-range outdoor access	1,006	44.5	5	5.0	2	4.9
Organic	515	22.8	5	5.0	0	-
I don't know	317	14.0	2	2.0	0	-
Total	2,259	100.0	100	100.0	41	100.0

Table 2. Mean perceived importance of broiler production aspects by citizens, poultry farmers and poultry veterinarians.[1]

Aspects broiler production	Citizen (n=2,259)	Farmers (n=100)	Veterinarians (n=41)	*P*-value
Broiler welfare	4.2	4.0	4.2	0.188
Broiler health	4.4^a	4.7^b	4.7^b	<0.001
System meets natural needs of broilers	4.0^a	3.5^b	3.4^b	<0.001
Environmentally friendly system	4.0^a	3.6^b	3.6^b	<0.001
Food safety broiler meat	4.5	4.5	4.7	0.399
Risk broiler farm to human health	4.2^a	3.7^b	3.8	<0.001
Taste of broiler meat	4.2	4.4^a	4.0^b	0.011
Cheap broiler meat	2.8	2.9	3.0	0.628
Broiler farmer income	3.5^a	4.5^b	4.1^c	<0.001

[1] 1=very unimportant, 5=very important.

[abc] Scores in a row with different superscripts are significantly different (1-way ANOVA and Post-hoc Games-Howell multiple comparisons test).

farmer income, different from farmers and veterinarians. The aspect risk broiler farm to human health is scored differently between citizens and farmers.

Perceived importance aspects vs best husbandry system

The mean perceived importance (Table 3) of broiler health, system meets natural needs of broilers and environmental friendly, increases with increasing extensity of the husbandry systems. The mean importance of cheap meat, famer income and broiler growth rate decreases with decreasing intensity of the husbandry systems. The mean importance of risk broiler farm to human health is lower for the conventional system than for the other three systems. The mean importance of food safety and taste does not differ among the husbandry systems.

Discussion

Most citizens view the free-range system with outdoor access as the best broiler husbandry system, whereas most farmers and veterinarians view the conventional system as the best system. Citizens that are female, between 25 and 34 years, high educated, lived during childhood in a town, have pets, eat meat once or less a week, and donate to an animal welfare or nature organisation chose more often for an organic system. The importance of most productions aspects is scored similar by farmers and veterinarians. Citizens however, score most production aspects differently from farmers and veterinarians. In agreement with previous research (e.g. Vanhonacker *et al.*, 2010) the three groups score food safety as more important than animal welfare. Food safety is scored equally important by the stakeholder groups. However, previous studies showed that citizens and experts perceive food safety risks differently (Jensen *et al.*, 2005). Also broiler welfare is scored equally important by the stakeholder groups, but of the indicators for animal welfare are scored differently. Citizens perceive the natural needs of broilers more important, and the production parameters broiler health and growth rate less important than farmers and veterinarians. These differences can be explained by different interpretations of the concept animal welfare, which is influenced by value based views on how the treat animals (Fraser *et al.*, 1997). Therefore, even though citizens, farmers and veterinarians perceive food safety and animal

Table 3. Mean perceived importance of broiler production aspects per perceived best husbandry system.[1]

Aspects broiler production	Conventional (n=242)	Free-range (n=303)	Free-range outdoor access (n=1,013)	Organic (n=520)	*P*-value
Broiler welfare	3.6[a]	3.9[b]	4.2[c]	4.6[d]	<0.001
Broiler health	4.1[a]	4.3[a]	4.4[b]	4.7[c]	<0.001
System meets natural needs broilers	3.3[a]	3.7[b]	4.1[c]	4.4[d]	<0.001
Environmentally friendly system	3.5[a]	3.8[b]	4.1[c]	4.5[d]	<0.001
Food safety broiler meat	4.5	4.6	4.6	4.6	0.385
Risk broiler farm to human health	3.8[a]	4.2[b]	4.2[b]	4.3[b]	<0.001
Taste of broiler meat	4.2	4.2	4.3	4.2	0.257
Cheap broiler meat	3.6[a]	3.2[b]	2.8[c]	2.2[d]	<0.001
Broiler growth rate	3.4[a]	2.8[b]	2.3[c]	1.8[d]	<0.001
Broiler farmer income	3.9[a]	3.7[b]	3.5[c]	3.3[d]	<0.001

[1] 1=very unimportant, 5=very important.

[abcd] Scores in a row with different superscripts are significantly different(1-way ANOVA and Post-hoc Games- Howell multiple comparisons test).

welfare as equally important, they might hold different value based views regarding both food safety and animal welfare.

The analysis of the relation between the perceived importance of production aspects and the best husbandry system shows that food safety and taste are valued equally important among all systems and consequently these aspects seem not determine the choice for the best system. People that view an outdoor system as best system perceive on the one hand the aspects concerning interests of animals and environment significantly more important, and on the other hand aspects concerning economic interests less important than people that perceive an indoor system as best husbandry system. This suggests involvement of moral values.

The results show that people perceive broiler husbandry systems and production aspects differently. Differences in perceptions have been explained by differences in knowledge, interests and moral values (Te Velde *et al.*, 2002). In society, people share moral values concerning humans, animals and environment (Cohen *et al.*, 2012). However, Cohen *et al.* (2009) explained that in a context the weighing of different moral values against each other can vary. In our study for example, all stakeholder groups view animal welfare as an important value. However, when farmers and veterinarians value animal welfare they will take into account the context and thus the economic interests of the farmer. In current broiler production it is for most farmers impossible to provide outdoor access. Therefore, farmers and veterinarians may value welfare aspects linked to outdoor access (e.g. system meets natural needs of the broiler) as less important than citizens. On the contrary, when citizens may value welfare they are more prone to all natural needs of broilers including outdoor access, without taking into account the context. Consequently, they value natural needs of broilers as more important than farmers and veterinarians. Likewise, in the case of the choosing the best husbandry system, farmers and veterinarians may take into account the context, as for example the production efficiency of outdoor systems (less eggs, more animal diseases). Citizens are less familiar with the context and may not take into consideration the constraints of an outdoor system and therefore view an outdoor system as best system. In conclusion, the participants might have valued, based on the same moral convictions, the broiler husbandry systems broiler production aspects differently.

In future research we will study the relation between moral values concerning humans and animals of relevance in the context of poultry production and the trade-off between animal welfare and public health risks. This knowledge may help to balance between the interests of humans and animals and contribute to a poultry sector that is more socially accepted.

References

Bokkers, E.A.M. and De Boer, I.J.M. (2009). Economic, ecological, and social performance of conventional and organic broiler production in the Netherlands. British Poultry Science 50: 546-557.

Cohen, N.E., Brom, F.W.A and Stassen, E.N. (2009). Fundamental moral attitudes to animals and their role in judgment: an empirical model to describe fundamental moral attitudes to animals and their role in judgment on the culling of healthy animals during an animal disease epidemic. Journal of Agricultural and Environmental Ethics 22: 341-359.

Cohen, N.E., Brom, F.W.A. and Stassen, E.N. (2012). Moral convictions and culling animals: a survey in the Netherlands. Anthrozoos: A Multidisciplinary Journal of The Interactions of People & Animals 25: 353-367.

Fraser, D., Weary, D.M., Pajor, E.A. and Milligan, B.N. (1997). A scientific conception of animal welfare that reflects ethical concerns. Animal Welfare 6: 187-205.

Harper, G. and Henson, S. (2001). Consumer concerns about animal welfare and the impact on food choice. The University of Reading, Reading, UK.

Jensen, K.K., Lassen, J., Robinson, P. and Sandøe, P. (2005). Lay and expert perceptions of zoonotic risks: understanding conflicting perspectives in the light of moral theory. International Journal of Food Microbiology 99: 245-255.

Kendall, H.A., Lobao, L.M. and Sharp, J.S. (2006). Public concern with animal well-being: place, social structural location, and individual experience. Rural Sociology 71: 399-428.

Kijlstra, A., Meerburg, B.G. and Bos, A.P. (2009). Food safety in free-range and organic livestock systems: risk management and responsibility. Journal of Food Protection 72: 2629-2637.

Swanson, J.C., Lee, Y., Thompson, P.B., Bawden, R. and Mench, J.A. (2011). Integration: valuing stakeholder input in setting priorities for socially sustainable egg production. Poultry Science 90: 2110-2121.

Te Velde, H., Aarts, N. and Van Woerkum, C. (2002). Dealing with ambivalence: farmers' and consumers' perceptions of animal welfare in livestock breeding. Journal of Agricultural and Environmental Ethics 15: 203-219.

Tuyttens, F., Heyndrickx, M., De Boeck, M., Moreels, A., Van Nuffel, A., Van Poucke, E., Van Coillie, E., Van Dongen, S. and Lens, L. (2008). Broiler chicken health, welfare and fluctuating asymmetry in organic versus conventional production systems. Livestock Science 113: 123-132.

Tuyttens, F., Vanhonacker, F. and Verbeke, W. (2014). Broiler production in Flanders, Belgium: current situation and producers' opinions about animal welfare. World's Poultry Science Journal 70: 343-354.

Vanhonacker, F., Van Poucke, E., Tuyttens, F.A.M. and Verbeke, W. (2010). Citizens' views on farm animal welfare and related information provision: exploratory insights from Flanders, Belgium. Journal of Agricultural and Environmental Ethics 23: 551-569.

One Health as a collective responsibility

M.F. Verweij[1] and F.L.B. Meijboom[2]*
[1]Wageningen University, Subdepartment Communication, Philosophy, and Technology, P.O. Box 8130, 6700 EW Wageningen, the Netherlands; [2]Utrecht University, Ethics Institute & Faculty of Veterinary Medicine, Janskerkhof 13a, 3512 BL Utrecht, the Netherlands; marcel.verweij@wur.nl

Abstract

In spite of the fact that in recent years many steps have been taken in the control of zoonotic diseases, we are still confronted with recent outbreaks of, for example Ebola and Avian Flu (H5N8) and with public debates on the preferred way to deal with zoonoses. Such debates can easily get polarised. Therefore, we argue that a more integrated approach is needed. In this paper we propose an integration on three levels. First, the One Health initiative could serve a fruitful approach to take the interrelatedness of human and animal health into account. Second, we argue that it might be fruitful to approach societal controversies about how societies should respond to zoonotic risks not by focusing on conflicting interests, but as conflicts between broadly shared values. Finally, an effective approach of and a discussion on zoonoses require a more sophisticated view on moral responsibilities, which goes beyond the level of the question 'who is to blame?' for a specific outbreak. In this paper we claim that although finding acceptable and well-justified policies will remain a difficult task, these three steps towards a pluralistic approach to thinking about prevention of and response to zoonosis outbreaks may create possibilities to overcome conflicts.

Keywords: responsibility, zoonoses, prevention, values

Introduction

In spite of the fact that in recent years many steps have been taken in the control of zoonotic diseases, we are still confronted with recent outbreaks of, for example Ebola and Avian Flu (H5N8) and with public debates on the preferred way to deal with zoonoses. Even if one agrees that culling should be used as a last resort in dealing with zoonotic diseases, many ethical questions remain when the focus shifts to prevention strategies. Such debates can easily get polarised – as happened for example in the case of the recent Q-fever outbreak in the Netherlands – which may easily undermine trust in agricultural, the food production and public health sectors (Evaluatiecommissie, 2010). Therefore, we argue that a more integrated approach is needed. This entails integration on three levels. First, the One Health initiative could serve a fruitful approach to take the interrelatedness of human and animal health into account. Second, we argue that, rather than interests, the plurality of values that underlie current debates should get a more central role in the debate. Finally, an effective approach of and a discussion on zoonoses require a more sophisticated view on moral responsibilities, which goes beyond the level of the question 'who is to blame?' for a specific outbreak. In this paper we elaborate on these three points that serve as stepping stones for future normative frameworks that help to assess current and innovative measures to prevent and control zoonotic diseases.

One Health and the need for ethical reflection

One Health is the idea that human and animal health and natural environment are connected, and that adequate protection against certain threats to human or animal health requires collaboration of a variety of different actors and scientific disciplines (Coker, 2011; Zinsstaga *et al.*, 2011). Zoonoses – infectious diseases that may spread from animals to humans and vice versa are a paradigm case for One Health. Infections occur through direct contact with (infected) biological material of other species, through vector organisms, food, or through the air. Some of these create major threats to public health. Ebola, SARS, H5N1 influenza, Q-fever, or West Nile virus, are examples of diseases that not only can result in severe illness of individual persons but may also cause large outbreaks that affect public health at large and may undermine societal life. Notwithstanding the advances of medical science, therapeutic options may be limited, as shown in recent cases of Ebola and MERS. Moreover, therapeutic options can further decline as a result of the development of antimicrobial or antiviral resistant strains of infectious agents. Indeed, some antibiotic-resistant infections like MRSA can be considered as zoonoses themselves, developing (among others) in livestock production and then spreading to human populations. Zoonosis infections in livestock farming and food production challenge public health but also affect animal welfare, consumer trust, and a sustainable animal and food production.

To contain the threats of infection, infectious disease control can involve drastic prevention and response measures that interfere with people's wellbeing, privacy and freedom, and which therefore require strong ethical justification and fair procedures to be acceptable and legitimate (Verweij, 2011). Public health legislation sets legal boundaries for legitimate interventions, but within those boundaries, public health officials, professionals, citizens and corporate agents are confronted with ethical questions: what precautions that affect people's lives are reasonable or morally required given certain (or often uncertain) risks (Krom, 2014)? This variety of potential value conflicts only increases or even exacerbates in the case of protection against zoonoses, given the value we attach to animal welfare, animal integrity, safety and trust in the food chain, and to preservation and protection of the natural environment (Akhtar, 2013; Wright *et al.*, 2010). One Health asks for collaboration between different perspectives, and, in the case of ethical reflection on prevention of zoonoses, it makes sense to combine arguments, concepts and ideas from animal ethics and public health ethics.

Interventions: from control to prevention

To control zoonotic diseases, the culling of animals which have been exposed, or which might have been exposed to a disease that is dangerous for human beings is still standard intervention. Especially in the case of disease with a high human health risk, such as H5N1 influenza, culling is considered by many to be an acceptable intervention. However, it is also one that raises a lot of criticism and public debate. Culling, however, is an *ultimum remedium*, and many precautions can and should be taken to prevent outbreaks of zoonoses and to detect infections as early as possible. Common ways of to prevent and respond to infections in modern agricultural practice include, for example following strict protocols for cleaning and disinfection of livestock houses as well as persons entering houses, minimisation of visitors, hygienic disposal of animal waste, etc. In addition, there is a variety of technical and societal innovations to prevent and respond to outbreaks, including the development of medical (human) and veterinary interventions like vaccination or (prophylactic) treatment and the breeding of more robust animals (Collins and Wall, 2004). Data collection in slaughterhouses could be used more systematically to track and trace sources of outbreaks in livestock. Further technological and scientific innovation, however might also facilitate syndrome surveillance in livestock, for example for influenza in poultry and swine). Moreover, such surveillance of possible symptoms in livestock veterinary data could be linked to systematic surveillance in human populations, e.g. in clinical data in family practices, to detect trends that might be early indications of an emergent outbreak.

Although these innovations often are very promising, maximising protection conflicts with other values (Meijboom *et al.*, 2009; Meijboom and Ohl, 2012) and each of the innovation strategies raises specific ethical questions, for instance concerning privacy, responsible use of big data, animal welfare, human quality of life, or sustainability. Furthermore, the effectiveness and reasonableness of specific precautions partly depend on precautions taken elsewhere in the food chain. Many parties can play some role, and share responsibility to contribute to prevention. This shows the need to discuss moral responsibilities next to the question of how to deal with conflicting values.

Dealing with conflicts: from interests to values

Zoonoses outbreaks raise controversies and debates as to how far government, farmers or food producers should go in reducing infection risks or otherwise take action to stop the spread of disease. These controversies are often framed in terms of conflicts of interests between stakeholders, such as farmers, retailers, different government departments (agriculture versus public health) and citizens (in their capacity as consumers or as persons living in high-risk areas). However, to deal with these conflicts it is important to explicate the values that underlie these conflicts. For this step we have two reasons. First, the debate is not only about benefits and stakes, but also about (the variety of) what actors consider to be important and valuable. This may help clarifying the moral conflicts at hand. For that matter, clarification may also imply showing that conflicts are much more complex than they appear at first sight. Conflicting values might be more fundamental than initially thought – certainly in the case of zoonosis, where concerns about society, individual well-being, biodiversity and animal welfare all come into play. Moreover, few actors will have only one, univocal stake in a controversial matter. This observation however also creates possibilities for overcoming conflicts – our second motive for shifting the perspective from interests to values. If we acknowledge that each different 'stakeholder' or moral agent may see a variety of values to be relevant – which are in effect often broadly shared within society – this opens ways to reconcile tensions between perspectives and seek reasonable compromises (Benjamin,1990).

The plurality of responsibilities

Societal controversies that arise in the context of livestock zoonotic outbreaks that threaten the health of people (whether as consumers or as people living close to farms) are easily framed in terms of victims and villains and questions about who is responsible for the loss of lives, health or quality of life. Understandably, citizens are seen as the most important victim; the animals that are culled are considered as another, still important victim. It is also tempting to see farms and government as the most important culprits, especially if governments seem to give some priority to protecting the economic interests of the agricultural sector over protecting the health of the public. Such framing in terms of victims and responsible agents only reinforces a polarised debate.

Responsibility is however a notoriously complex concept. Acknowledging this complexity may contribute to a depolarisation of discussions. One problem with talking about responsibility is that emphasising the responsibility of one party (often not oneself) suggests that others (including oneself) bear less responsibility, or none at all – and that creates tensions between stakeholders. Another problem is that in debates about responsibility, forward-looking and backward-looking conceptions of responsibility are often confused (Verweij and Houweling, 2014). We suggest that, with respect to the prevention and response to zoonotic risk, the focus should be on forward-looking responsibility: who should do what to prevent or respond to outbreaks? Normative judgments about backward-looking responsibility (who has played a causal role in the outbreak, and/or who is to blame for the consequences) are only one category of considerations that are relevant in determining who should take action to prevent or mitigate infection risks. Indeed, given that many parties are involved the food chain and in contributing to agricultural and food safety (governments, food safety authorities, farmers, transporters, slaughterhouses, food companies, retailers, consumers, etc.), it makes sense to frame the problem as a matter of how responsibility should be distributed among all stakeholders rather than as a question about who is responsible for zoonoses control.

David Miller (2001) has introduced a pluralistic approach to the question of how forward-looking (in his terms: remedial) responsibilities are to be distributed. Some parties may be seen as having some forward-looking responsibility because they formed a causal factor the emergence of an outbreak. Others may be considered as having responsibility to prevent and respond to risk because they, are in some way, be blamed for an outbreak. These (backward-looking) considerations however do not exhaust our reasons for allocating forward-looking responsibility. Some parties may have special capacity to prevent infection risk, and that might be an independent reason for seeing them as having a moral duty to contribute to the reduction or mitigation of risk. And yet another category of stakeholders could be seen as having a duty to bear some of the burdens of preventing or mitigating risk because they are benefiting from the conditions that have contributed to the zoonotic risk.

One of the strengths of Miller's approach is that the plurality of reasons makes it difficult for individual stakeholders to solely locate responsibility at other parties. As such it helps stakeholders to acknowledge responsibility as something they share. This not only concerns companies, professionals and government agencies in the food chain. Consumers also have reason to assume some responsibility: they benefit from practice that might contribute to risks: many citizens consume meat and other animal products, they often prefer the cheapest products, and many citizens attach positive value to living in the proximity to agricultural landscapes, where cattle are grazing in the fields. Consumers may have few possibilities to prevent zoonoses outbreaks, but given how they benefit from current agricultural practices that are a potential source for large-scale outbreaks, it might at least be reasonable that expensive precautionary measures will be reflected in the price of animal products.

Acknowledging the plurality of responsibilities is a first step in deliberation about what zoonosis prevention measures are reasonable, and what can be reasonably expected of each of the stakeholders. Such reflection leaves little room for stakeholders to flee from responsibility by pointing at responsibilities of others.

The plurality of values at stake

In practice prevention strategies invoke conflicts between several basic values: public health, privacy, environmental values, animal welfare, food safety, etc. (cf. Verweij, 2011; Verweij and Van den Hoven, 2012). Although it is tempting to focus on maximizing safety and minimizing environmental and health impacts as key parameters for prevention measurements, any responsible policy should acknowledge that risks cannot be ruled out completely, and that precautionary measures may come at the cost of other values. As a result, reflection on what risks are (un)acceptable is inevitable in responsible prevention and control strategies, and this involves a weighing of competing concerns. Such weighing requires that the values at stake are explicated and analysed. This is not an easy task given the plurality of values at stake. The current debate is characterized by a broad range of values, including animal welfare and integrity, urban and rural quality of life, privacy and sustainability. At face value, it seems that this plurality only complicates the opportunities to reach an effective strategy for the control and prevention of zoonosis. However, we argue that a fruitful prevention strategy should start in the acknowledgement of this plurality of values and actors.

The variety of values should not be considered as a complicating hurdle that needs to be taken if one wants to draft prevention strategies. It better is understood as a necessary part, or stepping stones to formulate ways to deal with risks, precautions and responsibility and to discuss and assess these proposals. First, this implies giving explicit room to the plurality of values in order to understand the normative background of (some of) the problems of and conflicts between actors. This does not immediately solve the issues at stake, but helps to get grip on it by better understanding. Furthermore, our proposal to approach zoonosis from an integrated perspective entails that the interrelatedness of valucs, such as human and animal health becomes more clear. If we start from a One Health perspective the assessment should not include an either/or discussion of health, but should take all values related to health and wellbeing of humans, animals and the environment into account. This may seem as just a further complication of the problem of prevention and control. However, prevention and control are complex by nature. This cannot and should not be 'solved' by denying the moral dimension or by any proposal to reduce the number of values that are considered as important for the debate. Such a reduction will in the end lead to new problems and questions (cf. Meijboom *et al.*, 2009). Finally, granting the plurality of values does not imply that we cannot longer come to agreement or consensus. Values often are shared by many stakeholders. For instance, safety, wellbeing and health are acknowledged as important values by farmers, retailers and public health professionals. Many interpersonal moral conflicts thus are, at the same time, intrapersonal conflicts for citizens, professionals, or governments as well (Benjamin, 1990). Such acknowledgement of the diversity of values and actors it is possible to start to build bridges and to formulate (new) prevention and control strategies.

Conclusion

We have suggested that it might be fruitful to approach societal controversies about how societies should respond to zoonotic risks not by focusing on conflicting interests, but as conflicts between broadly shared values. This may create possibilities to overcome conflicts. On the other hand, our analysis shows how these ethical problems are much more complex that they may appear at first sight. Many – probably irreducible – values are at stake. Many actors bear responsibility, and again, the reasons for assigning responsibility should not be reduced to one type of concern. Although we have suggested that our pluralistic approach to thinking about prevention of and response to zoonosis outbreaks may offer a basis for seeking reasonable compromises, establishing acceptable and well-justified policies will still be a difficult task.

References

Akhtar, A. (2013). The need to include animal protection in public health policies. Journal of Public Health Policy 34: 549-559.

Benjamin, M. (1990). Splitting the difference: compromise and integrity in ethics and politics. The University Press of Kansas, Lawrence, USA.

Coker, R., Rushton, J. and Mounier-Jack, S. (2011). Towards a conceptual framework to support one-health research for policy on emerging zoonoses. The Lancet Infectious Disease 11: 326-331.

Collins, J.D. and Wall, P.G. (2004). Food safety and animal production systems: controlling zoonoses at farm level. Revue Scientifique et Technique (International Office of Epizootics) 23(2): 685-700.

Evaluatiecommissie Q-koorts. (2010). Van verwerping tot verheffing. Q-koortsbeleid in Nederland 2005-2010. Den Haag, the Netherlands.

Krom, A. (2014). Not to be sneezed at. On the possibility of justifying infectious disease control by appealing to a mid-level harm principle. Department of Philisophy and Religious Studies, University of Utrecht, Utrecht, the Netherlands.

Meijboom, F. and Ohl. F. (2012). Naar een afwegingsmodel voor gezondheidsbeleid. Ontwerp van een instrument voor de identificatie van morele vraagstukken in het Nederlandse gezondheidsbeleid. In: De Mens Centraal? Ethische dilemma's bij gezondheidsbeleid met goede zorg voor dier en natuur. Centrum voor Ethiek en Gezondheid, Den Haag, the Netherlands, pp. 89-103.

Meijboom F.L.B., Cohen, N., Stassen, E.N. and Brom, F.W.A. (2009). Beyond the prevention of harm: animal disease policy as a moral question. Journal of Agricultural & Environmental Ethics 22(6): 559-571.

Miller, D. (2001). Distributing responsibilities. Journal of Political Philosophy 9(4): 453-471.

Verweij, M.F. (2011). Infectious disease control. In: Dawson AJ (ed.) Public health ethics, key concepts and issues in policy and practice. Cambridge: University Press, Cambridge, UK, pp.100-119.

Verweij, M.F. and Houweling, H. (2014). What is the responsibility of national government with respect to vaccination? Vaccine 32(52): 7163-7166.

Verweij, M.F. and Van den Hoven, M.A. (2012). Duurzame gezondheid haalt de angel uit het conflict tussen volksgezondheid, dier en natuur. In: De mens centraal? Ethische dilemma's bij gezondheidsbeleid met goede zorg voor dier en natuur. Centrum voor Ethiek en Gezondheid, Den Haag, the Netherlands, pp. 13-20.

Wright, N., Meijboom, F.L.B. and Sandøe, P. (2010) Thoughts on the ethics of preventing and controlling epizootic diseases. Veterinary Journal 186(2): 127-128.

Zinsstaga, J., Schellinga, E., Waltner-Toews, D. and Tannera, M. (2011). From 'one medicine' to 'one health' and systemic approaches to health and well-being. Preventive Veterinary Medicine 101: 148-156.

Animal husbandry and food production in China and Europe: a shared moral problem?

F.L.B. Meijboom[1] and J. Li[2]*
[1]Ethics Institute, Utrecht University, Janskerkhof 13a, 3512 BL Utrecht, the Netherlands; [2]Department of Science, Technology and Governance, China Agriculture University, 17 Qinghua Donglu, Beijing 100083, China P.R.; f.l.b.meijboom@uu.nl; lijianjun@cau.edu.cn

Abstract

In China and Europe many millions of animals are used for food production. For both regions animal food production is considered to be important for both the internal market, but also for export. In spite of these similarities there are many differences. First, while in Europe there currently is a lot of public attention to animal welfare, the moral value of animals and discussions on animal rights, the discussion on these topics in China seems to be relatively limited. This difference is remarkable and cannot only be explained by the fact that food security and food safety rather than food quality are in China more on the public agenda than in Europe. Traditionally, China has a much stronger philosophical tradition with regard to the human-animal relationship than the Western tradition. Nonetheless, current advocates of animal welfare or animal rights in China refer to modern Western philosophers rather than to the Chinese tradition. At the same time, recent discussions in Europe on animal ethics show that the proposals from the last century (e.g. Singer, 1995; Regan, 2004) still raise many theoretical and practical questions. This suggests that the East and the West can genuinely learn from each other in the case of the ethical evaluation of animal husbandry. In our paper we explore the Chinese traditions and the current animal welfare discussions in China. Next we draft ideas about how these traditions can serve as input for the debate on animal welfare in China and Europe.

Keywords: animal welfare, food production, China

Introduction

In China and Europe many millions of animals are used for food production. For both regions animal food production is considered to be important for both the internal market, but also for export. In spite of these similarities there are many differences. First, while in Europe there currently is a lot of public attention to animal welfare, the moral value of animals and discussions on animal rights, the discussion on these topics in China seems to be relatively limited. This difference is remarkable given the philosophical and moral tradition of reflections on animals in a very long history of Chinese civilization.

Although explicit references to animal rights or animal welfare in Chinese classical literature are rare, benevolence to animals or respect for animal life based on either human morality or the intrinsic value of nature can be recognized in the Chinese culture. Nonetheless, current advocates of animal welfare or animal rights in China refer to modern Western philosophers rather than to the Chinese tradition. That is to be regret for two reasons. First, it seems to disregard the precious spiritual heritage of China. This lack of attention for its own heritage, can be considered as trend in China and as a consequence of the pursue of survival and striving for economic power during the last 200 years. Second, recent discussions in Western philosophy-oriented animal ethics show that the proposals from the last century (e.g. Singer, 1995; Regan, 2004) still raise many theoretical and practical questions. This suggests that the East and the West can genuinely learn from each other in the case of the debate on animal ethics. In this paper, we first explore the Chinese discussion on animal welfare. Next, we shortly introduce the Confucian and Daoist views on the human-animal relationship. Finally, we discuss how these traditions can serve as input for the discussion on animal welfare in the context of food production.

Current changes in the Chinese debate on animals

Recently animal welfare has become part of public discussion in China. This is reflected in many references to animal use and animal cruelty on social media and by a number of public activities that have been organized in China. These activities aim to protest against many kinds of cruelty and animal welfare problems in China, ranging from fur farming to food production. These activities critically discuss whether (increasing) human welfare should come at the cost of reduced animal welfare. Activities have, for instance been focused on the protection of the welfare of the bears that are used by the Guizhentang Pharmaceutical company that keeps bears in order to extract bile for Chinese medicine. Also the Yulin Dog Meat Festival in Guangxi province of south-west China has been a target of fierce debates. These activities are partly inspired and initiated by international NGO's such as the Humane Society, World Wildlife Fund (WWF), and the International Fund for Animal Welfare (IFAW). However, also Chinese animal rights and welfare organizations play a central role. In the theoretical discussions on the arguments that underlie the attention to animal welfare one often refers to Western philosophy, especially Peter Singer's thoughts that were translated in Chinese in the 1990's. However, currently some Chinese philosophers and ethicists attempt to construct a new animal ethics that uses (elements of) Confucian animal ethics, Taoism ethics and Buddhism ethics.

Different terms, but same concepts?

Animal welfare has been a central concept in European discussion on animal policy, animal protection and ethics for many decades. Animal welfare is recognized by the European Member States as an important aspect that should be acknowledged in all kinds of activities including agriculture, research and technological development (Lisbon Treaty, 2009). Also the new European Directive on animal research stresses the importance of taking the interests of animals into consideration (Directive 2010/63/EU, art. 12). Likewise the Federation of Veterinarians of Europe (FVE) takes this development seriously and recently published its revised Code of Conduct and the Veterinary Act that explicitly includes ethical guidelines with respect to the health and welfare of animals (FVE, 2012). If we search for similar activities in the mainland of China, it seems that there are only small and limited steps to address questions of animal welfare (cf. Li, 2009). Even if one looks in the Chinese tradition, the term animal welfare cannot easily be recognized and the common Western distinction between animal welfare and animal rights is difficult to understand from a Chinese perspective (Lu *et al.*, 2013).

Nonetheless, it would be a too easy conclusion that Chinese tradition would lack thoughts on the position of animals and on our duties with regard to animals. Even though the term animal welfare lacks in Chinese literature, there are concepts that are or can be related to doing well towards animals. For instance, in Confucian ethics 'benevolence' ('Ren') is a central concept (Shen, 2014). This notion of 'Ren' implies that a man should care for his family members, his friends and other people. However, this care or love is not limited to humans. It should extend to everything on earth, which explicitly leaves room for care for animals. Confucians defend that everything on earth, including humans, animals, and society is part of a holistic system. This system of life and life creation includes the universal law ('Tao') (Shen, 2014). When participating in the life-creating-process of the holistic system, humans recognize this universal law. Once man recognizes this, then humanity will be his inherent nature and consequently benevolence will be his virtue. Therefore, 'benevolence' with respect to animals is more than an imperative that links to human interests only. We can have moral duties towards animals. At the same time, the source of normativity of this duty lies in the general requirements for self-improvement and the improvement of society of every man that is a central imperative in Confucian ethics. Therefore, we conclude that Confucian ethics can take animal interests seriously for its own sake, but always starts in human morality or moral needs. In general Confucians define the human in contrast to the animal (Roetz, 2010). In justifying the attitude of benevolence towards animals one does not need not to refer to notions of animal right or animal dignity, it rather requires a reference to the ideal human behaviour that is in harmony with nature, including the element of time. In this sense the tradition has been human centred (Cao, 2011; Roetz, 2010). However, even if we grant this, it is important to notice that in spite of this tradition its distinction between humans and animals provides room to go beyond a 'human interest only' account.

Next to Confucianism, 'Taoism' (or 'Daoism) is another important philosophical tradition in China. In Taoism nature is perceived as a well-ordered chaos that exists of many entities of equal importance and value (Roetz, 2010: 206). Taoist ethics focuses on the understanding of this reality of nature. This includes notions of increasing longevity, living morally, and building consciousness and attention one's diet. The content and interpretation of Taoists ethics can vary between the different schools, but in general it tends to emphasize four aspects: wu-wei (action through non-action), naturalness, simplicity, and spontaneity. Especially, Zhuangzi's perspective on naturalness of animals opens the normative possibility to address the value of animals for their own sake. From this the imperative follows to respect animals' naturalness, which encompasses the notions of animal integrity and autonomy. Zhuangzi argues that humans 'do not overcome what is of nature by what is of man' (Fall water of Zhuangzi), which implies that any loss of nature including the change or use of animals is to be evaluated negatively. Zhuangzi stressed the importance of freedom and autonomy for animal flourishing by a series of metaphorical stories, and he criticized the human tendency to distort the natural state of animals or their own innate nature (xing) and inner power (de) through domestication and instrumentalism. Zhuangzi also expressed his ideas on integrity and the inherent value of animals which should be respected by humans. In the Zhuangzi, we also find a discussion of the ritual practices and the recognition of animals as both independent agents and manifestations of the sacred, which entails that animals have their inherent value of life. This also emphasizes that humans are not allowed to violate or interfere with the life of animals. According to Zhuangzi's statement, humans should not distort their own innate nature, their own connection with the Dao, by interfering with the freedom and self-unfolding of animals.

From this short presentation of the Daoist tradition it follows that this tradition provides a clear basis to take the welfare of animals in to account and to formulate direct duties towards animals. It even provides the ground for an even more critical stance towards animal use in general.

Tradition in practice

The above presentation of the Chinese traditions is no more than a first step and it is beyond the scope of this paper to present and evaluate these traditions at full length. Nonetheless, it shows that the Chinese tradition has a potential to contribute to the discussions on animal welfare and animal ethics in general.

First, it shows for the Chinese debate it is possible to evaluate practices of animal use from an ethical perspective without reference to Western philosophy. This is important, because it shows that respect for animals and animal welfare are more than 'import products', but are considerations that fit in the Chinese philosophical traditions of Confucianism and Daoism. This may help in the process of creating more official support for the position of animals in food production.

Second, these traditions have the potential to improve the quality of the debate in China. As mentioned before the Western distinctions between rights and welfare do not have direct equivalents in the Chinese language and tradition. Furthermore, the Cartesian distinction between man, animal and nature that often form the background for Westerns discussions on the position of animals does not have direct links with the Confucian and Daoist perspectives. As a result the current discussion in China runs the risk of using concepts, terms and distinction that may not (yet) really be incorporated in the Chinese society. On the other hand, the debate may be frustrated by the fact that one appears to discuss animal welfare in univocal terms, while actually there are fundamental differences that underlie the used concepts and terms, but remain implicit. Especially with respect to animal welfare this is a serious risk. Already the European discussions on animal welfare show a huge variety of definitions and interpretations of the concept of welfare (cf. Haynes, 2011). In China this problem is multiplied by the cultural differences between China and the West. By elaborating on the traditional Chinese perspectives it is possible to clarify the (potential) differences and to explore the possibilities to find common ground for discussion.

Third, the Chinese traditions can serve as an incentive to reflect on the Western debates on animal ethics. For instance, the idea of a unity of humans and heaven (including the earth) and the relation between humans and animals as a continuum my not be absent in the Western tradition, but is not mainstream and challenges views that start from a strict(er) segregation between man and nature. Furthermore, the cognitive capacities that play a central role in the Western debate on moral status of entities is less important in (some of) the Chinese debates on the moral relevance of humans and animals. This can be a start to reflect (once again) on the rational capacities as the exclusive criterion for entering the moral circle.

Finally, given the fact that animals are kept worldwide and that animal products are part of global trade, animal welfare has a global dimension too. Therefore, it is crucial that the Chinese and the Western traditions are brought together in order find ways to discuss the moral position of animals and animal welfare.

Conclusion

If we turn to the question in the title of this paper, we can conclude that animal welfare with regard to animal food production is a shared moral problem. However, our answer has two dimensions. First, the problem is shared because both from a Western and a Chinese perspective we have good reasons to take the welfare of animals that are kept for food production seriously. It explicitly is not a Western idea only. There are clear indications that – in spite of the differences – the Confucian and Daoist traditions provide opportunities to take the interests of animal seriously and to define duties towards animals with regard to welfare. Second, it should be a shared problem because the short presentation of the Chinese traditions shows that traditional Chinese philosophy can contribute to the improvement of the Chinese debate on animal welfare, and can be an incentive to find ways to discuss animal welfare on a global level.

References

Cao, D. (2011). Visibility and invisibility of animals in traditional Chinese philosophy and law. International Journal for the Semiotics of Law 24: 351-367.

FVE (Federation of Veterinarians of Europe) (2012). Code of conduct & veterinary act. Available at: www.fve.org.

Haynes, R.P. (2011). Competing conceptions of animals welfare and their ethical implications for the treatment of non-human animals. Acta Biotheoretica 59: 105-120.

Li, P. (2009). Exponential growth, animal welfare, environmental and food safety impact: the case of China's livestock production. Journal of Agricultural and Environmental Ethics 22: 217-240.

Lu, J., Bayne, K. and Wang, J. (2013). Current status of animal welfare and animal rights in China. ALTA: Alternatives to Lab Animals 41(5): 351-357.

Regan, T. (2004). The case for animal rights. Updated with a new preface, The University Press Group Ltd, Berkeley, USA.

Roetz, H. (2010). On nature and culture in Zhou China. In: Vogel, H.U. and Dux, N. (eds.) Concepts of nature: a Chinese-European cross-cultural perspective. Brill, Leiden, the Netherlands, pp. 198-219.

Shen,V. (ed.) (2014). Dao companion to classical confucian philosophy. Dao companions to Chinese philosophy 3. Springer, Dordrecht, the Netherlands.
Singer, P. (1995). Practical ethics, 2nd edition. Cambridge University Press, Cambridge, UK.

Section 4. Aquaculture and fisheries

Pain perception in crayfish (*Astacus astacus*): empirical observations and ethical consequences

H. Vervaecke[1], E. Roelant[1], W. Hendrycks[1], T. Abeel[1], X. Vermeersch[2] and S. Aerts[1,3]*
[1]Odisee University College, Agro- and Biotechnology, Ethology and Animal Welfare, Hospitaalstraat 23, 9100 Sint-Niklaas, Belgium; [2]Institute for Agricultural and Fisheries Research, Technology and Food Science Unit, Brusselsesteenweg 370, 9090 Melle, Belgium; [3]Ethics@Arenberg, KU Leuven, Kasteelpark Arenberg 1, bus 2456, 3001 Leuven, Belgium; hilde.vervaecke@odisee.be

Abstract

In Belgium a public debate on pain perception in crustaceans was sparked upon a popular television program showing the dismembering and grilling of live lobsters by a top-chef. In crayfish little is known about pain perception, the transmission of nociceptive information to the brain that is associated with a more complex emotional component. We assessed the effect of electrical shocks on the noble crayfish (*Astacus astacus*), the reaction of crayfish to a newly introduced object and the differences in response towards familiar and unfamiliar objects. There were significant differences in activity among shocked and non-shocked crayfish. Noble crayfish that did not receive a shock did not approach the unfamiliar object during the first fifteen minutes after the introduction. The shocked animals approached the objects significantly more and with a lower latency time than the non-shocked crayfish, indicating a lower neophobia after experiencing a noxious stimulation. Crayfish outside their shelter after receiving a shock appeared disorientated as they were seen to bump their head against the novel object or were unable to enter their shelter. Nociception may capture the animal's attention or prevent it to respond to its innate distrust for the novel object. However, it was not confirmed that due to the neophobia the non-shocked crayfish would react differently towards a novel versus familiar object. The results are clear indications of nociception in crayfish and there is evidence of an altered state of attention. These preliminary results are relevant in the animal welfare debate and merit further investigation. If they can be confirmed, there is a strong incentive to re-evaluate a significant part of legislation and the animal welfare and ethics literature.

Keywords: nociception, animal welfare, stunning

Introduction

In Belgium a public debate on pain perception in crustaceans was sparked upon a popular television program showing the dismembering, slashing and grilling of live lobsters by a top-chef. The debate is prone to speculation as little is known about actual pain perception in crayfish.

Furthermore, it is necessary to differentiate between nociception and pain. Nociception is the process of physiological detection of noxious stimuli and the ability to respond towards these stimuli. As nociception does not involve conscious perception, there is no conscious suffering (Mather, 2001; Rose *et al.*, 2012). Pain on the other hand is defined 'as an unpleasant sensory and emotional experience associated with actual or potential tissue damage, or described in terms of such damage' (Broom, 2001).

In the paradigm of a novel-object-test, a decreased reaction towards a novel object is expected in animals that experience a temporary reduced environmental attention due to a nociceptive stimulus (Sneddon *et al.*, 2003). According to Braithwaite (2010), sensory information needs to be integrated at a higher order to assess the novelty of an object. An encounter with a novel object in a usually shy species that does not elicit attention, fear and avoidance can be indicative of a modulating effect of pain (Sneddon *et al.*, 2003). Until the interplay of the underlying mechanisms and the validity of the paradigm are firmly confirmed, results need to be interpreted with caution. We assessed the overall effect of electrical shocks on the noble crayfish (*Astacus astacus*) and determined how this affected the ability of the crayfish to focus their attention towards a newly introduced object. We also examined the difference in response towards a familiar object and an unfamiliar object.

Methods

In the experiment 24 two-summer-old pond raised noble crayfish (*Astacus astacus*) were used, obtained from a German crayfish farm, that had been kept in an indoor recirculation system for 120 days. They were housed in black 57 × 39 × 28 cm Samla tanks with a water temperature of 18-20 °C, each containing a plastic tube (32 mm diameter) for shelter. – Every crayfish was kept separately to prevent stress and cannibalism. Windows in the room allowed a natural day/night cycle. One third of the water was changed when necessary in order to maintain good water quality. The experiments were conducted in daylight between April and May 2013 and between 09.00 AM and 19:00 PM.

For the novel object test 12 crayfish of equal size were selected (6 males and 6 females). One female died, leaving 11 study animals. These crayfish were randomly assigned to two groups, a shock group and a non-shock group. For six days the crayfish were habituated to the presence of electrodes. Crayfish in the shock group received a shock of 4 V, 0.01 A, DC. This was the lowest benchmark to induce tail flicks reliably. Shocks were directly delivered by contact of the electrodes with the body of the crayfish using a VEPS613 Bench DC Top Power Supply. Crayfish were not restrained. Non-shock crayfish received a placebo treatment (contact with the electrodes without an electric shock).

The behaviours that were recorded were: rubbing (rubbing antenna and body against the side of the tray), grooming (rubbing with legs against antenna, legs, eyes, back and tail; rubbing antenna in claws of walking legs), stooping (pushing head and tail down and lifting abdomen from the substrate), tail flick (flicking the tail and darting backwards), recoil (rapid retrograde movement), autotomy (discarding own appendages), waving (slow waving of the legs while these are lifted off the ground). Stooping, tail-flicking, recoil, rubbing, waving, autotomy and grooming were considered as nociceptive behaviours. At the end of each minute, the frequency was counted. When a behaviour occurred on the transition of a minute, it was only scored once. Three behavioural states were scored 1/0 per minute: activity (locomotion outside the shelter), inactivity (lack of locomotion outside of the shelter) and sheltering (lack of locomotion inside the shelter). The total number of minutes with score 1 per behavioural state were summed for the analysis.

After a treatment was delivered, the crayfish were observed for 5 minutes. Then a novel object was carefully placed 10 cm in front of the crayfish. After this, observations of 15 minutes were conducted. After these observations the object was carefully removed. We recorded the distance between the crayfish and the object (<5 cm; 5-10 cm and >10 cm), their activity, inactivity or sheltering and nociceptive behaviours every minute as 1/0 score. On the first minute, 5th minute, 10th minute and 15th minute the occurrence of waving was recorded. Novel object presentation was repeated an additional 3 times at hourly-intervals. Novel objects consisted of a lid, a pincer, a stone and a plastic flowerpot. Objects were presented in a random order and no object was presented twice to the same crayfish. Testing occurred over the course of two days.

In the familiar object test we used 12 male crayfish. These crayfish were habituated to a specific object for 6 days. The day prior to testing the object was removed. Objects used were lids, plastic flowerpots and cups, stones, pincers and bowls. Only males were used during the familiar object test, because of there were no females available at that moment. Other animals were used to avoid double use of the same individuals and double exposure to the shocks, as these previous experiences could influence the results.

For the statistical analyses we used the data of the first 15 minutes on the one hand and on the other hand used more complex models for all data (first 15 minutes and additional 3 times at hourly-intervals). The latency time, defined as the time to approach the object closer than 5 cm is modelled with a Cox regression model with the factor shocked or not as the only predictor. The median number of minutes of the first 15 minutes in a certain position or doing a certain activity were compared between shock and non-shocked crayfish with a non-parametric Mann-Whitney test due to small numbers. The same analysis was also done for the number of grooms and wave frequencies observed in the first 15 minutes. To model all data (4 times 15 minutes observed) we use generalized linear mixed models with a random effect for crayfish correcting for repeated measurements of the same crayfish. For activity and position, outcomes were made binary (close to object or not, being active or not). As grooming and wave frequency are counts we used a Poisson regression model for these outcomes. In the model there were three predictors: shock (yes/no), time in minutes after shock and object (familiar or novel). First of all the significance of the three-way interaction was tested. If this was not significant, the significance of the two way interactions was investigated. If no interactions were significant, we studied the main effects.

Results and discussion

Shocked crayfish showed tail flicks and recoils as a response towards an electric shock. No autotomy was observed and stooping was only seen once.

After novel object presentation shocked crayfish had a significant lower latency time to approach the object (*P*-value Cox regression=0.003) than non-shocked crayfish and in the familiar object test there was also a significant lower latency time (*P*-value Cox regression=0.03) in shocked crayfish (Figure 1 and 2). Only shocked crayfish approached the novel object the first 15 minutes.

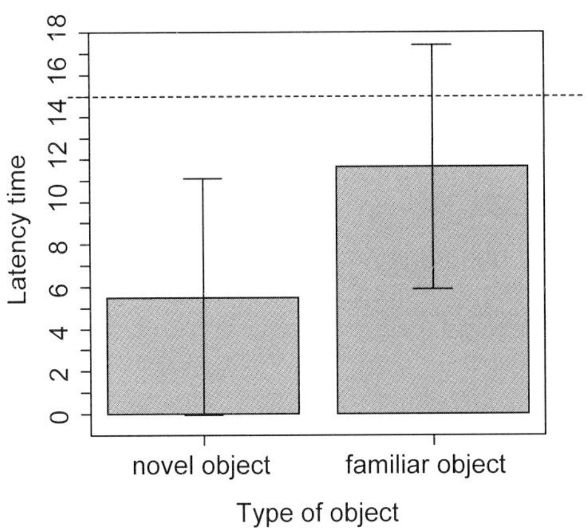

Figure 1. Average latency time in minutes to approach object (within first 15 minutes) in reaction to novel object (left) or familiar object (right) by shocked crayfish.
Note: None of the non-shock crayfish approached the (novel or familiar) object within 15 minutes (dashed line)

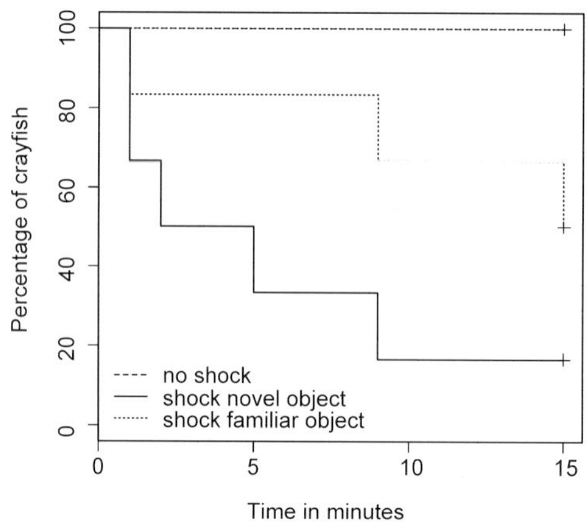

Figure 2. Percentage of crayfish that did not approach the object during the first 15 minutes in reaction to novel object or familiar object.

Shocked crayfish in the novel object test spent significantly more time close (Mann-Whitney U=2.5, *P*=0.015) to the object than non-shock crayfish. Non-shocked crayfish spent significantly more time at intermediate distance (between 5 and 10 cm) than shocked crayfish (*P* Mann-Whitney U=3.5, *P*=0.032). Shocked crayfish also had significantly more positional changes than non-shocked crayfish (Mann-Whitney U=3, *P*=0.019). A noxious stimulus might cause more wandering in search for a safe place which suggests a temporary loss of environmental awareness and short term memory. In both tests, crayfish that were not shocked did not spend any time in close proximity to the object for the first 15 minutes. After familiar object presentation there were no significant differences in positions towards the object, nor in positional changes.

In the generalized linear mixed model for the odds ratio of approaching the object or not, we found the interaction between shock and time (*P*=0.015) and the interaction between time and object (*P*=0.005) to be significant suggesting a different evolution over time between shocked and non-shocked and familiar versus novel object. Shocked crayfish showed a greater chance to spend time at close proximity of an object than non-shocked crayfish. This further suggests a loss of consciousness due to a noxious stimulus. Crayfish of the novel object test had a greater probability to spend time in close proximity than the crayfish of the familiar object test, something we cannot explain at this time. The results suggest that neophobia is lowered when crayfish receive a shock. The nociception may have taken up the majority of their attention, and they either ignored the novel object or could not divert attention to a fear response.

We also observed behaviours which suggest the occurrence of neophobia. Some crayfish approached objects, looked at them and then backed away or tail flicked away. Other crayfish moved around the novel objects in ways similar to how they avoid light beams (Yamane and Takahata, 2002) or left their shelter in the opposite direction of the objects. As these behaviours are indicative of aversion and have not been observed in shocked crayfish, they provide further evidence for the presence of neophobia and the reducing effect of nociceptive stimuli on neophobia. However, it was not confirmed that due to the neophobia the non-shocked crayfish would react differently towards a novel versus familiar object. This lack of difference in reaction may have to do with their general low activity during the day. Crayfish are most active during the night. Due to their nocturnal nature the present protocol may be inefficient to test the novel-object paradigm in non-threatened non-shocked crayfish under daylight conditions.

No significant differences in activity (Mann-Whitney U=4.5, *P*=0.058), inactivity (Mann-Whitney U=5, *P*=0.061) and sheltering (Mann-Whitney U=4.5, *P*=0.054) between shocked and non-shocked crayfish were found in the novel object test. There were also no significant differences among the two groups in activity, inactivity and sheltering in the familiar object test. In the generalized linear mixed model for activity (walking or not) we found the interaction of shock and time (*P*<0.001) and the interaction of time and object (*P*=0.007) to be significant. Shocked crayfish had a higher chance to show activity than non-shocked crayfish and novel object crayfish had a higher chance to show activity than familiar object crayfish but this difference among the groups diminishes over time. This suggests that a noxious stimulus might result in an increase of activity over a restricted time.

For the novel object the generalized linear mixed model for activity showed an interaction effect between shock, time and gender where shocked females had a significant higher probability to be active than shocked males (*P*=0.047). Non-shocked females had lower probability to be active than males. This suggests that females could be more sensitive to nociceptive stimuli than males what may be related to their higher vulnerability. Previous research found similar results (Appel and Elwood, 2009).

There was only a significant difference in grooming in the first 15 minutes between shocked and non-shocked crayfish (Mann-Whitney U=0, *P*=0.002) in the familiar object test with more grooming in the shocked group. For grooming there was only a significant difference (*P*<0.001) in the interaction between time and object in which the familiar object test crayfish always had a greater probability to groom than novel object test crayfish. Different studies have had different results on grooming (Appel and Elwood, 2009; Puri and Faulkes, 2010).

Although higher waving frequencies were observed in shocked crayfish, these difference were not significant. Non-shock crayfish however did not perform waving in the first hours. We also probably undercounted the waving frequencies as we only counted when one of the walking legs was waving. We noticed that some crayfish waved a lot with only their second maxilla. In a generalized linear mixed model with Poisson distribution for the number of waves we found that the interaction of time, type of object and shock to be very significant for waving (*P*=0.005). Shocked crayfish showed a far greater chance of waving than non-shocked crayfish in both experiments. This could be caused by increased stress which results in a higher need of oxygen. The function of this behaviour merits further exploration.

We also observed several odd behaviours. Some shocked crayfish in the novel object test exhibited wall climbing. Crayfish perform such behaviours to escape unfavourable conditions (Morrison and Calligan, 1998). One crayfish was observed to perform a meral spread (an elevation of the anterior portion of the body while spreading the major chelae) towards a novel object possibly a defensive behaviour towards a threat. The fact that wall climbing occurs in the presence of shelters further suggests a reduced consciousness.

One crayfish showed a stereotyped manner of walking in circles touching the walls or the object repeatedly at the same place. Some crayfish were observed bumping their heads into objects or making forward movements without forward motion. Others seemed to be unable to enter their shelters because they could not locate the entrance. These behaviours further suggest a loss of environmental awareness and orientation due to a noxious stimulus.

Conclusions and ethical significance

Throughout the different experiments we have come across results that make a strong case for the occurrence of nociception in crayfish. Nociception may capture the animal's attention thus preventing it to correctly focus and respond to its innate fear or distrust for the novel object. Seeing the unshocked animals maintaining a safe (intermediate) distance from newly placed objects while shocked animals seem unaware of the difference in their environment, is consistent with the paradigm of the novel-object-test. As expected, the attention towards the novel object was reduced in shocked crayfish. The results show that electrical shocks may influence the crayfish's attention. Further research using other than electric shocks as potentially painful stimuli will allow to exclude the mechanistic hypothesis that the brains' electrical signalling in the shocked crayfish was hampered by the electric shocks. In comparison to the study on nociception in hermit crabs by Appel and Elwood (2009), the intensity of the shock was lower and the duration was shorter, but the shock was applied in direct body contact. The lack of difference in reaction to novel and familiar objects by unshocked animals remains unexplained and may be due to the crayfish's memory capacity or an interference such as the loss of chemosensory information from the familiar objects. These preliminary results should be confirmed to examine the inclusion of crayfish and related decapods in the animal welfare debate. Positive results would entail a significant extension of the scope of the animal welfare and ethics debate. Furthermore, there would be an incentive to re-evaluate a significant part of the relevant legislation, such as that on animal experimentation, not to mention the techniques that are allowed when preparing crayfish and lobster in the kitchen. This study underlines the need for more research into the welfare of these animals.

References

Appel, M. and Elwood, R. (2009). Gender differences, responsiveness and memory of a potentially painful event in hermit crabs. Animal Behaviour 78: 1373-1379.

Braithwaite, V (2010) Do fish feel pain? Oxford University Press, Oxford, UK.

Broom, D. (2001). The evolution of pain. Vlaams Diergeneeskundig Tijdschrift 70: 17-21.

Mather, J. (2001). Animal suffering: an invertebrate perspective. Journal of Applied Animal Welfare Science 4(2): 151-156.

Morrison, S. and Callaghan, J. (1998). The emersion response of the Australian Yabby Cherax destructor to environmental hypoxia and the respiratory and metabolic responses to consequent air-breathing. Journal of Comparative Physiology B 168(5): 389-398.

Puri, S. and Faulkes, Z. (2010). Do decapod crustaceans have nociceptors for extreme pH? PLoS ONE 5(4). DOI: http://dx.doi.org/10.1371/journal.pone.0010244.

Rose, J.D., Arlinghaus, R., Cooke, S.J., Diggles, B.K., Sawynok, W., Stevens, E.D. and Wynne, C.D.L. (2012). Can fish really feel pain? Fish and Fisheries 15(1): 97-133.

Sneddon, L., Braithwaite, V. and Gentle, M. (2003). Novel object test: examining nociception and fear in the rainbow trout. The Journal of Pain 4(8): 431-440.

Yamane, S. and Takahata, M. (2002). Experimental modification of stepping course in spontaneously initiated locomotor behaviour in the crayfish *Procambarus clarkii* Girard. Journal of Comparative Physiology A 188(1): 13-23.

Genetic commons and codified commodities: exploring the role of intellectual property rights on genetic resources in Norwegian biotechnology innovation systems

E. Stien Grimsrud[1], *B. Aarset*[1], *A.I. Myhr*[2], *V. Skagemo*[3] *and I. Olesen*[3*]
[1]*NMBU School of Economics and Business, Norwegian University of Life Sciences, P.O. Box 5003, 1432 Ås, Norway;* [2]*Genøk Center of Biosafety, Forskningsparken i Breivika, P.O. Box 6418, 9294 Tromsø, Norway;* [3]*Nofima, P.O. Box 5010, 1431 Ås, Norway; ingrid.olesen@nofima.no*

Abstract

Exploitation of the extensive Norwegian marine resources has entered the era of modern biotechnology. The sources of knowledge of both bioprospecting and aquaculture are often genetic resources, considered to be common or public goods. Here, a selection of views on intellectual property rights and access regulation in certain areas of aquaculture and marine bioprospecting will be presented. With the aim to tailor intellectual property rights systems for genetic resources, we will argue that their specific biological features and implications as well as ability to promote good cooperation between industry and research institutions should be considered.

Keywords: aquaculture, marine bioprospecting, access regulation, private public cooperation

Introduction

Publicly funded biotechnological research and development programs have been key contributors to the scientific advances in the sectors of aquaculture and marine bioprospecting. In national and regional innovation systems, the need for industry-specific knowledge protection strategy regimes has not yet received much attention (Olesen *et al.*, 2007; Rosendal *et al.*, 2006, Tvedt, 2010, Tvedt, 2011). Innovation theory emphasizes the proprietary rights to knowledge production as one of the main drivers of commercial innovations in high technological sectors (Greenhalgh and Rogers, 2010; Lai, 1998).

In this paper, we present perspectives on how different knowledge protection mechanisms (intellectual property rights (IPR)) and access regulation in aquaculture and marine bioprospecting may affect innovation practices in the two sectors. The two innovation systems, aquaculture and marine bioprospecting, are examined by using survey response analysis and interviews with key actors, respectively. Our study of access regulation mechanisms in the biotechnical sector is conducted as a mixed methods study. 'Expansion' is one mixed methods approach useful for gaining a broader understanding of the topic at hand, across a large range of stakeholders (Alexander *et al.*, 2008). This research project explores several dimensions of intellectual property that can be observed in these systems and discusses their effect on performance of the innovation system.

Material and methods

Genetics of aquaculture

Fifty-five project leaders of aquaculture projects funded by the Fuge technology platform or the Havbruk research program received a web-based survey. The invited project leaders represented universities, governmental funded public research institutions, private research institutions and researchers in the industry. The aims for the research projects were to develop a more efficient and sustainable aquaculture industry and identifying biotechnological solutions for streamlining industry processes and meet the challenges of the aquaculture production. Design of the survey, choice of respondents, and collection of responses was conducted as a part of the project 'Stimulating sustainable innovation in aquaculture' (Project number 187970/S10 funded by the Research Council of Norway). Responses were obtained from 25 of 55 project leaders managing 34 projects with diverse funding, mostly public. Both basic research and projects with commercial objectives were represented here, and the respondents represented both public and private institutions and industrial companies.

Marine bioprospecting

Like other evolving technological industries, biotechnology is highly innovative. Intellectual property is seen by some as vital in knowledge creation and in order to be ahead of competitors (Greenhalgh and Rogers, 2010). Large innovation platforms in biotechnology have been formed and more are coming, such as the Salmon Genome, Aqua Genome and the Cod Genome Projects, in addition to Marbank, Mareano, MabCent and various marine bioprospecting platforms. This amounts to a voluminous bank of information with limited security.

The research questions we pursue here are:

- Which factors are dominating in IPR management strategies in Norwegian marine biotechnological innovation systems?
- What are their perceived effects on the performance of the systems?
- How are current IPR regimes in the systems perceived to attend to the public good of the marine commons and publicly funded knowledge production?
- Which concerns should be addressed by a future industry-specific IPR regime in the innovation systems?

Building on the findings of the questionnaire for aquaculture research (described above), in-depth interviews were carried out with 6 actors from a marine bioprospecting innovation system. Their input presents different views, experiences and opinions about IPR.

The case, Norwegian marine based biotechnology innovation consortium, was chosen on the basis of several criteria: Norwegian, biotechnological innovation with marine genetic resources, emphasis of IPR and composition of representatives (university researchers, TTO, knowledge bank, analysis platform and industry partners). The informants were mainly key actors within public institutions (including academia), but also industrial representatives were included.

To extend the perspective on the aquaculture industry from the results of the initial survey data, this case was chosen to compare and contrast the institutional similarities and differences between the industries, and how the innovation systems have adapted to them. This approach refers to the 'relevant case' (Yin, 1994).

Results

Genetics of aquaculture

Results indicate that within aquaculture research, patenting, publishing and contracts are preferred approaches to protect research results. A major part of the research results was not subjected to patenting, either because the results were not considered patentable, or because results were considered as part of the public good, or avoided due to the too demanding process of patenting.

Of the 25 respondents 13 answered that they had plans for securing IPR of their findings when they applied for funding. Product and Gene/Marker/Organism were the most common findings to secure. When the project was finished, 10 respondents answered that there was a need for protecting their findings through some type of IPR system.

The respondents were asked to rank from 1st to 4th choice of five different strategies regarding the legal arrangements in addition to 'I would rather publish' and 'I would rather keep it secret'. Here, patents got the highest rating among legal arrangements of protecting results. Publishing (which does of course not constitute legal protection, but is efficient for keeping the material in the public domain as 'prior art', and hence available for other researchers) came second, followed by contracts; however, the difference between them was small.

Fifty-four % of the respondents answered that they had experiences with IPR practice in earlier projects. When the respondents with experience were asked to describe the type of IPR used and whether the experience were positive or negative, most of them had experiences from patent applications. Their experience varied widely, and some researchers said that they had both positive and negative experiences with patent application. One said that he/she had experience with patent application and with patents from competitors stopping them from development of new products. The same respondent commented that their company was willing to spend money on research when they knew that IPR would be secured, and it was not so interesting to spend money if a cooperating research institute/university refused to license the findings, or wanted to publish immediately. Another researcher answered that the experience was generally negative, and that the aquaculture industry has been too small or immature to handle the patent issue.

Eleven respondents had experience from encountering others' patents that had affected their research. Some of these respondents mentioned that competing firms with a very broad patent can hinder development of for example new vaccines. Some of the respondents commented that some patents hamper research (i.e. if the patent blocks the development of a certain product there is no point in doing research), but on the other side, patents will stimulate to do research by the patent holder. Of the respondents 15 answered that their institution/enterprise holds an IPR, two answered that their institution/enterprise does not hold an IPR, and seven did not know. Those holding an IPR, had IPR on a product, process, method and/or gene/marker/organism. Of the IPRs 10 were within the field of animal health/disease, 6 within technology, 4 within nutrition, 3 within pharmacy, remaining within genetics, other fields or didn't know. The strategy or chosen agreement they have made regarding ownership, control and protection was contracts (3) and patents (1). Most respondents (21) answered that they did not have any plans for applying for a patent or that they had not applied for a patent, whereas only four project leaders answered that they had plans for applying for a patent or that they had applied. The four respondents were also asked to describe their reason for applying for a patent in terms of sustainability including both economic, environmental and social reasons. The economic reasons for patenting were 'protection of the invention', 'increased or more stable income, profit or funding for future research'. Further, one respondent agreed that a patent application would look good at scientific evaluations and one answered that they applied because of reduction of the losses due to non-authorised use of products. Of the environmental and social reasons only one respondent agreed on the incentives to solve environmental problems and two agreed on incentives to solve animal health/welfare problems.

Marine bioprospecting

As the industry is at an experimental stage, most of the funding is public. The informants considered the system as both dynamic and complex in terms of financing, interests, relations and institutions. One informant (from academia?) expressed that certain regional industry actors were prioritized when novel information was transferred. This implies that a limited number of industry actors directly involved in the consortia caused discoveries from research to be channelled directly to a specified actor. Another informant explained that industrial actors were actively promoting topics of research they needed from academia, which further emphasized a multilateral dynamic network.

Technological infrastructure, legal enforcement- and management of legal protection were expressed to be limiting factors to successful development. National actors are small and highly specialized in developing a discovery, whereas end markets are big international pharmaceutical corporations capable of carrying the complete process of innovation diffusion and enforcing legal protection of the intellectual property (IP). The generated knowledge is thus considered among many of the informants to yield only temporary innovative success at a regional and national level. A consensus of patents being the only fully applicable legal protection for commercialization was observed in the innovation system.

Most respondents expressed positive experiences of protecting inventions, particularly through patents. Similar to aquaculture, some respondents indicated the application procedure and the legal arrangement to be laborious and time demanding. Prior preparation and positioning of research as well as competence of the patenting procedure and legislation was considered important for filing efficiency.

Technological considerations to IP-strategy

Led compounds such as chemical molecules, enzymes and biological applications were considered most suitable to patent protection. Two informants preferred the application of complete organisms (low taxonomy class) for patenting in order to obtain sufficient protection. When searching for a licensee for commercialization, two informants explained commercial actors' strategy of shopping for wide-scoped patents that has better chances of innovation success, due to freedom to operate within the scope. It is therefore aimed at wide patents when deciding on research and protection strategy of findings (IPR).

Commercial development, testing and diffusion of products are time demanding in the end market. Patent rights are limited to 20 years. In the space of these years, the right-holder must develop product(s) and gain returns on R&D investments by the monopoly situation given by the patent. When development is time consuming, the industry actor will seek to do as much as possible before the patent application is filed. Several informants were concerned with this component of patent law.

Researchers indicated that commercialization policy of the innovation system delayed publication of publicly funded research results, even basal research results that could be considered significant to the discovery that could lead to an invention. One was concerned by this failure of knowledge production, but loyalty to the mission of the consortium – to keep the knowledge in regional industries – was considered more important. Commercial actors and the facilitator addressed the opposite concern; researchers' desire (and right) to publish results is a significant obstacle to commercialization.

Administrating actors said the consortia system was modelled on open innovation and the triple helix paradigm, as IPR was licensed to industrial actors in many ways. None of the informants was aware of inter-licensing of patents between industry actors. Only MTAs and the standard confidentiality agreement of the consortium were used to manage IP in the consortium. One informant claimed, 'Joint Ventures are not very applicable to biotechnology, as the research is specific and compound-oriented, and will not benefit two parties with different aims'.

Discussion

The low number of respondents in the survey and the specific case studied here limits the possibilities to draw firm conclusions and generalise. However, interesting perceptions from respondents can be discussed. From the case study, perceptions on innovation systems' performance indicators varied greatly. Some perceived the system as a mediator of knowledge resource management, while others perceived the platform as a facilitator of commercialisation. However, all informants seemed loyal to the commercial strategy of research within the consortium. It seems as if the complexity of institutional forces causes fragmentation of actors' perceptions of the mission and context of their innovation system. Hence, when discussing IPR, this system appears to function inefficiently as knowledge (IP) disclosure is delayed by legal arrangements and strategies of commercial progression within the patent period. Dislocation of knowledge is observed as linear relationships from academia to one specific industry actor, where failure of innovation rejects the invention as commercially viable. In the aquaculture survey, more respondents expressed their goal of research as altruistic in the sense of contributing knowledge to the public domain.

In the case study, all actors indicated that the actual innovation process in Norway was very short, only from discovery to pre-market testing and the realistic aim of innovation was sale of company or IPR to international industry actors with power to create a product and diffuse the innovation to a market. This response conflicts with the communication of the innovation system, which aimed to support a regional innovation cluster by dynamic technology transfer from publicly funded academia. In the aquaculture sector, most actors are either suppliers of products and services to the aquaculture industry or they use the IPR in their integrated production to supply seafood. Empirical evidence from both areas portrays many pitfalls of patent law in the innovation system. The main concern is patents hampering of research results publishing, and thus contributions to further research. Time between IPR agreements and patent filing prevents publication and further relevant research. Industry actors both within and outside the innovation system can evaluate the invention and suggest several plans for commercialisation before the duration of the grace period. This flips the constraint found in the case study that suggests that secrecy is vital before a commercial strategy is planned and legally arranged. Perhaps the limited time of grace period can even promote IPR handling efficiency.

No informants in the bioprospecting case study saw any alternative to patent protection. Two informants expressed that the innovation platform was based on open innovation, but there was little evidence of exchange of IP. Joint ventures were also expressed to be impossible in the innovation system, as the actors are technically specialised. None of the actors could recall any other type of licencing than exclusive-, and sole licencing. In the aquaculture survey, strict regulations of IPR was responded by some to hamper further research as others IP were experienced to block development of inventions, whereas IP-positioning in the case study of bioprospecting was considered an advantage to narrow the aim of research and make the innovation process more efficient. An ethical issue for discussion, but outside the scope of this paper, may be related to National public funds for research allocated for regional development to private actors with short-term aims of selling patents to large external MNC.

Two important bioinformatics knowledge banks are currently being formed in the two studied industries: (1) Salmon Genome Project, a complete genetic map of the salmon genome for aquaculture breeding and industry; and (2) Marbank, a national biobank of samples of a wide range of marine genetic organisms for utilization in marine biotechnological industries. When they are fully operational, these systems can be viewed symbolically as seas of (other place limited) available resources (knowledge) where the actors are 'fishing' for potential IPR. Protection by wide scoped patents increases total harvests per IP, but the by-catch, which can be of importance for others, may also be significant. As patent law is not tailored for biotechnology, this practice should be evaluated when designing the banks of bioinformatics.

Biological science differs from software development and source code. In software, copyright dominates, and the start-up cost for developing software is relatively low compared to biotechnology. An open source framework for inventions from marine biotechnology needs to create similar incentives as those created by patents, taking innovative conditions for the sector particularly into account. Joly (2007) reviews several open source-based biotechnological projects and the aspects arising of this application. His conclusions suggest a general negative view of open source applicability on innovation on genetic resources, but open source can complement and sometimes work best as an independent alternative to the innovation paths of proprietary rights incentive. The informants of the case study did not see any commercial alternative to patents as IP-protection in their field of biotechnology innovation. Joint ventures of innovation were also considered inappropriate due to the fragmented technological strategy of the actors.

Conclusion

It is concluded that when tailoring IPR systems for genetic resources, their specific biological features and implications as well as ability to promote good cooperation between industry and research institutions should be considered.

In conclusion, future access to intellectual property will be important to develop products and production technology because the aquaculture sector, as well as biotechnology in general, faces changes and challenges relating to social, economic and environmental issues.

Acknowledgements

This research was funded by the Research Council of Norway (Projects no. 187970/S10 and 208543/010). Valuable comments and suggestions from Dr. Kristin Rosendal are appreciated.

References

Alexander V.D., Thomas H., Cronin A., Fielding J., Moran-Ellis J. (2008). Mixed methods. In: Gilbert N. (ed.) Researching social life. Sage, London, UK, pp. 125-144.

Greenhalgh, C. and Rogers, M. (2010). Innovation, intellectual property, and economic growth. Princeton University Press, New Jersey, USA, 384 pp.

Joly, Y. (2007). Open source approaches in biotechnology: utopia revisited. Maine Law Review 59: 386-431.

Lai, E.L.C. (1998). International intellectual property rights protection and the rate of product innovation. Journal of Development Economics 55: 133-153.

Olesen, I., Rosendal, K., Tvedt, M.W., Bryde, M. and Bentsen, H.B. (2007). Access to and protection of aquaculture genetic resources: structures and strategies in Norwegian aquaculture. Aquaculture, 272: S47-S61.

Rosendal, K., Olesen, I., Bentsen, H.B., Tvedt, M.W. and Bryde, M. (2006). Access to and legal protection of aquaculture genetic resources: Norwegian perspectives. Journal of World Intellectual Property 9: 392-412.

Tvedt, M.W. (2010). Norsk genressursrett: Rettslige betingelser for innovasjon innenfor bio- og genteknologi, Cappelen akademisk forlag, Oslo, Norway.

Tvedt, M.W. (2011). Patent law and bioprospecting in Antarctica. Polar Record 47: 46-55.

Yin, R.K. (1994). Case study research: design and methods. Sage, London, UK.

Reconciling Haida ethics with Pacific herring management

M.E. Lam
University of British Columbia, Fisheries Centre, 2202 Main Mall, Vancouver, BC V6T 1Z4, Canada; University of New Mexico, Department of Biology, 167 Castetter Hall, MSC03 2020, 1, Albuquerque, NM 87131-0001, USA; mimibethlam@gmail.com

Abstract

Pacific herring have been a vital source of food, income, livelihood, and culture for Pacific Northwest indigenous communities for millennia. Today, the ecological and cultural value of herring, as forage fish for predatory marine fish, birds, and mammals within the marine ecosystem and traditional food for indigenous communities, are in direct conflict with the economic value of lucrative commercial fisheries for herring roe, exported primarily to Japanese markets, leading to a highly contested herring management regime. We examine Haida First Nation ethics and values in relation to existing management principles and practices for the Haida herring spawn-on-kelp (k'aaw) fishery in relation to the commercial herring roe and food and bait fisheries. We are developing an innovative practical ethics framework for analysis with Haida Gwaii community members to facilitate a collaborative Pacific herring governance solution. The practical ethics approach combines the ethical matrix, a conceptual tool that analyses the wellbeing, autonomy, and justice of different human and environmental interest groups, and rapfish, a semi-quantitative, rapid appraisal technique used to evaluate the sustainability of fisheries, adapted to the local Haida Gwaii herring fisheries context, with input from diverse stakeholders. By making values transparent, decision-making processes in high uncertainty/high stakes situations can be more easily negotiated among stakeholders to achieve just and sustainable outcomes for those affected.

Keywords: herring roe, ethical matrix-rapfish approach, indigenous culture, practical ethics, spawn on kelp (k'aaw), sustainability

Introduction

Pacific herring (*Clupea harengus pallasi*) have supported a Haida k'aaw (spawn-on-kelp) fishery for food and trade for millennia (Lepofsky and Caldwell, 2013). European colonization in the late 1800s and the ensuing industrialization of a commercial herring reduction fishery (1935-1967), however, severely depleted local herring populations (Jones, 2000). This was followed by a more carefully managed roe herring fishery beginning in 1972, but stocks still collapsed in the 1980s (Jones, 2000). Closed in major fishing areas since 2005 and 2008, Haida Gwaii and Central Coast commercial herring fisheries were re-opened in 2014 by the minister responsible for Canada's fisheries regulatory body. Local BC First Nations contested this decision, with serious conflict only narrowly averted by timely negotiations between the indigenous communities and the fishing industry. The ecological and cultural importance of herring, as forage fish to feed predatory marine fish, birds, and mammals within the marine ecosystem (Pikitch *et al.* 2012) and traditional food for indigenous communities (Jones, 2000), are in direct conflict with the economic value of commercial roe herring gillnet and seine fisheries. In partnership with the governments of other First Nations, BC, and Canada, the Haida Nation is committed to using Haida ethics (CHN, 2007), a holistic and ecosystem worldview, in the management of Pacific herring (Jones, 2000; Jones and Williams-Davidson, 2000) and Haida Gwaii's marine resources (Jones *et al.* 2010).

An evaluation of the sustainability and ethical status and regulation of the Pacific herring fisheries in Haida Gwaii is currently being conducted through ecosystem modelling of herring management scenarios and participatory community workshops. Our preliminary ecosystem modelling results indicate that DFO thresholds to open herring fisheries may be too low to support herring recovery and its provisionary role in marine ecosystems. Results of pilot research with Haida Gwaii residents, including Haida and commercial fishermen, members of the Haida Oceans Technical Team, and conservationists, suggest that they are reluctant to support the opening of a herring fishery until the entire food web can thrive. Partnering with the Council of Haida Nation, we are now exploring the integration of Haida values and ethics with ecosystem-based management and governance principles to facilitate collaborative solutions for Haida Gwaii herring fisheries. This work is in progress, so only the framework for analysis will be sketched out here; more detailed results will be given in the presentation.

Haida ethics and values

Haida Gwaii ('Islands of the People'), known also as the Queen Charlotte Islands, in Northern British Columbia (BC), Canada (Figure 1), are home to the Haida, indigenous people who have occupied the territory for millennia; today, in addition to a vibrant indigenous culture, the islands support both Haida and non-Haida livelihoods in productive fisheries and marine ecotourism. Haida Gwaii boasts a highly productive and diverse marine ecosystem, comprised of intertidal zones, kelp forests, shallow banks, and a continental shelf break (CHN, 2007). The marine productivity is due largely to oceanographic features, such as the Haida Eddies, created as the westerly North Pacific Current splits into northerly and southerly branches off the west coast of the islands, and the Hecate Strait Front, generated as opposing tidal flows from Dixon Entrance and Queen Charlotte Sound meet. The ecological diversity results mainly from its diverse bathymetric features (e.g. the shelf break, banks, reefs, estuaries, and fjords), produced by a combination of past tectonic and glacial action. Despite its land isolation, this ecosystem has long been affected by extractive human activities, such as the maritime fur trade and commercial whaling, which have depleted many ecologically important species, including sea otters and large whales. However, its waters remain home to many commercially important and charismatic species, supporting aboriginal, commercial, and recreational fisheries (Pitcher *et al.*, 2013a) and a growing marine ecotourism industry.

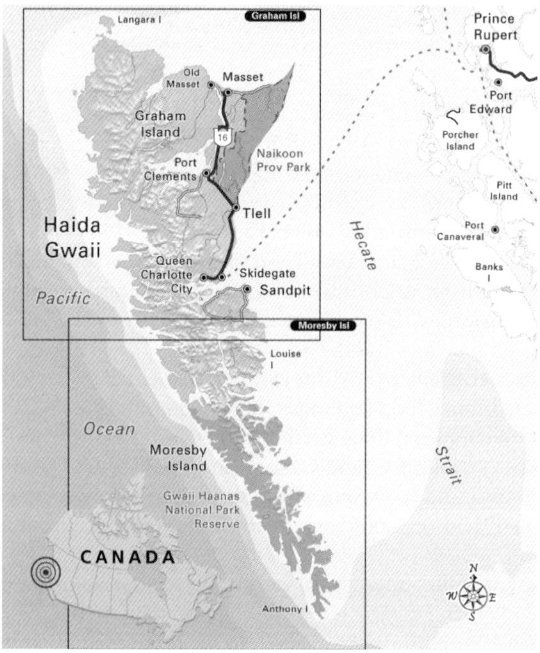

Figure 1. Haida Gwaii (Queen Charlotte Islands), British Columbia, Canada (http://tinyurl.com/kojgrcw).

The Haida have been marine resource owners and managers on Haida Gwaii for millennia, and continue to apply, as active partners with other indigenous, provincial, and federal governments, Haida traditional ecological knowledge and experience to marine-use planning (Pitcher *et al.*, 2013a; CHN, BC and MaPP, 2014) and integrated marine resource use and management on Haida Gwaii (Jones *et al.*, 2010). The Haida approach is place-based and guided by fundamental Haida ethics and values, such as respect, balance, and reciprocity (Table 1, Jones *et al.*, 2010). In the Haida worldview, all animate and inanimate beings are believed to have a spirit that must be treated with respect. Respect is taught for oneself and others, including one's clan, village, and nation, as well as one's environment (as in deep ecology). Fish and the food they supply are endowed with spirits that must be respected, which translates into a holistic ethical approach to natural resource management (Jones and Williams-Davidson, 2000). Offerings are made to honor the spirits of beings that are killed for food. To avoid offending the spirit of the food, food must not be refused when offered, wasted, or played with; thus, catch-and-release 'conservation measures' in recreational fisheries is considered disrespectful 'playing with one's food' in the Haida worldview. Another Haida guiding principle, balance, encompasses values, virtues, and activities. Violating this balance can lead to bad luck or loss, as in the eulachon run that was lost as a result of improper care (Jones and Williams-Davidson, 2000). The principle of reciprocity (giving and receiving), the sharing of wealth and equitable redistribution of limited resources, is expressed in decisions determining access

Table 1. Haida ethics and corresponding ecosystem-based management principles (Jones et al., 2010).

Haida ethical principle			Ecosystem-based management principle
Haida	English	Description	
1. *Yahguudang or Yakguudang*	Respect	Respect, for each other and all living things, is rooted in our culture. We take only what we need, we give thanks, and we acknowledge those who behave accordingly.	Precautionary approach
2. *Giid tll'juus*	'The world is as sharp as the edge of a knife'	Balance is needed in our interactions with the natural world. If we aren't careful in everything we do, we can easily reach a point of no return. Our practices and those of others must be sustainable.	Sustainable use
3. *Gina waadluxan gud ad kwaagiida*	Interconnectedness ('Everything depends on everything else')	This principle is comparable to an integrated approach to management.	Integrated management
4. *Isda ad diigii isda*	Reciprocity (giving and receiving)	Reciprocity (giving and receiving) is a respected practice in our culture, essential in our interactions with each other and the natural world. We continually give thanks to the natural world for the gifts that we receive.	Equitable sharing
5. *Gina k'aadang.nga gii uu tl' k'anguudang*	Seeking wise counsel	Our elders teach us about traditional ways and how to work in harmony. Like the forest, the roots of our people are intertwined. Together we consider new ideas and information in keeping with our culture, values, and laws.	Adaptive management/ best information
6. *'Laa guu ga kanhllns*	Responsibility	We accept the responsibility passed on (to us) by our ancestors to manage and care for the sea and land. We will ensure that our heritage is passed on to future generations.	Inclusive and participatory

to local fisheries and in the continuing practice of the potlatch to mark important life events, where the giving of gifts gives prestige and status to the giver and requires obligations of the recipient (Jones and Williams-Davidson, 2000; Jones *et al.*, 2010). The complementarity of Haida ethics and values and corresponding ecosystem-based management principles (Jones *et al.*, 2010) are being investigated to facilitate collaborative governance solutions (cf. Lam, 2014 for analysis of complementarity of traditional ecological knowledge and ecology).

British Columbia herring and Haida herring traditional k'aaw (spawn-on-kelp) fisheries

In 2013, BC commercial herring fisheries (spawn-on-kelp, roe, and food and bait) had a combined herring harvest of 22,100 tonnes, distributed as in Table 2 (BC, 2013). This 68% increase from the previous year reflected heightened market demand, particularly as a niche market export to Japan, as well as to China and the USA (BC, 2013). Herring roe products, including herring roe, herring meal and oil, generated the bulk of the total herring wholesale value of Canadian $ 42.3 million. Herring exports totalled Canadian $ 37.5 million in 2013 or 4% of BC's seafood exports value (BC, 2013). These harvests are much reduced from traditional and historical values (Jones, 2000), which has had severe implications on the Haida and other First Nations.

In April 2014, the traditional Haida community of Skidegate attended the first ceremonial feast of k'aaw, herring eggs spawned on kelp, in years (Figure 2), after Haida fishermen located herring spawn sites south of the village in Gwaii Haanas. Local herring populations in Skidegate Inlet, once the site of huge spawning populations supporting the traditional Haida herring k'aaw food fishery, were depleted by commercial overfishing in the 1950s and 1960s, from which they have never fully recovered. Herring return home to spawn and can be seriously depleted by overfishing, not included in Canadian fisheries management. Current DFO management of herring populations in Northern BC has not been well received by local communities. Coastal First Nations believe it favours a corporate commercial herring fishery and ignores herring's ecosystem value to the marine food web, including as food to recovering whales, seals, salmon, and seabirds. Pilot work with Haida Gwaii residents reveal that they are reluctant to support the opening of a herring fishery until the entire food web can thrive. To address their belief that the current level of herring recovery is insufficient, we are performing ecosystem modelling to determine acceptable herring levels that can support the ecosystem. In the 2014 commercial herring fishing season, the Minister of Fisheries opened the Haida Gwaii and Central Coast herring fisheries, against the advice of DFO's own senior scientists. Potentially serious conflict between industrial fishermen and Haida and Heiltsuk First Nations was only narrowly averted by negotiations between the First Nations and the industry (Secher, 2014).

Table 2. BC commercial herring fisheries production for 2013 (BC, 2013).

Fishery	Herring harvest		Landed value		Wholesale value	
	'000 tonnes	% total	Canadian $ millions	% total	Canadian $ value	% total
Spawn-on-kelp	0.07	0.3	1.8	12.7	2.0	4.7
Roe herring	14.0	63.4	10.5	73.9	35.0	82.7
Food and bait	8.0	36.2	1.9	13.4	5.3	12.5
Total	22.1	100	14.2	100	42.3	100

Figure 2. Kyra Russ (Haida Raven clan) enjoying her first k'aaw feast in Skidegate, April 2014 (photo: M.E. Lam).

Practical ethics approach

We adopt a practical ethics approach to investigate how Haida ethics and values may be integrated with ecosystem-based management principles to facilitate collaborative Pacific herring governance solutions[2]. Documenting Haida and non-Haida community values and ethics as they relate to herring is a critical first step to reducing conflict over the Pacific herring fisheries. Through participation with Haida Gwaii communities and consultation with Canadian government and industry representatives, we aim to identify sustainable and ethical practices for collaborative herring fisheries governance. Participatory workshops with diverse stakeholders, including traditional and commercial Haida herring fishermen, Haida elders and youths, commercial fishermen, tourism operators, conservationists, fisheries managers, and scientists, are planned for April 2015 in the traditional Haida communities of Skidegate and Massett. Adapting the ethical matrix (Mepham 2000, 2008) and rapfish (Pitcher *et al.*, 2013b) into a combined methodological approach (Lam, 2013) yields a powerful theoretically-based method for empirical ethical analysis and evaluation of fisheries that may facilitate collaborative decision-making in policy and governance. We build on this combined ethical matrix-rapfish approach by embedding it within an innovative practical ethics approach in the community workshops. The sustainability and ethical analysis framework being developed employs fishery management scenarios of various thresholds to open the herring fishery that are being explored with ecosystem modelling; the current DFO threshold is 20%, although forage fish science (Pikitch *et al.*, 2012) suggests a 40% threshold. The participatory process will include three steps with value elucidation and ranking exercises of these fishery management options: (1) identify values through narrative, both Haida (displayed in Table 3) and ecosystem-based; (2) rank the values; (3) select the fishery management scenarios that instantiate those values. Ways to reconcile, if needed, the preferred fishery management scenario with the current DFO herring fishery management regime will also be explored.

Conclusions

Indigenous communities, such as the Haida First Nation in BC, are actively collaborating with government, industry, and NGO representatives to steer modern natural resource management towards a more precautionary, just ecosystem-based approach consistent with indigenous ethics and values. By

[2] Funded by University of British Columbia Peter Wall Institute for Advanced Studies Solutions Initiative Grant

Table 3. Preliminary sustainability and ethical analysis framework.

Haida value (embedded in narrative)	Fishery management scenario				
	Unrestricted fishing	Open fishing when current biomass 25% of unfished biomass	Open fishing when current biomass 50% of unfished biomass	Open fishing when current biomass 75% of unfished biomass	No fishing
Respect					
Responsibility					
Interconnectedness					
Balance					
Seeking Wise Counsel					
Reciprocity					

working with diverse stakeholders in the Pacific herring fisheries, including Haida Gwaii communities, government, and industry, to elucidate, rank, and incorporate cultural and scientific values and ethical principles in herring governance, we aim to facilitate a way out of the conflict that has mired herring management in BC for decades. This innovative practical ethics and sustainability approach towards collaborative herring governance aims to contribute to the goal – common to all parties – of a herring recovery that supports the Haida spawn-on-kelp food fishery, profitable commercial herring fisheries, herring conservation, and ecosystem health. Collaborative herring governance is expected to result in improved protection of the Haida Gwaii marine ecosystem for present and future generations and more sustainable local fish populations and supported human communities.

We believe that by making values transparent, decision-making processes in high uncertainty/high stakes situations can be more easily negotiated to achieve just and sustainable outcomes for affected communities.

References

British Columbia (BC) (2013). British Columbia seafood industry 2013 year in review. Available at: http://tinyurl.com/ofjk6ke.

Council of the Haida Nation (CHN) (2007). Towards a marine use plan for Haida Gwaii. Available at: http://tinyurl.com/nznnnnh.

Council of the Haida Nation (CHN), BC (British Columbia) and Marine Planning Partnership (MaPP) (2014). Haida Gwaii draft marine plan version 2.1. Available at: http://tinyurl.com/orjkcc4.

Jones, R. (2000). The herring fishery of Haida Gwaii: an ethical analysis. In: Coward, H. Ommer, R. and Pitcher, T. (eds.). Just fish: ethics and Canadian marine fisheries. Social and economic papers no. 23. Memorial University of Newfoundland, St. John's, Newfoundland, Canada.

Jones, R., Rigg, C. and Lee, L. (2010). Haida marine planning: first nations as a partner in marine conservation. Ecology and Society 15(1): 12.

Jones, R. and Williams-Davidson, T.L. (2000). Applying Haida ethics in today's fishery. In: Coward, H., Ommer, R. and Pitcher, T. (eds.). Just fish: ethics and Canadian marine fisheries. Social and economic papers no. 23. Memorial University of Newfoundland, St. John's, Newfoundland, Canada, pp. 100-115.

Lam, M.E. (2013). Comparing the ethics of capture fisheries and aquaculture. In: Roecklinsberg, H. and P. Sandin (eds.) The ethics of consumption: the citizen, the market, and the law. Wageningen Academic Publishers, Wageningen, the Netherlands, pp. 305-310.
Lam, M.E. (2014). Building ecoliteracy with traditional ecological knowledge: do, listen, and learn. Frontiers in Ecology and the Environment 12(4): 250-251.
Lepofsky, D. and Caldwell, M. (2013). Indigenous marine resource management on the northwest coast of North America. Ecological Processes 2(1):1-12.
Mepham, B. (2000). A framework for the ethical analysis of novel foods: the ethical matrix. Journal of Agricultural and Environmental Ethics 12: 165-176.
Mepham, B. (2008). Bioethics: an introduction for the biosciences. Second edition. Oxford University Press, Oxford, UK.
Pikitch, E.K., Rountos, K.J., Essington, T.E., Santora, C., Pauly, D., Watson, R., Sumaila, U.R., Boersma, P. D., Boyd, I.L, Conover, D.O., Cury, P., Heppell, S.S., Houde, E.D., Mangel, M., Plagányi, É., Sainsbury, K., Steneck, R.S., Geers, T.M., Gownaris, N. and Munch, S. B. (2012). The global contribution of forage fish to marine fisheries and ecosystems. Fish and Fisheries 15: 43-65. DOI: http://dx.doi.org/10.1111/faf.12004.
Pitcher, T.J., Kumar, R., Lam, M.E. and Surma, S. (2013a) Spatial modelling of the Haida Gwaii ecosystem: a pilot project contributing to Haida Gwaii marine spatial planning. Report to the Haida oceans technical team, 77 pp.
Pitcher, T.J., Lam, M.E., Ainsworth, C., Martindale, A., Nakamura, K., Perry, R.I. and Ward, T. (2013b). Improvements to the 'rapfish' rapid evaluation technique for fisheries: integrating ecological and human dimensions. Journal of Fish Biology 83(4): 865-889.
Secher, K. (2014). Without better science, BC's herring crisis could resurface. The Tyee, Vancouver, Canada. Available at: http://tinyurl.com/nw8wa58.

Section 5. Transparency and certified markets

Which forms of nudging food decisions are ethically acceptable?

M. Korthals
Emeritus Professor Applied Philosophy, Free University Amsterdam; Wageningen University, Wageningen; Noolseweg 28, 1251 GP Laren, the Netherlands; michiel.korthals@wur.nl

Abstract

Thaler and Sunstein's Nudge (TSN) (2008) introduced the concept of nudge as '... any aspect of the choice architecture that alters people's behaviour in a predictable way without forbidding any options or significantly changing their economic incentives. To count as a mere nudge, the intervention must be cheap and easy to avoid'. All contexts in which food preferences are transformed in decisions to buy food are prestructured more often than not in the interest of sellers; nudges claim to structure the context in the self-enlightened interests of buyers. The aim of this paper is to analyse the phenomenon of nudging and to assess its ethical acceptability. First, several definitions of nudging will be discussed and one will be proposed that takes into account that choice contexts in supermarkets or canteens let many people buy their food items quite routine like on the basis of easy accessible things like tasteful looking wrappings. Choice contexts are always already framed, and it is then ethically relevant that restructured choice contexts indeed express norms generally seen as desirable that respect certain conditions, like no restrictions on choice and respecting the consumer. Secondly, ways will be discussed in which non-willed choices inhibit individuals to buy the things they really want. Thirdly, a distinction is made between several types of nudges that can stimulate alternative willed choices instead of unwilled choices. Finally, an assessment is made of ethical criteria in evaluating seven aspects of nudging.

Keywords: normative and descriptive nudges, not really willed choices, choice contexts, ethical deliberation, decision making, buying food items

Introduction

Nudging can mean giving information about your choices (descriptive social norm or nudge) or that of others (injunctive norm) or changing the default choice context into a context more adapted to justifiable normative objectives like sustainable food production and healthy eating (TSN, 2008). Nudging is everywhere: the British government started a nudging program; the Dutch government has also interest in nudging consumers towards more healthy and sustainable behaviour. However, not much is known about the effectiveness of nudges in changing purchasing attitude of consumers, such as in groceries. In the meantime, the ethical criticisms increase in number and content. You hear terms as 'Nanny state' and 'paternalism'. Scientific and ethical input in the debate is therefore necessary. In this paper I will focus on the main ethical problems and give some answers on the basis of ethical theories and ethical considerations. I will show that nudging in general is inevitable and some in cases even ethical desirable.

The main part of this paper circles around seven ethical issues: why should one nudge; who nudges whom; when; where; about what are the nudges; how does the nudge work? First, a definition of types of nudging will be proposed, which is in so far necessary, that in the discussion until now several definitions are circulating, often as a kind of straw puppet to make a point about the ethical unacceptability of the nanny state or the right to choose for your own unhealthiness. Moreover, before the main part, a discussion is necessary of four ethical considerations and assumptions about making a choice in social contexts, which mostly is not a totally isolated and independent activity. We conclude with a list of criteria of ethically acceptable nudges based on these considerations.

What is nudging?

TSN give several definition of a nudge, but some aspects often return: nudges do not allow for coercion, they are in the nudgees' interest (and improve their lives), options are available and not eliminated, and they can be useful in situations of choice context, where people are confronted by temptations and complexity with many options to choose from. They also add that those contexts should not happen too frequently and should give sufficient feedback (TSN, 2008: 247). The additions are surprising, because the authors use a lot of examples in daily life, like the cafeteria example, that happen very frequently; also feedback is not always necessarily the case; for example, it could be that a person wants to buy fair trade stuff, but does not always wants to bother about it, so he wants to make it a form of routine, and a nudge can help in this case. Anyhow, the following is a good circumscription of a nudge:

> A nudge ... is any aspect of the choice architecture that alters people's behaviour in a predictable way without forbidding any options or significantly changing their economic incentives. To count as a mere nudge, the intervention must be cheap and easy to avoid. (TSN, 2008: 6).

This choice context is always already structured in daily life in one way or another in someone's particular interest (Selinger, 2010); however, nudges are those stimulating contexts that are in one's own interest. It is necessary to keep in mind that nudges are seen as instruments that improve a choice context in the interest of the choicer (nudgee) and should not come in the place of other (policy) instruments like taxing products or state ordained laws that prescribe the context (House of Lords, 2011).

Four assumptions of nudging; philosophical considerations

Nudging assumes that making decisions and acting accordingly is not a deductive affair, placed in a contextfree space, free from others and without routines and habits. These four assumptions require some clarification.

Firstly, starting point of nudging is the assumption that acts of human beings cannot deductively be derived from their internal ethical decision making process, but are elements in a world characterized by bounded rationality. This concept is the alternative to the 'clear' and 'conscious' rationality as a distanced and objective essence of the human being that is so characteristic of the anthropology of Descartes and neo-classical economy. This concept, elaborated by H. Simon, takes into account the humans are acting in a world with a 'fuzzy logic', in which dynamics, emotions and other (important and meaningful) events play a role. Human beings do not control themselves and their context totally and have to live with weakness of will, character flaws, complex contexts and limitations of their cognition. Our rationality is contextually bounded. It is interesting to see that many ethicists accept this bounded rationality, but do evaluate it differently. For Kant, you have to fight against it, for Nietzsche, it is exactly the core characteristic of the super human. Bounded rationality implies also that humans cannot always live consciously with respect to everything in the world; you have to rely on routine, be it (ethically) consciously chosen routines. Habits as such are not to be evaluated negatively. The anthropology of on the one hand American Pragmatism and Continental Philosophy (of f.i. Gadamer) on the other hand corresponds with this notion.

Secondly, the concept of context is implicated by the concept of nudge, which implies that the living world of individuals is always already prestructured, not in a determining way, but at least in an influencing way. For conscious, rationally justifiable choice architecture this is the starting point to analyze and evaluate the default choice situation. The context of buying food stuffs is still full of mysteries, and yet one of the most important contexts that influence the choice of foodstuffs and therefore how people live. Mostly, it consists of a particular arrangement of things and signs (like assortments in a supermarket), which reflects the policy of the management to let you buy as much as profitable (for the shop) as possible.

Thirdly, when one speaks about choice architecture, freedom of choice in the sense of positive relations to others is also at stake. According to the famous definitions of Berlin (2006), there are two concepts of freedom: negative and positive. The negative covers freedom from... (mostly constraints), the positive freedom to...(opportunities, in the interest of the person). However, many people have difficulties, to say the least, to identify their interests in their openness to others. Sen (2010: 275) argues:

> Trying to bring people to see their own interest comes often rightly so to be seen as a form of paternalism; otherwise it is also the case that many people adapt to circumstances that make life less bearable for them (think about women oppression and the 'inactive acceptance of a subjugated role without complaint or dispute').

Most people, probably all, 'adapt to circumstances' that in the end make life less bearable: from overweight people buying and eating too much risking to become obese to ethically conscious people that plan to buy organic meat but succumb to a cheap offer. Sen rightly stresses the abuse of this fact, in the sense that others are cleverly exploiting this but also in the sense that others, often in a beneficent mood, assume that they know better the 'real' interests of others. The dilemma is either to refrain from assisting people trying to get rid of harmful situations or to assist them and run the risk to impose interests on people that are not really their own interests. It is impossible to remain constantly vigilant and to act continuously on the basis of your own objectives when the context is not favorable. There are contexts, that 'seduce' people to act unsustainable and to buy salty, fatty and sweetie things, which are indeed against the interest of a person. Nudging starts with these quite fundamental interests people have: to stay healthy, to assist people with their own sustainable, animal friendly, or fair trade choices. Because nudging never implies the destruction of alternatives, constraint (manipulation) is not applicable to a nudge.

Fourthly, the bounded rationality of meaning life is connected with routines, and processes of deroutinization and reroutinization. Individuals constantly develop and maintain routines and habits in interaction with social rules and institutions. The critical reflection and change of established routines may be based on a prior conscious awareness (as in: now I want to change my favorite supermarket for my groceries) but can also mean simply to connect to the habits of some others. These routines and habits are actively maintained, not only by the agent itself, but also by various 'points of support' in the area. If at some point my routine or habit runs into difficulties, for example because I do not like that one supermarket anymore, I must build new routines. There is an ongoing process which involves three phases: first a phase of routinization interspersed by a phase of deroutinization and then a phase of new decision making with respect to a particular route, feedback is important here (as Thaler and Sunstein indicate); never all routines at once can be questioned. Especially with regard to food purchases consumers take very little time to at first just orienting themselves in the supermarket before they start buying; apparently, the vast and broad food supply and increasingly convenience-orientation encourage consumer behaviour in direction of continuous routinization. Questioning a routine and new decision making reverts back to a new routine. So, routines and habits are necessary 'devices: people cannot constantly be aware of all the old and new elements of their surroundings and how they should cope with these. A fully conscious life would make anyone crazy.

Five types of not really willed choices

The assumptions together make it clear that individuals in acquiring food are a dynamic mix of conscious and non-conscious consumers. They are prone to routines that harm their interest or do not express their conscious choices. People make not always really willed choices. Following Bovens (2009), one can distinguish several types of agency that can be made subject of a nudge, because they all result in the category: not really willed choices. My hypothesis is that these types of agency of not really willed choices can be redirected into willed choices by different types of nudges. I come to five not really willed choices:

1. (Willed) ignorance. Sometimes ignorance is not intentional. For instance, many consumers think that an A grade on eggs means that they are healthy or had the approval of a rabbi (The Wallstreet Journal, 2010). This lack of knowledge or lack of expertise restricts our access to information. A lot of devices are indeed so constructed that they don't give us information about our use of them. We don't know how much water, electricity, gas, etc. we spend in opening taps or pushing buttons (here 'descriptive norms' are helpful). It is interesting to see that TSN (2008: 160) call even labeling of a product a form of nudging. However, many forms of ignorance are intentional, because it serves the interests of the particular person very well that he or she is ignorant. Interest is taken here in the broad sense of a kind of resistance to a cognitively accepted food choice that willingly rejects the information given.
2. Inertia. 'It may be the case that we do have sufficient knowledge, but somehow inertia gets the best of us. We are absorbed in our daily activities and simply put off filling out the forms until we forget. Moreover trying to give serious attention to everything leads to depletion of attention. In this case a default option kicking in for the lazy, the forgetful or the weary may be welcome. ' (Bovens, 2009: 210)
3. Perception of small effectiveness. There is some well validated research about the perception of consumers on their impact, which they mostly estimate as quite low (called perceived consumer effectiveness, PCE). This low perception regards their involvement, certainty, values, and social norms on attitudes and intentions towards sustainable food products (Cronan *et al.*, 2005). Perceived low importance is empirically tested and it is found to significantly impact both ethical judgment and behavioural intention.
4. Weakness of will (Levy, 2010). We have often good knowledge that creamy, sweet things are not good for our overall health but they are so easy to get! The easy choice is often not our own choice and our will is weak. Bovens (2009: 211): 'We are weak-willed in choosing the proper steps towards our long-term goals. By structuring the choice situation it becomes easier to correct for such weak-willed actions, because temptation will have less of a pull on us'.
5. Individual benefits, social costs. It may well be the case that a particular individual choice is not socially beneficial. Bovens (2009: 211): 'I may see no individual benefit in giving to a charity. In a Tragedy of the Commons, society may be better off if I decide to refrain from, say, adding a fishing boat to over-fished waters or drilling for oil in an oil well that is already quite depleted. But unless I am being compensated or find alternative employment, I may be worse off in the short term'.

Ethical deliberation: criticisms and responses

There are many misunderstandings and many objections against nudging. One misunderstanding is that nudging means manipulation, implying cutting of alternatives (Grüne-Yanoff, 2011: 5). However, nudges should always include a choice between alternatives and the choice context can never be neutral. One more or less similar objection argues that nudges are not about the preferences of the nudgee; but with a nudge the nudgee gets the same options, but in a different 'colour', which better fits societal acceptable ideas. A more serious concern is the Slippery Slope argument. Some (Rizzo and Douglas, 2008) draw a number of stations along a slippery slope, of which the first is a very soft intervention, for example a yes, unless clause in an organ donation contract instead of a no, but clause, and the last one due to the servile acceptance of citizens comprises forceful external control over their lives. This argument suggests an automatic process of degrading peoples' responsibility that cannot be stopped and where the future is nothing else than a linear extrapolation of the past. However, this suggestion is very doubtful. Moreover, who will object to descriptive nudges? A more serious objection is about the possibility of an abusive strategy of nudging (TSN, 2008: 238). However, there are some measures that can reduce the risk of abuse, for example transparency, no restriction of the number of choice options, and clear delineation which choice context are allowed to be nudged and which not.

Both TSN (2008: 240) and Bovens (2009) mention one feature of the human condition, the right to be wrong. Only by making mistakes people can learn to improve their ability to choose. However, how many mistakes must people make to learn? In particular with respect to food intake it seems as if you always too late and that is well supported by empirical evidence; many people that always succumb to the sweetie food stuffs in the supermarket simple are not able to resist the current choice context. Nudging gives consumers in super markets and canteens a little help to stick to their intentions by changing the choice contexts, or giving them information in how far they really did comply with their own intentions. A similar objection stresses moral character and the possible effect of the decrease of phronesis, the competence to reflect ethically on important life issues. However, by reflecting upon the choice context, and by taking into account that the human beings have to build and to maintain routines in order to make more time for being conscious or even reflexive when necessary, a fragmented self is not developed or encouraged.

Nudging is accused by some of being paternalistic, in particular in connection with the state ('nanny state'). Often, a distinction is made between three types of paternalism: hard paternalism, soft paternalism and maternalism (Haÿry, 1991). The first type comprises direct coercion and denies autonomy of choice (the law is often seen as an example), the second one involves cutting off some options for action or, softer, giving unwanted information (like: 'smoking kills'). Maternalism covers psychological control by addressing moral feelings like guilt. However, when the second and third types are used in criticizing nudging, they presuppose a fully rational agent in a totally free context, wrongly as I have shown.

Discussion and recommendations: what, why, when, how, where, who and to whom

We started initially with these seven questions; on the basis of our considerations they indeed generate an exhaustive list of the main aspects of nudging. Therefore I will discuss the ethical issues according to the seven issues.

1. What. Nudges must cover clear choices, that are frequent made, but are not too complex by controversies or high stakes of the nudgee. The descriptive social nudges (giving information about your buyings) are ethically easier than the normative social nudges (giving the socially accepted appreciation).
2. Why. People aren't formal rational subjects, but rooted in contexts, with their own depletable resources (even attention is subject to depletion). Nudging assumes that one alternative is clearly better (in the interests of the person). However, only in cases that cannot or should be regulated by law of taxation. So, governments should consider in how far tax/commercial or legal measures are more appropriate.
3. When. During the choice process in buying food stuffs in supermarkets, canteens, groceries.
4. How. A nudge does not eliminate alternatives and therefore afterwards reflection must be possible. This implies that the nudge must in principle give the possibility afterwards to ask, even when realizing the nudge: did I still like my decision? There must be a possibility to again reflect upon the differences between short term and long term interests, some space to talk about the choice architecture.
5. Where. Nudges can be used in choice contexts that are frequently and routinely visited, like groceries, canteens and larger institutional food settings.
6. Who. Nudges: grocery managers, canteen managers, managers of institutional restaurants, or (indirectly) government agencies can become nudgers, however, on the basis of accessible guidelines. To make the nudger accountable, it is recommendable that in the phase of designing the nudge

this process is documented and the document is accessible for everyone. Moreover, in the phase of implementation these documents should also in principle be accessible. This elaborates Bovens (2009: 217) idea: 'every nudge should be such that it is in principle possible for everyone who is watchful to unmask the manipulation' (but remember, a nudge should not manipulate).

7. Whom. In principle everyone, however in this paper we discussed in particular consumers or Managers, CEO of grocery stores, supermarkets. Nudges should not mean unacceptable curtailment of freedom of consumers; they should not discriminate and they should not unduly appeal to guilt feelings of nudgees.

Conclusion

Nudges must in principle be transparent for the persons to be nudged, both in its design as in its end effects, they must respect choices, not restrict choices, they must be efficient and efficacious, and the nudger must be accountable.

Some issues regarding social justice and respect for people deserve more attention. It seems plausible that some people are more resistant than others to nudges. Furthermore, to avoid stigmatization, a nudge should not cover issues where only certain groups of people are involved. Finally, nudging should not intrude the emotional household of consumers by producing unnecessary shame, guilt or fear.

Not only armchair ethics is necessary in trying to understand and evaluate the phenomenon of nudging. In another paper I will report about empirical research complementing this ethical project in which the interviewees were asked in how far nudged people do feel stimulated by particular nudges.

References

Berlin, I. (2006). Two concepts of freedom, in idem, political ideas in the Romantic age. Chatto, London, UK, pp. 155-208.

Bovens, L. (2009). The ethics of nudge. In: Grüne-Yanoff, T. and Hansson, S.O. (eds.) Preference change. Springer, Dordrecht, the Netherlands.

Cronan, T., Leonard, L. and Kreie, J. (2005). an empirical validation of perceived importance and behaviour intention in IT ethics. Journal of Business Ethics 56(3): 231-238.

Grüne-Yanoff, T. (2011). Old wine in new casks: libertarian paternalism still violates liberal principles. Social Choice and Welfare 38: 635-645.

Haÿry, H. (1991) The limits of medical paternalism. Routledge, London, UK.

House of Lords (2011) Behaviour change. The Stationery Office Limited, London, UK.

Levy, N. (2010). Weakness of will. Philosophy and Phenomenological Research 82(1): 134-155

Rizzo, M. and Whitman, D. (2008). Little brother is watching you: new paternalism on the slippery slopes. New York University of Law, New York, USA.

Selinger, E. and Powys Whyte, K. (2010). Knowledge and trust in choice architecture. Knowledge, Technology, Policy 23 (3-4): 461-482.

Sen, A. (2010). The idea of justice. Penguin, London, UK.

Thaler, R. and Sunstein, C. (2008). Nudge. Penguin, London, UK.

The Wallstreet Journal (2010). Eggs' 'grade a' stamp isn't what it seems.

Fair, local and environmentally sound? Corporate social responsibility in Norwegian food production

B.K. Myskja, S.G. Carson and L.Ø. Ursin*
Department of Philosophy and Religious Studies, Norwegian University of Science and Technology, 7491 Trondheim, Norway; siri.granum.carson@ntnu.no

Abstract

Norwegian farming is among the most subsidized and protected areas of food production in the world. In a world where issues of food safety and food security are increasingly important, the social responsibility of central actors in the food chain – producers and the main grocery chains – becomes more pressing. As a response, these actors move from implicitly assuming social responsibilities implied in laws, regulations and ethical customs, towards explicitly expressing social responsibilities as a matter of company-specific values. In this paper, we discuss the ethical values expressed in the Norwegian food discourse, and the social responsibility of central food producers and retailers. We go into the case of Tine, a farmer-owned dairy cooperative with market regulator function, and analyse the challenges of Tine in dealing with and expressing their social responsibility.

Keywords: corporate social responsibility, food ethics, Norwegian agriculture

Introduction

To compensate for the lack of arable land and the challenging topography, Norwegian farming is among the most subsidized and protected areas of food production in the world, with close collaboration between the authorities and the major food producers, owned by farmer cooperatives. This structure gives the major food producers much power and control, with a triple role as interest organization for the farmers, commercial actor and political tool. In a world where issues of food safety and food security are increasingly important, the social responsibility of central actors in the food chain becomes more pressing. In this paper, we discuss the ethical values expressed in the Norwegian food discourse, and the social responsibility of central food producers and retailers. As an example, we analyse Tine a dairy producer owned by a farmer cooperative.

Food production in one of the world's most subsidised food markets

Farming in Norway is a challenging task. Only 3% of the land is arable (Heie, 2003), and only limited parts of this are suitable for 'large-scale' industrialized farming, which in the Norwegian context would mean dairy farms with approximately 100 cows The rest consists in small-scale farms requiring high input of labour or are only suitable as grazing land for cattle or sheep. Due to climatic conditions, the season is short and there is high risk of damage to harvest. Adding to this, Norway has one of the world's highest costs of living, and next-to-full employment. Thus, it is difficult for others than the large farmers to produce an average level of income based on competition with products from countries more suitable for industrial farming. In order to compensate, Norwegian farming is among the most subsidized and protected areas of food production in the world (OECD, 2013), and the authorities generally fight all attempts at reducing the high import tariff (Gaasland, 2009). The motivation behind this regime is complex, consisting in several interacting factors. The most obvious one is future food security, even if the relative significance of Norwegian land-based food production for the national market is decreasing due to an expanding, industrialised global food industry, cheap and efficient transport, and an increasing use of food that cannot grow in cold climate. As long as the international supply is more than sufficient at affordable prices, Norwegian land-based food production is not justified by concerns for contemporary food security. However, the future is less certain. Spending resources on maintaining a high degree of self-sufficiency may be a sound policy in order to prepare for a potential future food crisis. There are also other issues justifying the subsidy structure, such as maintaining traditional rural population structure and cultural landscape. Subsidies and import tariffs are combined with a strong semi-corporatist structure of Norwegian land-based food production, with close collaboration between the authorities and the major food producers, owned by farmer cooperatives (Terragni, 2004).

Explicating social responsibility in the Norwegian food sector

Historically, corporate social responsibility (CSR) has not been a big issue for Norwegian companies. A reason is that many of the activities falling into the broad category of CSR in Norway are firmly rooted in laws, regulation and societal expectations. Applying the distinction between 'explicit' and 'implicit' CSR (Matten and Moon, 2008), we could say that a majority of Norwegian companies have traditionally not communicated their sense of social responsibility explicitly. However, in a time of economic globalization, where the state's control function is played down, many Norwegian companies become increasingly explicit regarding their role in society, and start expressing their social responsibilities as a matter of stakeholder communication and brand building (Carson *et al.*, 2013).

The Norwegian food sector is, even to a larger extent than other market areas, characterized by strong institutionalization built up around the interplay between a few dominating industry actors, political institutions such as import restrictions, price regulation, and annually negotiated agreements. The dominating industry actors Tine and Nortura play the triple role of interest organization for the farmers, commercial actor, and implementer of Norwegian agricultural politics. A precondition for mastering this role is that these corporations retain a significant market share, in Tine's case of about 81% (Listhaug, 2014), as they are obliged to buy products from farmers as well as deliver to processing plants. The question of the social responsibility of these central actors in the food chain becomes more prominent when issues of food safety and food security rise on the agenda. The strong institutionalization is recognized in the fact that the basic system is taken for granted and not questioned, and that the actors enjoy high legitimacy and significant latitude (Kjærnes *et al.*, 2010). Traditionally, consumers have been more or less invisible in the Norwegian food market, in any other way than as an anonymous source of demand of safe and reasonable products of standard quality. Consumers are generally satisfied with the supply of Norwegian food, and the producers enjoy a high level of trust and loyalty (Halkier *et al.*, 2007).

However, in spite of a political consensus on the continuation of traditional agricultural- and food policy in Norway – i.e. maintaining a high level of regulation and subsidies and even on increasing Norwegian food production in the years to come (Meld. St. 9, 2011-2012) – there are a number of challenges to the status quo of the Norwegian food sector, such as international agreements on trade (WTO, EU) leading to increasing competition from abroad, as well as an increasing importance attached to consumer interests over producer interests in the political debate (Farsund, 2014). In addition, with new systems of traceability and quality control, the responsibility for the safety of Norwegian food is gradually redirected from the direct control of the authorities to the industry actors themselves (e.g. ISO; hazard analysis critical control points (HACCP)). The safety and security of food products gradually goes from being secured by authorities and implicitly assumed by consumers, to becoming a matter of product differentiation, in other words something the producers use in order to distinguish certain products from other, similar products, e.g. through explicit branding. Ethical, environmental and social values are increasingly used for product differentiation as well, as we have seen in imported goods such as fair trade coffee and rain forest certified teak furniture.

The development towards self-regulation represents a challenge for the Norwegian food sector, since the high trust consumers place in Norwegian food products traditionally has been tightly connected with the comprehensive external control systems (Terragni, 2004) and become more vulnerable when the commercial sector takes over some of the control. The role of the cooperative-owned food producers thus becomes more complex. As a political tool, the primary social responsibility of these organizations is to protect the interests of the farmers, or more generally to maintain Norwegian food production. The strategy has been to offer reasonably priced, standard-quality products, while the health and safety of the products has been controlled and guaranteed by the authorities. Self-regulation and increasing international competition creates a new situation where food producers must communicate explicitly on issues of quality, health and safety.

'Enjoy Norway'

Modern food production and distribution is characterised by an increased relegation of control to the consumer, through a competitive food market with a number of different products and brands, and strong demands on labelling. In the Norwegian food sector, there are a number of so-called collective labels, where either a public body or a private initiative decides on and controls the standard, with the intention of influencing production and consumption. For commercial actors, the label becomes a way to differentiate their products by communicating social and environmental responsibility, or simply the origin of the product. The 'Keyhole' label is meant to promote healthy food choices, the 'Debio' label signals organic production, while the 'Enjoy Norway' label simply serves to help consumers pick out Norwegian products. In the area of information beyond that which is required by law, social responsibility issues are raised, e.g. in relation to issues of environmentally sound food production, animal welfare, food as a political issue and food as a cultural value.

Norwegian food producers have tried to combine these values in the promotion of Norwegian food as inhabiting all qualities – local, sustainable, environmentally sound, animal friendly and fair. The original trademark 'Godt norsk' ('Good Norwegian') carried associations both that these products were well-tasting, morally and qualitatively good products of Norway, but also that Norwegian products in general are 'good' in this multi-faceted way. This did hit a note with the consumers, who generally considered Norwegian farming to be 'good'. Norwegian farming has not suffered any major food scandals and the geographical conditions means that the majority of farms are still relatively small, and therefore perceived as unsuitable for the kind of industrial farming associated with poor animal welfare, high levels of imported feed and non-sustainable practices (Storstad, 2007). This strategy of marketing 'Norwegian' as particularly good is threatened from two sides: standard-quality, Norwegian products meet internal competition from regional products alluding to the values associated with local, 'short-travelled' food, and at the same time external competition on price and quality from foreign products gradually perceived as equally good as the 'good Norwegian' bulk products (Jacobsen, 2004). Still, the focus on the quality of Norwegian products has been a persistent and (so far) relatively successful strategy. The special character of the protected Norwegian food market is undoubtedly at least part of the explanation for this success.

In 2006, the 'Godt norsk' label was replaced, as the use of it was perceived as arbitrary as only some producers chose to use it, and then for a limited range of products (Folsland, 2003). It was also ridiculed as conceited and nationalistic. The new label 'Nyt Norge' is less value-laden, although it also represents an attempt at combining the national with quality, as it translates as 'Enjoy Norway' or 'Savour Norway'. Their information page (http://www.matmerk.no/nytnorge) states that the label is a guarantee that the products are made from Norwegian raw materials, produced by companies localised in Norway and coming from farms following the KLS quality system. This labelling reinforces the idea of Norwegian food production as something different from food production in other countries.

The continued and state-supported focus on the national as a guarantee of quality and purity is part of the explanation why products from organic farming have been less important in the Norwegian food markets. People have believed that Norwegian farming is better in terms of sustainability, animal welfare and intensity, and 'good enough' when compared with organic food (Storstad and Bjørkhaug, 2003). Although the 'Enjoy Norway' information never states explicitly it is of higher quality, purer and more environmentally sound than products from other parts of the world, this is still the general impression. In order to retain market shares for the Norwegian dairy, meat and egg production and fulfil their role as market regulators, this national emphasis is reasonable. However, when the market regulating producers and grocery chains join forces with the government in creating a combined quality control system and product promotion campaign for Norwegian products, such as 'Godt norsk' or 'Nyt Norge', this affects the accessibility of other products, since the promotion of competing products face a serious uphill struggle. This is probably most important concerning niche products associated with food ethics, as they compete on the same market of ethical consumerism. As long as the association of 'Norwegian' with pure and natural is successful, the national labelling alludes to at least some of the values promoted in local and organic food.

Tine – responsibility in dairy production

Tine, Norway's by far largest producer, distributor and exporter of dairy products, began as an association of Norwegian dairy farmers in 1881, and is today owned by 13,000 Norwegian farmers (Tine, 2014a). Historically, the cooperation of dairy farmers was established to secure the sales and a stable and optimal price of the farmer's products, and the system of cooperation between the farmer's cooperatives and the government described above, is still fully in use. Tine's triple role of being a cooperative of Norwegian farmers, market regulator and commercial actor is however heavily challenged by principled criticism from the perspective of liberal trade. The three roles are governed by demands and objectives that are pointing in different directions. Firstly, being owned by the dairy farmers, Tine has 'the best possible price for milk' as a stated objective for their activity (Tine, 2014b). The price level of milk has direct consequences for the viability of farms, and thus the interest of farmers. As a farmer cooperative, Tine would work for a high price of dairy products by supporting a policy of high import tax on dairy products and favourable adjustments of milk quotas. Secondly, in their role as market regulator, Tine's task is to receive milk from all farmers for an equal price, and to supply competing domestic dairy companies with milk at an equal price without making profits from this. The milk price and quantity should create an equal starting point for Tine and competing domestic dairies in the market for dairy products in Norway. Thirdly, as a commercial actor in this market, Tine's strategy is to offer products that are competitive on price and quality.

Tine – fair, local and environmentally sound?

How should Tine handle the responsibility given by its triple role as a co-operative of Norwegian farmers, market regulator and commercial actor? The answer to this question depends both on Tines' ability to fulfil its role obligations, and the legitimacy and coherence of the interests that shape the objectives of these roles. The advertising budget of Tine is one of the largest of all Norwegian firms (Kampanje, 2012). The protectionism from foreign products and the dominating domestic position of Tine in the Norwegian market can and has been challenged together with the environmental and social aspect of shielding Norwegian farming. In order to be convincing, Tine's advertised message depends on several factors, and importantly on Tine's ability to appear as fair, local and environmentally sound.

Regarding fairness, Tine has been met with criticism from competitors and consumers for utilizing their dominant market position to obstruct the establishment of a fair domestic market for dairy products, most markedly in the conflict over access to markets with the competing Synnøve Finden-company in 2004 (Norges høyesterett, 2011). Tine's central position in shielding the Norwegian market from import has also led to criticism of blocking the interests of both farmers from other countries (News in English, 2014) as well as Norwegian consumers and their interest in minimizing the household expenditure on food.

Regarding local identity, Tine finds itself in a precarious position. In the role of being a farmer's co-operative with the best possible milk price and high import tariffs as clear interests, both international and local goods present a potential threat to Tine's interests of upholding Norwegian milk farms. The introduction of strong regional and local Norwegian brands, or strong international organic brands, might weaken the association of 'Norwegian' with 'quality', 'purity' and 'naturalness'. If it is the local qualities that are associated with the quality of the products, and not the national qualities, there would be no reason to buy Norwegian instead of products from small-scale products from abroad. Tine commercials are focusing on the qualities of Norwegian milk. Matmerk, the administrator of the 'Enjoy Norway' label, is partly financed by Tine. Tine now offers milk from Norwegian regions, and milk from ecological farms, but the main message is still that Norwegian milk as such is 'maybe the finest milk in the world'.

The environmental aspect of Tine's shielding Norwegian farming has also been questioned, as the Norwegian climate and topography makes farming difficult. Farming practices with low effectivity that needs to be kept up by governmental measures might be environmentally unfavourable (McWilliams, 2010). In addition, Tine's policy of securing the viability of farming in all parts of the country, even in remote districts where production volumes and yields are low, can be questioned from an environmental perspective. From a comparative point of view, it is more effective to grow crops in the lowlands in the southern half of Norway – or indeed in countries south of Norway. However, it is difficult to make a comprehensive assessment of the environmental impact of farming (cf. Bioforsk Øst, 2013; Roer *et al.*, 2013). It follows that it is hard to assess the costs and benefits of protecting Norwegian farming throughout the country by subsidies and toll barriers.

Concluding remarks

Norwegian customers want Norwegian dairy products, and there is broad support for keeping Norwegian farming viable (ABCnyheter, 2014). The traditional co-operative 'bulk product regime' of Norwegian agriculture policy is, however, challenged by several developments: new public management and the emphasis on the autonomous consumer, the steady push towards liberal trade policies in the negotiations with the EU and in the WTO, and the wave of regional and local food. Following new public management principles (Christensen and Lægreid, 2001), our case study, the dairy producer Tine, has divided its enterprise into subdivisions along the lines of the company roles of being co-operation, market regulator and commercial actor. The ideal of the autonomous consumer presumes the possibility of choice and the enablement of the consumer to make informed food choices. This ideal is a driver for the restructuring of collective and monopolist provision of goods and services into market solutions. For Tine this implies a strengthening of their identity as a commercial actor.

Being a commercial actor among other actors on the one hand directs Tine away from the company's hard tied implicit social responsibilities as an integral part of Norwegian agricultural policy. On the other hand, this new role makes more explicit Tine's social responsibility in its relations with the customers.

References

ABCnyheter (2014) Trygg mat viktigst – ingen interesse for kutt i støtten. Available at: http://tinyurl.com/pkqm6jx.

Bioforsk Øst (2013). Økte avlinger kan gi miljøgevinst. Available at http://tinyurl.com/pd8wrlo.

Carson, S.G., Hagen, Ø. and Sethi, S. P. (2013). From implicit to explicit CSR in a Scandinavian context: the cases of HÅG and hydro. Journal of Business Ethics. DOI: http://dx.doi.org/10.1007/s10551-013-1791-2.

Christensen, T. and Lægreid, P. (2001) New public management: the effects of contractualism and devolution on political control. Public Management Review 3(1):73-94

Farsund, A.A. (2014). Norsk jordbruk i krysspress mellom nasjonal og internasjonal politikk. Norsk statsvitenskapelig tidsskrift (2).

Folsland, J. (2003). Godt Norsk ikke godt nok. Nationen. Available at: http://tinyurl.com/p3erct8.

Gaasland, I. (2009). Agriculture versus fish – Norway in WTO. Food Policy 34: 393-397.

Halkier, B., Holm, L., Domingues, M., Nielsen, A. and Terragni, L. (2007) Trusting, complex, quality conscious or unprotected? Journal of Consumer Culture 7(3): 379-402.

Heie, K. (2003). Viable small scale farming in Norway. International Farm Management Congress. Available at: http://tinyurl.com/q7brx87.

Jacobsen, E. (2004). Norsk mat i et åpent marked. Report. National Institute for Consumer Research. Oslo, Norway, 60 pp.

Kampanje (2012). Åtte reklamemilliarder i spill. Available at http://tinyurl.com/nzdflk8.

Kjærnes, U., Borgen, S.O., Borch, A. and Lavik, R. (2010): Tillit til kjøtt – endringer og utfordringer i det norske markedet. Sifo Oppdragsrapport nr. 14. Available at: http://tinyurl.com/pzattpe.

Listhaug, S. (2014). Regjeringens 100 dager er over – hva nå? Available at: http://tinyurl.com/nn7h2jr.

Matten, D. and Moon, J. (2008). 'Implicit' and 'explicit' CSR: a conceptual framework for a comparative understanding of corporate social responsibility. Academy of Management Review 33(2): 404-424.

McWilliams, J. (2010). Just food: where locavores get it wrong and how we can truly eat responsibly. Back Bay Books, New York, USA.

Meld. St. 9 (2011-2012). Landbruks- og matpolitikken – Velkommen til bords. Available at: http://tinyurl.com/qjn36nz.

News in English (2014) EU blasts Norway's protectionism. Available at: http://tinyurl.com/nhmsvp4.

Norges høysterett (2011) Prøving av Konkurransetilsynets vedtak om ileggelse av overtredelsesgebyr. Available at: http://tinyurl.com/qatcyrd.

Organisation for Economic Co-operation and Development (OECD) (2013). Producer support estimates (subsidies). Agriculture and food: key tables from OECD, No. 1. Available at: http://tinyurl.com/nc4dg7b.

Roer, A.G., Astrid Johansen, A., Bakken, A., Daugstad, K., Fystro, G. and Strømman, A. (2013). Environmental impacts of combined milk and meat production in Norway according to a life cycle assessment with expanded system boundaries. Livestock Science 155(2): 384-396.

Storstad, O. (2007). Naturlig, nært og trygt. En studie av hvordan forbrukertillit til mat påvirkes av produksjonsmåte og matskandaler. Norwegian University of Science and Technology, Norway.

Storstad, O. and Bjørkhaug, H. (2003). Foundations of production and consumption of organic food in Norway: common attitudes among farmers and consumers? Agriculture and Human Values 20: 151-163.

Terragni, L. (2004). Institutional strategies for the production of trust in food in Norway. National Institute for Consumer Research, Oslo, Norway, 83 pp.

Tine (2014a). Årsrapport 2013. Available at: http://tinyurl.com/orleslf.

Tine (2014b) Visjon og verdier. Available at: http://tinyurl.com/ndd659h.

Section 6. Feeding the people

Requirements of safety and acceptability in food security definitions

H. Siipi
Department of Behavioural Sciences and Philosophy, University of Turku, Assistentinkatu 7, 20014 Turun Yliopisto, Finland; helsii@utu.fi

Abstract

Enhancing global food security requires hefty changes in ways of food production, distribution and consumption in the near future. The way the term 'food security' is understood partly determines what kind of changes in diets and in ways of consumption can be seen as the duties of individuals and societies. It is quite generally agreed that food security should not be understood as a simple demand versus supply calculation. A large amount of calories available for an individual or society is not sufficient for food security if the food is not safe and/or lacks necessary nutrients. Moreover, the existence of safe and nutritious food is valueless if those who have access to it consider it inedible or otherwise unacceptable. In this paper, I analyse the requirements of safety and acceptability in mainstream definitions of food security. I argue that, in order for the requirements of safety and acceptability to imply and be consistent with the duties of well-off individuals and societies to considerably change their diets and ways of consumption, they should be interpreted in the following way: (1) the requirement of safety should broadly include indirect health effects of food; (2) the requirement of acceptability should merely be understood as a right to avoid food that is against the overriding values, beliefs and/or interests of an individual or a society in question.

Keywords: global food security, changing ways of consumption, duties, meanings of terms

Introduction

Because of population growth, climate change and other challenges (e.g. overexploitation of many marine ecosystems), enhancing, or even maintaining, the current levels of global food security requires hefty changes in the near future. The changes needed concern food production, distribution, and also the ways of consumption. (Brown and Funk, 2008; Godfrey *et al.*, 2010) In his list of solutions for famine, Julian H.J. Cribb (2011) goes as far as suggesting a 'reinvention of world diet to one that is lighter, healthier, kills fewer consumers and treats less heavily on the planet'.

Supposing that the goal of enhancing food security can imply moral duties for individuals to modify their personal diets and for societies to adjust their ways of consumption, what kind of changes can be morally required? The question is approached by spelling out different interpretations of the following two components of the food security: food safety and food acceptability. The idea is that the way in which 'food security' is understood, at least partly, determines what kind of duties to societies and individuals its enhancement can be taken to imply.

Components of food security

Different kinds of groups and even individuals can be assessed with respect to their level of food security. Historically food security is often taken to concern countries, but with globalization and global food system the interest in national food security is fading (Schmidhuber and Tubiello, 2007). Current approaches include food security of cities and other smaller areas, households, and even individuals (Conway, 2012; Pinstrup-Andersen, 2009). Through global food systems the global food security has become a central viewpoint. In this paper the interest is in individuals' and societies' contribution to global food security through their consumption choices and ways of consumption.

Several definitions for 'food security' can be found. Following Anderson *et al.* (1990) the United States Department of Agriculture (2014) understands food security 'as access of all people at all times to enough food for active healthy life and includes as minimum: (1) ready availability of nutritionally adequate and safe foods; (2) assured ability to acquire acceptable foods in socially acceptable ways'. According to the Food and Agriculture Organization of the United Nations (FAO, 2014), food security 'exists when all people, at all times, have physical and economic access to sufficient, safe and nutritious food to meet their dietary needs and food preferences for an active and healthy life'. FAO together with International Fund for Agricultural Development and World Food Program (2014) further conceptualized food security by setting four dimensions to it: availability (i.e. quantity, quality, and diversity), access (affordability and allocation), utilization, and stability.

As can be seen from the above definitions, food security should not simply be understood as calculation of demand and supply. Even large amount of calories available for an individual or society is not sufficient for food security, if the food is not safe and/or lacks necessary nutrients. Moreover, the existence of safe and nutritious food does not guarantee an individual's access to it, and an individual's access to food is valueless if she considers that food inedible or otherwise unacceptable. (Pinstrup-Andersen, 2009; Schmidhuber and Tubiello, 2007). The requirements of safety and acceptability are, thus, integral to food security.

Safety

The term 'safe' has narrow and broad meanings. In its narrowest sense, the safety of food is a negative requirement of food not to contain harmful materials such as toxic bacteria or foreign allergens. In a broader sense the term 'safe' refers also to the presence of beneficial elements and necessary nutrients. When the nutritional adequacy of food is listed as a separate requirement in a food security definition, the term 'safe' can be understood as a narrow negative requirement (cf. e.g. FAO, 2014).

However, the health effects of food are not limited to its nutritious values (Scrinis, 2008). Food also affects our well-being through its aesthetic, social, cultural, and symbolic meanings. Celebrating an important happening by sharing a delicious meal with friends, for example, enhances one's mental and physical well-being beyond the nutritional qualities of food. Moreover, ways of food production, distribution and consumption affect us indirectly through their social and environmental impacts. (Thompson and McDonald, 2013) To simplify a little, favoring food with a large carbon footprint may bring about climatic changes that are risky to human health and well-being.

Should the expression 'safe food' be understood to include, in addition to the direct health effects of food, also its indirect effects on human well-being? Economic, ecological and social arrangements that affect the availability of food and individuals' access to it are often taken to be important parts of food security assessments. Schmidhuber and Tubiello (2007), for example, understand the 'availability' dimension of food security to refer to an 'overall ability of the agricultural system to meet food demand. Its subdimensions include the agro-climatic fundamentals of crop and pasture production and the entire range of socio-economic and cultural factors that determine where and how farmers perform in response to markets'. The question presented here, however, differs from their approach in this paper's focus on the component of safety and its interpretation.

Broadening the meaning of 'safe food' in the suggested way to include environmental and social impacts of food would imply that certain foods (or even diets) commonly consumed by well-off individuals and societies fail to meet the requirements for food security. In other words, the suggested understanding implies a low level of food security to a household which has good economic situation and easy access to wide variety of food and which favors food that happens to have bad environmental and/or social consequences (e.g. beef or exotic fruits produced by child labor). The risk of this kind of understanding is that it directs the focus away from original targets of the term 'food security' – that is from the individuals, households and societies with poor access to food. On the other hand, the suggested understanding of food safety is not just compatible with the changes required for enhancing global food security, it also directly motivates and necessitates them.

The expression of 'safe food' is also ambiguous in referring either to a real (or expert) estimation of food safety or to consumer perceptions of the safety of the food available. Perceived food safety means individuals' or societies' views on the health effects of the food available. It may be in accordance with an expert estimation of food safety, but it is also possible that lay persons consider food safe when it is not, or that lay persons take safe food to be unsafe. It seems clear that perceived food safety is not sufficient food security. Food security requires that the food available really is safe (or safe enough). The crucial question concerns the necessity of perceived safety to food security. To what extent and how should the mistaken views regarding safety of food be taken into account? Can individuals and societies have a duty to learn and change their mistaken views regarding safety of food? As the perceived unsafety of food may offer an individual or a society one reason to reject that food, these questions connect food safety to the wider question of food acceptability.

Acceptability of food

As noted by Per Pinstrup-Andersen (2009), 'the addition of food preferences changes the concept of food security from mere access to enough food to access to the food preferred'. The requirement of accordance with preferences implies that two individuals or societies with equal access to the same kind of food may differ with respect to their levels of food security provided that their preferences differ. Thus, in an imagined situation in which the main source of protein is horse meat, the level of food security in Italy would be much higher than in Great Britain (horse meat is widely consumed in Italy, whereas most British people consider horses inedible).

But what kinds of preferences are relevant to the requirement of food preferences? Surely my family's love to fresh strawberries and blueberries cannot imply a lower household food security level for us in the winter when they are not available in Finland. According to Per Pinstrup-Andersen (2009), 'as long as term 'preferences' is interpreted to mean foods that are socially and culturally acceptable and consistent with religious and cultural values, rather than a broader interpretation to mean household or individual preferences for caviar instead of sorghum, there may not be a problem with addition of preferences'.

It is easy to accept Pinstrup-Andersen's view that the requirement of food preferences should not be interpreted as a requirement of individuals' or societies' access to their favorite foods. Yet, despite this clarification, a dilemma arises. If the requirement of food preferences is, as in the above quote from Pinstrup-Andersen (2009), interpreted as consistence with cultural values, how could the goal of enhancing food security ever justify a duty to bring about changes in food culture and in the current ways of consumption (e.g. Cribb, 2011; Grote, 2014; Godfrey *et al.*, 2010)? In other words, if the requirement of food preferences is taken seriously, can the goal of enhancing food security ever justify a duty for a society to change its food culture or an individual to change her general way of consumption?

One might suggest solving the dilemma in interpreting the requirement of preferences in a weak and negative way. According this line of understanding, all sources of food and food products that are not against the cultural or religious values of a society (or an individual) and that are not found unacceptable or inedible in it (or by her) are considered as preferred and acceptable. Under this interpretation, the goal of enhancing food security could, for example, justify a duty of Western people to cut their meat consumption considerably (or even to give up eating meat). Even though many Western consumers prefer meat as a protein source, and even though meat can in the traditional Western food culture be considered as the centerpiece of meals, there are numerous other protein sources (soya beans, eggs, nuts etc.) that are not against Western cultural values and not considered unacceptable or inedible.

However, as weak as this interpretation of the requirement of preferred foods may seem, it may still be too strong – that is too strong for justifying demands for such changes in food cultures that are needed for the enhancement of global food security. Under the provided interpretation, the goal of enhancing food security could never justify a duty to give up views or cultural values regarding inedibility or unacceptability of food. Thus, since, for example, most Western individuals and societies are reluctant to eat insects and find eating them disgusting (Van Huis *et al.*, 2013), Western individuals and societies cannot under this interpretation have a moral duty to replace (or even try to replace) meat in their diet with insects – even though insects have been found (Van Huis *et al.*, 2013) both from environmental and health perspectives to be among the best candidates to replace meat in the traditional Western diet.

How should this implication of suggested interpretation of food preferences be met? If the suggested interpretation is accepted, it implies that a Western individual or society working towards replacing meat with insects (in order to enhance global food security) is acting in a supererogatory way, doing something good beyond duties. At the same time, a Western consumer working towards enhancing global food security by replacing meat with soya, eggs, nuts and other products that are already accepted as part of the Western diet can be seen to act according to her duty. If this implication is not found acceptable, further adjustments to the requirement of food preferences are needed.

In order to further adjust the requirement of food preferences, we need to be able to conceptually distinguish between cases in which it is and it is not appropriate to require individuals and societies to change their views regarding what is acceptable, inedible and against their cultural or religious values. The ultimate goal of the requirement of preferences is, of course, to exclude any judgments of good level of food security in situations where, for example, members of a devout Muslim society would have to eat pork to fulfil their need for protein. How does this imaginary case differ from a requirement of Western individuals and societies to start to consume insects in order to enhance global food security?

My suggestion is that the expressions 'cultural or religious values' and 'found unacceptable or inedible' should be understood to refer to values and beliefs that connect to the type of interests Gary Comstock (2013) calls 'overriding'. Overriding values, beliefs and interests are, according to him, the ones that are central to individual's worldview and identity. They are determined by the decisions regarding the kind of life one wants to live and chooses to live. Overriding values connect to the individual's or society's lifestyle. They can be considered as strong values in a sense that their violation is serious moral wrongdoing that harms something central to one's life (for further discussion on the strength of values see Poli, 2012). In a devout Muslim society, the views of inedibility and unacceptability of pork rise from values that can be seen as overriding in the described sense. The Western view of inedibility of insects, in contrast, does not usually connect to overriding values, interests or beliefs. Rather the view rises from a mere habit of associating insects with harm, pest and spreading disease (Van Huis *et al.*, 2013). This is not to say that for some Western individuals the inedibility of insects may result from overriding values. It may also well be that for some Muslims the view of inedibility of pork does not connect to their overriding values.

Thus, in order to make the definition of food security compatible and consistent with moral duty of individuals and societies to change their food culture and ways of consumption even to a considerable extent, I suggest interpreting the requirement of preferences as follows: all sources of food and food products that are not against the overriding cultural or religious values of an individual or society and that are not found unacceptable or inedible due to the overriding interests, values and beliefs, should be interpreted as preferred and acceptable in that society or for that individual.

Conclusions

Safety and acceptability are important components of food security. It is crucial to avoid situations in which a level of food security is estimated as good even though the food to which an individual or a society has access is risky and/or unacceptable for them. At the same time, the requirements of safety and acceptability should be interpreted and understood in ways that are compatible with the requirements and duties of well-off individuals and societies to adjust and change their ways of consumption and even food culture. In order to achieve this goal, it is suggested: (1) that the component of food safety should be understood broadly to include the indirect health effects of food; and (2) that the component of acceptability should merely be interpreted to refer to a right to avoid food that is against the overriding values, beliefs and/or interests of an individual or a society.

References

Anderson, S.A. (ed.) (1990). Core indicators of nutritional state for difficult to sample populations. The Journal of Nutrition 120: 1557S-1600S.

Brown, M.B. and Funk, C.C. (2008). Food security under climate change. Science 319 (5863): 580-581.

Conway, G. (2012). One billion hungry: can we feed the world? Cornell University Press, Ithaca, USA, 439 pp.

Comstock, G. (2013). Research ethics: a philosophical guide to responsible conduct of research. Cambridge University Press, Cambridge, UK, 305 pp.

Cribb, J.H.J. (2011). Food security: what are the priorities? Food security 3: 123-125.

Godfrey, H.C., Beddington, J.R., Crute, I.R., Haddad, L., Lawrence, D., Muir, J.F., Pretty, J., Robinson, S., Thomas S.M. and Toulmin, C. (2010) Food security: the challenge of feeding 9 billion people. Science 327: 812-8.

Food and Agriculture Organization of the United Nations (FAO). (2014). Introduction to basic concepts of food security. Available at: http://tinyurl.com/pvfae4j.

Food and Agriculture Organization of the United Nations (FAO) and International Fund for Agricultural Development and World Food Program. (2014). State of food insecurity in the world: strengthening the enabling environment for food security and nutrition. Available at: http://tinyurl.com/l6b26mm.

Grote, U. (2014). Can we improve global food security? A socio-economic and political perspective. Food Security 6: 187-200.

Pinstrup-Andersen, P. (2009). Food security: definition and measurement. Food Security 1: 5-7.

Poli, R. (2012). Nicolai Hartmann. In: Zalta, E.N. (ed.) The Stanford encyclopedia of philosophy. Fall 2012 edition. Available at: http://tinyurl.com/ojcm5y3.

Schmidhuber, J. and Tubiello, F.N. (2007). Global food security under climate change. Proceeding of the National Academy of Sciences of the United States of America 104: 19703-19708.

Scrinis, G. (2008). On the ideology of nutritionism. Gastronomica. The Journal of Food and Culture 8: 39-48.

Thompson, D.B. and McDonald, B. (2013). What food is 'good' for you? Toward a pragmatic consideration of multiple values domains. Journal of Agricultural and Environmental Ethics 26: 137-163.

United States Department of Agriculture. (2014). Food security in U.S. – measurement. Available at: http://tinyurl.com/kbtybfv.

Van Huis, A., Van Itterbeeck, J., Klunder, H., Merterns, E., Halloran, A, Muir, G. and Vantomme, P. (2013). Edible Insects: future prospects for food and feed security. Food and Agriculture Organization of the United Nations, 187 pp. Available at: http://tinyurl.com/d3dykve.

Food security and ethics: the first world hunger

T. Kortetmäki
University of Jyväskylä, PL 35, 40014 Jyväskylän yliopisto, Finland; teea.kortetmaki@jyu.fi

Abstract

Food security is the main policy objective of food systems. FAO (2013) defines food security as a situation when 'all people, at all times, have physical, social and economic access to sufficient, safe and nutritious food that meets their dietary needs and food preferences for an active and healthy life'. Despite of the increasing importance of food security and food politics, ethical discourse in the field has been scarce. Even more this is the case concerning the first world food security. This paper explores the three major challenges the first world hunger poses to ethics: (1) access to food; (2) local culture and food sovereignty; and (3) sustainability. These challenges are 'wicked problems' and require ethical theorising for discussing and finding the solutions. Through the discussion of these problems this paper shows that whereas traditional, distribution oriented theories of justice are too narrow in front of the aforementioned challenges, the multidimensional theory of justice offers a promising conceptual framework for approaching the wicked problems.

Keywords: food justice, first world hunger, food sovereignty, sustainability

Introduction

According to the FAO's (2013) official definition, food security has four elements: availability (of food), access (to food), utilisation (the quality and safety), and stability (of supply). By the very definition food security should also be sustainable: this is implicit in the definition words 'at all times'. Food security is often linked with food sovereignty or 'the right of peoples to define their own food and agriculture; to protect and regulate domestic agricultural production and trade in order to achieve sustainable development objectives; to determine the extent to which they want to be self reliant' (FAO, 2005).

Sustainable food security is has been discussed in politics, but the related ethical discourse has been rather scarce. In his classic 'The spirit of the soil', Thompson (1995) discussed the sustainability and food ethics comprehensively. Food security and sovereignty has been treated by Schanbacher (2010), and the food security and its tensions with sustainability have been brought out by Sandler (2014), Lang and Barling (2012) and EurSafe 2010 book (Casabona *et al.*, 2010). Yet in all these works, the first world food security is rarely mentioned. The most comprehensive treatment of the first world hunger has been made by Silvasti and Riches (2014); their book however touches not so much the ethical side of the debate but mostly focuses on the national-level policy discussion.

In this paper I discuss the first world food security, making two main arguments. Firstly, the first world food security challenges are 'wicked problems': this highlights the importance of ethics in dealing with them. Secondly, these problems can better be tackled with the broad view of justice, as opposed to the narrower distributional paradigm of justice.

Wicked problems

Three issues related to first world food security are especially interesting from the ethical viewpoint: access to food, food sovereignty, and sustainability. These are characteristically 'wicked problems'. Wicked problems differ from tame problems (found in many natural sciences) in several terms. Their solutions are not true-or-false, but good-or-bad; there are no tests for finding the solutions; the set of possible solutions cannot be exhaustively described; and there is no room for learning by trial-and-error but every solving attempt affects the situation. Every problem is essentially unique and can be considered as a symptom of another problem. (Norton 2005: 132) Wicked problems tend to have competing and conflicting interests at stake. Due to the lack of solutions that could be found by calculations or other means, wicked problems necessitate ethical considerations. Next I will shortly explain what makes the above mentioned three problems wicked – and at the same time very pressing issues for the ethicists.

Access to food is often the real reason for food insecurity: the global dietary energy supply would exceed the needs was it equally distributed. Access to food is a wicked problem due to several reasons. Firstly, there is no single answer to how to improve the access in the first hand: should we distribute food, the means for accessing it, or promote other related capabilities? Should we target individuals, families, or certain groups? There are no methods for calculating the best solutions. Moreover, the problems related to access to food are context-dependent and unique everywhere. Lacking the access to food in Finland is different from facing the food insecurity in Spain, let alone in the developing countries.

Food sovereignty posits the second wicked problem. Roots of this wickedness have been already identified by Sandler (2014), although he does not label them wicked. They are: (1) Promoting food security may ignore the (individual or) local autonomy and food sovereignty. There is no clear answer to whether this is justifiable (and in which cases if so). (2) On the other hand, exercising food autonomy and sovereignty may entail practices that undermine sustainable food security, and there is no clear answer how these cases should be dealt with. Tensions arise between food sovereignty and other environmental and social values.

Sustainability forms the third wicked problem. The wickedness lies in the fact that currently many of the most food secure states exercise unsustainable food system practices, undermining the future food security. General affluence and the low proportion of food poverty in wealthy states is grounded on the economic growth and low prices, benefiting from the poverty of developing nations. This concerns particularly the cheap labour, cheap raw material, resource exploitation and land grabbing. This is economically and socially unsustainable. Environmental unsustainability of the agriculture and food systems does not need a detailed explanation here: carbon footprints, water and energy use and food waste of agriculture and food systems are widely known.

The answers to these problems are not 'true-or-false': there is a competition between conflicting goods and interests (cf. Norton 2005: 137), and modeling or scenario-making tools can provide but predictions. Ethics is needed to frame the questions, evaluate the weights of different risks and possibilities and help to set the right questions. But what kind of philosophical tools are most fit for this task? In what follows, I shortly introduce two influential approaches of justice: distribution oriented and multidimensional. I argue that the distributional approach provides a too narrow viewpoint for dealing with the first world hunger, whereas the multidimensional theory provides a more useful framework for this task. I compare these two approaches in the light of the first world hunger and the cases of access to food and food sovereignty. Sustainability is a crossing theme in both issues.

Narrow and broad views of justice

By the traditional conception of justice I refer to distributive justice, an umbrella term covering various theories of justice. The proponents of this approach argue that justice is first and foremost about distributing goods and bads (or benefits and harms): John Rawls, David Miller, Richard Rorty, and Ronald Dworkin are the well-known proponents of this idea. Many of their critics have followed the same school, taking distribution as the only core of justice.

However, some political theorists, especially Nancy Fraser (accompanied by for example Iris Young and David Schlosberg), have argued that the distributive paradigm is truncated and insufficient. Fraser does not reject distribution as an element of justice, but calls for understanding justice more broadly, having three irreducible dimensions: distribution, recognition and participation (Fraser, 2009). These dimensions are in reality intertwined but can be analysed separately.

Distribution dimension refers to the (re)distribution of goods and bads. Often this means the economic distribution and equality. Recognition concerns the institutionalised cultural patterns that promote or hamper equality: even though two persons may be equal on economic basis, the other one may encounter inequality in terms of discrimination (due to e.g. religious beliefs). The third dimension of justice, participation, involves political representation. It is particularly about the questions of political inclusion and exclusion. This dimension also asks how the different claims of justice can be considered justified and who defines the procedures by which the considerations are made. Regarding all these three dimensions, the aim is to promote equality of the society members: their participatory parity in all dimensions.

To sum up the difference between the two approaches, distributional theories hold that injustices are either reducible into questions of (often economic) maldistribution, or if not, they should not treated as issues of injustice. Distributions define whether the societies (or their institutions) are just or not. Opposing this, the broad theory argues that we must adopt multiple dimensions of justice in order to capture the whole nature of injustice and the structures that reproduce it. The famous example made by Fraser and Honneth (2003: 34) is an African-American Wall Street banker, enjoying equal wage and benefits with his colleagues but being unable to get a taxi to pick him up due to cultural discrimination and misrecognition. Whereas this is not an issue of justice for the distributional approaches, the broad view of justice takes it such. Next I will discuss the differences of these two approaches in the context of first world food security and the issues of access to food and sustainability.

Access to food in the developed world

Food insecurity in the first world has received only little attention in the ethical considerations. Although smaller in scale than the the developing world challenges, food insecurity in the affluent states is similarly tragic for those facing it. According to the FAO (2013) report, there are around 15-16 million food insecure people in the developed countires, although there are reasons to doubt that the real numbers are significantly higher (Silvasti and Riches, 2014).

Several factors make the first world hunger distinct from hunger in the developing world:

- In affluent nations, the problem is even less in the availability of food: a significant amount of citizens are actually obese.
- Food insecurity in these countires indicates serious failures in the food policy or social and public policy (Silvasti and Riches 2014: 5).
- The problem is not only with the access to food, but increasingly in the utilization, resulting in the combination of malnourishment and obesity among the poorest.
- The food system is unsustainable in many affluent states. In other words, even the achieved level of food security is not sustainable.

The practices of food charity and food banking have become a common way to respond to the first world hunger. Distributional theories of justice find this a generally acceptable solution if it corrects the food inequality. Naturally, unequal income distribution is usually the preferable remedy, but food charity can fix the holes in safety nets. In this view, the institutional status or form of redistribution does not matter in terms of justice, if it corrects the maldistribution.

Even if we leave aside the empirical fact that food charity often fails in providing enough food (and correcting the maldistribution), the reasoning described above ignores important questions of justice. The first of them is that currently the food aid in affluent countries is based on charity. It is a gift, not an entitlement: Consequently, no claims of justice can be placed to charitable practices. In the charity breadline the rights and human dignity is lost (Silvasti and Riches 2014: 9-10).

Some distributional theorists hold that rights are a distributable good and argue for a right to food: they can criticise the charity problems mentioned above. Rights entail corresponding obligations and allow making claims of justice. However, this says not much. The solution for ensuring food distribution could still be some kind of food bank, only with the exception that governments (or other institutions) guarantee the sufficiency of food aid. Whereas making food banks a service and not charity would increase the stability of supply and restore some of the human dignity, problems would still remain. The gravest is the stigmatising effect. Even if made a right, food banks for 'those at the bottom' still maintain the stigmatising divisions. Another problem is that merely institutionalising the current system would still leave the clients without sovereignty and autonomy. A person with a cardiovascular disease may have to accept foods that are unhealthy for him to satisfy his need for energy; or a vegetarian person may have to choose between not having enough protein or giving up his moral principles and accepting the meat offered by the food bank (even though vegetable proteins are widely considered as ecologically, ethically, and in health terms more sound – and, usually, cheaper as well). Distribution-oriented justice theories do not consider these cases as injustice, but I contend (along with the broad approach) that they should be conceived such.

The broad approach recognises that in the current welfare system resolutions often have unintended consequences. Fraser and Honneth (2003: 64-65) notes that the redistribution policies tend to stigmatise, constituting divisions such as 'taxpayers' versus 'welfare-receivers'. To avoid such consequences (as long as non-stigmatising measures like basic income are not adopted), measures of recognition should be taken along with the distributional aid instead of mere redistribution. What would this mean in the context of first world food insecurity? The first step would be to publicly acknowledge the insufficiency of current society's safety net (and try to patch it). This should be paired with making the food aid as an issue of basic right – and not just right to food, but right to healthy and culturally acceptable food. This raises the question whether food aid should be combined with the operation of normal grocery stores, bringing the breadline customers back to 'normal clients': in addition, this would open up the possibility to increase the sovereignty of those in need of food aid, leaving them for example the freedom to choose between certain sources of protein or the ability to choose more vegetables instead of meat if their health needs are better met by that choice.

Broad approach also covers the problem of participatory deficiency, which usually is not deeply dealt within the distributional approaches. The relevant questions (cf. Fraser 2009: 17-20) are: who are included in the community of justice? Who can participate in decision making? Are there obstacles that impede participatory parity of our fellow people, even if they are 'in principle' allowed to participate? Who are framed as the 'affected ones' concerning the questions of food policy and food security? Food charity approaches are mostly of top-down nature: charity organisers and recipients are different parties and barely meet outside the ten-second encounter in the breadline. Opening the chances for participation in designing the food security policies and practices would be more just for those who are (possibly) affected by these decisions; it could also help to improve the food sovereignty of those otherwise excluded.

Food sovereignty and the first world

If both food security and food sovereignty are rights, the aim should be to promote them sustainably hand in hand as much as possible. This conflicts with the UN mainstream strategies for improving food security: developmental growth, agricultural reform, and global market integration tend to decrease sovereignty (Schanbacher, 2010: 18-22). It is arguable that these mainstream policies harm food sovereignty and misrecognise the local values. Whereas the distribution oriented theories of justice lack specific tools for dealing with this problem, the broad view provides a useful framework for addressing this tension. Ignoring the food sovereignty violates both the recognition and representation aspects of justice. The claims for recognition and representation of local communities with regard to promoting food security are, in this view, viable claims of justice.

However, the challenge becomes more complex in the first world context. I will now shortly discuss two problems that relate to the tension between food sovereignty and food security in the first world: (1) 'redistributing' food power to locals does not guarantee promoting equal food security to all society members; and (2) food sovereignty may threaten food security or the sustainability of food production.

Firstly, plain (re)distribution of food power does not deal with the deeper question of framing: who are included and who are excluded in the local deliberation? Who are agreed to have knowledge that counts? Discrimination of certain minorities may prevail. Moreover, it is not clear what claims are justified under the umbrella of sovereignty. In Finland there has been discussion on making vegetarian school lunches more common in public schools. However, the opposition has been strong, arguing that the children are entitled to have meat in a proper school lunch. On the other hand, the epistemic deficiencies give rise to the question whether the locals are in the best position to make decisions that truly promote food security and nourishment. This concerns especially the obesity and malnourishment related problems.

Furthermore, some local practices of food production and consumption may hamper sustainable food security. In the first world, various consumption patterns and values of food are environmentally or socially unsustainable. For example, viewing meat eating as a symbol of increasing standard of living is an example of cultural practice that seriously risks the sustainability of global food production (as well as human health). Another example is the Finnish coffee culture: Finland's per capita coffee consumption is highest in the world, and coffee is considered a basic foodstuff that should be received at a low price (meaning high social and environmental costs for the producing countries). Is there any way to differentiate between the justifiable and unjustifiable practices that are exercised in the name of local food sovereignty?

The notion of recognition as justice is useful here. Fraser has set double criteria on evaluating whether a claim for recognition is justified or not. In order to be justified, the changes following from recognition: (1) must improve the equality of people in the society (in e.g. preventing some form of subordination); and (2) must not create or worsen other disparities of participation or inequalities (Fraser and Honneth, 2003: 38-40). This can be translated to food context very well. Food sovereignty deserves recognition when: (1) acknowledging the sovereignty increases the equality of the claimants in the society; and when (2) it does not create or maintain other inequalities, local or global. For example the claims of recognition of local food culture that subordinate women or are environmentally unsustainable (therefore jeopardising the global equality) are not considered as deserving recognition but the opposite.

Conclusions

Promoting sustainable food security involves many problems that are 'wicked', unsolvable with empirical research or decision making models. Therefore food security requires ethical theorising to help frame the questions and evaluate the acceptable risks and clarify problems of interest conflicts. This theorising must not limit itself into the 'more traditional' cases of the developing world hunger; the first world hunger must be addressed as well, as its existence implies serious failures in the welfare policies of the affluent states.

Food security justice and the question of first world hunger reaches far beyond the plain question of distributing dietary energy more equally. In this light, the traditional distributional approaches of justice remain narrow in their perspective to food security. A broader conception of justice, taking justice to have at least three dimensions (distribution, recognition and representation) provides a more fruitful framework for discussing the wicked problems of food security. The thorough discussion about the first world hunger and justice must start.

References

Casabona, C.M.R., Epifanio, L.E.S. and Cirión, A.E. (2010). Global food security: ethical and legal challenges. EurSafe 2010 congress. Wageningen Academic Publishers, Wageningen, the Netherlands, 532 pp.

Food and Agriculture Organization of the United Nations (FAO) (2005). The state of food and agriculture. Available at: http://tinyurl.com/m42j2fm.

Food and Agriculture Organization of the United Nations (FAO) (2013). the state of food insecurity in the world: the multiple dimensions of food security. Available at: http://tinyurl.com/njzqzjp.

Fraser, N. (2009). Scales of justice: reimagining political space in a globalizing world. Columbia University Press, New York, USA, 224 pp.

Fraser, N. and Honneth, A. (2003). Redistribution or recognition? A political – philosophical exchange. Verso, London, UK, 276 pp.

Lang, T. and Barling, D. (2012). Food security and food sustainability: reformulating the debate. The Geographical Journal 178(4):313-326.

Norton, B.G. (2005). Sustainability. University of Chicago Press, Chicago, USA, 607 pp.

Nussbaum, M. (2011). Creating capabilities. Harvard University Press, Harvard, USA, 237 pp.

Thompson, P.B. (1995). The spirit of the soil: agriculture and environmental ethics. Routledge, London, UK, 196 pp.

Sandler, R.L. (2014). Food ethics: the basics. Routledge, London, UK, 216 pp.

Schanbacher, W. (2010). The politics of food: the global conflict between food security and food soverneignity. Praeger, Santa Barbara, USA, 148 pp.

Silvasti, T. and Riches, G. (2014). First world hunger revisited. Palgrave Macmillan, London, UK, 280 pp.

Section 7. Traditional and local food production

A view to a (staged) kill? The perception of game bird shooting among different Danish stakeholders: hunters, landowners and the general public

C. Gamborg[1], F.S. Jensen[2] and P. Sandøe[1]*
[1]Department of Food and Resource Economics, University of Copenhagen, Rolighedsvej 25, 1958 Frederiksberg, Copenhagen, Denmark; [2]Department of Geosciences and Natural Resource Management, University of Copenhagen, Rolighedsvej 23, 1958 Frederiksberg, Copenhagen, Denmark; chg@ifro.ku.dk

Abstract

Is it acceptable, in principle, to shoot wild animals as a sport? The ethical controversy around recreational hunting is not only of academic interest but plays an important role in public and political discussions concerning the legitimacy of hunting. For the hunter, hunting is typically imbued with a perception of deeper meaning. For ardent non-hunters it may be hard to understand the meaning of enjoying nature by shooting at other living beings. This lack of mutual understanding makes it difficult to create a dialogue. However, the issue is not just about whether or not to accept hunting but also the specific forms of hunting. The aim of the paper is to explore the attitudes at stake in debates about a controversial form of hunting: shooting of birds that are farm reared and subsequently released for the purpose of being hunted. The paper is based on a nationally representative survey (web-based questionnaires) of different Danish stakeholders. Results show that a majority of the general public is primarily against the practice; a majority of hunters are pro, and landowners somewhere in between. Reasons may be found in relation to views on wildlife, naturalness and what constitutes 'proper' hunting. Moreover, differences in expression of one's fundamental values toward wildlife may help to explain differences found in views. This type of recreational hunting challenges traditional hunting/anti-hunting arguments, such as hunting being acceptable when based on sustained yield 'harvest' principles, and raises a discussion about the ethical acceptability of providing food from 'the wild'.

Keywords: conflict, ethics, food, hunting, naturalness

Introduction

The practice of hunting wild animals for food or fur is as old as humankind. Hunting is still widely practiced but the purpose has widened to cover wildlife management, recreation and sport. To the extent that the hunting is not a life necessity it increasingly gives rise to moral concerns (Cahoone, 2009; Causey, 1989; Loftin, 1984) and challenges deeply held values (Morris, 2013). For the recreational hunter, hunting is typically imbued with a perception of deeper meaning. For non-hunters it may be hard to understand what the point is supposed to be of enjoying nature by shooting at other living beings (Dahles, 1993; Wood, 1997). This lack of mutual understanding makes it difficult to create a dialogue on hunting ethics. It is increasingly discussed whether hunting as recreational activity or a sport should be considered morally distinct from other forms of hunting, such as subsistence hunting (Curnutt, 1996; List, 2004). Thus, it is being questioned whether it is acceptable to kill animals for recreational (and not 'survival') purposes; and whether hunting amounts to animal cruelty under such circumstances (Humphreys, 2010).

In a way the controversy around recreational hunting becomes a struggle about moral justification (Humphreys, 2014). It entails a number of ethical discourses associated with the killings, often framed as issues of animal rights or animal welfare (McLeod, 2007). Such discussions are not only of academic interest but play an important role in public and political discussions concerning legitimacy of hunting

(Fischer *et al.*, 2013). However, there is not much research on actual views, attitudes and underlying motivations of some of the key parties involved in recreational hunting – such as the hunters, the landowners and the general public – and the research that exists is often not representative of (national) populations (Gamborg and Jensen, 2014).

It seems a fair assumption that many people – if they think at all about 'game bird' shooting – imagine that the birds come from the wild. In several European countries, such as Norway, Sweden, Poland and Romania, game species are relatively abundant and recreational hunting allegedly takes place based on sustained yield 'harvest' principles – i.e. not based on farm reared and released game. However, in a number of countries, notably Spain, Denmark and the UK game bird shooting takes place on birds, such as pheasant and mallard, which are purpose-bred in cages, sheds and pens and subsequently released for hunting (Mustin *et al.*, 2012). Release can take place in many different ways and to different extents – from the rear and release of (few) birds for hunters' own recreational hunting to major releases of game birds for hunting/shooting on a commercial basis (Sanchez *et al.*, 2009). In the game bird/restocking industry, habitats are created which are conducive to the breeding of wild birds; birds are reared in enclosures; the birds are given supplementary feeding and provided with water once released; and natural predators are (often) controlled. In short, management is intensive in order to maintain high numbers of birds for 'driven shooting' where the game birds are driven towards the hunters by e.g. 'beaters'. There is a considerable commercial element (in the UK it provides an annual turnover of 1.6 billion £ (Canning, 2014), and in Denmark game birds represent two-thirds of the economic turnover related to game. Although perfectly legal in these countries, it is ostensibly contested (Animal Aid, 2012). The aim of this paper is to map out the attitudes of the general public, landowners and hunters to recreational hunting as such and the specific form of game bird shooting on reared and released birds.

Methods

The paper is based on a nationally representative survey from Denmark of the general public (n=1,001), landowners (n=1,207) and hunters (n=1,130). The respondent numbers (n) indicate fully completed questionnaires. Data were collected in late 2012/early 2013 as three separate web-based questionnaires prepared in SurveyExact. The general public questionnaire was sent out to a representative sample of the Danish population aged 18-79 years through a polling firm (AnalyseDanmark) which collected responses until a representative sample of one thousand respondents (in terms of age and sex) was reached. The landowner survey was based on a random sample from the Danish Central Business Register (CVR) among owners of agricultural and forestry holdings. In November 2012 2,127 selected landowners received a letter with a request to complete the online questionnaire via a personal link. After about 6 weeks a reminder was sent to those who had not yet replied. The survey achieved a total of 1,207 responses (48 via printed questionnaires that were sent to respondents upon request), yielding an overall response rate of 56.7%. For the hunter survey 2,000 hunters were randomly selected from the national Danish hunting license register among persons younger than 80 years old who had taken out a hunting license in 2011. In October 2012 the selected hunters received a letter with a request to complete the online questionnaire via a personal link. After about 5 weeks a reminder was sent to the non-respondents. The survey achieved a total of 1,130 responses (19 via printed questionnaires that were sent to respondents upon request), yielding an overall response rate of 56.5%.

The three surveys consisted of a pool of common questions regarding attitudes to hunting and game management and questions related to wildlife value orientations (cf. Fulton *et al.*, 1996). There was no distinction made in the surveys between release of pheasant, mallard or partridge, which are the three game birds species in Denmark. Partly the questions were in the format of reactions to statements relating to different aspects of rear and release of game birds. Respondents could indicate their level of (dis)agreement on a 5 point Likert scale. In the following, only descriptive statistics is presented in this paper although statistical analysis (chi-square tests) have been carried out.

Results and discussion

In general, as surveys have consistently shown, the importance of wildlife to the general public for recreational purposes is high in Denmark (and most other Western countries in fact). When Danes in 2008 were asked in a national representative survey, 'What changes in nature/the landscape around your town/residence would make you make more visits into nature?' the option 'more wildlife' (so there is a greater chance of seeing wildlife) was number one among 25 listed changes, with 40% of the respondents sharing that opinion (Jensen and Cleemann, 2015). The importance of experiencing wildlife is also demonstrated in two forest preference studies completed in the 1970s and 1990s where meeting wildlife like roe deer and pheasant were among the most preferred experiences in connection with a forest visit (Jensen, 1999).

So, how do respondents view taking the lives of wildlife for recreational purposes? To the general question of whether it is okay, in principle, to shoot wild animals as a pastime – formulated as 'What is your attitude to (recreational) hunting – all in all?' – 26% of the general public are somewhat or very negative; 30% indifferent, and 44% somewhat or very positive. Among landowners, not surprisingly only 10% are negative, 16% indifferent and 73% are positive. So, there is a significant support to recreational hunting in general. Respondents were also asked whether there may be circumstances concerning the way hunting is undertaken which played a role in their overall hunting attitude. The majority of the listed forms of hunting (e.g. at sea or in groups) did not affect respondents' attitude. The biggest effect, though, was found with regard to whether the hunt takes place on a natural wild stock or on farm reared and released game birds, where 48% of the general public and 37% of the landowners responded that hunting the farm reared game birds affected their hunting attitude negatively.

It is worth mentioning that some of the features typically associated with commercial shooting of game birds are also found to affect attitudes to hunting negatively. These features, for example, concern the type of lease of a hunting area and the number of hunters participating in the specific hunt/shooting. While, for example, only 9% of the public believe that single-man hunting affects their attitude negatively, the figure is 25% for hunting clubs. The corresponding figures for landowners are 4 and 23%. It also appears that attitudes are affected by the length of the hunting agreement: leases for a whole hunting season are viewed more favourably than one day shooting leases by all three groups. Looking at the overall attitude toward game birds shooting based on restocking, just over half the population are somewhat or very negative. Only a quarter is slightly or very positive, while a fifth is indifferent. For hunters, the situation is almost the reverse: 60% are somewhat or very positive about the release and a quarter is negative. The landowners are slightly more positive (45%) than negative (37%). So, there is a marked difference in the attitude to recreational hunting in general and the specific form of game bird shooting.

To get a better understanding of how the public, hunters and landowners think about rear and release practices solely for hunting, respondents were presented with 20 statements about the practice, based on an analysis of current public and stakeholder specific debates, and cf. Manfredo (2008). These statements consisted of 3 clusters of (to some extent overlapping) 'themes': animals (wildlife); nature (naturalness) and people (culture).

With regard to the first theme, animals: several statements dealt with game birds and their behaviour. 76% of landowners and 72% of the hunters agree that it is acceptable that released animals are fed for a period after the release. 55% of the public also agree to this. The answer to that question can be considered to be related to how released game birds are perceived: as a form of livestock or as non-reared wildlife. 47% of hunters and 49% of the landowners agree that one can observe a difference in behaviour of reared and released pheasants to wild ones. Only 30% of the public agree, but 49%, not surprisingly, answer 'do not know'. Approximately 20% of hunters and landowners are indifferent. 53% of the population, 51% of hunters and 50% of the landowners say they agree that released birds could be characterized as 'livestock without fences'. Only 26% of hunters disagree with this statement. So, this raises an ethical question of whether that changes the responsibilities we have toward the game birds.

With regard to the second theme, nature: part of the statements were about the impact on nature or how 'natural' or 'unnatural' rear and release of game birds is considered to be. One of the recurrent arguments put forward by hunters and landowners in favour of game bird shooting is that the whole set up of rearing and subsequent release has a positive impact in that the release helps to relieve the extent of hunting on wild populations (e.g. of pheasants). 52% of landowners and hunters state that they agree with this line of argument, while about 20% of these two groups answer they are indifferent. 41% of the public agree, but here it is worth noting that 21% answered 'neither nor' and 24% 'do not know', which probably indicates that a substantial number of the respondents are lacking knowledge or experience about this particular aspect. 61% of the hunters and 54% of landowners say that they agree to the statement that the practice of shooting reared and released game birds leads to increased motivation for nature conservation measures among hunters and forest owners. Only 31% of the public agree (30% answer 'do not know'). The opposite view, which is an argument that has been advanced in the debate among restocking opponents, and in particular argued by nature conservation groups and ornithologists, is that the actual so-called nature conservation measures in connection with the release cause a 'distortion' of nature. 42% of the members of the public agree, 26% answer 'do not know' and 19% are indifferent, probably (again) reflecting a lack of specific knowledge about the effects (or to be seen as an indication that it is a bit unclear what is meant by 'distortion' of nature). 35% of hunters and 32% of landowners disagree with the statement. Another variant on the theme of 'distortion' or unnaturalness is the argument that released pheasants can occur in such large quantities that it appears 'unnatural'. 69% of hunters and 61% of landowners actually agree with this (but still perhaps continue the practice of restocking for shootings). 42% of the general public also agree that it is 'unnatural', but 34% respond 'do not know' – again reflecting a probable lack of knowledge/experience or lack of clarity about what precisely to associate with 'unnatural' (in the questionnaire, the term was deliberately placed in inverted commas to indicate a somewhat ambiguous term). So, here is an ethical discussion of what kind of nature is to be preferred and the associated degree of perceived naturalness.

With regard to the third theme, people, an often heard argument among proponents of the restocking practice is that shooting of the reared and released game birds is important for economic development in rural areas. This is an argument of necessity: protecting rural livelihoods. However, only 28% of landowners and 27% of the hunters agree with this, and only 14% of the public (though it should be noted that 34% of public answer 'do not know'). A relatively large percentage of hunters (19%) and landowners (24%) answer 'neither nor' or 'strongly disagree'. For the statement that the shooting of released game birds is a centuries-old tradition that should continue, 46% of the landowners and 59% of hunters agree (in varying degrees). 32% of the public disagree with this, but about as many 'do not know'.

Perhaps they do not know whether it is an old tradition or whether the practice should continue. When it comes to another type of reason with pronounced ethical content – the claim that the release of game birds is only to satisfy hunters' need to shoot many birds for pleasure – 57% of the public and landowners agree, and, somewhat surprisingly, 51% of the hunters also agree. To the related claim that reared and released game birds are easier to shoot than the wild birds, 47% of the hunters, 43% of landowners and 45% of the public agree (not surprisingly 35% of the public answer 'do not know'). The latter also seems to entail a discussion among 'ranks' of what constitutes 'proper' hunting.

Generally, it seems there is more or less agreement among the three stakeholder groups on a number of specific issues relating to releasing and shooting of game birds – yet different overall conclusions about the acceptability of the practice are drawn. How come? Before trying to answer that question it is worth noting a high proportion of 'either-or/don't know' in the responses of the general public. One obvious explanation is lack of interest in the subject. If that is the case the opinion may be marked but not necessarily strongly held. If it is due to lack of knowledge it is important to be aware that more knowledge will not necessarily serve to change attitudes, as seen in e.g. debates about genetically modified crops. Marked differences in overall attitudes to shooting of released game birds between landowners and hunters and the general public could also be explained in terms of differences in underlying values.

Conclusion

Is it acceptable, in principle, to shoot wild animals as a sport? And is it more or less ethically problematic when it is a 'staged' kill of animals which have been farm reared and subsequently released into the wild? Distinctive differences in extent and legislation are at play across Europe concerning this form of hunting (two extremes: in the UK, 50 million pheasants and partridges are reared and released every year whereas in the Netherlands the practice was banned in 2004).

The results presented in this paper show that in Denmark, there is broad agreement among the general public, hunters and land owners that recreational hunting of 'surplus' wildlife (i.e. non-reared and released) is acceptable, but release of game birds for shooting is a dividing issue both within the groups and between the groups. In the general public there is a majority against the practice; a majority of hunters are in favour, and landowners somewhere in between. Reasons may be found within different views on how the reared and released birds should be viewed (as 'wild' or 'domesticated' animals; how (un)natural the practice is seen to be; and whether recreational hunting in the form of game bird shooting is considered 'proper' hunting. Differences may at another level also be explained in terms both of knowledge and shifts in underlying values, partly towards wildlife. Here, further analysis and work need to be done. Ultimately, the observed division might come to influence individual's views on the support or opposition to other forms of hunting – and hunting as such.

References

Animal Aid (2012). Background notes on the pheasant shooting industry. Available at: http://tinyurl.com/orxmcq7.

Canning, P. (2014). The UK game bird industry – a short study. Prepared for DEFRA. Available at: http://archive.defra.gov.uk.

Cahoone, L. (2009). Hunting as a moral good. Environmental Values 18: 67-89.

Causey, A.S. (1989). On the morality of hunting. Environmental Ethics 11: 327-343.

Curnutt, J. (1996). How to argue for and against sport hunting. Journal of Social Philosophy 27: 65-89.

Dahles, H. (1993). Game killing and killing games: an anthropologist looking at hunting in a modern society. Society and Animals 1: 169-184.

Fischer, A., Kereži, V., Arroyo, B., Delibes-Mateos, M., Tadie, D., Lowassa, A., Krange, O., and Skogen, K. (2013). (De) legitimizing hunting – discourses over the morality of hunting in Europe and eastern Africa. Land Use Policy 32: 261-270.

Fulton, D.C., Manfredo, M.J. and Lipscomb, J. (1996). Wildlife value orientations: a conceptual and measurement approach. Human Dimensions of Wildlife 1: 24-47.

Gamborg, C. and Jensen, F.S. (2014). Recreational hunting – a comparative study of attitudes to rear and release among Danish hunters, land owners and the general public. The International Forestry Review 16: 53.

Humphreys, R. (2010). Game birds: the ethics of shooting birds for sport. Sport, Ethics and Philosophy 4: 52-65.

Humphreys, R. (2014). The argument from existence, blood-sports, and 'sport-slaves'. Journal of Agricultural and Environmental Ethics 27:331-345.

Jensen, F.S. (1999). Forest recreation in Denmark from the 1970s to the 1990s. The research series, vol. 26, The Danish Forest and Landscape Research Institute, Hørsholm. Denmark, 117 pp.

Jensen, F.S. and Cleemann, M.F. (2015). Friluftsliv i landskabet 2008 (20). Forandringer i naturen/landskabet. Videnblade Planlægning og Friluftsliv. Forest & Landscape, Frederiksberg, 2 pp.

List, C. (2004). On the moral distinctiveness of sport hunting. Environmental Ethics 26: 165-167.

Loftin, R.W. (1984). The morality of hunting. Environmental Ethics 6: 241-250.

Manfredo, M.J. (2008). Who cares about wildlife? Social science concepts for exploring human-wildlife relationships and conservation issues. Springer, New York, USA. 228 pp.

McLeod, C.M. (2007). Dreadful/delightful killing: the contested nature of duck hunting. Society and Animals 15: 151-167.

Morris, S.P. (2013). Challenging the values of hunting: fair chase, game playing, and intrinsic value. Environmental Ethics 35: 295-311.

Mustin, K., Newey, S., Irvine, J., Arroyo, B. and Redpath, S. (2012). Biodiversity impacts of game bird hunting and associated management practices in Europe and North America. Report. James Hutton Institute, Aberdeen, Scotland, UK, 72 pp.

Sanchez García-Abad, C., Alonso de La Varga, M.E., Prieto Martín, R., Gonzalez Eguren, V. and Gaudioso Lacasa, V.R. (2009). A view of game bird farming in Spain. Aida-itea 105(3): 169-183.

Wood, F. (1997). The delights and dilemmas of hunting. The hunting versus anti-hunting debate. University Press of America, Lanham, USA, 237 pp.

Assess the impact of traditional and localized food under economic recession

G. Vlontzos[1*], M.N. Duquenne[2] and S. Niavis[2]
[1]University of Thessaly, Department of Agriculture Crop Production and Rural Environment, Fytoko street, 38446 Volos, Greece; [2]University of Thessaly, Department of Planning and Regional Development, Pedion Areos, 38334 Volos, Greece; gvlontzos@agr.uth.gr

Abstract

The aim of this paper is to examine the changes on food consumption patterns of households, following an economic crisis and recession. By applying the Theory of Planned Behaviour (TPB), this paper identifies objective and subjective factors influencing consumers to purchase, or not, traditional and localized food. Furthermore, there is an assessment of the impact of environmental and social criteria for food selection, like CO_2emissions, locality of production, healthy nutrition habits, etc. Quite enlightening is the research on the spatial basis. The results of this approach signify the different consumption trends dominating the urban, semi-urban, and rural areas. There are also some implications for marketing and public policy makers to further strengthen consumption of traditional and locally produced foodstuffs, such as promotion of their nutritional characteristics, support of local economies, encouragement for adoption of healthy eating by children, as well as enjoyment of cooking.

Keywords: traditional food, theory of planned behaviour, food safety, local production

Introduction

Consumption of Traditional Food Products (TFPs) is the topic of many scientific researches, trying to indentify consumer perceptions, behaviours, as well as interactions with a series of developmental mechanisms. Focusing on the product image it was evident that exploiting the impact of it, leads to quite successful implementations of marketing strategies (Guinard *et al.*, 2001). It is quite important though from the very beginning to clarify what TFP is. Up to now there are several approaches, applying a series of criteria or methodologies to shape a definition, with the most important ones to be cultural, territorial, focus on traditional ingredients, composition and processing (Jordana, 2000; Trichopoupou *et al.*, 2007; Vanhonacker *et al.*, 2010; Weichselbaum *et al.*, 2009). Having in mind the importance of the final consumer's beliefs on the same issue, it is quite noteworthy for the food industry to follow the already proven as best practices, which are high quality and high value image of TFPs, trying on the same time to improve convenience, increase the notion of exclusivity, develop the healthy image of them, and reassure the environmentally friendly profile of them (Almli *et al.*, 2011;. Guerrero *et al.*, 2009; 2010) makes up a ten dimension qualitative food concept, consisting of sensory, health, elaboration, heritage, variety, habit, origin, simplicity, special occasions, and marketing. Despite the fact of the existing controversy between the terms innovation and tradition, there is continuous demand for innovative products, even in the sector of TFPs (Gellynck *et al.*, 2008; Jordana, 2000). Such changes refer to quality aspects, modern packaging techniques, reassurance for health and safety issues, guarantee of origin, and convenience improvements (Kühne *et al.*, 2010). Further attempts to clarify the impact of this controversy, led to application of the sorting task methodology, which suggested, as innovative techniques, the improvement of nutritional value of TFPs, as well as the development of the variety of choices and stressing the national/local origin of them (Guerrero *et al.*, 2012). Perhaps one of the most important characteristics of TFPs is the level of consumers' satisfaction. Their acceptability is based upon the quality characteristics, and the symbolic connotations they raise (Sanzo *et al.*, 2008). Such a dietary pattern seems to be a sustainable alternative for westernized societies, after many decades of adopting diets responsible for cardiovascular disease and cancer (Kearney, 2010). Having as motives the health attribute, among others, it is realized that there is a negative association between weight control and TFP consumption in regions where they are quite fatty, like Norway and Poland (Pieniak *et al.*, 2009). Field researches in different places lead to controversial results, with Asian societies and Canadian Inuit adults, to adopt rapidly westernized diets, by far different from their traditional meals (Pingali, 2006; Hopping *et al.*, 2010). In Greece there was an attempt to assess the consumer perceived quality for traditional meat supply chain. The critical points for choosing meat are the intrinsic quality cues, and the place of origin (Krystallis *et al.*, 2007). Regarding the consumers' perception of meat quality, significant criteria are the taste, in accordance with colour and leanness. There are also quite discrete consumer groups focusing on other issues too, like labels, brand name, safety certifications etc. (Krystallis and Arvanitoyannis, 2006).

Greece for six continuous years is facing a quite intense economic crisis aggravating, among others, purchasing power of consumers and unemployment rates. This new economic and social status quo modified in many cases food consumption habits (Duquenne and Vlontzos, 2013). Due to the fact that the country holds a quite important heritage of food products, it is essential to identify the impact of this crisis on TFPs. Therefore, a field research was conducted on a national level, during the period September-October 2014. Data have been collected from 225 respondents selected through a stratified random sampling. This sample covers a wide field of Greek urban and rural areas. For the purpose of the analysis, the very elderly consumers, more than 85 years old, as well as the youngest, less than 18 years old, were excluded from the analysis, in order to avoid some specific constraints, especially as regards health aspects and certain diets. The sample includes 98 men and 127 women with mean age 41 years (SD=16.5).

Results

Food-related perceptions

In order to measure how the respondents link food consumption to the attainment of life values and quality, 34 items were subjected to assessment through a seven-point scale, from 'totally disagree' to 'totally agree'. The Principal Component Analysis with varimax rotation allows to extract nine (9) dimensions of food-related lifestyle. Pleasure procured by foods and health issues are clearly the two first consumers' concerns, with food origin (localization, proximity to the place of residence) to be also a determining aspect of consumers' choice (Table 1). On the contrary, cooking methods – especially those refereeing to novelty – are not perceived as major aspects.

Table 1. Food consumption as regards attainment of life values.

Food-related lifestyle	Number of items	Eigen values	% of variance	Cronbach a
Pleasure	5	7.955	23.398	0.883
Health aspects	6	3.592	10.566	0.845
Foods' origin	5	2.571	7.561	0.781
Taste aspects	3	1.943	5.714	0.740
Ways of shopping	4	1.766	5.195	0.724
Attitude against obesity	3	1.488	4.377	0.810
Consumption habits	2	1.232	3.625	0.793
Simple methods of cooking	3	1.118	3.289	0.704
Looking for new cooking methods	3	1.099	3.232	0.600

Some significant differences are observed between men and women. Systematically, women scored higher aspects related to: health, obesity risk and pleasure. It also appears that, compared to men, simple ways of cooking is more important for women (Table 2).

The income level is partly influencing the consumers' perception. More precisely, some significant differences are observed as regards the first group of respondents, those with the lowest level of income (less than 500€ per respondent income per month). In most cases, their scores are significantly lower for items directly related with health and obesity comparatively to respondents with higher income (Table 3). Consequently, we can consider that the crisis effectively contributes to place to a secondary position, health aspects as well as nutritional value of foods. Even if the level of statistical significance is not very intense, the impact of TFPs tends to be less important for respondents with low income, suggesting, at least indirectly, that prices' consideration are most important.

Respondents living in rural areas scored systematically higher the origin of foods, giving therefore more importance to traditional and localized food, comparatively to urban population. Regarding urban population, taste aspects have systematically higher values (Table 4).

Attitudes against traditional and localized food products.

Most of the respondents indicated that they purchase occasionally (46%) and frequently (41%) traditional food products produced locally. Quite interesting is though those consumers purchasing rarely such foods are mainly urban residents. It is interesting to mention that there is no strong relation between this purchasing attitude and income level, except consumers with very low income (less than 500€) whose scores as regards purchasing frequency are on average quite lower (Table 5). Consequently, even in this period of deep economic crisis, living standards cannot be considered as a determinant factor.

Table 2. Attainment of life values for men and women.

Food-related lifestyle	Men (n=98)		Women (n=127)	
	Mean	SD	Mean	SD
Pleasure	5.2	1.4	5.5[a]	1.4
Health aspects	6.0	0.8	6.3[b]	0.8
Foods' origin	5.1	1.3	5.3	1.2
Taste aspects	5.6	1.2	5.8	1.1
Ways of shopping	5.6	1.2	5.9[c]	1.0
Attitude against obesity	5.2	1.5	5.6[d]	1.3
Consumption habits	4.5	1.7	4.5	1.6
Simple methods of cooking	5.5	1.2	5.8[e]	1.0
Looking for new cooking methods	3.4	1.5	3.6	1.4

[a] Mean difference significant at *P*<0,044**.

[b] Mean difference significant at *P*<0,013**.

[c] Mean difference significant at *P*<0,063*.

[d] Mean difference significant at *P*<0,019**.

[e] Mean difference significant at *P*<0,060*.

Table 3. Attainment of life values as regards consumers' income.

Food-related lifestyle	Average monthly income's levels							
	<500€ (n=99)		500-1000€ (n=67)		1000-1,500€ (n=38)		>1,500€ (n=15)	
	Mean	SD	Mean	SD	Mean	SD	Mean	SD
Pleasure	5.2	1.4	5.6	1.2	5.4	1.6	5.6	1.5
Health aspects	5.9	1.0	6.3[a1]	0.6	6.2	0.8	6.4[a2]	0.6
Foods' origin	5.0	1.4	5.5[b]	1.1	5.4	1.2	5.3	1.2
Taste aspects	5.7	1.1	5.8	1.2	5.6	1.2	5.9	1.0
Ways of shopping	5.8	1.2	5.9	1.1	5.7	1.1	5.5	1.1
Attitude against obesity	5.2	1.5	5.5	1.3	5.7[c1]	1.2	5.9[c2]	1.0
Consumption habits	4.3	1.7	4.7	1.6	4.9[d]	1.6	4.5	0.9
Simple methods of cooking	5.5	1.1	5.8	1.1	5.6	1.0	6.0	1.1
Looking for new cooking methods	3.4	1.5	3.7	1.4	3.3	1.4	3.5	0.6

[a1] Income <500€ versus income 500-1000€, F=2.710, $P<0.007$***.

[a2] Income <500€ versus income >1,500€, F=3.207, $P<0.076$*.

[b] Income <500€ versus income 500-1000€, F=5.685, $P<0.018$**.

[c1] Income <500€ versus income 1000-1,500€, F=3.606, $P<0.058$*.

[c2] Income <500€ versus income >1,500€, F=2.949, $P<0.089$*.

[d] Income <500€ versus income 1000-1,500€, F=3.840, $P<0.052$*.

Table 4. Attainment of life values as regards consumers' place of life.

Food-related lifestyle	Rural (n=75)		Semi-urban (n=45)		Urban (n=105)	
	Mean	SD	Mean	SD	Mean	SD
Pleasure	5.43	1.52	5.07	1.22	5.47	1.35
Health aspects	6.23	0.73	5.73[a]	0.81	6.25	0.88
Foods' origin	5.63	1.09	4.78[b1]	1.22	5.14[b2]	1.30
Taste aspects	5.56	1.28	5.35	1.20	6.02[c]	1.00
Ways of shopping	5.83	0.90	5.49	1.26	5.87	1.18
Attitude against obesity	5.48	1.44	5.41	1.16	5.38	1.42
Consumption habits	4.55	1.73	4.06	1.66	4.66	1.56
Simple methods of cooking	5.69	0.89	5.13[d]	1.19	5.82	1.04
Looking for new cooking methods	3.34	1.55	3.59	1.47	3.62	1.27

[a] Rural versus semi-urban. F=12.076, $P<0.001$***.

[b1] Rural versus semi-urban. F=15.810, $P<0.0005$***.

[b2] Rural versus urban. F=7.156. $P<0.008$***.

[c] Rural versus urban. F=7.409, $P<0.007$***.

[d] Rural versus semi-Urban. F=8.493, $P<0.004$***.

Table 5. Traditional and localized purchasing frequency by income level.

	Average monthly income's levels							
	<500€ (n=99)		500-1000€ (n=67)		1000-1,500€ (n=38)		>1,500€ (n=15)	
	Mean	SD	Mean	SD	Mean	SD	Mean	SD
Purchasing frequency (from 1: 'very rarely' to 7: 'very frequently')	4.8	1.47	5.2[a]	1.55	5.3[b]	1.33	4.8[c]	2.01

[a] Lower income (<500€) versus income (500-1000€). F=3.126, *P*<0.079*.

[b] Lower income (<500€) versus income (1000-1,500€). F=2.639, *P*<0.107.

[c] Lower income (<500€) versus high income (>1,500€). No significant difference.

Furthermore the educational level appears with a quite 'surprising' impact: the respondents having basic education levels are purchasing traditional food products more frequently (Table 6).

As regards the motives to purchase traditional and local foods, we observe an especially high correlation with purchasing frequencies. More precisely, three motives: (1) good consumption behaviour; (2) healthy attitude; and (3) economic concerns, have clearly a positive impact on the decision to purchase more or less frequently this type of product (Table 7). It is also necessary to underline that most of the respondents do not perceive traditional and local foods as expensive ones, as there are no significant price differences between TFPs and imported ones. Only 15% of them consider that their income does not allow them to consume this type of foods. Consequently they rarely purchase traditional and local foods even if they admit that this consumption behaviour is a good one, especially in terms of health. Among these consumers, the vast majority (60%) are experiencing a very fragile economic situation, with a monthly income less than 500€.

Table 6. Traditional and localized purchasing frequency by education level.

	Education levels							
	Primary (n=19)		Lower secondary (n=26)		Upper secondary (n=79)		University (n=101)	
	Mean	SD	Mean	SD	Mean	SD	Mean	SD
Purchasing frequency (from 1: 'very rarely' to 7: 'very frequently')	5.9	1.47	5.2[a]	1.55	5.3[b]	1.33	4.8[c]	2.01

[a] Primary level versus lower secondary. F=4.361, *P*<0.043**.

[b] Primary level versus higher secondary. F=8.499, *P*<0.004***.

[c] Primary level versus university level. F=4.7056, *P*<0,032**.

All other comparisons not significant.

Table 7. Purchasing motives by purchasing frequency.[1]

Motives	Rarely (n=29)		Sometimes (n=104)		Often (n=92)	
	Mean	SD	Mean	SD	Mean	SD
Good consumption behaviour	4.3	1.93	5.4	1.19	6.4	0.87
Healthy attitude	3.9	1.69	4.9	1.42	5.9	1.22
Economically advantageous	4.4	1.82	4.6	1.61	5.9	1.30

[1] For each one of these three motives, all the mean differences are significant with $P<0.001$***. The only one exception concerns the economic motive: the difference between the 1st (rarely) and 2nd (sometimes) groups has a $P>0.10$.

Conclusions

This research identifies the most important characteristics of consumers' behaviour for TFPs. While it was expectable low income consumers not to choose traditional foodstuffs, because of their weak purchasing power, it is quite surprising the fact that well educated consumers with relatively high incomes do not include in their nutritional habits TFPs very often. The main obstacle for this is the lack of trust regarding food safety issues, as well as significant difficulties on finding them near their place of residence. These consumer trends are controversial to European nutrition trends, creating scepticism about the future of production channels focused on TFPs. It is a first priority issue for policy makers to implement a promotion strategy to encourage TFPs consumption, based on the competitive advantages of them on both health and developmental argumentation, and promote the ethical issues accompanying consumption of TFPs.

References

Almli, V.L, Verbeke, W., Vanhonacker, F., Næs, T. and Hersleth, M. (2009). General image and attribute perceptions of traditional food in six European countries. Food Quality and Preference 22(1): 129-138.

Duquenne, M.N. and Vlontzos, G. (2013). The impact of the Greek crisis on the consumers' behaviour: some initial evidences? British Food Journal 116(6): 890-903.

Gellynck, X., Molnár, A. and Aramyan, L. (2008). Supply chain performance measurement: the case of the traditional food sector in the EU. Journal on Chain and Network Science 8: 47-58.

Guerrero, L., Claret, A., Verbeke, W., Enderli, G., Zakowska-Biemans, S., Vanhonacker, F., Issanchou, S., Sajdakowska, M., Signe Granli, B., Scalvedi, L., Contel, M., Hersleth, M. (2010). Perception of traditional food products in six European countries using free word association. Food Quality and Preference 21(2): 225-233.

Guerrero, L., Claret, A., Verbeke, W., Vanhonacker, F., Enderli, G., Sulmont-Rossé, C., Hersleth, M., Dolors Guàrdia, M. (2012). Cross-cultural conceptualization of the words traditional and innovation in a food context by means of sorting task and hedonic evaluation. Food Quality and Preference 25: 69-78.

Guerrero, L., Guardia, M.D., Xicola, J., Verbeke, W., Vanhonacker, F., Zakowska-Biemans, S., Sajdakowska, M., Sulmont-Rossé, C., Issanchou, S., Contel, M., Luisa Scalvedi, M., Signe Granli, B., Hersleth, M. (2009). Consumer-driven definition of traditional food products and innovation in traditional foods. A qualitative cross-cultural study. Appetite 52(2): 345-354.

Guinard, J. X., Uotani, B. and Schlich, P. (2001). Internal and external mapping of preferences for commercial lager beers: comparison of hedonic ratings by consumers blind versus with knowledge of brand and price. Food Quality and Preference 12(4): 243-255.

Hopping, B.N., Mead, E., Erber, E., Sheehy, C., Roache, C. and Sharma, S. (2010) Socioeconomic indicators and frequency of traditional food, junk food, and fruit and vegetable consumption amongst Inuit adults in the Canadian Arctic. Journal of Human Nutrition and Dietetics 23: 51-58.

Jordana, J. (2000). Traditional foods: challenges facing the European food industry. Food Research International 33(3-4): 147-152.

Kearney, J. (2010). Food consumption trends and drivers. Philosophical Transactions of the Royal Society 365: 2793-2807.

Krystallis, A. and Arvanitoyannis, I.S. (2006). Investigating the concept of meat quality from the consumers' perspective: the case of greece meat science 72(1): 164-176.

Krystallis, A., Chryssochoidis, G. and Scholderer, J. (2007). Consumer-perceived quality in 'traditional'food chains: the case of the Greek meat supply chain. Appetite 48(1): 54-68.

Kühne, B., Vanhonacker, F., Gellynck, X. and Verbeke, W. (2010). Innovation in traditional food products in Europe: do sector innovation activities match consumers' acceptance? Food Quality and Preference 21: 629-638.

Pieniak, Z., Verbeke, W., Vanhonacker, F., Guerrero, L. and Hersleth, M. (2009). Association between traditional food consumption and motives for food choice in six European countries. Appetite 53(1): 101-108.

Pingali, P. (2006). Westernization of Asian diets and the transformation of food systems: implications for research and policy. Food Policy 32: 281-298.

Sanzo, M.J, Belén del Río, A., Iglesias, V. and Vázquez, R. (2003). Attitude and satisfaction in a traditional food product. British Food Journal 105(11): 771-790.

Trichopoulou, A., Soukara, S. and Vasilopoulou, E. (2007). Traditional foods: a science and society perspective. Trends in Food Science & Technology 18(8): 420-427.

Vanhonacker, F., Verbeke, W., Guerrero, L., Claret, A., Sulmont-Rossi, C., Raude, J., Granli, S., Hersleth, M. (2010). How European consumers define the concept of traditional food: evidence from a survey in six countries. Agribusiness. DOI: http://dx.doi.org/10.1002/agr.20241.

Weichselbaum, E., Benelam, B. and Soares Costa, H. (2009). Traditional foods in Europe. In European Food Information Ressource (EuroFIR) consortium, EU 6th framework food quality and safety thematic priority. Contract FOOD-CT-2005-513944. Available at: http://www.eurofir.net.

Transforming a conflict into a market creation for Dutch chicken meat

J.M. Bos[1], P.H. Feindt[2] and B. Gremmen[1]*
[1]Department of Philosophy, Wageningen University, P.O. Box 8130, 6700 EW Wageningen, the Netherlands; [2]Department of Strategic Communication, Wageningen University, P.O. Box 8130, 6700 EW Wageningen, the Netherlands; jacqueline.bos@wur.nl

Abstract

In the late 1990s, animal scientists from Wageningen University started to develop the very idea of an 'intermediary' chicken meat product that has more welfare quality than broilers in mainstream production systems but less than broilers in organic production systems. This was recognised as a market opportunity by seemingly opposing groups of actors in the Dutch chicken meat supply chain: industry, NGO's and retail. Their shared mission was to develop an intermediary chicken meat product that included animal welfare attributes and in turn, was affordable for consumers. However, in the process of collaboration, consensus was rarely reached, and fragile when as it was. At the same time, cooperation and communication continued, often un-problematically. How was this possible? In this paper, we present a case study and focus on different groups of actors transforming a societal conflict on animal welfare into collaboration and market creation. We use the concept of boundary object as the approach to explain both the process and the product. The case study on the transition of a societal conflict into market creation for Dutch chicken meat brings up at least two issues that require attention in the process of collaboration and market creation. These issues pertain to negotiating the shared meaning of an intermediary product, and the organisation of interactions between groups of actors that maintain their differences in the process. The boundary crossing feature of an intermediary product that positions itself on both animal welfare and price, makes it multi-interpretable, and therefore, acceptable to otherwise possibly conflicting interpretations. The notion of an intermediary product made this struggle explicit. The bridging and conflicting concerns at the boundary of the different groups of actors offered a new horizon on the transition towards sustainable practices, in our case in the broiler sector. The organisation of interactions that led to continuity and consistency in the process of collaboration and market creation for Dutch chicken meat are summarized in three lessons learned.

Keywords: animal welfare, boundary object, meat supply chain, groups of actors

Introduction

Despite the widely held notion that the issue of animal welfare, like other issues such as climate change, can be characterised as an intractable problem (Lazarus, 2008; Levin *et al.*, 2012; Rittel and Webber, 1973), several solutions for animal friendly food production have been explored. In the Netherlands, government and multiple stakeholders in livestock production introduced the idea of an 'intermediary segment' that positions itself between mainstream and organic products (ANF, 2009a; 2009b). In the late 1990s, animal scientists from Wageningen University already started to develop the very idea of an 'intermediary chicken' that has more welfare quality than broilers in mainstream production systems but is cheaper than broilers in organic production systems (Van Horne *et al.*, 2002; 2003). This concept was recognised as a market opportunity by seemingly opposing groups of actors in the chicken meat supply chain: industry, NGO's and retail. Their shared mission, although initially not made explicit, was to develop an intermediary chicken meat product that included animal welfare attributes and in turn, was affordable for consumers. However, in the process of collaboration, consensus was rarely reached, and fragile when as it was. At the same time, cooperation and communication continued, often un-problematically. How was this possible? In this paper, we focus on groups of actors transforming a societal conflict on animal welfare into collaboration and market creation. We use the concept of boundary object as the approach to explain both the process and the object. This paper seeks to answer the following questions: How were societal conflicts transformed into market creation for animal friendly products? How were different groups of actors able to collaborate while at the same time maintaining their disagreements? To explore these questions, we present a case study on the transformation of a societal conflict on animal welfare between 1997 and 2012.

Material and methods

We adopt in this study a case study methodology, because case studies are well suited for exploratory research, rich in context and enabling for tracing of processes and causal mechanisms (Yin, 2014). Our perspective practices a mechanism-oriented style of theorizing which aims to uncover the social mechanisms that underlie certain processes. We have chosen a case on the transformation of a societal conflict on animal welfare, because animal welfare is a topic that has received attention for a long time in Europe, as evidenced by the wide range of claims of animal-friendly production on products in stores across Europe (Schmid and Kilchsperger, 2010). Moreover, as an ethical issue, animal welfare is highly ambiguous because human perceptions of animal welfare may diverge strongly from what the animals themselves experience (Bracke and Hopster, 2006).

The case study is based on multiple methods and data sources, which allows for triangulation (Yin, 2014). First, from the perspective of communication science, a media analysis was conducted of newspapers and professional magazines on the coverage of 'intermediary segment'. This served as a first approximation of public attention for the issue and provided an indication of the attention peaks, i.e. the relative importance of 'intermediary segment' in the public sphere (Downs, 1972). Full articles of five major Dutch national newspapers and five professional magazines related to the agro-food sector were electronically accessed through LexisNexis. In a first step, all articles were sampled which contained the keyword 'intermediary segment' (n=309). In a second step, all articles that did not report on agriculture, food production and sustainability were excluded by manual interpretative content analysis. Of the 221 articles that remained, 92 articles were published in the upswing phase, 73 articles in the peak, and 56 articles in the downswing phase. For each of these phases, the individual articles were uploaded in 'Wordle' to create wordclouds. The wordclouds display the most frequently used words in the largest letter type, and this indicates what the dominant topics were. Because wordclouds do not reveal in what way these topics were discussed, we looked for concepts, actors linked to the intermediary segment, general tone and main concerns in each individual article (n=221) by manual interpretative content analysis.

Second, we used policy documents, articles, research reports and public sources such as websites and blogs to identify the dominant topics in the discussions over time, as well as the nexus of interactions between the different groups involved in the intermediary segment. Third, we conducted, recorded and transcribed five interviews (one was held in 2012 and four were held in 2014); one interview for each group collaborating in the Dutch chicken meat product development between 1997 and 2012 (identified by the informants). We interviewed representatives from a feed company, an animal protection organisation, a large national chicken meat processor, a farmers' organisation, and a large national retailer. Data from the interviews were analysed using a narrative approach called re-transcription that filters the organisations' storylines and practices, and others involved in the conflict, enabling analysis of competition across storylines (Dodge, 2014; Feldman *et al.*, 2006). Third, all the information acquired from media analysis, desk research and interviews was summarised and presented as a whole narrative of the case.

We provide our own narrative on the transformation of a societal conflict on animal welfare, by putting the boundary object at the centre of our conceptual framework. A boundary object refers to a multi-interpretable concept – or object – that is both plastic enough to adapt to the local needs and the constraints of the several parties employing them, yet robust enough to maintain common identity across sites. They can be more or less tangible, and hence can be more a less an object in the true sense (Star and Griesemer, 1989). Boundary objects incorporate ambiguity (Bowker and Star, 2000) and have interpretative flexibility which means that they can mean different things to different people depending on contextual factors, and that these different interpretations, uses and views of the object are valid in different contexts (Klerx *et al.*, 2012; Star and Griesemer, 1989).

Results

The experience of animal epidemics such as Bovine Spongiform Encephalopathy, Food and Mouth Disease and Avian Influenza, and the massive culling of animals that these generated, fuelled a new and heated public debate at the beginning of the 20th century about the welfare of farm animals and European agriculture's 'licence to produce' (Bock and Van Huik, 2007). Coupled with a growing environmental awareness and concerns about health and quality of diet led many people question modern agricultural practices. There was genuine interest among the pioneers of organic farming in creating livestock systems that better fulfil the animal needs than the mainstream livestock systems. However, an increasing supply of organic products was not reaching the European consumer as organic marketed; a situation to which production oriented rather than market oriented organic schemes contributed (Latacz-Lohmann and Foster, 1997). Thus, there was some kind of urgency and at the same time an opportunity to fill the gap between organic and mainstream products. In 2009, a covenant between Dutch government, agro-food industry, NGO's and retail led to a similar idea that was coined as intermediary segment (ANF, 2009a). Against this background, a process of collaboration and market creation for Dutch chicken meat unfolded. This process essentially consisted of three stages.

In the first stage (1997-2003), the legitimacy of the Dutch broiler chicken sector was contested by 'outsiders' who gradually coalesced into a research project investigating an intermediary product, i.e. a product that has more animal welfare quality than broilers raised in mainstream production systems, but less than broilers raised in organic production systems. The research primarily focused on fast growing broiler breeds and broiler health problems, such as ascites and footpad lesions, and permanent hunger of broiler parents. Novel broiler husbandry systems were also investigated and test-panels considered the taste and texture of chicken meat. As relative outsiders, a feed company, an animal protection organisation, a farmers' organisation, and a group of retailers became involved in the research. They had to deal with the problem that intermediate chickens have much higher environmental impact, and the welfare aspect of the growth rate of the breed is very debatable, as it is the balanced breeding programme that counts. Throughout the research project, they became aware of each other's vision by talking among themselves and exploring new concepts. Scientific knowledge about broiler health and welfare, broiler husbandry systems and meat quality provided a basis to actually develop a new chicken meat product.

In the second stage (2003-2007), a supply chain was developed for Dutch chicken meat product that included animal welfare attributes and was also affordable. In line with the results of the research project, animal welfare was in the first place conceived as a slower growing broiler breed. At the same time, different groups of actors made suggestions from their own perspective. The animal protection organisation insisted on a covered free range area in the barn, a view that matched with the retailers' focus on consumer perception and product image. For the food processor, a covered free range area in the barn would lead to a higher cost price, unless they could maximise the utilisation of the carcass. Furthermore, the farmers' organisation and the food processor both pointed to the occupancy rate in the barn and the extended life-time before slaughtering as important cost price increasing aspects. On the other hand, a better diet and an extended life-time before slaughtering would lead to a better taste of the meat as well as a higher level of animal welfare; a view that was shared between the feed company and the animal protection organisation.

In the process of partnership building and relationships between the actors, a sense of realism and sound arguments about what is technically feasible (or not) came to the forefront. The animal protection organisation was willing to deviate from their organic ideal. Similarly, the farmers' organisation was willing to take their responsibility instead of protesting against policies. The feed company insisted on a product with a national image that would build trust for the consumer. The animal protection organisation found a partner in their opponents, a feed company and a farmers' organisation, to persuade retailers to adopt a new chicken meat product in their assortments. In 2006, a memorandum of agreement with four retailers resulted in the launch of a new product brand called Volwaard. By placing a Better- Life hallmark on the packages, the animal protection group explicitly communicated their commitment.

In the third stage (2007-2012), the largest national retailer decided to exchange the product brand Volwaard for their own sustainability home brand Pure-and-Honest. For the retailer, this step had two advantages. First, they got the freedom to purchase the intermediary chicken meat from any other party than from the closed Volwaard supply chain. This led to an increase in turnover each year (between 2009-2012) for this product category. Second, they could develop a new intermediary chicken meat product together with a large national chicken meat processor. For this product, called Free-Range, the slower growing broiler breed criterion remained the same and the reduction of antibiotics came more to the forefront. Interestingly, the free range criterion for animal welfare – in terms of a covered free area in the barn – was abandoned. Although the animal protection organisation was principally against the direction towards less animal welfare product attributes in the Free-Range initiative, the conversations between the actors remained intact. Moreover, the Free-Range product offered an acceptable outcome, 'not in the least, because a more radical NGO challenged it as not being (sufficiently) animal friendly. This helped the animal protection organisation to enforce corporate commitment to incremental improvement of animal welfare' (Bos *et al.*, 2013).

Discussion and conclusion

The current study on the transition of a societal conflict into market creation for Dutch chicken meat brings up at least two issues that require attention in the process of collaboration and market creation of an intermediary product. These issues pertain to negotiating the shared meaning of an intermediary product, and the organisation of interactions between groups of actors that maintain their differences in the process. We discuss these subsequently.

The boundary crossing feature of an intermediary product that positions itself on both animal welfare and price, makes it multi-interpretable, and therefore, acceptable to otherwise possibly conflicting interpretations. In livestock farming, animal welfare can be examined from various disciplinary angles and different problem definitions (Fraser, 2008; 2009). Animal scientists and veterinarians study animal welfare from the animals' point of view, dealing with the animal, its health and behaviour as influenced by certain treatments and living conditions (e.g. Hiemstra and Ten Napel, 2013; Tuyttens *et al.*, 2008). Organic livestock production follows the basic philosophy for organic farming and departs from the concept of naturalness. For the animals, this includes access to outdoor areas and freedom of choice that allows each animal to express its individual preferences (Lund, 2006; Vaarst and Alrøe, 2012). In the case study, the different groups of actors struggled with the various meanings of animal welfare that could be aligned with the notion of affordability: was it a slower growing broiler breed, was it the reduction of antibiotics, was is the extended lifetime before slaughtering and/or was it a covered free range area in the barn? The notion of an intermediary product made this struggle explicit. Starting off from a more organic perspective, the Dutch chicken meat initiative resulted in a position closer to mainstream livestock production that deals with the animal, its health and behaviour.

The bridging and conflicting concerns at the boundary of the different groups of actors offered a new horizon on the transition towards sustainable practices, in our case in the broiler sector. How were different groups of actors able to collaborate while at the same time maintaining their differences? The organisation of interactions that led to continuity and consistency in the process of collaboration and market creation for Dutch chicken meat can be summarized in three lessons learned. First, putting the operational decisions on the intermediary product at the forefront of the conversations while leaving the principal differences between the groups at the background. Second, building trust by making small steps forward and subsequently by positive utterances about the results of these small steps. Third, being explicit about things that need to be changed and remaining an open dialogue about animal welfare, prices, volumes and the process itself. Boundary objects may as well create a sphere of engagement in which the different groups of actors can elucidate and reflect upon the ambiguity of the boundary object itself, in our case the process of collaboration and market creation for Dutch chicken meat. In line with earlier findings of Metze and Van Zuydam (2013), 'this may also result in a credible and tangible new boundary object through which participants can alter dominant constellations at the individual level, at the group level and for society'.

References

Bock, B. and Van Huik, M. (2007). Animal welfare: the attitudes and behaviour of European pig farmers. British Food Journal 109: 931-944.

Bos, J.M., Blok, V. and Van Tulder, R. (2013). From confrontation to partnerships: the role of a Dutch non-governmental organization in co-creating a market to address the issue of animal welfare. International Food and Agribusiness Management Review 16 (Special Issue A): 69-75.

Bowker, G.C. and Star, S.L. (2000). Invisible mediators of action: classification and the ubiquity of standards. Mind, Culture, and Activity 7: 147-163.

Bracke, M. and Hopster, H. (2006). Assessing the importance of natural behaviour for animal welfare. Journal of Agricultural and Environmental Ethics 19: 77-89.

Dodge, J. (2014). Civil society organizations and deliberative policy making: interpreting environmental controversies in the deliberative system. Policy Sciences 47: 161-185.

Downs, A. (1972). Up and down with ecology: The issue attention cycle. Public Interest 28: 38-50.

Dutch Ministry of Agriculture, Nature and Food quality (ANF) (2009a). Convenant marktontwikkeling verduurzaming dierlijke producten: tussensegmenten. (Covenant Market development Sustainable Livestock Products: Intermediary segment). Ministry of Agriculture Nature and Food, the Hague, the Netherlands.

Dutch Ministry of Agriculture, Nature and Food quality (ANF) (2009b). Uitvoeringsagenda duurzame veehouderij. (Implementation Agenda Sustainable Livestock). Ministry of Agriculture Nature and Food, the Hague. the Netherlands.

Feldman, M.S., Khademian, A.M., Ingram, H. and Schneider, A.S. (2006). Ways of knowing and inclusive management practices. Public Administration Review 66: 89-99.

Fraser, D. (2008). Animal welfare and the intensification of animal production. In: Thompson, P. (ed.) The ethics of intensification. Springer, the Netherlands.

Fraser, D. (2009). Assessing animal welfare: different philosophies, different scientific approaches. Zoo Biology 28: 507-518.

Hiemstra, S.J., and Ten Napel, J. (2013). Study of the impact of genetic selection on welfare of chicken bred and kept for meat production. Final report of a project commissioned by the European Commission (DG SANCO 2011/12254).

Klerkx, L., Van Bommel, S., Bos, B., Holster, H., Zwartkruis, J. V. and Aarts, N. (2012). Design process outputs as boundary objects in agricultural innovation projects: Functions and limitations. Agricultural Systems 113: 39-49.

Latacz-Lohmann, U. and Foster, C. (1997). From 'niche' to 'mainstream' – strategies for marketing organic food in Germany and the UK. British Food Journal 99: 275-282.

Lazarus, R.J. (2008). Super wicked problems and climate change: restraining the present to liberate the future. Cornell Law Review 94: 1153.

Levin, K., Cashore, B., Bernstein, S. and Auld, G. (2012). Overcoming the tragedy of super wicked problems: constraining our future selves to ameliorate global climate change. Policy Sciences 45: 123-152.

Lund, V. (2006). Natural living – a precondition for animal welfare in organic farming. Livestock Science 100: 71-83.

Metze, T. and Van Zuydam, S. (2013). Pigs in the city: reflective deliberations on the boundary concept of agroparks in the Netherlands. Journal of Environmental Policy & Planning: 1-18.

Rittel, H.W. and Webber, M.M. (1973). Dilemmas in a general theory of planning. Policy Sciences 4: 155-169.

Schmid, O. and Kilchsperger, R. (2010). Overview of animal welfare standards and initiatives in selected EU and third countries. Final Report Deliverable 1.2. Research Institute of Organic Agriculture (FiBL), Frick, Switserland.

Star, S.L. and Griesemer, J.R. (1989). Institutional ecology, 'translations' and boundary objects: amateurs and professionals in Berkeley's Museum of Vertebrate Zoology, 1907-39. Social Studies of Science 19: 387-420.

Tuyttens, F., Heyndrickx, M., De Boeck, M., Moreels, A., Van Nuffel, A., Van Poucke, E., Van Coillie, E., Van Dongen, S. and Lens, L. (2008). Broiler chicken health, welfare and fluctuating asymmetry in organic versus conventional production systems. Livestock Science 113: 123-132.

Vaarst, M. and Alrøe, H. (2012). Concepts of animal health and welfare in organic livestock systems. Journal of Agricultural and Environmental Ethics 25: 333-347.

Van Horne, P., Van Harn, J., Van Middelkoop, J. and Tacken, G. (2003). Perspectieven voor een alternatieve kuikenvleesketen. LEI – Animal Sciences Group Wageningen University, the Hague, the Netherlands.

Van Horne, P., Van Middelkoop, J. and Tacken, G. (2002). Bouwstenen voor een marketingconcept van een welzijnsvriendelijk gehouden vleeskuiken. LEI – Animal Sciences Group Wageningen University, the Hague, the Netherlands.

Yin, R.K. (2014). Case study research: design and methods, Sage publications, Thousand oaks, USA.

A shorter agro-food value chain is a more ethical one? Case of AgroTransilvania cluster

H.F. Arion, I.C. Muresan and I.D. Chiciudean*
Department of Economic Sciences, University of Agricultural Sciences and Veterinary Medicine of Cluj-Napoca, 3-5 Manastur St., Cluj-Napoca, 400372 Cluj, Romania; felix.arion@usamvcluj.ro

Abstract

The more localized is a food systems and the shorter is the food supply chain, the more ethical is seen by scholars, by authorities, by small producers and by dedicated consumers. The main reason why it is preferred a short food system is that it assures a sustainable development of each of the three pillars: economic, social and environment. The paper discusses an example of a short supply chain of local producers of the AgroTransilvania Cluster (Romania), a professional association playing an important role on the agri-food sector of the region. The results and discussions are based on three successive researches (a review of most representative source of information related to the agrifood sector and consumer behaviour from the studied area; of face-to-face discussion with members of the AgroTransilvania Cluster and of a survey of all the 81 communities of Cluj County). The most important common concern which showed up was the difficulty of finding local products on local markets. AgroTransilvania Cluster created two shops located onto the most important shopping areas of the Cluj city. Instead of their economic success, a number of ethical issues raised and are still waiting for an answer from the economic, social and environmental perspective.

Keywords: local food systems, cooperation, short food supply chains

Introduction

The new market orientation of the agriculture in the frame of the common agriculture policy intends to maximise the use of the economic potential of local agriculture and, implicitly, of short food supply chains emphasising, as 'local food supply services have tremendous potential' according to Dacian Ciolos, former European Commissioner for Agriculture and Rural Development (Ciolos, 2012). But such orientation cannot be viable without an adequate adaptation of EU legislation on food hygiene. John Dalli, former European Commissioner for Health and Consumer Policy, frankly expressed the opinion that the legislation on food hygiene does not favour industrialised methods of food production. The single transparent hygiene policy, so-called 'hygiene package' created in January 2006, stated that food operators assume full and prime responsibility for the safety of the food they produce. The provision is applied at all levels of the food chain, from primary production to delivery to the final consumer (Dalli, 2012).

The common agricultural policy (CAP) orientation is an obvious react to fulfil the need of the consumers. Saba and Messina (2003) observed the change in consumer attitudes in relation to modern farming practices, by increasing the challenges for the new food products, the ones with a higher respect for the environment and for the health. Scholars consider that local food systems are the base of many innovations in the food sector in the last decade (Nurse *et al.*, 2010), as the preferences of consumers are more and more heterogeneous and diverse. Consumers gradually become more interested in the origin of products consumed and in the production methods (Roininen *et al.*, 2006), so shorter short food systems can offer them the feel of a more accurate information about the food they consume, only by bringing consumers closer to the origin of food and even involving the direct contact between producers and consumers (Feagan, 2007).

Paloviita (2010) considers that actual local products are nothing but a retro innovation in business food. Their unprecedented development in the last decade has been achieved as a result of the consumers' behaviour, more orientated for sustainability and the development of the food sector requires viable alternatives to global supply chains. The author also notes that current researches related to food sustainability are not strictly limited to economic issues, but it is ethically extend to other areas, such as: environment and socio-cultural environment.

It can be observed cases of authorities trying to encourage consumers to buy local products as much as possible in order to develop a sustainable consumption. The initiative of the local authorities in the United Kingdom has as final result a positive attitude of the consumers for local food. The consumers perceived local food as being authentic, having higher level of quality, freshens, nutrition, taste and safe (Chambers *et al.*, 2007). Zepeda and Reid (2004) also identified a positive attitude of consumers towards local products. They believe that through the consumption of local products it can be developed the local economy and it can be improved the protection of the environment.

Seyfang (2008) identifies four major positive consumption positive motivational perceptions on consumption of local food products: benefits to the environment (by reducing costs of packaging, by reducing pollution generated by long distances transport, by increasing the diversity); benefits to the economy (by supporting the local farmers, by supporting the agricultural cooperatives, by maintaining money into the local economy flows); social benefits (by knowing the origin of foods and their production process, by preserving the traditions and cultural heritage, by direct contacts between consumers and producers, by their good taste and refinement) and personal benefits (by tasting better, by being safer). Nurse *et al.* (2010) identifies, also, the main factors that motivate any consumer of local products: economic factors related to environmental, social and related social responsibility.

Based on the above mentioned considerations, the aim of this paper is to analyse the ethic of a particular short supply chains, the one created by AgroTransilvania Cluster (in Transylvania, Romania). The reason of creating the short supply chain was to increase the economic efficiency of producers by taking advantage of the changing consumer behaviour in favour of local products 'consumption, and it did not directly planned to improve other factors as environment or social responsibility. It can be said then that the producers betrayed the consumers? Or they indirectly improved social and environmental issues attributed them by the consumers?

Material and methods

AgroTransilvania Cluster is a professional association of entities from the Cluj County, and not only, created to support the development of agri-industrial sector, by increasing the competitiveness of the association itself and its members individually, both on national and international market, based on a common development strategy. The cluster includes various members with direct/indirect implication into agri-food sector: input suppliers (tools, machineries, software, etc.), producers, processors, manufacturers, distributors, universities, consultants and administrative authorities. The main objective of the association represents the integration of all these entities into a sustainable and competitive value chain that could bring notoriety and visibility to Cluj local products.

The mission of the cluster is to become an integrative pole of sustainable development of agri-industrial sector in Transylvania (northwest of Romania) and to support the sustainable competitivity of agri-industrial sector. The AgroTransilvania Cluster is following six strategic objectives that have to be accomplished through several operational objectives (increase of sustainable competitivity of agri-industrial sector in Transylvania; encouraging the set-up and/or the development of associative forms; integration of producers and/or associative forms into the value chain; the increase of qualitative and quantitative representatives of cluster; support the development of local and regional initiative; encouraging the research, innovation and technological transfer in the agri-industrial sector).

AgroTransilvania Cluster succeeded to create a full value chain (Figure 1), integrating diverse inputs suppliers, agricultural producers, food processors, commercial center for agro food products and stores, involving regional and local authorities, catalytic organisation and a commercial bank. The association collaborates, as well, with 2 universities from Cluj that are participating to research projects and prepare high qualified specialists in agro-food domain. The public institutions provide logistic, human and administrative support and representation when negotiating.

Results discussed into this paper are based on three successive researches. It started by analysing the most representative references and other source of information related to the agrifood sector and consumer behaviour from the studied area. Beside this, there was conducted a face-to-face discussions with all members of the cluster and, during 2013, were visited all the 81 communities of Cluj (75 communes, 2 cities and 4 municipals) in order to conduct a survey for observing the perceptions of authorities (mayors, vive-mayors), of the representatives of the agribusiness companies (between 2 and 5 for each companies on each community) and of the individuals (randomly selected based on the information offered by authorities) about the local agri-food sector development. All information was consolidated

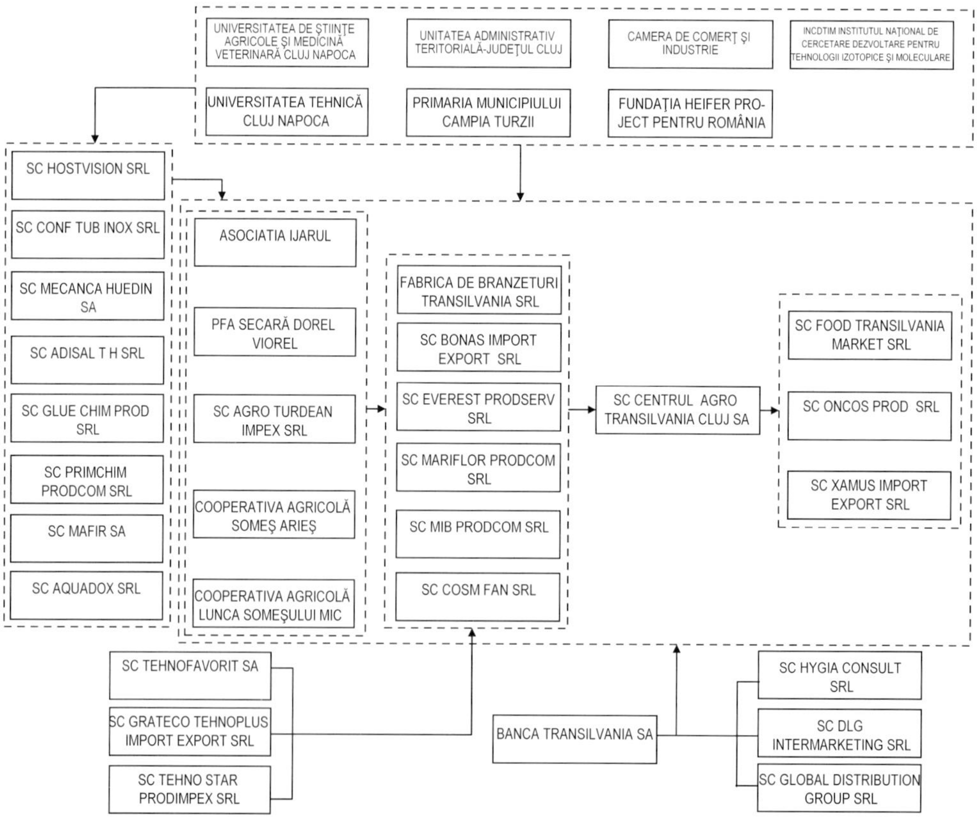

Figure 1. Value chain of AgroTransilvania cluster.

both in a SWOT Analysis and in a PEST Analysis used to develop the strategy for shorting the value chain of local producers by creating short marketing chain for local products.

Results and discussions

The producers, as well as consumers, demonstrate originality and openness to develop more innovative food marketing solutions. These solutions include direct sales arrangements; large and complex markets, use of new technologies in order to fit peoples' daily lives. As result, the source for food is neither linked to a single location (or a limited area), nor a particular time. Hence, including all supply chains that are not a conventional retailer outlet in the category of 'alternative food network' is perhaps no longer sufficient. It is necessary a deeper analyse in order to observe the depth and the creative/innovative relationships created through new consumer-producer partnerships.

It was observed that in Cluj there are a large number of locally produced agri-food products. Many of them are recognised, both by consumers and by producers, as being representative for the region and having a high potential to be successfully involved into a short supply chain. The main problem that appears for these specific local producers is the lack of possibility of penetrating the markets, being too focused on production process and having not enough entrepreneurial information, abilities and knowledge to play on a competitive market. Like most of the small producers, they have low productivity, reduced turnover and they chaotically use direct selling strategies among other marketing channels (Bertazzoli, 2010). The decisions are adopted based on experience of the owner and on opinion leader's feelings, and not on market research.

They raised an ethical issue regarding the difficulty to access the consumers, mainly because of the diverse interests of the middlemen that control the marketing chain. Studies revelled that, beside some notable success stories of local producers to reach themselves to consumers, most of local products are traded by important retailers and diverse middlemen, which do not allow them to increase their involvement on the value chain of their products. (Renting *et al.*, 2003). In Cluj – the studied area –, like in entire Romania, in fact, it can be observed the highly bipolarity phenomenon of the agricultural structures; the agricultural exploitations under 2 hectares representing almost three quarters of the total farms. It was observed that the power of negotiation of individual producer is extremely limited, so they have to accept the imposed trade conditions.

On the other side, the consumers complained about the difficulty of buying local products, as the products are not available on the places they are regularly buy the agri-food stuff. They affirmed they need to make supplementary effort (travel, time, money and so on) to get into direct contact with the producers if they intend to consume local products. According to Venn e*t al.* (2006) consumers' style and standards of living force them to look for agri-food products on different location, space and time, fact that complicates the process of buying. So, it was raised other ethical issues, now from the consumers' perspective: it is correct for them to expect the products to come to the places where they are buying, or should they find solutions, on their expenses, to buy the products, as they are offering them supplementary advantages, as was mentioned before?

Starting from those two ethical issued, AgroTransilvania Cluster was pushed to find solutions able to offer positive solutions in the same time. Firstly, it was find out that there it is no unique, perfect answer. Anyway, arguing that the consumers require local products on the place they are regularly buying, the obvious solution was to penetrate both the main commercial centres areas in Cluj by opening in each of this a shop (one in December 2013, the other one in November 2014, the next one, bigger and more complex, is scheduled for opening in spring 2015. Each of the commercial centres areas are daily visited by an average of 25,000-30,000 potential consumers, with a large range expectation for food products. As it was previously observed that producers complained about high marketing costs, the new created shops are managed with minimum administrative costs. This is a result of the negotiation of AgroTransilvania Cluster with commercial centres' managers, the main reason being the possible advantages for the commercial centres by attracting those consumers which are looking for local products on short supply chain.

The short supply chain is assured by the fact that producers themselves created and run the shops, keeping their identity under the umbrella of AgroTransilvania Cluster, so no traders interfere on their way to final consumers. AgroTransilvania Cluster used its communication skills and abilities to offer the producers a higher degree of confidence level, fact that could lead to increase the interest for business development (Kneafsey *et al.*, 2013).

The development of this new food supply chains, is anyway, a possible way to get together all four types of evolution mentioned by Marsden *et al.* (2000) as being significant for its success: temporal, spatial, demand and associational.

Created at the beginning more as an advertising action for the producers that an economical one, in time, the shops proved to be efficient for the producers. The experience gained is used to create a code of good practices for local producers, designed to meet the needs of all actors from the marketing chain. The shops created by AgroTransilvania Cluster could by assimilated as being 'shop purchasing directly from producers', according to the report elaborated by EHNE, a farmer's union of the Basque Country, Spain. Their level of compromise adopted by producers is medium and the level of compromise adopted by consumers is low (Galli *et al.*, 2013), so they have the potential to be feasible in time. Obviously, it is compromise a solution that with proved advantages, but nevertheless, it not excluded ethical considerations.

Ethical considerations

The economic success of AgroTransilvania Cluster, and its members contributed, indirectly, to the satisfaction of the consumers but this short supply chain. Anyway, a number of ethical topics still exist and they cannot be neglected. Firstly, the concept of super/hypermarket is seen by orthodox adepts of short supply chain, by its own concept, a contradiction of fair trade, so they purposely avoid to buy in those shopping areas for not encouraging the globalisation and standardization of food products, no matter by the products available there. In their opinion, the real consumer that encourages short supply chain will buy only from producers specialised shops, no matter the additional costs. This assumption get to the second ethical problem: in this case a large share of consumers (mainly urbanised ones) will have to possibility to get to a specific range of local products (for instance for the simple reasons on not having time for shopping them in selected areas). It arises obvious negative consequences of not being encouraged the local economy, of consuming of other products that travelled long distances and so one. The third ethical matter raised forms this statement is the fact that pollution generated by transport is one of the main issues for the advocates of the short supply, it is necessary to prevent its hypocrite approach: it will generate less pollution to transport the products to one selling place that to bring all consumers to the producer's residence, or to send every products to each of the consumer that is asking for?

Conclusions

It can be concluded that there is a general trend to encourage sustainable local consumption by short supply chains. Short supply chains are considered to have multiple benefits at the macro level that consumers understand and they resonate. So the trend towards local consumption occurs as a response to global consumption and to people's desire to reconnect with their food source, to know where it comes from, how it is produced. This gives them the feeling of food safety. Anyway, if the short supply chains are created only to take advantage of the consumers' expectations and not for having any other ethical reasons (environment, local economy, social welfare, etc.) there will be no correct communicating to the consumers. The consumers indirectly, expect the local producers to offer them these services, too. Without a further and deeper comparative research on the ethical behaviour of the actors of the marketing chain there is no fait to assume that a short supply chain is offering is the best alternative to consumers' indirect needs.

References

Bertazzoli A., Ruggeri, A. and Samoggia, A. (2010). Short supply chain: analysis of the competitiveness of organic horticultural farmers at Italian regional level. In: Proceedings of 118th EAAE seminar rural development: governance, policy design and delivery, Ljubljana, Slovenia, 25-27 August 2010, pp. 771-783.

Chambers S., Lobb, A., Butler, L., Harvey, K. and Traill, W.B. (2007). Local, national and imported foods: a qualitative study. Appetite 49(1): 208-213.

Ciolos, D. (2012). Local farming and short supply chains: enhancing the local dimension of the common agricultural policy. Conference 'Local agriculture and short food supply chains', 20 April 2012, Brussels, Belgium. Available at: http://tinyurl.com/p5mlh3p.

Dalli, J. (2012). Local farming and short supply chains. Conference 'Local agriculture and short food supply chains', 20 April 2012, Brussels, Belgium. Available at: http://tinyurl.com/p5mlh3p.

Feagan R. (2007). The place of food: mapping out the local in local food systems, Progress in Human Geography 31(1): 23-42.

Galli, F. and Brunori, G. (eds.) (2013). Short food supply chains as drivers of sustainable development. Evidence document. Document developed in the framework of the FP7 project FOODLINKS (GA No.265287). Laboratorio di studi rurali SISMUNDI, Pisa, Italy, 7 pp.

Kneafsey, M., Venn, L., Schmutz1, U., Balázs, B., Trenchard, L., Eyden-Wood, T., Bos, E., Sutton, G. and Blackett, M. (2013). Short food supply chains and local food systems in the EU. A state of play of their socio-economic characteristics. Publications Office of the European Union, Luxembourg.

Marsden, T., Banks, J. and Bristow, G. (2000). Food supply chain approaches: exploring their role in rural development, Sociologia Ruralis 40(4): 424-438.

Nurse, G., Onozaka, Y. and McFadden, D.T. (2010). Understanding the connections between consumer motivations and buying behaviour: The case of the local food system movement, Southern Agricultural Economics Association Meeting, Orlando, USA.

Paloviita, A (2010). Consumers' sustainability perceptions of the supply chain of locally produced food. Sustainability 2(6): 1492-1509.

Renting, H., Marsden, T.K. and Banks, J. (2003). Understanding alternative food networks: exploring the role of short food supply chains in rural development. Environment and Planning A 35(3): 393-411.

Roininen, K., Arvola, A. and Lahteenmaki, L. (2006). Exploring consumer perceptions of local food with two different qualitative techniques: laddering and word association. Food Quality and Preference 17: 20-30.

Saba A. and Messina, F. (2003). Attitudes towards organic foods and risk/benefit perception associated with pesticides. Food Quality and Preference. 14: 637-645.

Seyfang, G. (2008). Avoiding Asda? Exploring consumer motivations in local organic food networks. Local Environment 13(3): 187-201.

Venn, L., Kneafsey, M., Holloway, L., Cox, R., Dowler, E. and Tuomainen, H. (2006). Researching European 'alternative' food networks: some methodological considerations. Area 38(3): 248-258.

Zepeda, L. and Leviten-Reid, C. 2004. Consumers views on local food. Journal of Food Distribution Research 35(3): 1-6.

Section 8. Consumer values and habits

Know your pork – or better don't: debating animal minds in the context of the meat paradox

J. Benz-Schwarzburg[1] and C. Nawroth[2]*
[1]Department of Ethics and Human-Animal Studies, Messerli-Research Institute Vienna, Veterinaerplatz 1, 1210 Vienna, Austria; [2]Institute of Animal Breeding and Husbandry, Martin-Luther University Halle Wittenberg, Theodor-Lieser-Strasse 11, 06120 Halle, Germany; judith.benz-schwarzburg@vetmeduni.ac.at

Abstract

Ethologists and philosophers increasingly understand (some) animals as individuals with complex socio-cognitive abilities. However, strong animal rights claims linked to such abilities have been pushed for great apes and dolphins, but not yet for any farm animal species. We hypothesize that the reason for this is not because farm animals lack morally relevant abilities but because granting them extended rights seems a profound step considering the markets involved. Pigs are used intensively for consumption which renders them an interesting case for investigating the interface between mind attribution, moral status and practical treatment of animals. Here, research results from comparative cognition are being discussed in the light of ethical, psychological and sociological theories: Do pigs have complex socio-cognitive abilities? To which debates on animal mind and moral status do such abilities link? Does the knowledge of such abilities change our perception and treatment of animals? What prevents us from translating ethical claims into practice? The case of pigs reveals that the acknowledgment or withdrawal of mind and moral status takes place within psychological and sociological frameworks constituting the so-called meat paradox, where the denial of mind and moral status allows, maintains or even fosters processes like an objectification or de-individualisation of animals. It seems that we do not want to get to know animals as 'cognitive relatives' – at least as long as they are considered food animals.

Keywords: pigs, cognitive abilities, animal rights, meat consumption

Introduction: animal cognition and the ethical implications

Since more than 20 years philosophers and biologists have been debating animal rights, including rights comparable to human rights for great apes. In the face of these animals' complex social, emotional and cognitive capacities, as well as their presumably complex needs three rights have been demanded for them: a right to life, to freedom and not to be tortured (Cavalieri and Singer, 1993). This challenges the husbandry, well-being and killing of great apes in zoos, circuses and laboratories as well as their well-being and survival in the wild. These claims have only been extended to cetaceans so far (The Helsinki Group, 2010; White, 2007). When dealing with the matters of farm animals, ethicists seem far more focused on a welfare perspective (e.g. Rollin, 2003) than on the link between cognition, human rights and animal rights. Maybe the reason for this is not because farm animals lack morally relevant abilities but rather because thinking about the ethical implications makes us specifically uncomfortable. Let's take for example the case of pigs. Billions of farm animals are being slaughtered each year worldwide with pig meat representing 37% of global meat output (FAO, 2010: 32). There are few other species in the world which we use equally intensively, produced in industrial units (Steinfeld *et al.*, 2006). According to the US Department of Agriculture, global pork production for 2014 continues its decade long expansion and is projected at a record 108.9 million tons, due to lower expected feed costs and a growing demand. The global trade of pork has expanded by about 25% in just five years (USDA, 2013). What if pigs have much more sophisticated abilities than many people think? What ethical implications arise?

Pigs' socio-cognitive abilities

Pigs form matrilinear family groups (with male offspring separating eventually in bachelor groups) and they show complex social interaction like collective defence of group or offspring and cross-fostering (Newberry and Wood-Gush, 1985; Stolba and Wood-Gush, 1989). The families forage together for food which is distributed patchily both in space and time. Given this social lifestyle, according to Mendl *et al.* (2010), pigs would likely benefit from having sophisticated socio-cognitive abilities. In fact, cognition research showed that they can discriminate and remember individuals, they remember for a long time where food is hidden, where they have already searched for and retrieved food, and they preferentially return to sites containing a greater value of food (Mendl *et al.*, 2010). Broom *et al.* showed that pigs who have been introduced to mirrors learn how to use them to locate food hidden behind barriers (2009: 1037). 'Naïve' pigs looked behind the mirror instead (Gieling *et al.*, 2014). Held and colleagues showed that pigs apply sophisticated social tactics: in an arena with hidden food, non-informed pigs who did not know the food's location, were 'able to exploit the knowledge of others by following them to a food source' (Held *et al.*, 2000: 569). In competitive trials where the dominant pig (who would chase them away from the food) was naïve with regard to the location of the food 'the subordinates were more likely to show food-directed behaviour when the chances of arriving at the food source ahead of their exploiters were higher' (Held *et al.*, 2002a).

Though there is a lot we do not know yet about pigs, it can be assumed in the light of current directions in cognition research that future experiments will for example turn to pro-social tendencies, cooperative behaviour or inequity aversion in pigs. The latter means that a subject resists an inequitable reward if his/her partner gets a better pay-off. Several experiments showed inequity aversion in nonhuman primates (e.g. Brosnan and De Waal, 2003) and dogs (e.g. Range *et al.*, 2009). As pigs show sophisticated social behaviour, inequity aversion might play a role in their social interactions as well. Abilities like these are highly relevant for discussions about fairness and morality in animals. It is a hotly debated topic in biology and philosophy if and which animals share such abilities with us (Andrews, 2013; Bekoff and Pierce, 2009; Broom, 2003; De Waal, 2003; 2009; Pluhar 2013; Rowlands, 2012). For most of these authors welfare and/or rights implications would follow. Marc Rowlands, for example, grants animals who are able to act according to moral reasons (for example on the basis of empathy) the status of moral subjects. This status is subordinate to the status of moral agents (fully aware and fully responsible humans) but it clearly deserves moral respect which again asks for more than mere welfare considerations (Rowlands, 2012).

Implications: from welfare to rights?

Animal welfare science argues on the basis of animals' complex ethological needs. According to Duncan and Petherick (1991) 'animal welfare is dependent solely on the mental, psychological and cognitive needs of the animals concerned'. Learning and memory capacities for example are assumed to have an impact on the animal's capacity to cope with housing conditions (Gieling *et al.*, 2011; Held *et al.*, 2002b, Mendl and Paul, 2008). Many welfare problems in (intensive) pig farming have been described and discussed, often in connection with husbandry conditions (for a summary see European Food Safety Authority/Panel on Animal Health and Animal Welfare, 2007).

It has been recognized that links between cognition and welfare also matter because welfare constitutes an important factor for production success (Held *et al.*, 2002b; Mendl and Paul, 2008; Mendl *et al.*, 2001). Furthermore, according to Buller and Roe (2012), 'animal welfare is increasingly being commodified by various foodchain actors' as it is 'a 'value added' component or distinctive selling point for food products, brands or even particular manufacturers and retailers'. However, this development is also problematic. Greater emphasis is being placed either on those welfare elements that lend themselves to commodification processes or on those that respond to consumer interpretations of what 'good' welfare might be (Buller and Roe, 2012: 131).

Other reasons for the importance of the connection between cognition and welfare than their positive impact on production and sale have to be taken into account. Most importantly, it can be asked if animal cognition and welfare matter from the perspective of the animals, and to what amount they matter. Is 'good welfare' good enough for animals with complex socio-cognitive abilities (Benz-Schwarzburg and Knight, 2011: 20-21)?

In contrast to welfare accounts, animal rights accounts (e.g. Francione, 2000; Regan, 2004; Wise, 2002) don't argue within the logic of established production systems but ultimately question it. Francione states that a reformist approach challenges how we exploit animals while not challenging the reasons why these animals are bred, kept and slaughtered in the first place (Francione, 2007; 2009). If this is a valid critique, a very dominant aspect of the human-pig relationship cannot be satisfactorily discussed within a welfare account: the killing of pigs for meat consumption.

What prevents us from transferring ethical claims into practice?

The human-pig relationship constitutes the prime example of what has been termed 'the meat paradox' defined as the inconsistency between the appreciation for animals and enjoyment of their meat (Bastian *et al.*, 2012; Loughnan *et al.*, 2010). The denial of morally relevant qualities, specifically the denial of mind and suffering in animals, has been claimed to be a 'particularly useful strategy' whenever people want to reduce the conflict between eating meat and their moral concern for animals. Empirical studies support this claim (Bastian *et al.*, 2012: 248; Bilewicz *et al.*, 2011; Bratanova *et al.*, 2011; Loughnan *et al.*, 2010). In fact, farm animals are rated significantly lower with regard to complex cognition but significantly higher with regard to edibility than pets and wild animals. In dogs and great apes this relation is exactly reverse: they are rated high with regard to cognitive abilities but low with regard to edibility (Bastian *et al.*, 2012, Figure 9: 249). Unfortunately, the study of Bastian *et al.* (2012) did not include the perception of pigs.

Such studies into humans' perception of animals are highly relevant from an ethical point of view. More and more scholars point to this. According to Bastian *et al.* (2012) meat eating is a morally significant behaviour – even though it is rarely conceptualized as a moral choice. But if we start to investigate it as such it can provide insights into how cognitive and motivational processes obscure moral responsibility (Bastian *et al.*, 2012: 247). Francione describes the different treatments given to companion vs farm animals as a 'moral schizophrenia' (Francione, 2000). Melanie Joy claims that we care about animals but also want to eat them – which is a moral dilemma: 'our values and behaviours are incongruent, and this incongruence causes us a certain degree of moral discomfort' (Joy, 2011: 18). In order to alleviate this discomfort consumers establish a belief system which provides them with justifications for eating animals and which builds on a range of complicated defence mechanisms.

But meat eating is not only constructed on a personal, psychological level but also on the level of society. It is a social norm and '(s)ocial norms aren't merely descriptive – describing how the majority of people behave – they are also prescriptive, dictating how we ought to behave' (Joy, 2011: 105). Joy names the belief system that coordinates us to eat certain animals 'carnism' (Joy, 2011: 30). As a socially constructed belief system, carnism is – much as veganism and vegetarianism – an ideology. It renders the consumption of some animals as normal, natural and necessary (Joy, 2011: 96). Joy assumes that most people care for animals and want to avoid intense and unnecessary suffering in others, including suffering in animals (Joy, 2011: 33). Therefore, carnism restores to psychic numbing, a psychological process by which we disconnect, mentally and emotionally, from what we know (Joy, 2011: 18) which again relies on a set of specific defence mechanisms: denial, avoidance, routinization, justification, objectification, de-individualization, dichotomization, rationalization and dissociation (Joy, 2011: 19). Joy highlights three of these mechanisms – objectification, de-individualization, and dichotomization – as the 'Cognitive Trio' of carnism, having a profound effect on our perception of animals. These mechanisms are not normatively neutral and they link directly to the denial of mind.

Farm animals like pigs are objectified as a 'something' rather than a 'someone' through language (e.g. referred to as production units). According to Stewart and Cole (2009) the discursive positioning of pigs renders them objects of consumption, objects which are also highly invisible (as we rarely meet living farm animals). An objectification of animals is legitimized through institutions, legislation, and policies. We literally turn them 'into a piece of live stock' by buying, selling, trading or exchanging them (Joy, 2011: 117-118). Animal rights philosophers criticize exactly the status of animals as such things which belong to humans while not having any rights themselves (Francione, 2000). The denial of mind is a core element of objectifying an animal as it strips it off any abilities which characterize it as an individual subject coming close to or even being a nun-human person.

The second mechanism highlighted by Joy is the de-individualization of animals as they are not perceived as individuals but seen as 'abstractions', as herds, flocks, or a 'bunch' of living things which is an oxymoron in itself (Joy 2011: 118). Again, denying animals complex minds serves stripping them of individuality. They disappear into the group.

The third key mechanism within the cognitive trio of carnism is dichotomization. The term captures the idea, that animals are grouped in categories like 'edible' – 'inedible' but also 'intelligent' – 'not very smart' or 'cute' – 'less attractive' (whereas 'not very smart' seems to be connected with edible animals). Again, such categories are not normatively neutral, they allow for establishing prejudiced hierarchies. Words such as 'hoggish', 'swine' and 'pigsty', and phrases like 'greedy as a pig' associate the pig with greed and dirt (Malcolmson and Mastoris, 2003). Pigs served and still serve as images of the ugly and freakish (Bondeson, 2006) and they are main figures in anti-Semitic strategies because they rank lower than most other animals (Fabre-Vassas, 1997). Thus, pigs seem to be rather on the pejorative side of dichotomization processes.

A forth mechanism of carnism which Joy does not specifically highlight is the commodification of animals which renders living beings not just into objects but into specific kinds of objects: commodities. Objectification, commodification and commercialization are connected: 'To objectify something is implicit in treating it as a market commodity, but what is disturbing about objectifying a person or organism is not so much the exchange of money as it is the notion that a subject, a moral agent with autonomy and dignity, is being treated as if it can be used as an instrument for the needs or desires of others without giving rise to ethical objections' (Radin, 1991, cited by Schrecker *et al.*, 1997: 62). The happy pig in commercial ads, literally slicing itself into pieces of pork or the body of pigs divided into pieces of consumable meat are just two examples where the pig is advertised as a commodity, ready to consume, even in notably sexist ways where the objectification of a living being follows narrations similar to the objectification of women (Adams, 2010). Pigs being among the most heavily consumed animals are also being heavily commercialized in advertisements.

If mind attribution, moral status and the treatment of pigs are interlinked via the complicated psychological and sociological mechanisms described by Joy and others, a broader question arises: What is the role of scientific knowledge about animal cognition in the changing human-animal relationship?

The denial of mind seems to be a defence mechanism which provides a rational explanation for a mental classification which is not rational (Joy, 2011: 128). It uses the traditional role of science as a generator of objective, true knowledge but it operates within the social matrix of carnism. The modus operandi of carnism is to distort, rather than to report reality (Joy, 2011: 129). Thus, if actual scientific knowledge on pigs' complex social cognition aims to inform and even change the human-pig relationship, it has to compete not only with consumers' individual perception of pigs but also with the social matrix. In other words: actual scientific knowledge would have to replace socially constructed ideology (that is the idea of dichotomized, de-individualized, objectified and commodified pigs) in order to allow identification and empathy to set in. This is a largely unheeded, major challenge for ethologists and philosophers who hope to change human attitudes by providing knowledge about animals' complex minds.

References

Adams, C.J. (2010). The sexual politics of meat (20th anniversary edition): a feminist-vegetarian critical theory. Continuum, New York, USA.

Andrews, K. (2013). Ape autonomy? Social norms and moral agency in other species. In: Petrus, K., Wild, M. (eds.) Animal minds and animal ethics: connecting two separate fields. Transcript, Bielefeld, Germany, pp. 173-195.

Bastian, B., Loughnan, S., Haslam, N. and Radke, H.R.M. (2012). Don't mind meat? The denial of mind to animals used for human consumption. Personality and Social Psychology Bulletin 38(2): 247-256.

Bekoff, M. and Pierce, J. (2009). Wild justice: the moral lives of animals. University of Chicago Press, Chicago, USA.

Benz-Schwarzburg, J. and Knight, A. (2011). Cognitive relatives yet moral strangers? Journal of Animal Ethics 1: 1-36.

Bilewicz, M., Imhoff, R. and Drogosz, M. (2011). The humanity of what we eat: conceptions of human uniqueness among vegetarians and omnivores. European Journal of Social Psychology 41: 201-209.

Bondeson, J. (2006). Freaks: the pig-faced lady of Manchester Square and other medical marvels. History Press Tempus Publishing, Gloucestershire, UK.

Bratanova, B., Loughnan, S. and Bastian, B. (2011). The effect of food categorization on the perceived moral standing of animals. Appetite 57: 193-196.

Broom, D.M. (2003). The evolution of morality and religion. Cambridge University Press, Cambridge, UK.

Broom, D.M., Sena, H. and Moynihan, K.L. (2009): Pigs learn what a mirror image represents and use it to obtain information. Animal Behaviour 78: 1037-1041.

Brosnan, S.F. and De Waal, F.B. (2003). Monkeys reject unequal pay. Nature 425 (6955): 297-299.

Buller, H. and Roe, E. (2012). Commodifying animal welfare. Animal Welfare 21(S1): 131-135. DOI: http://dx.doi.org/10.7120/096272812X13345905674042.

Cavalieri, P. and Singer, P. (1993): The great ape project: equality beyond humanity. St. Martin's Griffin, New York, USA.

De Waal, F. (2003). Good natured: the origins of right and wrong in humans and other animals. Harvard University Press, Cambridge, USA.

De Waal, F. (2009). Primates and philosophers: how morality evolved. Princeton University Press, Princeton, USA.

Duncan, I.J. and Petherick, J.C. (1991). The implications of cognitive processes for animal welfare. Journal of Animal Science 69(12): 5017-5022.

European Food Safety Authority/Panel on Animal Health and Animal Welfare (2007). Scientific opinion of the Panel on Animal Health and Welfare on a request from the Commission on animal health and welfare aspects of different housing and husbandry systems for adult breeding boars, pregnant, farrowing sows and unweaned piglets. The EFSA Journal 572: 1-13. Available at: http://tinyurl.com/ld3wbas.

Fabre-Vassas, C. (1997). The singular beast: Jews, Christians, & the pig. Columbia University Press, New York, USA.

Food and Agriculture Organization of the United Nations (FAO) (2010). Agribusiness handbook: poultry, meat and eggs. FAO, Rome, Italy. Available at: http://tinyurl.com/7nfwh46.

Francione, G.L. (2000). Introduction to animal rights: your child or the dog? Temple University Press, Philadelphia, USA

Francione, G.L. (2007). The four problems of animal welfare: in a nutshell. Available at: http://tinyurl.com/qetjtyw.

Francione, G.L. (2009). New welfarism fails on its own terms. Available at: http://tinyurl.com/pnx6bqr.

Gieling, E.T., Nordquist, R.E., and Van der Staay, F.J. (2011). Assessing learning and memory in pigs. Animal Cognition 14: 151-173.

Gieling, E.T., Mijdam, E. and Van der Staay, J.F. (2014): Lack of mirror use by pigs to locate food. Applied Animal Behaviour Science 5.

Held, S., Mendl, M., Devereux, C., and Byrne, R.W. (2000). Social tactics of pigs in a competitive foraging task: the 'informed forager' paradigm. Animal Behaviour 59(3): 569-576.

Held, S., Mendl, M., Devereux, C. and Byrne, R.W. (2002a). Foraging pigs alter their behaviour in response to exploitation. Animal Behaviour 64(2): 157-165.

Held, S., Mendl, M., Laughlin, K., and Byrne, R.W. (2002b). Cognition studies with pigs: livestock cognition and its implication for production. Journal of Animal Science 80: E10-17.

Joy, M. (2011). Why we love dogs, eat pigs, and wear cows: an introduction to carnism. Conari, San Francisco, USA.

Loughnan, S., Haslam, N., and Bastian, B. (2010). The role of meat consumption in the denial of moral status and mind to meat animals. Appetite 55: 156-159.

Malcolmson, R. and Mastoris, S. (2003). The english pig: a history. Bloomsbury, London, UK.

Mendl, M., Burman, O., Laughlin, K., and Paul, E. (2001). Animal memory and animal welfare. Animal Welfare 10: 141-159.

Mendl, M., Held, S. and Byrne, R.W. (2010). Pig cognition. Current Biology 20 (18): R796-R798.

Mendl, M. and Paul, E.S. (2008). Do animals live in the present? Current evidence and implications for welfare. Applied Animal Welfare Science 113: 357-382.

Newberry, R.C. and Wood-Gush, D.G.M. (1985). The suckling behaviour of domestic pigs in a semi-natural environment. Behaviour 95: 11-25.

Pluhar, E. (2013). The nonhuman roots of morality. In: Petrus, K., Wild, M. (eds.) Animal minds & animal ethics: connecting two separate fields. Transcript, Bielefeld, Germany, pp. 197-227.

Radin, M. J. (1991). Reflections on objectification. Southern California Law Review 65(1): 341-354.

Range, F., Horn, L., Viranyi, Z., and Huber, L. (2009). The absence of reward induces inequity aversion in dogs. Proceedings of the National Academy of Sciences USA 106: 340-345.

Regan, T. (2004). The case for animal rights. University of California Press, Berkeley; Los Angeles, USA.

Rollin, B. (2003). Farm animal welfare: social, bioethical, and research issues. Iowa State Press Wiley-Blackwell.

Rowlands, M. (2012). Can animals be moral? Oxford University Press, Oxford, UK.

Schrecker, T., Elliott, C., Hoffmaster, C.B., Keyserlingk, E.W., and Somerville, M.A. (1997). Ethical issues associated with the patenting of higher life forms. Westminster Institute for Ethics and Human Values, McGill Centre for Medicine, Ethics and Law.

Steinfeld, H., Gerber, P., Wassenaar, T., Castel, V., Rosales, M., and De Haan, C. (2006). Livestocks long Shadow. FAO, Rome, Italy. Available at: http://tinyurl.com/96csxh.

Stewart, K. and Cole, M. (2009). The conceptual separation of food and animals in childhood. Food, Culture and Society 12(4): 457-476.

Stolba, A. and Wood-Gush, D.G.M. (1989). The behaviour of pigs in a semi-natural environment. Animal Science 48: 419-425. DOI: http://dx.doi.org/10.1017/S0003356100040411.

The Helsinki Group (2010). Declaration of rights for cetaceans: whales and dolphins. Available at: http://tinyurl.com/a26snyl.

United States Department of Agriculture (USDA) (2013). Livestock and poultry: world markets and trade. Biannual report by the USDA, foreign agricultural service. Available at: http://tinyurl.com/mtqvtfu.

White, T. (2007). In defense of dolphins: the new moral frontier. Willey-Blackwell, Oxford, UK.

Wise, S.M. (2002). Drawing the line: science and the case for animal rights. Perseus Books, Cambridge, UK.

Analysis of potentialities for the consumption of fresh strawberry tree fruits: preliminary results

M.F. Oliveira and J. Franco*
Instituto Politécnico de Coimbra, Escola Superior Agrária de Coimbra/CERNAS, Bencanta 3045-601 Coimbra, Portugal; foliveira@esac.pt

Abstract

The main purpose of this article is to evaluate consumers' perception of fresh *Arbutus* fruit and its market potential. This undervalued product from the local flora. Being a fruit that is not yet the target of marketing outside of its production area, an appraisal is needed to assess the perception of the market for fresh *Arbutus* fruit. To achieve part of these goal 164 surveys were carried out and validated in two commercial surfaces in Coimbra region, Portugal with experiment of product in the place. The results revealed that 80% of the surveyed knew the *arbutus* and 39% were very satisfied with proof of the fruit and 56% expressed satisfaction. The most appreciated attribute was the taste. About 87% of respondents revealed that willingness to pay (WTP) was below 1.5€/125 g. Respondents frequently mentioned the fact that the *arbutus* to be the fruit of youth and nostalgia ('saudade').

Keywords: marketing, value chain, *Arbutus*, consumer, willingness to pay

Introduction

The strawberry tree (*Arbutus unedo* L.) is a plant of Mediterranean origin known and eaten in ancient Greece, but ignored by the Romans and grown in consociation with forest species. Traditionally the population linked the effects of excessive consumption of the fruit with unwell and drunkenness symptoms. *Unedo* means that only one should be eaten. Nevertheless its food and medicinal value lead to the making of research studies so that it can be considered the *Arbutus* fruit a more soft fruit, able to be consumed without fear. *Arbutus* tree also has an important role in the protection of forest areas, as firebreak, increases biodiversity in pine forest stands areas that are particularly prone to plagues and sensible to climate change. *Arbutus* tree is a direct source of income due to the fruit and indirect due to its role in forest protection, playing a significant role in the diversity and resilience of local economies.

In recent years the *Arbutus* tree has gained importance as fruit tree orchards planted in order to produce fruit for brandy production. Growers are interested in diversifying the ways to consume the fruit. The *Arbutus* fruit is an edible and flavorful fruit when completely mature, very rich in fiber and minerals when compared with other soft fruits (Table 1).

The carotenoids, lycopene and the β carotene are important in the *Arbutus* fruit and total of phenols vary between 386-1973 mg GAE/100 g. Fructose is more important sugar, ranging from 3.6-14.5 mg/100 g, which gives it a sweet and pleasant taste when fully mature. Several compounds with antioxidant properties were found as organic acids such as: fumaric, lactic, malic, suberic and citric. Some authors consider the *Arbutus* fruit one of the richest fruit in proanthocyanidins and the consumption of mature fruits will induce health benefits (Ayaz *et al.*, 2000; Alarcão-E-Silva *et al.*, 2001; Oliveira *et al.*, 2011; Pallauf *et al.*, 2008). The *Arbutus* fruit consumption is traditionally recommended for the treatment of gastrointestinal problems, bladder, kidney, skin and cardiovascular disorders. However, in spite of containing higher amounts of nutrients and bioactive compounds when compared with other cultivated species, the consumption of fresh *Arbutus* fruit is limited. These fruits have the disadvantage of sometimes being considered indigestible (which can be related to their content of cellulose: 3 g/100 g),

Table 1. Composition of Arbutus versus blackberry per 100 g edible weight.

Not minerals items	*Arbutus* fruit[1]	Rapberry[2]	Minerals items	*Arbutus* fruit[1]	Rapberry[2]
Energy (kcal)	61.48-135.95	25.0-43.92	Potassium (mg)	79.72-323.14	160.0-189.0
Carbohydrates (g)	14.11-31.55	5.1-6.24	Calcium (mg)	40.54-104.12	41.0-44.0
Total fiber (g)	10.04-22.27	3.1-3.16	Magnesium (mg)	9.56-45.85	23.0-30.0
			Vitamin C (mg)	122.0-262.7	15.0-17.0

[1] Ruiz-Rodríguez *et al.* (2011).
[2] Frutas and Hotalizas (2014).

which can cause an 'undefined sense of malaise'. Molina *et al.* (2011) report that although they have not found scientific references to its ethanol content, the consumption of the *Arbutus* fruits in an advanced state of maturation is popularly seen as causing drunkenness symptoms. It is important to clarify this question, since tradition and beliefs regarding the product are crucial for its consumption. In the analysis of food demand trends, we must consider that consumers are all different and their motivations are: nutritional needs, health, pleasure, convenience motive and food security in the sense of food quality and preservation of the environment, particularly in societies with greater available income. Moser and Raffaelli (2011) state that the choice to purchase and willingness to pay (WTP) for fresh fruits and vegetables is mainly driven by personal attributes such as health or experiential quality. The results show that the local attribute has increased its importance when compared to organic, certification, and origin. Consumers identify local with quality-assured, making the local attribute an important tool for the communication of other attributes, such as, nutritional quality, sustainable production and the support of the local economy. Communication is an important tool although, Wills *et al.* (2009) concluded that food labeling can help consumers make decisions about their lifestyle. However food labeling did not prevent the increasing obesity and consumers have expressed confusion about how this information is transmitted. Local attribute can be a tool for the dissemination of the *Arbutus* fruit but labeling can have a contradictory effect. There is not research work about the fresh consumption of the fruit and the existing market is restricted to local fairs and integrated in very short commercialization channels with reduced economic capacity. It is fundamental to appraise the market perception and the WTP, defined as 'the maximum price that a buyer is willing to pay for a certain amount of product' (Wertenbroch and Skiera, 2002). The WTP for food is intrinsically connected to the shaped of quality perception through certain attribute or it tries to observe the value that consumers attach to certain attributes such as information about the mode of production or the introduction of an attribute. This assessment is carried out by applying different methodologies (Angulo and Gil, 2007). In this work, these attributes cannot be evaluated in order to be scientifically validated because consumers do not have the fruit at their disposal and they cannot compare it to other experiences. To build a market for the consumption of the *Arbutus* fresh fruit requires that consumers recognize the quality, making the quality definition an essential element for this study. The quality concept has two dimensions: an objective dimension linked to technical product attributes and a subjective dimension which refers to the consumer perception, there is a difference between the acknowledged quality by those who produce it and quality that is recognized by the consumer who is the market 'The relationship between the two is at the core of the economic importance of quality: only when producers can translate consumer wishes into physical product characteristics, and only when consumers can then infer desired qualities from the way the product has been built, will quality be a competitive parameter for food producers' (Grunert, 2005: 371). There is a horizontal dimension of quality perception (before and after purchase) and a vertical dimension that includes a set of intrinsic or extrinsic characteristics and the way consumers find if these attributes are desirable by linking them to drivers of human behaviour (Grunert, 2005). These concepts are important because for *Arbutus* fruit the horizontal dimension is not know, through experimentation of the product in a market situation. We aim to observe the vertical dimension, considering the short time that some consumers contact with the fruit. The inference in the WTP cannot be compared with other studies presented in the literature, since there is no experience with the product at least in the market situation and in cases where it exists is not seen as a fruit, rather something that is consumed in rural areas for enjoyment, and an easy to pick up product. This work assesses consumers' perception of fresh *Arbutus* fruit and its market potential.

Material and methods

Taking into consideration the different quantitative and qualitative research techniques of markets in food studies, we decided to conduct a simple and short survey to the consumer after tasting the *Arbutus* fruit. Reactions of respondents' were observed in situ after delivery of a brochure on the product and the work developed in project. The survey is standard market research technique, but is not immune to the interviewer' interference. The location and time will be reflected in the answers that only a larger study may clarify. As Gofton (1997) refers 'The best way to learn about marketing research, however, is to do it'. The survey was conducted in the perspective of an evolutionary process of introducing a new product in consumer's habits. The survey design took into account the consumers' time availability and the wiling they have to taste the product. The fruits were harvested in the region of Penacova/Coimbra (Portugal) the day before their presentation at the Shopping surface. The selected fruits had an orange-red color, weighing ≥5 g, with an equatorial and longitudinal diameters ≥1.6 and 1.5 mm respectively, firmness 16-25% (Agrosta 100 and tip 25), with total soluble solids of 18 to 25° Brix and acidity of about 8.5 g/l malic acid. The *Arbutus* fruits tested were taken from wild *Arbutus* tree selected but not treated with any plant protection products and without fertilization. This fact was mentioned during the test of *Arbutus* fruit and the surveyed were asked about the desire for change in fruit by applying farming techniques. The study was conducted in two commercial surfaces in 2013 in Coimbra region, Portugal. On 21 November 2013 were performed 104 surveys (63%) in the Commercial Surface Hypermarket Jumbo Dolce Vita in Coimbra city, between 9 am and 15 pm. On 12 December 2013, 60 surveys (37%) were conducted in the Supermarket Intermarché in Condeixa city between 9 am and 14 pm. The fruit harvested for presentation at the Jumbo Hypermarket, had greater diameters and brighter colors than those presented at the Intermarché Supermarket. This difference was due to differences in the harvest season and not in the harvest region.

Results and discussion

Data collection was analyzed using SPSS 21 software. To measure the strength of association between variables we used the Pearson correlation coefficient (r) that is applied when the variables follow a normal distribution. We consider for analysis of correlation the follow scale: very high: 0.9-1.0; high 0.7-0.89, medium: 0.4-0.69, weak: 0.2-0.39 and a very weak <0.2. In some situations the variables did not follow a normal distribution. The Spearman coefficient (rho) is used as an alternative and is not sensitive to asymmetries in the distribution, it is not required the presence of the Normal Distribution. To characterize the social characteristics of the sample it was limited to age and gender (Table 2). Questions on the economics characteristics of the surveyed had were not carried since they could cause apprehension and biases consumer's perception during the fruit proof.

Respondents were inquired about their prior knowledge of the *Arbutus* fruit. A small number had heard about it but had never seen or tasted the fruit. In these cases it was considered that they did not hold knowledge. About 80% of the surveyed had previously tasted or seen the fruit and only 20% of the surveyed never tasted or seen the fruit. There is a weak negative correlation but significantly different from zero, significant at the 1% level between knowledge and age (r=-0.291; rho=-0.0281). The

Table 2. Age and gender of surveyed.

Age (years)	No. of survey	(%)	Cumlative (%)	Gender			
				No. survey		(%)	
				Male	Female	Male	Female
≤18	6	4	4	5	1	83	17
>18-25	20	12	16	9	11	45	55
>25-40	22	13	29	5	17	23	77
>40-60	50	31	60	16	34	32	68
>60	66	40	100	29	37	44	56
Total	164	100		64	100	39	61

magnitude of the Pearson correlation coefficient show that nearly 8% of the variance of the 'knowledge' variable is explained by age. Knowledge increases with the age. There is no significant relationship between this variable and others. Some respondents reported previous experience with the fruit, but not to bias the response, it was pointed out the need to limit the answer to the moment when the fruit was displayed (Table 3)

About 39% of the surveyed were very satisfied and 56% satisfied. Of those surveyed who had no prior contact with the fruit, 27% of surveyed were very pleased with the test and 73% expressed satisfaction.

Table 3. Degree of satisfaction by degree of knowledge, age and gender.

	Degree of satisfaction (number)					Degree of satisfaction(%)				
	Very satisfied	Satisfied	Unsatisfied	Very unsatisfied	Total	Very satisfied	Satisfied	Unsatisfied	Very unsatisfied	Total
Total	64	91	3	6	164	39	56	2	4	100
Knowledge of *Arbutus*										
Yes	55	67	3	6	131	42	51	2	5	100
No	9	24	0	0	33	27	73	0	0	100
Gender										
Male	19	40	2	3	64	30	63	3	5	100
Female	45	51	1	3	100	45	51	1	3	100
Age										
≤18	1	5	0	0	6	17	83	0	0	100
18-25	4	16	0	0	20	20	80	0	0	100
25-40	7	13	1	1	22	32	59	5	5	100
40-60	21	26	1	2	50	42	52	2	4	100
>60	31	31	1	3	66	47	47	2	5	100

The female gender showed a higher degree of satisfaction, the group aged ≤18 years were satisfied and for the higher age groups the answers are in line with global results. Despite the differences in percentages there are no statistically significant relationships between the degree of satisfaction, age, gender and level of knowledge. It was requested to the surveyed that they indicated the three attributes of the fruit that were most appreciated by descending order of preference (Table 4). The respondents elected in first place the taste followed by color. The texture was the first attribute for 20% of the respondents. Taste was also chosen more frequently in 2nd, followed by texture. There are two statistically significant correlations (at 5% level) between color, other attribute and gender. In the analysis of statistically significant correlations (color/gender) and (other/gender) we observe that women were more sensitive to color, and 40 of the surveyed (24% of the total and 49% of females surveyed) selected this as the 1st attribute Some of the surveyed correlated the red color with the richness of the fruit in antioxidants before having been delivered the flyer with the fruit characteristics. In contrast the male gender revealed more drifters in what concerns the attribute they appreciate most (gender/color: r=-0.173, rho=-0.173; gender/other: r=0.642, rho=0.660)

From the standpoint of marketing, the results show a significant relationship between gender and the color attribute and between gender and other attributes. This may be used to take action, using the fruit color in its promotion

The surveyed was asked in an open question if they would be interested in changing some of the fruit attributes. About 40% of the surveyed responded affirmatively and 58% would not change any attribute. The majority of respondents answered that they would like to change something in the fruit having indicated they would like a more intense taste. About the willingness to purchase the fruit if it were on the market, 87% of respondents would be willing to purchase it. There is a weak positive correlation

Table 4. Attributes preferences and analysis between attributes.

Position	Texture			Color			Flavour			Taste			Other		
	No.	%	% Cum.	No.	%	% Cum.	No.	%	% Cum	No.	%	% Cum.	No.	%	% Cum.
1°	33	20.1	26.8	53	32	41	6	4	11	65	40	44	1	1	
2°	47	28.7	65	38	23	71	16	10	39	52	32	80	5	3	5
3°	43	26.2	100	37	23	100	34	21	100	30	18	100	13	8	32
Total	123	75		128	78		56	34		147	90		19	12	100
Not selected	41	25		36	22		108	66		17	10		145	88	
Total	164	100		164	100		164	100		164	100		164	100	

Gender	Color (no)				Color (%)				Ohter (no.)				Other (%)			
	1°	2°	3°	Total in 1°	1°	2°	3°	Total in 1ª	1°	2°	3°	Total in 1°	1°	2°	3°	Total in 1ª
Male	13	18	16	47	25	47	43	28	1	4	2	7	100	80	15	14
Female	40	20	21	81	75	53	57	49	0	1	11	12	0	20	85	0
Total	53	38	37	128	100	100	100	41	1	5	13	19	100	100	100	5

(at the 5% level) between degree of satisfaction and the willingness to purchase (r=0.308; rho=0.239), e.g. 9.5% of the willingness to purchase is explained by the degree of satisfaction. No other statistically significant associations were found.

The surveyed that demonstrated the willingness to purchase the fruit were asked about the WTP. This question is central to the purpose of the project work and for the *Arbutus* farmers, but it is a sensitive issue due to the fact the response is influenced by the interviewer and by the economic and social crisis that Portugal has been experiencing in the last five years. This question was performed considering a transparent package of 125 g used for the sale of soft fruit. Packaging was presented to the respondent for a better perception in terms of packing capacity. This package was selected because the fruit is sensitive to touch and pressure. Moreover this package is transparent allowing its visualization. Of the surveyed who had showed willingness to purchase, 61% revealed a WTP for *Arbutus* fruit between 1-1.5€ (exclusive) and about 26% of the surveyed revealed a WTP less than 1€. In the correlation analysis it is observed that there is a very weak correlation (significant at the 5% level) between the variable age and the variable WTP (r=-0.202; ro=-0.252). No other statistically significant associations were found. If we consider the relationship between the WTP and the age group, the group aged ≤18 years revealed WTP between 1-1.5€. The WTP for the groups between 25-40 years old and 40-60 years old show the same WTP (1-1.5€) as we can observe in Table 5.

We analyzed WTP in relation to the place where the survey and was no association was observed between the location of the survey and the results obtained, there are no differences between the survey conducted in an urban hypermarket and a smaller shopping center (supermarket) in less urban area such, as the city of Condeixa. Respondents frequently mentioned the fact that the *Arbutus* is a fruit of youth and nostalgia ('saudade'). It should be noted that some of the surveyed to determine WTP were influenced by the facts of knowing the *Arbutus* tree and recognizing the difficulty in harvesting the fruit

Table 5. Willingness to pay analysis.

Price in € per 125 g	Surveyed		Validated answers		Age					
	No.	(%)	(%)	Cumulative (%)	≤18	18-25	25-40	40-60	>60	Total
<1	38	23.2	26.2	26.2	0	4	3	9	22	38
1-1.5	88	53.7	60.7	86.9	6	11	12	28	31	88
1.5-2	16	9.8	11	97.9	0	5	2	7	2	16
≥2	3	1.8	2.1	100	0	0	1	1	1	3
Total	145	88.4	100		6	20	18	45	56	145
Not answered	19	12								
Total	164	100								

Price in € per 125 g	Percent distribution by age groups (%)						Percent distribution total of survey (%)					
	≤18	18-25	25-40	40-60	>60	Total	≤18	18-25	25-40	40-60	>60	Total
<1	0	20	17	20	39	26	0	11	8	24	58	100
1-1.5	100	55	67	62	55	61	7	13	14	32	35	100
1.5-2	0	25	11	16	4	11	0	31	13	44	13	100
≥2	0	0	6	2	2	2	0	0	33	33	33	100
Total	100	100	100	100	100	100	4	14	12	31	39	100

for commercial purposes. The surveyed also mentioned that the fruit is available during walks in the countryside without the need to pay. Respondents were sensitive to the leaflet delivered after the survey has been conducted and highlighted the fact that the fruit has antioxidant properties, is a Portuguese fruit, local and known in the Coimbra region for the production of brandy ('aguardente de medronho').

Conclusion

As previously stated these results are provisional: the surveyed revealed to be satisfied with the fruit, were sensitive to color, and related the color to its antioxidant value. They were sensible to the fact that it is a local fruit. It is also found that the delivery of the flyer helped in this relation. However, when defining a price for the product this value was much lower compared to the soft fruits that are currently in the market. Some of the surveyed mentioned the fact that to eat it the *Arbutus* fruit can cause inebriation or malaise. Laboratory results obtained to date failed to find relationship between the consumption of fruit and the perceived unwell; may be a belief (creed) and not an event. It is therefore important to take this variable into account as it may affect WTP. It will also be important to conduct a brief analysis about the potential of the *Arbutus* fruit as a fruit free from synthetic inputs. This first evaluation doesn't allow us to say that it is profitable crop for production of fruits for fresh consumption, considering the current crop conditions of *Arbutus* tree, the yields obtained and the price that the consumer is willing to pay. Further studies should be conducted at the urban areas of Lisbon and Porto which concentrates a large number of consumers that can justify the logistical cost that this fruit may have. It will be necessary to conduct interviews to the stakeholders of fresh fruit channels. This is one of the steps for the diversification and development of the strawberry tree (*medronheiro*).

Acknowledges

This work was performed under two design tasks supported by Foundation for Science and Technology project '*Arbutus unedo* plants and products quality improvement for the agro-forestry sector'. Tanks to Vivina Carreira for her support with the English.

References

Alarcão-E-Silva, M., Leitão, A., Azinheira, H.G. and Leitão, M. (2001). The Arbutus berry: studies on its color and chemical characteristics at two mature stages. Journal of Food Composition and Analysis 14: 27-35.

Angulo, A. and Gil, J. (2007). Risk perception and consumer willingness to pay for certified beef in Spain. Food Quality and Preference 18: 1106-1117.

Ayaz, F, Kucukislamoglu, M. and Reunanen, M. (2000). Sugar, non-volatile and phenolic acids composition of strawberry tree (*Arbutus unedo* L. var. *ellipsoidea*) fruits. Journal of Food Composition and Analysis 13: 171-177.

Frutas & Hotalizas (2014). Available at http://tinyurl.com/ot4r4wu.

Gofton, L.(1997). Marketing research. In: Padberg, D., Riston, C., Albisu, L. (eds) Agro-food marketing. CAB International, New York, pp. 225-252.

Grunert, K. (2005). Food quality and safety: consumer perception and demand European Review of Agricultural Economics 32 (3): 369-391.

Molina, M., Pardo-De-Santayana, M., Aceituno, L., Morales, R. and Tradío, J. (2011). Fruit production of strawberry tree (*Arbutus unedo* L.) in two Spanish forests. Forestry 84 (4): 419-429.

Moser, R. and Raffaelli, R. (2011). Credence-based attributes: a review. International Food and Agribusiness Management Review 14(2): 121-142.

Oliveira, I., Baptista, P., Albino, B. and Pereira, J. (2011). *Arbutus unedo* L. and its benefits on human health. Journal of Food and Nutrition Research 50(2): 73-85.

Pallauf, K., Rivas-Gonzalo, J., del Castillo, M., Cano M. and Pascual-Teresa, S. (2008). Characterization of the antioxidant composition of strawberry tree (*Arbutus unedo* L.) fruits. Journal of Food Composition and Analysis 21: 273-281.

Ruiz-Rodríguez, B-M., Morales, P., Fernández-Ruiz, V., Sánchez-Mata, M-C., Cámara, M., Díez-Marqués, C., Pardo-de-Santayana, M., Molina, M. and Tardío, J. (2011). Valorization of wild strawberry-tree fruits (*Arbutus unedo* L.) through nutritional assessment and natural production data. Food Research International 44: 1244-1253.

Wertenbrock, K. and Skiera, B. (2002). Measuring consumers'willingness to pay at the point of purchase. Journal of Marketing Research 39: 228-241.

Wills J., Schmidt, D., Pillo-Blocka, F. and Cairns, G. (2009). Exploring global consumer attitudes toward nutrition information on food labels, Nutrition Reviews 67(1): 102-106.

Simple plain fare or exquisite eating – is simplicity really a consumer virtue?

P. Sandin[1*] *and P. Moula*[2]
[1]*Department of Crop Production Ecology, Swedish University of Agricultural Sciences, P.O. Box 7043, 75007 Uppsala, Sweden;* [2]*Department. of Philosophy and History of Technology, Royal Institute of Technology, Brinellvägen 32, 10044 Stockholm, Sweden; per.sandin@slu.se*

Abstract

This paper discusses the proposed consumer virtue of material simplicity, as advocated by Joshua Colt Gambrel and Philip Cafaro. We argue that the introduction of simplicity as a virtue is unnecessary and arbitrary. We also argue that there are reasons to doubt whether simplicity is really a virtue, and using counterexamples, we argue that the arguments presented by Gambrel and Cafaro are unconvincing in this respect. Finally, some considerations for further research are suggested.

Keywords: virtue ethics, consumption, temperance, frugality

Introduction

We all have to eat, and most readers of this text are likely to buy substantial parts of what we eat at stores or restaurants. We are also likely to buy some other things, some of them essential to our survival, some of them contributing to a reasonably comfortable life, and perhaps some outright luxuries. In a word, we are consumers.

Today we are aware that some types of consumption might have severely detrimental consequences for human health, animal welfare, and the environment, and that something has to be done to prevent or mitigate these consequences. Therefore, to an increasing degree, our consumer actions and decisions are subject of scrutiny and evaluation from an ethical point of view. Sometimes, the focus is on responsibility (e.g. Meisch, 2013; Raterman, 2012). Explored to a less degree, but at least as interesting, is the focus on the consumer's moral personality. What are the character traits that a consumer ought to have, or not have, and why? What are the consumer virtues and vices?

To our knowledge there is as of yet no comprehensive approach to the virtues of consumers. The development of such an approach would be highly commendable. However, it is a project beyond the scope of the present paper. So the task here is more limited. In this paper, we will take on recent work by Joshua Colt Gambrel and Philip Cafaro, who make the case that material simplicity within the sphere of our consumer decisions is a virtue (Cafaro and Cambrel, 2013; Gambrel and Cafaro, 2010). They argue that material simplicity, the disposition to act appropriately with regard to our consumption choices and to have a conscientious and restrained attitude toward material goods, is a virtue because it 'furthers human and nonhuman flourishing' (Gambrel and Cafaro, 2010: 95). Against this, we will argue first that it is to be doubted whether the introduction of the virtue if simplicity is fruitful, and second that it is not obvious that simplicity is virtue at all.

Before going further it is appropriate to distinguish between virtue ethics on the one hand and virtue theory on the other. Virtue ethics is an approach in normative ethics which emphasizes the virtues in contrast to duties, rules or consequences. In virtue ethics the notion of a virtue is central to understanding conceptions of the good life and rightness; virtues do not only play an instrumental role in order to achieve some higher good. In contrast, virtue theory is a field of theory about what virtues are and why. This field does not have to be virtue ethical itself. A deontologist like Kant had his theory of virtues, and more recently Julia Driver (2012) and Dale Jamieson (2007), among others, have proposed consequentialist justification of virtues. In this paper we are within a virtue ethical framework, and Gambrel's and Cafaro's proposed virtue of material simplicity is justified by virtue ethical considerations.

Let us also observe that one could argue that it is impossible to live virtuously in a consumer society, or at least a consumerist one, and that virtue would require one to opt out of such a society to live self-sufficiently (Hursthouse, 2007). On this view, there would be no room for consumer virtues. It is trivially true that all sorts of societies do not allow the exercise of virtue, at least not all virtues. Consider attempting to exercise justice in a dictatorship, or caring virtues in a Spartan warrior society. (The relationship between virtues in individuals and states, communities or societies is an issue as old as virtue ethics itself, e.g. Plato's (1945) 'Republic' (Book IV:427-434; cf. Sandin, 2012), but we will not dwell on it here.)

Gambrel and Cafaro (2010: 93) however, allow that at least some virtues are compatible with consumption in a consumer society and we will follow them in this.

What is a virtue?

How does one find a virtue, and how does one recognize it as a virtue? This question is not as straightforward as it seems. There are at least two relevant meta-ethical views here, which are not mutually exclusive. (cf. Sandler, 2007, Chapter 1)

According to the first view, virtues are 'out there' to be discovered. They exist as character traits in individuals who are habituated to do and feel the right thing for the right reasons and with the right state of mind. To discover them, one would have to proceed in the following manner: First, identify a virtuous individual (say Jesus, Gandhi, or Rachel Carson). Second, observe his or her character traits – 'In situations like X, Jesus/Gandhi/Carson consistently did Y for reasons Z and with state of mind W'. Now you have found a virtue (or virtues) of relevance in situations like X.

On the second view, virtues are invented. You start with a conception of the good – that is, a set of ideas about what actions should be performed, for what reasons, with what state of mind, and so on. From this conception of the good a virtue, a character trait that promotes the good, is derived and then you might go on to, for instance, cultivate the virtue in yourself or others. This appears to be, for instance, Kant's view, and it is the view of consequentialist virtue theorists like Dale Jamieson (2007).

Neither view is necessarily virtue ethical (though they may be), but virtue theoretical. And both views are compatible with the introduction of new virtue terms. Suppose that I have studied the character traits displayed by some role model and realized that she behaves in a commendable way for which there is no appropriate term. A concrete situation where this might happen is when society or technology has developed in an unforeseen way so that established virtue terms do not adequately cover the discovery. In that case, naming the virtue is warranted. Or imagine a consequentialist virtue theorist arguing that one ought to cultivate a disposition to perform some type of actions that lead to an increased amount of well-being in the world, and that there is no obviously corresponding virtue term – perhaps not surprisingly since traditional virtue terms have not been developed in a consequentialist context (compare the remarks in Parfit, 1987: 453f).

Gambrel and Cafaro's position is an instance of the second, virtue ethical view. They proceed from their conception of the good in the form of human and nonhuman flourishing (Gambrel and Cafaro, 2010:

95), a good which simplicity supposedly furthers. There are, however, two principal objections to the introduction of simplicity as a virtue in this context.

Is simplicity necessary, and is it a virtue?

First, it can be questioned whether it is necessary to introduce it, or whether the work simplicity is supposed to do can be done at least as well by other means, such as already existing virtues.

In order to introduce a new virtue, or a new virtue term, there should be some positive reason for doing so. There can be a practical purpose for introducing a new virtue, for instance. The introduction of simplicity as a virtue could be justified if it facilitated moral education of good consumers. It could also serve a more theoretical purpose if it would give action-guidance in cases were there otherwise would be any. Introducing unnecessary virtues (or virtue terms) would, arguably, complicate matters and for instance make moral education more difficult because unnecessarily complex. Here it might be worth comparing with Regan's set of desirable characteristics for ethical theories: consistency, scope, precision, conformity to reflective intuition, and simplicity (which in this case means the quality of being uncomplicated, not to be confused with the virtue of material simplicity (Regan, 1983: 131-40, 146f). They are parallel to Kuhn's quality criteria for scientific theories (Kuhn, 1977).

Furthermore, virtues are often thought to be mean between vices. For instance, courage is the mean between cowardice and recklessness. Cafaro and Gambrel propose that simplicity, being a complex virtue, 'appears to be a mean along several axes' (Gambrel and Cafaro, 2010: 90f, emphasis added). However, this makes it hard to see what the reason is for introducing a new virtue, the extension of which is superimposed arbitrarily on a number of already existing virtues Gambrel and Cafaro do not systematically relate the new virtue to those virtues – for instance temperance, frugality, prudence, self-control and wisdom – even though they mention them as overlapping with simplicity. Temperance, they think, is perhaps not suited for affluent societies. It was developed in societies marked by comparative scarcity and it might indeed become more directly relevant again: 'perhaps as people's consumption options narrow, 'temperance' and 'frugality' say all that needs to be said' (p. 105). But the fact that it is difficult to be temperate or to live frugally in an affluent society does not justify introduction of a new virtue. Quite the contrary. (for a discussion of frugality in present-day society, see Nash 1995. Interestingly, Nash also follows the tradition of emphasizing the political aspects of the virtues. He also mentions simplicity, which he conceives of differently than Gambrel and Cafaro).

Secondly, and more seriously, it can be questioned whether simplicity, as characterized by Gambrel and Cafaro, really is a virtue.

It is not obvious what consumer decisions would be in accordance with, or contrary to, the virtue of simplicity. This is not surprising – virtues, unlike deontological rules, do not always prescribe or prohibit particular actions. The rightness of actions is context-dependent to a large degree, and many virtue ethicists, but not necessarily virtue theorists, have been reluctant to present specific action-guiding criteria (Hursthouse, 1999.)

However, by presenting some examples Gambrel and Cafaro do give a clear indication of what the virtue of simplicity would demand of us as consumers. They depict a case of a philosopher who is 'happy to eat 'foie gras' and drink Dom Perignon champagne when the opportunity presents itself, or jet around the world to take part in symposia on art, science, literature, or philosophy at expensive resorts'(Gambrel and Cafaro, 2010: 103f). Such persons have 'fallen prey to their own poor consumption decisions, undermining their health, financial wellbeing, personal relationships, or careers' (Gambrel and Cafaro,

2010: 104). The authors are thus making the claim that such consumer behaviour is detrimental to the agent's well-being in different ways.

However, this is weak. Moderate consumption of Dom Perignon, just as participation in symposia on art, etc., could in fact to a large degree *benefit* my career and boost my personal relationships, while at least not undermining my health, and not the condition of the environment either. A virtuous person will not necessarily have her flourishing damaged if she eats exquisite dishes or spends a weekend at an exclusive resort. A virtuous person will, however, do it with moderation and sincerely appreciate it when she does. The reasons for this are two. The virtuous person ought to feel some resistance towards engaging in behaviour that collectively leads to morally problematic outcomes, and a virtuous person will appreciate things of value and that includes some material things.

This is something which might be problematic for the proposed account of material simplicity. This virtue is supposed to lead to greater appreciation of things other than material ones, while at the same time giving the agent who possesses simplicity greater appreciation of material things – not taking them too lightly, so to speak. Appreciation is certainly a good quality to have, towards the right things in the right amount and so on, but if it is desirable and good to have greater appreciation for material things, why then should one not make use of expensive resorts and other exclusive things if one is lucky enough to have the opportunity? If I am to appreciate things more, wouldn't it be wrong for me to have a negative attitude toward more refined things? Obviously, consumption of some things, such as cigarettes and other things that damage the consumer's health or the environment is both unreasonable and wrong. (Travel to resorts might, but need not, belong in this category.) This is true also of consumption of goods that have been produced through immoral means, for instance involving slave labor or animal suffering. 'Foie gras' would be an example of this. But these things are wrong because of the nature of these things and the process in which they are made. They are not wrong to consume because consumption is wrong. It should also be noted that some material things which are expensive or luxurious are not necessarily worse with regard to human and animal well-being or other relevant considerations in the environment. Buying a new car is a case in point. Many people live in societies where being without access to a car is not a real option (of course, there are several of us who can manage perfectly well with buses, the underground, bikes, or car pools). Suppose that you actually need a car. In this case it is hard to see that buying a new, fancy car is contrary to the flourishing of human and non-human entities. It might well be the other way around. You are safer and more comfortable in the new car, and, being a hybrid, it is significantly less hostile to the environment than your previous car. It is also less damaging to anyone or anything you might accidentally hit, since many modern cars are designed with energy-absorbing bonnets and windshields for improved pedestrian safety. It might be that abstaining from some of these luxuries would be a good thing – but Gambrel and Cafaro seem to think that the very use of them is a sign of vice.

We propose a distinction to be made between one's actual consumption and one's attitude with regard to consumption. There are some choices that will lead to worse health, deteriorating personal relationships, and so on, and having a disposition against making those choices seems justified. But that kind of disposition will not cover all the cases that Gambrel and Cafaro want it to cover. Some, but obviously not all, cases of drinking champagne, staying at expensive resorts and enjoying other finer things will be beneficial for human and sometimes animal flourishing.

We believe this distinction is important. One could still claim that there are benefits with introducing the virtue of material simplicity, to be more conscious about one's consumer choices and to appreciate things more in general. However, if we are right in that it is nothing necessarily wrong with drinking champagne and enjoying it at an expensive resort, we must ask what the new virtue supposedly brings to the table. It is hard to see how simplicity is something more than an arbitrary mix of the existing virtues of temperance, frugality, prudence, self-control and wisdom, etc. Both the theoretical and practical case for introducing the virtue then seems weak.

Conclusion

Gambrel and Cafaro do say a lot of sensible things about consumption, how we ought to consume and what type of consumers we ought to be. This is laudable. However, we have argued, their proposed virtue of simplicity appears superfluous and arbitrary, since other more established virtues do the same work at, and furthermore there are reasons to ask whether simplicity is really a virtue – at least the arguments presented by Gambrel and Cafaro are unconvincing in this respect.

Despite this, we believe that virtue theory and virtue ethics can have a place in ethical assessment of consumption. However, against the background of the discussion above, we propose the following:
- In order for a new virtue to be introduced, it should add something to existing virtues. Preferably, it should satisfy desirability criteria akin to those of theories, among them the property of being easy to apply or learn.
- The introduction of new virtues should not be arbitrary.

The further development of these themes will have to be topics for further research.

References

Cafaro. P. and Gambrel, J. (2013). The virtue of simplicity. In: Röcklinsberg, H. and Sandin, P. (eds.) The ethics of consumption: the citizen, the market and the law. Wageningen Academic Publishers, Wageningen, the Neteherlands, pp. 24-28.

Driver, J. (2012). Consequentialism. Routledge, New York, USA.

Gambrel, J. and Cafaro, P. (2010). The virtue of simplicity. Journal of Agricultural and Environmental Ethics 23: 85-108.

Hursthouse, R. (1999). On virtue ethics. Oxford University Press, Oxford, UK.

Hursthouse, R. (2007). Environmental virtue ethics. In: Walker, R.L. and Ivanhoe, P. J. (eds.) Working virtue. Oxford University Press, Oxford, UK, pp. 155-171.

Jamieson, D. (2007). When utilitarians should be virtue theorists. Utilitas 19: 160-183.

Kuhn, T. (1977). The essential tension: selected studies in scientific tradition and change. University of Chicago Press, Chicago, USA.

Meisch, S. (2013). Green food consumption: whose responsibility? In: Röcklinsberg, H. and Sandin, P. (eds.) The ethics of consumption: the citizen, the market and the law. Wageningen Academic Publishers, Wageningen, the Netherlands, pp. 160-165.

Nash, J.A. (1995). Toward the revival and reform of the subversive virtue: frugality. Annual of the Society of Christian Ethics: 137-160.

Plato (1945). The republic of plato. Oxford University Press, London, UK.

Parfit, D. (1987). Reasons and persons. Oxford University Press, Oxford, UK.

Raterman, T. (2012). Bearing the weight of the world: on the extent of an individual's environmental responsibility. Environmental Values: 417-436.

Regan, T. (1983). The case for animal rights. University of California Press, Berkeley, USA.

Sandin, P. (2012). A collective virtue approach to agricultural ethics. In: Potthast, T. and Meisch, S. (eds.) Climate change and sustainable development: ethical perspectives onn land use and food production. Wageningen Academic Publishers, Wageningen, the Neteherlands, pp. 151-153.

Sandler, R. (2007). Character and environment. Columbia University Press, New York, USA.

'Space, pasture, water, grain (and) love': dairy cattle needs according to citizens visiting a dairy farm

B.A. Ventura[1], M.A.G. von Keyserlingk[2], H. Wittman[2] and D.M. Weary[1]*
[1]Animal Welfare Program, Faculty of Land and Food Systems, University of British Columbia, 2357 Main Mall, Vancouver, BC V6T 1Z4, Canada; [2]Integrated Studies in Land and Food Systems, University of British Columbia, 2205 East Mall, Vancouver, BC V6T 1Z4, Canada; bethventura2@gmail.com

Abstract

Citizen engagement on issues around farm animal welfare is important for the livestock industries to sustain their social license to produce. One barrier to this engagement is that non-farming citizens are sometimes perceived as ignorant of farming practices and their views are dismissed. Despite recent interest in understanding citizen views on farm animal welfare in Europe, little is known about North Americans' views. We surveyed the views of 67 citizens before and after a self-guided tour of a working 500-head dairy farm in Agassiz, British Columbia. The mixed-methods survey used quantitative and qualitative questions designed to explore the range of perceptions and concerns about dairy cattle welfare. Of those who responded, 60% were female, 96% consumed dairy products and 93% reported themselves as either somewhat or not knowledgeable about dairy farming. Content analysis revealed that, before visiting the farm, participants considered the following elements (in decreasing frequency) as necessary for dairy cattle to have a good life: fresh food and water, pasture access, gentle human care, space, shelter, cleanliness, fresh air and sunshine, social companions, absence of stress, health, and safety from predators. In general, the farm visit appeared to mitigate some concerns (e.g. provision of adequate food and water, gentle human care) while reinforcing others (e.g. lack of pasture and outdoor access, early cow-calf separation). These results illustrate a multi-dimensional conception of animal welfare that links with values of other livestock industry stakeholders. The results also indicate that efforts to educate citizens may address some concerns but are also likely to create awareness of other issues relating to the welfare of farm animals.

Keywords: attitude, knowledge, value, engagement, animal welfare

Study context

The post-WWII demographic shift of citizenry from rural to urban and semi-urban areas has separated – in place and in activity – the majority of the North American populace from food production and regular contact with livestock (Kendall *et al.*, 2006). This in turn has meant that the actors (including the animals) that contribute to animal production have become largely invisible in the day-to-day life of most North Americans. Concurrently, society has become increasingly critical of the livestock industries, including animal welfare in these systems (Mench, 2008).

The livestock industries must address these criticisms to sustain their social license to produce (Brom, 2000; Benard and De Cock Buning, 2013). Inclusion of non-farming citizens' views in policy may help shift production practices into greater alignment with mainstream social values (Guehlstorf, 2008; note: we focus here not on the general public's role as consumer, but on the citizen, who as a part of civil society contributes to a social consensus of how livestock agriculture should ideally be carried out (Aerts 2013)). This democratization may help legitimize standards that are then developed and improve societal trust in livestock production (Guehlstorf, 2008).

Industry stakeholders sometimes dismiss citizens' concerns about agriculture on the basis of their perceived ignorance of production practices, as has been documented among farmers in both Europe (Benard and De Cock Buning, 2013; Hubbard *et al.*, 2007) and North America (Spooner *et al.*, 2012). This assumption is reminiscent of the informational deficit model of public understanding (referred to commonly as the knowledge deficit), which assumes that public concerns can be corrected through education on 'the facts,' bringing public opinion into line with 'expert' views (Hansen *et al.*, 2003).

We argue that there are limitations to this type of thinking, particularly when it is used to define the frame of reference for engaging with the public on contentious issues surrounding livestock production. Providing people with more information does not always translate to higher acceptance of practices (Hansen *et al.*, 2003), in part because the deficit model fails to recognize the role that values play in attitude formation. People may object to practices that insiders or experts advocate, not because they lack knowledge of relevant facts, but because the practices conflict with their own values (Hansen *et al.*, 2003).

That said, there is some evidence (at least in Europe: Boogaard *et al.*, 2006; 2008; Ellis *et al.*, 2009) that many citizens lack experience with and knowledge of agricultural practices. Factual knowledge and emotional experiences, in addition to values, contribute to people's frames of reference (Te Velde *et al.*, 2002), so it is important that studies of citizen attitudes toward farm animal production be grounded in experiences (including farm visits) that provide urban citizens with some lived context from which to evaluate production practices (Boogaard *et al.*, 2008; 2011; Krystallis *et al.*, 2009).

Objectives and methods

Our objectives were to: (1) understand the range of perceptions, concerns and values held about dairy cattle welfare among a convenience sample of North American citizens; and (2) explore the effect of a dairy farm visit, as a form of information provision, on the plasticity of people's concerns and values about dairy cattle welfare.

We conducted surveys of citizens attending an 'open house' at a working 500-head free-stall dairy farm in Agassiz, British Columbia (n=67 participants, with an 85% completion rate of the 'after' portion). Before touring the farm, participants answered a 5-10 minute survey to gauge their baseline perceptions (i.e. top of mind impressions of the dairy industry), concerns about dairy production, and values (i.e. what a good life for dairy cattle means). Fraser *et al.* (1997)'s conceptualization of animal welfare into three distinct but interrelated aspects – biological functioning (how an animal physically functions), affective states (how an animal feels), and natural living (the degree to which an animal can live a natural life) – was used as the framework to organize the diverse values that participants held regarding a good life for dairy cattle. Participants also completed a five-question quiz on basic dairy husbandry practices. After a 1 hour self-guided tour of the farm, including visits to eight educational stations covering dairy husbandry practices like calf care and cow nutrition, respondents completed another 5-10 minute survey designed to capture any changes in their concerns and values, and answered the quiz questions again.

Preliminary results

Analysis is ongoing, but we summarize some initial findings below.

Content analysis of participants' responses showed that before their tour participants considered the following elements (in decreasing order of frequency) as necessary for a good life for dairy cattle: fresh food and water, pasture access, gentle human care, space, shelter, cleanliness, fresh air and sunshine, social companions, absence of stress, health, and safety from predators. Their predominant welfare concerns, which reflected the value judgments over which elements they considered to be important for a good life, included issues like inadequate or inappropriate nutrition, lack of pasture access and indoor confinement (related concerns being overcrowding and behavioural restriction), and abusive treatment of cattle, in addition to other issues like antibiotics and hormone use and the intensification of cattle farming.

Different participants contextualized their concerns around different aspects of animal welfare. For example, some described pasture access as being important to cows because it allowed them to consume a biologically appropriate diet (i.e. referencing values for biological functioning), while others emphasized the importance of cows being able to engage in the natural behaviour of grazing (i.e. referencing values for natural living).

As anticipated, the farm visit mitigated some concerns; for example, many people expressed satisfaction with this farm's feeding practices and what they observed to be a high level of attention to the cattle. Other concerns, like lack of pasture access, were reinforced. The farm visit also elicited new concerns, likely reflecting an increase in knowledge regarding certain practices. For example, before participants toured the farm, many were unaware of the common practice of early separation and individual rearing of dairy calves. In the post-tour survey, participants showed much greater awareness of this practice. This change in knowledge was accompanied by an increase in participants' reported concerns regarding early separation.

Implications

The current study shows that a farm visit can assuage some citizen concerns about farm animal welfare. However, these results also indicate that certain concerns may be reinforced or even elicited when people become more aware of common practices on farms, suggesting that certain practices may conflict with citizens' values. We conclude that farm visits – by revealing the complexity of farming – may be a useful tool for public engagement and may ultimately help livestock production harmonize with evolving societal norms.

References

Aerts, S. (2013). The consumer does not exist: overcoming the citizen/consumer paradox by shifting focus. In: Rocklinsberg, H. and Sandin, P. (eds.) The ethics of consumption: the citizen, the market and the law. 11th Congress. of the European Society for Agricultural and Food Ethics (Eursafe), Uppsala, Sweden, pp. 172-175.

Benard, M. and de Cock Buning, T. (2013). Exploring the potential of Dutch pig farmers and urban-citizens to learn through frame reflection. Journal of Agricultural and Environmental Ethics 26: 1015-1036.

Boogaard, K., Boekhorst, J.S., Oosting, J. and Sørensen, T. (2011). Socio-cultural sustainability of pig production: citizen perceptions in the Netherlands and Denmark. Livestock Science 140(1-3): 189-200.

Boogaard, K., Oosting, J. and Bock, B. (2006). Elements of societal perception of farm animal welfare: a quantitative study in the Netherlands. Livestock Science 104(1-2): 13-22.

Boogaard, K., Oosting, J. and Bock, B. (2008). Defining sustainability as a socio-cultural concept: citizen panels visiting dairy farms in the Netherlands. Livestock Science 117(1): 24-33.

Brom, F.W.A. (2000). Food, consumer concern and trust: food ethics for a globalizing market. Journal of Agricultural and Environmental Ethics 12: 127-139.

Ellis, K., Billington, K., McNeil, B. and McKeegan, D. (2009). Public opinion on UK milk marketing and dairy cow welfare. Animal Welfare 18: 267-282.

Fraser, D., Weary, D.M., Pajor, E.A. and Milligan, B.N. (1997). A scientific conception of animal welfare that reflects ethical concerns. Animal Welfare 6: 187-205.

Guehlstorf, N.P. (2008). Understanding the scope of farmer perceptions of risk: considering farmer opinions on the use of genetically modified (GM) crops as a stakeholder voice in policy. Journal of Agricultural and Environmental Ethics 21(6): 541-558.

Hansen, J., Holm, L., Frewer, L., Robinson, P., and Sandøe, P. (2003). Beyond the knowledge deficit: recent research into lay and expert attitudes to food risks. Appetite 41(2): 111-121.

Hubbard, C., Bourlakis, M. and Garrod, G. (2007). Pig in the middle: farmers and the delivery of farm animal welfare standards. British Food Journal 109(11): 919-930.

Kendall, H., Lobao, L., and Sharp, J. (2006). Public concern with animal well-being: place, social structural location, and individual experience. Rural Sociology 71(3): 399-428.

Krystallis, A., De Barcellos, M.D., Kügler, J.O., Verbeke, W. and Grunert, K.G. (2009). Attitudes of European citizens towards pig production systems. Livestock Science 126: 46-56.

Mench, J.A. (2008). Farm animal welfare in the U.S.A.: farming practices, research, education, regulation, and assurance programs. Applied Animal Behaviour Science 113(4): 298-312.

Spooner, J.M., Schuppli, C.A. and Fraser, D. (2012). Attitudes of Canadian beef producers toward animal welfare. Animal Welfare 21(2): 273-283.

Te Velde, H., Aarts, N. and van Workum, C. (2002). Dealing with ambivalence: farmers' and consumers' perceptions of animal welfare in livestock breeding. Journal of Agricultural and Environmental Ethics 15: 203-219.

Misperceived climate friendliness of organic food and consumer willingness to pay for actual greenhouse gas emission reduction

S.B. Olsen, T. Christensen, S. Denver, A. Dubgaard and N. Kærgård*
Institute of Food and Resource Economics, University of Copenhagen, Rolighedsvej 23-25, Frederiksberg, 1958 Copenhagen, Denmark; tove@ifro.ku.dk

Abstract

Organic farming practices are generally associated with lower environmental impacts than conventional farming but this does not generally hold true in a climate context. A recent meta-analysis shows that organic milk production often has higher greenhouse gas (GHG) emissions per kilo milk produced compared to conventional systems. Organic beef production has, on the other hand, lower GHG emissions per kg compared to conventional whereas GHG emissions from organic pork production are higher. Yet, a recent survey among Danish consumers demonstrated that a great majority of the respondents perceived organic food production to be more climate friendly than conventional food production. In the present investigation it's assumed that consumers also have access to foods from production systems with a documented climate action plan to reduce GHG emissions. We test the hypothesis that misperceptions about the climate friendliness of organic production negatively affect consumers' willingness to pay (WTP) for food products with documented GHG emission reductions. Secondly, we aim to identify segments of consumers that are meaningful in terms of predicting which types of consumers are likely to be interested in climate change programs. A web survey with 390 respondents was conducted in Denmark in September 2013. The data comprise attitudinal questions and a choice experiment where consumer preferences for different varieties of beef were elicited. The beef varied with respect to price, whether it was organic, and whether it was produced at farms with contracts to reduce emissions of GHG. On average, consumers who perceived organic farming to be more climate friendly than conventional farming also had a higher WTP for beef from farms with documented reduced GHG emissions. Thus, the survey did not confirm the hypothesis that misperceived assumptions about the climate friendliness of organic farming might reduce consumers' WTP for documented initiatives to reduce GHG emissions. A latent class analysis indicates that it is important to take a more detailed look at consumers since there is a lot of heterogeneity in consumers' preferences. Hence, if reduced GHG emissions are to be pursued by increased consumption of climate friendlier food then marketing efforts and information strategies need to be targeted specifically to different segments of the population.

Keywords: organic food, climate mitigation strategies, consumer behaviour, choice experiment, latent class

Introduction

Based on Lancaster's theory of consumer demand, food products can be seen as differentiated goods where the consumer's utility is derived from attributes embodied in the good, such as nutritional value, taste, appearance, etc. (Lancaster, 1966). Standard economic assumptions of utility maximization imply that only private good characteristics of a product would be considered as relevant by consumers. However, the demand for organic food products seems to be motivated also by public good characteristics in terms of negative as well as positive externalities associated with the production of organic food. Among other things this raises the question of to what extent consumers' perceptions of differences between conventional and organic practices when it comes to the creation of externalities affect their food choice.

Many consumers associate organic production methods and organic products with a range of positive characteristics such as being healthier, better for the animals and better for the environment and wildlife (Aertens *et al.*, 2009; Christensen *et al.*, 2014; Denver and Christensen 2010). Some of the benefits are guaranteed by the organic standards such as no use of pesticides and no use of most food additives. Other benefits are not guaranteed by the standards but nevertheless commonly perceived to exist. For instance, many consumers believe that organic products are always better tasting, healthier and better for the climate. The study by Denver and Christensen (2014) found that consumers who believe that organic food contains benefits beyond those guaranteed by the standards have a significantly higher willingness to pay (WTP) for organic food. More generally, the findings in Christensen *et al.* (2014) indicate that a great majority of the Danish consumers trust that the Danish organic label guarantees that the organic standards are met and that they often expect more from organic food than from non-organic food. This overall trust in organic farming methods is not unequivocally backed up by scientific facts. A recent meta-analysis of climate impacts of various production systems shows, that in most cases organic milk production has higher greenhouse gas (GHG) emissions per kilo milk produced compared with conventional systems. In meat production, organic beef production was found to cause lower GHG emissions compared to conventional per kg of meat produced whereas GHG emissions from organic pork production were higher than conventional pork production (Tuomisto *et al.*, 2012). In a detailed study on consumer segments related to organic products, Tveit and Sandøe (2011) found that the Danish consumers roughly can be divided into two segments of approximately the same size. One of the segments is positive towards organics while the other is negative or indifferent. Even more interestingly, though, they found that around 13% of the consumers – consumers who have an orientation towards ethical consumption which extends far beyond the current organic rules – were responsible for almost half of the total Danish organic consumption.

In the present paper we test the hypothesis that the belief that organic production is more climate friendly than it actually is negatively affects consumers' WTP for documented GHG emission reductions. Secondly, we aim to identify meaningful segments of consumers in terms of predicting which types of consumers are likely to be interested in climate programs or environmental programs.

Material

A web survey with 390 respondents was conducted in Denmark in September 2013 among a large consumer panel that is owned and maintained by the market research institute Userneeds. The data comprise questions on attitudes and consumption patterns of organic vs non-organic food items. Also, a choice experiment was included in the questionnaire where consumer preferences for different varieties of beef were elicited.

Method

A stated preference approach is used as there is presently in Denmark no specific marketing of any products from farms with documented GHG emission reductions. More specifically, a choice experiment is used to estimate consumers' valuation of specific characteristics of beef production. The underlying assumption is that the value to a consumer of a package of beef depends on the specific attributes of the package (Lancaster, 1966). In the present case, the characteristics in question comprise whether the beef is produced organically, whether specific concern for the climate has been addressed in the production, whether the farmer has committed himself to raise young Jersey bull calves and use them for beef production instead of euthanizing the new-born calves, and finally the price of the package of beef. A detailed description of the attributes and their levels can be seen in Table 1.

Table 1. Information on attribute and attribute levels as presented to respondents.

Attributes	Attribute levels and text provided to respondents
Organic production	No – conventional production (reference) Yes – the package of beef has a label certifying that the beef is produced according to the Danish rules for organic production. This means that no artificial fertilizer or pesticides have been used in the field; more strict rules concerning use of medicine for cows; no use of feed based on genetically modified organisms; cows are on grass at least 6 hours per day during the summer months; more space per cow is required in the stables.
Climate friendly	No – conventional production (reference) Yes – the package of beef has a label certifying that the beef is produced in a climate friendly way. The farmer has committed himself to formulate a climate action plan for his farm and implement initiatives that will reduce the CO_2-emissions from the farm with at least 15% within a five-year period. This may be reached by building new windmills or solar panels, by saving energy or by increasing CO_2 storage through afforestation. The impact of the farm climate action plan will be the same on organic and conventional farms
Jersey live	No – conventional production (reference) Yes – the package of beef has a label certifying that the farmer has committed himself to raise young Jersey bull calves and use them for beef production instead of euthanizing the new-born calves. Jersey cows are relatively small and since the bull calves are not as effective meat producers as calves from the heavier dairy breeds, they are typically euthanized right after birth.
Price levels	Six levels ranging from 25-75 DKK modelled as a continuous variable

Well aware that other attributes such as expiration date, fat content, appearance, package size and type of shop might be more important to a consumer in a real shopping situation, we chose to focus on differences in the appearance of only four attributes. This was done by specifying to the respondents that 'everything else is equal across alternatives'.

A consumer's utility of a package of beef is assumed to be a linear combination of the four attributes in question where s_{kij} represents the WTP for the *k*'s attribute for the *j*'s alternative in the choice set as evaluated by respondent *i* (included as independent variables in the econometric estimations) as well as an alternative specific constant A_j capturing the systematic variations that are not captured by the chosen independent variables. Also, a random error term capturing all unsystematic variation in data ε_{ij} is included.

$$U_{ij} = \beta_{ASCj} ASC_j + \beta_{1i} s_{ij} + \beta_{2i} s_{2ij} + \ldots + \beta_{ki} s_{kij} + \varepsilon_{ij}$$

The full factorial design comprised of 48 alternative combinations of beef which due to fatigue concerns was considered to be too many to include in the final choice experiment. Hence, a fractional factorial design was used, based on only 24 of the alternatives, resulting in 12 choice tasks in total. Focus group testing as well as previous Danish food choice experiments has indicated this to be a suitable number of choice tasks per respondent. The software package Ngene 1.0 was used to select these, based on a D-efficient experimental design, utilising prior knowledge of respondents' preferences obtained in a pilot study. Consequently, all respondents were asked to go through the same 12 choice sets. The software package Biogeme (Bierlaire, 2003) was used to carry out a latent class analysis (LCA) in order to analyse to what extent it would be possible to identify different segments of consumers differentiated by their WTP for the individual attributes.

Results

Asked about which quality attributes of meat the consumers found most important for deciding whether to choose organic vs non-organic meat, the most chosen attributes was 'price' closely followed by 'animal welfare', 'taste' and 'health'.

Confronted with statements concerning the climate and animal welfare friendliness of organic production, two out of three respondents agreed or partly agreed that organic farming is better for the climate. As many as three out of four agreed or partly agreed that organic farming is better for the animals. Almost one out of five wrongly thought that organic standards forbid killing Jersey calves while almost every second respondent wrongly believed that the organic standards include requirements to reduce GHG emissions.

As a reference case, we estimated a model with a multinomial conditional logit model with only one type of consumers. This provided a simple relationship between the average consumer's WTP for the individual attributes. Relative to the price of 25 DDK for a pound of minced beef, we found a WTP a price premium of 51% for organic production, 15% for having a GHG emission reduction plan and 23% for promising to keep Jersey calves for meat production.

Analysing the data obtained in the choice experiments indicated that consumers' perception of the climate friendliness of organic farming did not affect their WTP for beef from farms with documented reduced GHG emissions. On the contrary, the segment of consumers buying organics and believing that it is better for the climate, were found to be also willing to pay for additionally documented GHG emission reductions.

In order to obtain more details on these average numbers, more consumer segments were included in the analysis. In particular, a latent class with seven classes was found to be able to capture variability between classes without being too complicated to interpret. In all seven classes, there is a negative price coefficient which indicates that all respondents associate lower utility with increasing price. In the first three classes that account for slightly more than half of the respondents, respondents show WTP for the beef being produced organically, Jersey live and a GHG emission plan. In all three classes, respondents show higher WTP for organic production than for Jersey live and a GHG emission plan. Respondents in class 1 and 2 tend to consider having a GHG emission reduction plan and assuring to not kill the jersey calves to have roughly the same value. Therefore, they mainly differ from each other in price sensitivity which is higher among respondents in class 1. Respondents in class 3 can, on the other hand, be characterized by considering a climate plan to be more important than raising Jersey calves.

The preferences in the remaining four classes are more diverse. Respondents in class 4 are only willing to pay for organic production while respondents in class 5 are only willing to pay for raising Jersey calves. Class 6 has positive marginal valuation of organic production as well as for having a GHG emission reduction plan but not for raising Jersey calves. Finally, class 7 was characterized by being the only class where the respondents were not willing to pay for any of the proposed attributes and even considered organic production as negative.

Discussion

Consumers' WTP to mitigate climate change associated with food consumption is also addressed in Heerwagen *et al.* (2014) and Koistinen *et al.* (2013). Koistinen *et al.* (2013) used a choice experiment to analyse how Finnish consumers' preferences for different quality attributes in minced beef and pork were affected by information on the carbon footprint of beef vs pork. The information sheet indicated that the carbon footprint from beef was higher than that from pork. They also found that WTP for beef decreased when respondents were provided with information on carbon foot prints. Similar to our findings, the estimated impact of the carbon foot print information was not large. Heerwagen *et al.* (2014) found that people include food-related behaviour such as the purchase of organic food and reduced meat consumption as ways to mitigate climate change but that people generally perceive other behavioural modifications such as reduction of car usage and household heating as more important strategies to mitigate climate change.

The results of our survey as well as those of Heerwagen *et al.* (2014) and Koistinen *et al.* (2013) indicate that there is a significant WTP price premiums in order to reduce the negative externalities associated with food production. It is not surprising in itself that people are interested in supporting policy programmes that aim to reduce negative externalities and where implementation and individual contributions to the programme are guaranteed through legislation. However, our result is stronger in that it indicates that individuals voluntarily are willing to pay a price premium on environmentally friendly products even though they are not guaranteed a specific environmental outcome due to the risk of free riding. We suggest that there is a need for further research to establish to what extent free riding may prevail in spite of the findings that a considerable number of respondents indicate that they are willing pay voluntarily for reducing the externalities of food production – preferably interdisciplinary research that involves other disciplines such as psychology and sociology.

There is always risk of hypothetical bias in this type of stated preference studies. Hence, it's uncertain to what extent WTP estimates from a stated preference survey provide a correct picture of the actual behaviour by consumers when it comes to accepting price premiums on environmentally friendly goods. Some respondents may answer according to how they believe they ought to behave, rather than how they actually behave when choices are made between more or less environmentally friendly food products. It requires further research to establish to what extent the stated WTP price premiums on environmentally friendly products that are elicited in the present study are likely to comply with actual consumer behaviour.

Conclusions

The case study results indicate that overall Danish consumers are willing to pay for climate friendly food either as an individual label or as an extended organic label. On average, consumers who perceive organic farming to be more climate friendly than conventional farming also have a higher WTP for beef from farms with documented reduced GHG emissions. Misperceived assumptions about organic farming generally being more climate friendly do not seem to reduce consumer WTP for additional initiatives to reduce GHG emissions.

A latent class analysis indicates that it is important to take a more detailed look at consumers than simply analysing the preferences of an average consumer. The latent class analysis indicates that four out of seven segments value a GHG emission plan. However, differences in the relative WTP for the production being organic, climate friendly or animal friendly indicates that in order to catch the interest of consumers from different segments, a targeted marketing or information strategy is needed. Further analyses of consumer characteristics should be pursued as they might shed light on how to proceed in that direction.

References

Aertens, J., Verbeke, W., Mondelaers, K. and Van Huylenbroeck, G. (2009). Personal determinants of organic food consumption: a review. British Food Journal 111(10): 1140-1167.

Bierlaire, M. (2003). BIOGEME: a free package for the estimation of discrete choice models. Proceedings of the 3rd Swiss Transportation Research Conference, Ascona, Switzerland.

Christensen, T., Olsen, S.B., Kærgård, N. and Dubgaard, A. (2014). Spørgeskema om økologisk forbrug. 'Questionnaire on organic consumption.' IFRO documentation 2014/3. Institute of Food and Resource Economics, University of Copenhagen.

Denver, S. and Christensen, T. (2010). Dokumentation af valgeksperiment og spørgeskema, concepts. 'Documentation on choice experiment and questionnaire, concepts.' Institute of Food and Resource Economics, University of Copenhagen, Copenhagen, Denmark.

Denver, S. and Christensen, T. (2014). Consumers' grouping of organic and conventional product; implications for the marketing of organics. Journal of Food Products Marketing, 20(4): 408-428.

Heerwagen, L.R. Andersen, L.M., Christensen, T. and Sandøe, P. (2014). Can increased organic consumption mitigate climate changes? British Food Journal 1168: 1314-1329.

Koistinen, L., Pouta, E., Heikkilä, J., Forsman-Hugg, S., Kotro, J., Mäkelä, J. and Niva, M. (2013). The impact of fat content, production methods and carbon footprint information on consumer preferences for minced meat. Food Quality and Preferences 29: 26-136.

Lancaster, K.J. (1966). A new approach to consumer theory. Journal of Political Economy 74(2): 132-157.

Tuomisto, H.L., Hodge, I.D., Riordan, P. and Macdonald, D.W. (2012). Does organic farming reduce environmental impacts? A meta-analysis of European research. Journal of Environmental Management 112: 309-320.

Tveit, G. and Sandøe, P. (2011). Økologiske Fødevarer – hvor bevæger forbrugerne sig hen? 'Organic food – where are the consumers heading?' Centre for Bioethics and Risk Assessment, University of Copenhagen, Copenhagen, Denmark.

Ethical challenges of Romanians' consumer preference related to satiety induced by commercial breads

R. Suharoschi
Department of Food Science, Faculty of Food Science and Technology, University of Agricultural Sciences and Veterinary Medicine of Cluj-Napoca, 3-5 Manastur St., Cluj-Napoca, 400372 Cluj, Romania; ramona.suharoschi@usamvcluj.ro

Abstract

Bread represents one of the principal fundamental food of the Romanians' consumer meals. Several studies related to whole-wheat and oat products diet shown an increased satiety. This studies shown that the effect on energy consumed was inconsistent with no effect on energy intake while other shown a decrease food energy intake. Moreover, a study regarding whole cereals (oat and wheat) doesn't find any correlation between food energy intake and satiety (Berti *et al*., 2005). The main aim of this study is to analyse Romanian consumer's preferences on three different commercial breads, and to add more knowledge and understanding into how consumers perception of satiety value of bread influence their choice of a variety of analysed bread and to get a better insight of consumer vocabulary and perceptions about bread-related satiety. The study provides insights into Romanians' consumer attitudes toward food safety, nutrition and health issues at which we add other characteristics related to satiety induced by three different types of commercial breads. This paper offer a frame for a debate on the ethical challenges related to consumer perception applying the ethical grid as a theoretical matrix to explore and argue with the study of ethics concerns.

Keywords: consumer behaviour, satiety perception, fully-informed choice, ethical grid, bread

Introduction

The patterns of bread consumption vary widely within the EU, but most countries have an average consumption of 50 kg of bread per person per year. The consumption of bread in Romania is in the first place in EU with 74 kg on average per year (American International of Baking (AIB International)). The expenses for the foodstuffs and beverages (including food expenses out-of-house: restaurants, bistros, food-to-go, fast food, catering, home ordering, etc.) were on average 100€ per months per family (less than 40€ per months per person), and represented 43.8% of total consumer expenditures, 34.8% of budgetary consumption expenditure.

Bread and bakery products (8€ per person) take up 15.2% at the national average of the total expenditure on food and beverages. A significant proportion of expenditure for the purchase and holding of the meat and meat products (25.3%), milk, dairy, cheese and eggs (14.8%). Expenditures for the purchase of three groups of food-bread and bread-like products, meat and meat products; milk and dairy, cheese and eggs is 55.3% of the total expenditures for the purchase of food and alcoholic beverages. The main part of the households' food intake as a whole is represented by the food of vegetal origin. Almost 80% of the calories and about 60% of the amount of proteins are provided by the food of vegetal origin. The residence did not have a too high influence on calorie weight of vegetal origin (on urban area 74.4% and 76.4% of rural area) while the animal protein intake was 50.9%, and 44.5% respectively (The National Institute of Statistics (INS)). Amounts of food consumed by a person reflects a way of more real effect of the intricate factors of influence at the level of the household income: the size of the family income, number of children, residence, etc. By region of residence, the average consumption of bread and bread-like products was 8.8 kg/person in urban areas and 9.6 kg per capita in rural areas. By region, the largest average consumption of bread and bread-like products was registered in the southwest region (Oltenia) (11.4 kg per capita), with lowest grade of education (8.58% population with HE from total population of the region), and with highest population over 60 years (55.97%) compared to the lowest in the northeast region (7.4 kg per capita) with highest grade of education (17.60% population with HE from total population of the region) and lowest population over 60 years 44.03% (INS).

Theoretical approaches to satiety (fullness feeling) understanding in the context of food consumption

Appetite control is a complex event under the combined control of different signals nutritive, sensory, cognitive, and physiological response to consumption of food (Blundell *et al.*, 2010). At the phenomenological level, it expresses feelings of hunger and fullness (satiety). Considering that satiety is a subjective sense of a decrease in the rational of crib, hunger is an unpleasant situation that was unloaded by the ingestion of food (De Graaf, 2011). Foods are distinguished based on their degree to which they induce the feeling of satiety, and composition among other factors (Holt *et al.*, 1995). There is a large volume of researches showing that the acute satiating effects of food can be influenced by their energy density and macronutrient components (Van Kleef *et al.*, 2012) as well as non-nutrient components such as specific fibre (Wanders *et al.*, 2011). Blundell and collaborators (2010) provide a substantial review and recommendations on the methodology for evaluating the satiating effects of food (objective, reasonable and reliable). Food producers have recognized the need to increase fibre in current and new products, but there are challenged since not all fibres are equally satiating (Klosterbuer *et al.*, 2012). Several studies related to whole-wheat and oat products diet shown an increased satiety, the effect on energy consumed was inconsistent with no effect on energy intake (Solah *et al.*, 2007), while other studies shown a decrease food energy intake (Melanson *et al.*, 1999). Moreover, a study regarding whole cereals (oat and wheat) doesn't find any correlation between food energy intake and satiety (Berti *et al.*, 2005).

The aim of this paper is to present state of knowledge on the subject of Romanian consumers' perception of bread consumption. In order to serve consumers' needs and wants, it is crucial that those involved in the market are informed in as much detail as possible about the perception of bread eating, especially since free gluten diet and free-gluten products have left their market niche (Engleson and Atwell, 2008; Dunn, 2014). Over the past few years, various studies focusing on individual aspects of bread consumer preference have been published. Currently, the only comprehensive literature review on bread consumption deals with personal determinants of consumer preference in terms of sensory quality (Heenan *et al.*, 2008), freshness perception (Heenan *et al.*, 2009) or perception of bread quality (Gellynck *et al.*, 2009). None of these studies are not questioning the ethical challenges of the research. The present contribution, therefore, gives an insights of the state of research on the topic, including basic and additional characteristics (nutritional, health, and satiety), product labelling, product innovations and the consumer fully-informed choice.

Methodology

The main aim of this study is to analyse the Romanian consumer's preferences on three different commercial breads (multi-grain bread (MG); whole-meal bread (WM); white bread (WB)) and to add more knowledge and understanding into how consumers perception of satiety value of bread influence their choice of a variety of analysed bread and to get a better insight of consumer vocabulary and perceptions about bread-related satiety. An adapted version of the Food Frequency Questionnaire was applied (19 questions). The questionnaire design was an open-ended type of questions (dichotomous questions), and the measurement were done by using Likert response scale (bipolar scale – opinion questions). The satiety induced by the different type of bread was assessed between and before each of the meals.

Subjects

The consumer preference on bread behaviour pathway were assessed on 102 healthy-volunteers (29 men and 73 women representing 71% of female population compare with INS 51.3% female population), normal-weight, non-smokers were recruited from the urban general population (large and small towns of the Transylvania), with the age ranging 18-40 years, with an intermediate and high level of education. The Fullness Feeling (FF) (Satiety Index (SI)) correlated to Energy Value (total energy of the diet (kcal)), and assessed on six healthy volunteers, normal-weight, non-smokers.

Hypothesis

The bread with barley, oats and seeds has a greater FF (SI), with a lower Glycaemic Index (GI). A blinded randomized study was used in order to assess the three commercial breads (available in Romanians' hypermarket) – a within-subjects repeated-measures designed (tested during 3 different days, at the same time, under similar conditions as possible). The volunteers eat the tested bread at the same main meal. The Questionnaire (Visual Analogue Scale (VAS) and Satiety Labelled Intensity Magnitude (SLIM)) assessed: hunger (from 0: 'extremely hungry' to 100: 'not at all'), fullness (from 0: 'not at all' to 100: 'extremely full'), desire to eat (from 0: 'very weak' to 100: 'very strong'), quantity of eaten foods (from 0: 'nothing at all' to 100: 'very large amounts'), and thirsty (from 0: 'very weak' to 100: 'very strong'). The volunteers follow the same 'eatwell plate' menu (breakfast (8.00 AM) (2 subjects with 50 g MG; 2 subjects with50 g WM; 2 subjects with50 g WB)); snack (11.00 AM) 1 fruit; lunch (2.30 PM): regular meal + 1 slice (25 g) MG/WM/WB; dinner (7.30 PM): salad + 1 slice (25 g) MG/WM/WB.

Ethical grid

In order to analyse the ethical challenges of the research, it was used the theoretical framework of an ethical grid (EG). The EG, based on the work of Seedhouse and Flinders, is a tool designed to help the user understand the ethical issues that they are facing. The grid was developed in the context of healthcare (Seedhouse, 1998). It provides a structure to guide the moral reasoning and ethically thinking of the user, in order to enable systematically respond and issues. The research becomes more ethically and can be presented the decisions in such a way that the decision-making process is transparent and can be discussed. The EG approach to ethical dilemmas as being like a network where each idea is connected to all the others by all manner of routes. The debate of the ideas in the framework of EG are overlap and interconnected. The ethical grid present four concentric layers within a square. The four layers in conformity with Seedhouse represent four aspects of comprehensive ethical analysis: external, consequential, deontological and individual (Seedhouse, 1998; Stutchbury and Fox, 2009). Blue boxes represent the core rationale, the red boxes represent the deontology layer with a focus on duties and motives, the green boxes represent the consequential layer with attention on outcomes, and the black boxes are of major importance and include factors often omitted representing the level of external considerations (a persons' legal right to fully-informed choice being able to effectively takes any decision). It is also necessary to assess the research degree of risk the (allergy, intolerance, potential risk of contaminants ingestion – like acrylamide). The common questions are:
- Will physical harm occur?
- Will there be emotional damage?
- Is there a danger that an unfortunate precedent will be set?
- Is there a chance that the proposed research will produce unintended consequences?

There is an associated demand on the EG user to try to ensure that whatever study is made is carried out with maximum efficiency and effectiveness.

Limits of the ethical grid

The limit of the EG is straightforward. The grid cannot direct the tasks, nor can decide which tasks are the most important. The grid can enhance deliberation, but the responsibility lies with the user not with the grid. The questions asked need to be very specific – the study need an informed consent?

Possible questions that could be posed by each level of the framework are: What are the values, norms and roles in the environment in which the study is conducted and are they likely to be challenged by this research? What is the relationship between the group/individual the researcher work with? How does it affect the participant(s)? How the research might be viewed by others? How will the language used be interpreted? What are the responsibilities to the people paying for or supporting this research (local authority, university, external bodies)? Has the research worked respecting the ethical guidelines? Is the researcher aware of his/her rights and responsibilities through to publication? Have the researcher made efficient use of the resources available to me, including people's time? Are there any risks to anyone as a result of this research? Have the researcher treated all participants fairly?

There were no major differences in the results of the questionnaire VAS area changed for the three products assessed. However, subjects reported significantly less hungry sensation before lunch compared to their hunger before breakfast when not eaten bread. Top scores on satiety were found in the case of multigrain bread. The results obtained are in conformity with the literature; the oatmeal bread induced an increased satiety (due to β-glucan effect on glucose absorption and satiety) (Benini *et al.*, 1995). The study limits are the reduced sample size, the wide variability of the result, and the absence of a sensory analysis.

Discussion

Essentially this model assumes that from their perception of actual information (e.g. the stated product attributes), consumers infer functional (e.g. what the product does functionally) and psychosocial (e.g. emotional) outcomes. This product related perceptions then form the basis for further cognitive elaboration processes to infer functional and psycho-social benefits. At a final level of cognitive elaboration, these outcomes are understood in the context of relevant goals that the consumer wants to achieve or to avoid. Industry has been encouraged to take an active role in offering products that facilitate healthy choices. The media constantly seek to tempted consumers with reports of 'novel', 'interesting' and 'unusual' functional aspects of foods/bread. The EFSA has established procedures for evaluating scientific evidence in support of health claims about foods. In this environment, there is enormous interest in developing and marketing functional foods to manage satiation, satiety and body weight (Blundell, 2010).

Evidence that consumers understand health claims related to foods, in general

The study did not afford enough evidence to suggest that consumers can make an informed choice about bread products based on their understanding of the science of satiety and energy balance. The study makes a distinction between consumer demand for bread, approachability to eat bread, and understanding of health claims related to satiety and weight control in perspective of bread consumption. Currently, it is a great demand for foods with satiety and weight control functionality and consumers clearly respond to the market offerings available. However, consumer understanding is not evident, and, in fact, there is more evidence for consumer naivety in relation to weight control bread-like products, in general (Cleland *et al.*, 2002). The consumer understanding related to bread eating habits related to health claims, satiety and weight control, can be either subjective (i.e. perceptive) or objective. Evidence for both attitudes is very limited. Results from this study indicate that bread labels do not significantly impact consumer's perceptions of bread quality, although consumers may have uncorroborated tenets about the healthfulness and potential positive impact of assessed bread. The recommendations to add ethical value to the study consist on a trained panel enabled an objective understanding of bread satiety and health claims. Clearly the application of this approach to predict satiety of a newly developed bread products will benefit the industry but also the consumer.

As concerns Romanians' eating behaviour culture is based on bread home-made tradition, the tendency of Romanians' population is to address specific home-made bread recipes presenting a subjective approach. About foods (e.g. bread), diets (e.g. gluten sensitivity) and weight control, consumers seem less interested in evidence than they are in the promise of market strategy prospects. While some argue that this is the result of unprincipled trading and manufacturers, it is clear from the popular media that obesity holds a certain fascination for the public. Many western adults have an intense, emotional and personal interest in managing their weight. These findings showed that consumers do not effectively use existing information relating to satiation, satiety and appetite to control their weight.

Conclusion

Currently, the existing evidence is insufficient to support the most likely bread candidates for satiety claims, and few claims have yet been approved regarding the whole grain-rich products made from fibres such as oat fibre. The study show that is little evidence consumers understands such claims. The scientific knowledge of the consumer is limited. Bread consumers seem more alert to the promise of weight control solutions than knowledgeable about how to use them in the context of effective weight control behaviours. Future progress depends on more accurate product assessments, greater consumer understanding of related claims, and scientific evaluation of consumer understanding. Confronting dilemmas with ethical implications is an integral part of any research involving people (even healthy volunteers) acting decisively, which include dealing directly or indirectly with numerous moral issues that are linked to the entire process of producing, distributing, making-decision and consuming food (Brom, 2000; Heenan *et al.*, 2008; Mepham, 1996; Zwart, 2000). However, nowadays food ethics is far more diverse. This lead to increasing separation between the production and consumption of food as part of the most general process named 'the fragmentation of society' (Miller *et al.*, 1990). We raise an ethical dimension added to the experimental design referring to the consumer/citizen/people right to make a fully-informed choice as possible about the bread they eat (Hepting *et al.*, 2014; Lapinski *et al.*, 2014). Based on this ethical issues is important that the research decision that we make to have a justifiable moral basis and that the process of making those decisions to be transparent. The ethical challenges of the research will arise at both macro and micro levels. At a 'macro' level the main ethical concern will be to gather enough data in order to be able to draw valid conclusions; and at a 'micro' level the ethical attention should be paid to consider the details of how we might conduct individual interviews in respect with informed consent. Another ethical aspects is related to the needed of the population to be fully-informed about the bread ingredients (as far as the main commercial bread are using flour with food additives (known as 'processing aids'), premix, or frozen dough and/or bread. Mass-media should do the consumer education in terms of the information about food ingredients and food labelling (EU Regulation N°. 1169/2011, 2011). For example, on the Romanian market the dark bread is made from white wheat flour with artificial food colorants in 60%. That raise an ethical concern questioning the honesty and fairness of producers, as far the bread market is about 1 Billion Euro, and bread has the lowest VAT (9%) comparing with 24% of all other food. It looks like the main concern for producers being the suspicious profit.

References

AIB International – American International of Baking (2011). Meet the opportunities of the Romanian baking industry. AIB International study conducted in 2011. Available at: http://tinyurl.com/oxhjb96.

Benini, L., Castellani, G., Brighenti, F., Heaton, K.W., Brentegani, M.T., Casiraghi, M.C., Sembenini, C., Pellegrini, N., Fioretta, A., Minniti, G., Porrini, M., Testolin, G. and Vantini, I. (1995). Gastric emptying of a solid meal is accelerated by the removal of dietary fibre naturally present in food. Gut 36: 825-830.

Berti, C., Riso, P., Brusamolino, A. and Porrini, M. (2005). Effect on appetite control of minor cereal and pseudocereal products. The British Journal of Nutrition 94: 850-858.

Blundell, J. (2010). Making claims. Nature Reviews Endocrinology 6(1): 53-56.

Blundell, J., De Graaf, C., Hulshof, T., Jebb, S., Livingstone, B., Lluch, A., Mela, D., Salah, S., Schuring, E., Van der Knaap, H., Westerterp, M. (2010). Appetite control. Methodological aspects of the evaluation of foods. Obesity Reviews 11(3): 251-270.

Brom, F.W.A. (2000). Food consumers concerns and trust: food ethics for a globalizing market. Journal of Agricultural and Environmental Ethics 12(2): 127-39.

Clark, M.J. and Slavin, J.L. (2013). The effect of fiber on satiety and food intake. Journal of the American College of Nutrition 32(2):1-12.

Cleland, R.L., Gross, W.C., Koss, L.D., Daynard, M. and Muoio, K.M. (2002). Weight-loss advertising: an analysis of current trends. Federal Trade Commission staff report.

De Graaf, C. (2011). Trust worthy satiety claims are good for science and society. Appetite 57(3): 778-783.

Dunn, H.R. (2014). Beyond bread: gluten-free potential in beer and pizza. Available at: http://tinyurl.com/omuebkt.

Engleson, J. and Atwell B. (2008). Gluten-free product development. Cereal Foods World 53(4): 180-184.

European Union Regulation N° 1169/2011. Official Journal of the European Union 18 (L 304).

Gellynck, X., Kühne, B., Van Bockstaele, F., Van de Walle, D., Dewettinck, K. (2009). Consumer perception of bread quality. Appetite, 53(1): 16-23.

Heenan, S.P., Dufour, J.P., Hamid, N., Harvey, W. and Delahunty, C.M. (2008). The sensory analysis of fresh bread. Food Research International 41: 989-997.

Hepting, D.H., Jaffe, J., Maciag, T. (2014). Operationalizing ethics in food choice decisions. Journal of Agricultural and Environmental Ethics 27: 453-469.

Holt, S.H.A., Brand Miller, J.C., Petocz, P. and Farmakalidis, E. (1995). A satiety index of common foods. European Journal of Clinical Nutrition 49(9): 675-690.

Klosterbuer, A.S., Thomas, W. and Slavin, J.L. (2012). Resistant starch and pullulan reduce postprandial glucose, insulin, and GLP-1 but have no effect on satiety in healthy humans. Journal of Agricultural and Food Chemistry 60: 11928-11934.

Lapinski, M.K., Funk, J.A., Moccia, L.T. (2014). Recommendations for the role of social science research in One Health. Social Science and Medicine 128: 1-10.

Melanson, K.J., Westerterp-Plantenga, M.S., Saris, W.H.M., Smith, F.J. and Campfield, L.A. (1999). The American Journal of Physiology: Regulatory, Integrative, and Comparative Physiology 277: 337-345.

Mepham, B. (ed.) (1996). Food ethics. Routledge, London, UK.

Miller, J.P., Cassie, J.R.B. and Drake, S.M. (1990). Holistic learning: a teacher's guide to integrated studies. Ontario Institute for Studies in Education, Toronto, Canada.

Seedhouse, D. (1998). Ethics: the heart of healthcare. Wiley-Blackwell, Chichester, UK.

Solah, V.A., Crosbie, G.B., Huang, S., Quail, K., Sy, N., Limley, A. (2007). Measurement of satiety of wheat-based bulgur by intervention and sensory evaluation. Cereal Foods World 52(1): 15-19.

Stutchbury, K. and Fox, A. (2009). Ethics in educational research: introducing a methodological tool for effective ethical analysis. Cambridge Journal of Education 39(4): 489-504.

The National Institute of Statistics – INS. (2013). Romania, repere economice si socio-regionale: statistica teritoriala. The country report. Available at: http://tinyurl.com/pyzgkgm.

Van Kleef, E., Van Trijp, J.C.M., Van den Borne, J.J.G.C. and Zondervan, C. (2012). Successful development of satiety enhancing food products. Critical Reviews in Food Science and Nutrition 52(7): 611-628.

Wanders, A.J., Van den Borne, J.J.G.C., De Graaf, C., Hulshof, T., Jonathan, M.C., Kristensen, M., Mars, M., Schols, H.A., Feskens, E.J. (2011). Effects of dietary fibre on subjective appetite, energy intake and body weight. Obesity Reviews 12(9): 724-739.

Zwart, H. (2000). A short history of food ethics. Journal of Agricultural and Environmental Ethics 12(2): 113-26.

Technologies in organic farming: consumers' values and acceptance

J. Lassen
Department of Food and Resource Economics, University of Copenhagen, Rolighedsvej 25, 1958 Frederiksberg, Copenhagen, Denmark; jlas@ifro.ku.dk

Abstract

In organic farming a dilemma is posed by the heavy reliance on nutrients from conventional livestock farming. For Danish organic plant producers the influx of conventional nutrients accounts for up to 70% of their nutrients. Facing this problem, Danish organic farmers' organizations have decided to phase out their use of conventional manure before 2021. This, however, raises a number of questions about consumers' acceptance of the alternative technologies that have been proposed to close the nutrient gap. Drawing on qualitative interviews with Danish organic consumers, this paper first discusses what, from a consumers perspective, characterizes the technologies consumers associate with organic production. This part of the analysis shows that by and large consumers regard organic technologies as the opposite of conventional farming. Second, consumers' perceptions of solutions suggested to close the nutrient gap by recycling sewage sludge, and in particular the criteria applied when these technologies are assessed, are analysed. This part of the analysis shows how organic consumers base their assessment of alternative strategies and technologies primarily on concerns about environmental risks (pollution) and consequences for human health. Broader ideas about ecosystems and the recycling of nutrients between the agricultural sector and the urban population are notably absent. On the basis of these findings the paper concludes by discussing the relationship between the consumers' values that guide what is perceived as an acceptable organic technology and the ideological principles behind organic production systems as they are formulated by the International Federation of Organic Agriculture Movements (IFOAM).

Keywords: agriculture, perceptions, sewage sludge, nutrient gap

Introduction

In 2008, following years of internal discussions, the two Danish societies for organic farmers, Økologisk Landsforening (ØL) and Sektion for Økologi (SØ), made a decision in principle to phase out use of conventional manure in organic plant production and straw in organic animal production by 2021. On the one hand, this would resolve a problem faced not only by organic producers in Denmark, but also in countries around the world (Oelofse *et al.*, 2013): namely, the fact that organic production, in its present form, is closely intertwined with the conventional agricultural sector it sees itself as an alternative to. Thus organic agriculture as a whole cannot be seen as living up to the ideal of recycling, one of the core elements in the principles of ecology laid out by the International Federation of Organic Agriculture Movements (IFOAM) (2014).

The seriousness of the challenge is illustrated by the fact that conventional manure, mostly from the largely criticized and intensive pig production, accounted for about 70% of the total input of nutrients in Danish organic plant production in 2010. The discussions in LØ and SØ leading up to the 2008 decision were heated, not least because phasing out conventional nutrients would be a much smaller challenge to organic dairy producers – who by and large have plenty of nutrients – than to plant producers. In addition, suitable alternative nutrient sources and technologies were not easy to identify.

Some of the technologies put forward to fill the nutrient gap following the 2008 decision represent a radical vision of recycling. They involve returning nutrients which leave the agricultural sector in the shape of food products to be consumed in the cities to organic plant production in the shape of sewage sludge. It is, however, unclear how such a change in organic agriculture would be received by consumers of organic food. Since the organic sector is both a social movement and business producing organic food products, it needs to balance between its own values and consumers values. This paper aim to help the organic movement handling this balance by scrutinizing the arguments organic consumers give when they are asked whether or not a technology is relevant to organic production. Thus the values organic technologies must meet to be acceptable from a consumer's point of view are revealed.

Method

The paper draws on the empirical results of five focus groups of 7-8 persons (total n=39) conducted in September 2012. Across the groups, participants were recruited paying respect to educational background, age and gender. In addition, only organic consumers were recruited. These were defined as consumers who had either purchased an organic product in the two weeks preceding the recruitment interview or agreed that organic farming should be promoted.

The interviews were carried out in Copenhagen (two groups); Kolding, a provincial town (two groups) and Christiansfeld, representing a rural area (one group). All participants were recruited through an analysis company in a topic-blinded manner. Each focus group discussion lasted approximately 2 hours and organised around four themes as follows. (1) Organic foods in an everyday perspective. (2) Organic production in a societal perspective. (3) Organic technologies. (4) Actors responsible for solving the nutrient problem. This paper is based primarily on an analysis of themes 2 and 3, where the participants were encouraged to discuss organic technologies.

In theme 2 participants were asked to write down any keywords they associated with organic agriculture. They were reminded that the organic products they had been discussing in the previous theme originated from a production system. This was followed, in theme 3, by a section where participants were asked to reflect on the technologies they associate with organic plant production. Drawing on previous research on the importance of contexts for expressed themes and values (Korzen and Lassen, 2010), the intention of these sections was to encourage the participants to discuss technologies rather than products, and thus illuminate what in general, and specifically for plant production, characterises an organic technology.

The second part of theme 3 involved a more structured approach, where participants were first introduced to, and subsequently discussed, the problem of the nutrient gap, and the decision to close it. Following this, participants were shown six cards suggesting different technological solutions to the problem. These included three sewage sludge-related solutions that will be analysed in this paper: use untreated sewage sludge; use composted sewage sludge, and use phosphate extracted from sewage sludge. The intention of this section was to illuminate the interviewees' perceptions of the different solutions to the problem of phasing out conventional nutrients, and in this way to facilitate an analysis of the values drawn on in such assessments.

The focus group interviews were recorded and transcribed verbatim. Stephen Toulmin's approach to the analysis of argumentation (Toulmin, 2003) was applied in order to identify and analyse the core arguments that participants were deploying when they responded to questions about organic technologies. Thus the analysis revealed the values underlying the expressed perceptions.

What is an organic technology?

It is striking that in the general discussions of organic technologies interviewees most often characterise organic agriculture by saying what it is not. Accordingly, organically relevant technologies are often identified by references to their counterparts in conventional production systems: organic production methods are those not using pesticides, not using artificial fertilizers or not relying on genetic manipulation, and so on. Occasionally there is an explicit reference to conventional agriculture, but most often the comparison is tacit or implicit – as in the following exchange, where Carsten defines organic agriculture: 'The first that comes into my mind, when I think of the organic, that is of course the farmer doesn't spray – and I suppose this is also true for market gardening. And they neither use straw shortener or other crop sprays; and this must be positive for nature – for plants and animals in nature. This was actually what came into my mind'.

This is one of the relatively few examples where organic technologies are justified with reference to nature and the environment – in this case pesticides and the use of straw-shortening chemicals are presented as positive organic technologies because of the benefits for plants and animals. By contrast, the avoidance of pesticides and other auxiliary chemicals is more often discussed within a frame of health: organic technologies here are favoured because the products are not as contaminated with chemicals or because the working environment is free from chemicals. Alberte, for example, argues in favour of organic vegetable production in greenhouses, justifying it with reference to consumer as well as worker health: '... the first (thing) I wrote was better working environment, as Anne says, because of the working conditions and all the poisons, sprayed in the greenhouses. It is not fair that people are exposed to this – and I'm sure that less (poison) will end up in our body if we eat things other than these sprayed vegetables'.

Quite often links between organic technologies and, say, pesticide-free agriculture are followed by positive accounts of alternative pest control. These are both referred to in general with terms like biological pest control or (less often) by specifying the alternatives – as happens here, when David talks about alternatives like predatory mites, revealing a high degree of insight into (organic) agricultural technologies: 'And that makes me think about – and that's definitely something they use in greenhouses – these biological pest controls, that is predatory mites. And then I think about this awful quantity of fruit... with deceases, because it will be affected'.

The auxiliary chemicals closely connected with conventional production, but not with organic production systems, are often discussed within a critical frame of efficiency and unnaturalness. At the core here is the notion that virtues of conventional agriculture are speed and efficiency, and the idea that this is achieved by artificial – that is, not natural – means such as chemical sprays, genetic manipulation, and so on. By contrast, organic agriculture, because of its limited use of chemicals and high-tech solutions, is seen as more natural. Core features of organic production here seem to be that plants and fruit should not be harvested until they are ripe, that growth (or breeding) should not be accelerated, but follow the pace of nature. Time, or rather slowness, is seen as an important aspect of organic agriculture. So, organic technologies are perceived as technologies which embody naturalness and induce slowness. This is expressed by Ejvind when he explains why he dislikes the way conventional agriculture fiddles with things. His argument in favour of organic production is supported via a link between the slowness of organic agriculture and a more natural and therefore desirable production system: 'Really, (in organic farming) it is something natural, coming out of the soil, without anybody fiddling too much. ...like a real seed, you could say, coming out of the soil, when you water it. It's not something that has been grafted, where they have used straw shorteners or whatever they do to make them grow faster... I suppose that that's what it is about (in conventional agriculture), and that is not a very nice idea...'

Following the pace of nature in organic production is often seen as a good thing in itself, a positive aspect that needs no elaboration or backing – 'a nice idea', as Ejvind express it. It is, however, striking that on the few occasions on which participants say why naturalness is desirable, the explanations they offer are only once linked to the idea that following the speed of nature ensures that plants are more robust in coping with pests. Typically, the explanations have instead to do with a perceived and positive association between naturalness and better health, improved taste, or being more filling, although some interviewees were also concerned about reduced shelf-life.

Closing the gap: recycling nutrients

Considering that one of the pillars, or ideals, of organic plant production is the recycling of nutrients and abandonment of artificial fertilizers, these very issues were taken up by surprisingly few participants during sections of the discussion dealing with organic technologies in general. One of the groups omits to address the issue at all, and in the others, soil enrichment and closing the nutrient cycle are typically addressed in passing, and never discussed in detail. In groups where nutrients were discussed within the framework of nutrient recycling, it is striking that the participants never take up nutrient recycling beyond the farm gate. That is, they only address farm-level circuits such as using manure, harvested weed and crop rotation as a means of providing nutrients.

However, some like Bjarne, below, express uncertainty about whether these strategies are actually applied in organic plant production. Bjarne sees organic agriculture as a practice moving towards what he understands as sustainable agriculture, or in other words an agricultural practice not exhausting the fertility of the soil: 'It would be clever, in the long view, if organic agriculture would be what you consider as truly sustainable, so we don't exhaust (the soil)'.

Hardly any of the participants were aware of the reliance of organic production on conventional manure. Although some received this information with a shrug, most reacted with disappointment or disillusion. Thus Birgit said: 'This makes me think, well, maybe it doesn't really make a difference buying organic products, if the difference is so small. Although I would still buy some products organic'. Some clearly see organic farming as *the* good and sustainable alternative – as a sort of golden standard meeting the highest requirements. From this perspective, organic farming's reliance on conventional farming is shocking and makes it hard to see the point of organic farming.

While some participants are more positive than others, discussions of the three alternative sludge technologies show that none of the focus groups embraced these solutions. The solution least accepted was the use of untreated sludge, followed by composted sludge. Extracted phosphorous was seen as most acceptable. Despite this, and although there was frequent criticism of the present reliance on conventional nutrients, several of the groups ended up agreeing that none of the sludge-based technologies are better alternatives.

In discussions of the pros and cons of the technologies most arguments are related to the risks, to nutrients as a scarce resource, and to discomfort at the idea of mixing human lavatory waste with food. It is thus remarkable that during these discussions, following the presentation of the dilemma, the issue of recycling is only taken up once in one group. This happened when Alice, towards the end of the discussion of sludge-based technologies, said: 'I believe (the soil) has been exhausted since the fifties. We need to supply some nutrients. After all we are removing products and we take away energy: we need to return it. It is too easy; it doesn't work properly. It's like the heathland in the 50'ies'. Considering that the sewage-based solutions are clearly technologies that close the cycle, it is surprising that this topic is not taken up in any of the other groups.

The only arguments presented invoking the flow of nutrients revolved around scarce resources. To some participants the utilisation of an unexploited resource is an advantage of the sewage technologies. Thus Anja says: '...there must be a surplus (of sludge) somewhere; what happens to that? If they just dump it in the sea, it doesn't make a difference. You might as well make use of it (in organic farming)'. As in the discussions above, organic farming is not seen as part of a larger system where energy and matter ideally circulate. It is regarded as an enterprise, like conventional farming, with inputs and outputs – the basic difference lying in the quality of the inputs, and hence also the outputs.

Most arguments about the sludge-based technologies relate to their risks. While some participants felt confident that once the use of sewage sludge is approved for organic farming, we must assume that the authorities have examined the risks, several expressed concerns about risks. When these risks were appealed to in order to argue for or against the technologies, health risks seem to predominate. But it was often not clear whether the interviewees were talking about health or environmental risks – as can be seen in this discussion between three participants: Allan: 'I get some horror visions: what is in (the sludge)?' Astrid: 'Sludge' – just thinking about it!' Alice: 'It's full of hormones'. Allan: 'Yes, hormones and heavy metals!' Alice: 'Contraceptive pills...' Anne: 'It's disgusting'. Environmental risks were particularly important in assessments of phosphorus extraction, because that process requires high temperatures and thus large quantities of fossil fuels.

Many of the negative reactions to the sludge-based technologies relate to the fact that they mix categories. In line with the views of Mary Douglas (Douglas, 1966), the technologies are often seen as more risky and thus more unacceptable because they mix sludge from, and therefore associated with, the lavatory with organic production associated with food. This is indirectly expressed in Anne's last remark in the quote above; but is sometimes also expressed more clearly, as happens in the following discussion. Else: 'Is cowpat any better that human shit?' Emma: 'Yes, definitely! There are all the medical residues'. Esben: 'There are also medicinal residues in cowpat, just so you know it...!' Emma: 'I find human faeces more repulsive than cow(pat)'.

Finally it should be noted that economic concerns surfaced in the arguments about the technologies. These related to rising consumer prices, as well as concerns about the economic performance of the Danish organic sector.

Discussion

Although values and perceptions are at different levels, as values are more stable and perceptions more likely to change depending on the context, the analysis of consumer's perceptions above points to a number of core values which consumers connect with organic technologies.

First, it should be noted that consumers base their assessments on a number of values which are also visible in the principles for organic farming stated by IFOAM (2014). These shared values include risks to health, to the environment and to workers (the latter all included in IFOAM's principle of health). Although it is not always stated directly, consumer concern about risks is closely related to a demand for caution. The feeling is that knowledge in the area is limited, and that accordingly we should exercise caution – another key value principle stated by IFOAM in their principle of care.

Second, the organic consumers do at some points diverge from IFOAM's principles. It is less surprising that the principle of fairness, stating that organic agriculture should build in fairness, in the common environment and life opportunities, is not a value people use to decide whether a technology is acceptable or nor. After all, fairness is not a major issue in debates over Danish organic production. It is, however, striking that recycling, a key element in the principle of care, is more or less absent as well: at this point consumers diverge from one of the pillars of organic agriculture.

Third, organic consumers seem to base their assessments of organic technologies on a hybrid value best described as 'naturalness/slowness'. This value and its components are not expressed directly in the organic principles.

Fourth, and finally, these differences between official organic principles and consumer belief should be heeded: they show that decision makers in the organic food system need to consider the new technologies they want to introduce into organic agriculture in the future with care.

References

Douglas, M. (1966). Purity and danger. An analysis of the concepts of pollution and taboo. Routledge, London, UK.

International Federation of Organic Agriculture Movements (IFOAM) (2014). Principles of organic agriculture. European Commission.

Korzen, S. and Lassen, J. (2010). Meat in context. On the relation between perceptions and contexts. Appetite 54(2): 274-281.

Oelofse, M., Jensen, L.S. and Magid, J. (2013). The implications of phasing out conventional nutrient supply in organic agriculture: Denmark as a case. Organic Agriculture 3(1): 41-55.

Toulmin, S. (2003). The uses of argument. Cambridge University Press, Cambridge, UK.

Fitness for which purpose? A quality assessment of a Norwegian public health intervention in nursery schools

H. Nilsen
Centre for the Studies of the Sciences and the Humanities, University of Bergen, P.O. Box 7800, 5020 Bergen, Norway; helene.nilsen@svt.uib.no

Abstract

This paper seeks to explore Silvio Funtowicz and Jerome Ravetz's concepts of fitness for purpose and fitness for function as an analytical framework for evaluating a public health intervention in nursery schools in Bydel Grorud, Oslo, Norway. It argues that reducing the practical and ideologically based problem of social health inequity to a manageable technical problem of diet and exercise carries with it a range of potential pitfalls and may cause the original purpose of intervention to be lost from view.

Keywords: public health ethics, nutrition, prevention, children's health, post-normal science

Introduction

NCD's (non-communicable diseases, popularly known as 'lifestyle diseases') has gained increasing focus on a global level during the past decade, and are often seen in relation to lifestyle, dietary habits and physical activity. There is broad understanding that the solution to this modern health problem is to 'prevent more in order to repair less', encompassing both socio-economic and public health benefits. This is reflected in the Scandinavian context, where the focus is also revolving around social health equity as an important area of action, reflected in the Nordic Societies for Health and Well-being conference in Trondheim 2014. In Norway, the issue of social health equity has become an area of priority, as is reflected in the White paper 'Prescription for a Healthier Norway' (Norwegian Ministry of Health and Care Services, 2002-2003). In this report, Oslo is emphasized as a particularly relevant region for action, as this is the city with the highest diversity, and highest degree of inequity.

This paper will seek to explore a particular public health intervention directed at nursery-schools in the area of Bydel Grorud, Oslo.

Materials and methods

This paper will draw upon Post Normal Science (PNS) as methodological framework. PNS was developed by Silvio Funtowicz and Jerome Ravetz as a model of enquiry in cases where 'facts are uncertain, values in dispute, stakes high and decisions urgent' (Funtowicz and Ravetz, 1991). More specifically this paper will find its point of departure in Funtowicz' and Ravetz' conception of 'quality as fitness for purpose' in order to perform a critical analysis of the intervention report. In their work from 1992, 'The Good, the True and the Post-modern', Ravetz and Funtowicz make a distinction between internal and external quality. Internal quality is defined as 'assessed within the field of practice' in accordance with criteria such as dexterity, craftsmanship and creativity (Funtowicz and Ravetz, 1992: 964). External quality is defined 'by a relationship with a broader community of users, has its own criteria and modes of assessment. It is conceived in such terms as fitness for purpose, reliability and economy' (Funtowicz and Ravetz: 964).

The concepts of internal and external quality, are related to Ravetz' distinction between scientific, technical and practical problems (Ravetz, 1996). A technical problem, according to Ravetz, is defined as 'a statement of a function to be performed, whose means are to be established as the conclusion of an argument, with a plan for its accomplishment' (Ravetz, 1996: 323). A practical problem, on the other hand, is defined as a 'statement of a purpose to be achieved, whose means are to be established as the conclusion of an argument, with a plan for its accomplishment' (Ravetz, 1996: 339). The central tenet of this perspective, then, is the distinction between the function, or goal, of a technical problem, and the purpose of, or the ideology informing, a practical problem in terms of how these problem framings affect their potential solutions. This holds particular relevance where welfare is concerned: 'The correct identification of the distinction between technical and practical problems is of great importance for understanding the human aspects of modern physical science and technology. If one's language, in the term 'science', forces one to assume that work with physical things has an autonomous system of final causes, and that by contrast work with people is automatically concerned with welfare, then there is no framework in which these problems can be analysed, much less solved' (Ravetz, 1996: 338).

Pointing to the potential pitfalls of failing to recognize this distinction, Ravetz' emphasizes the danger of immediate technical problems replacing the purpose of a decision in the execution of a project, which in turn may transform 'welfare invention' to 'welfare engineering' (Ravetz 1996: 342).

This article will seek to explore and evaluate a public health intervention in Norwegian nursery schools in the light of quality as fitness for purpose by examining the policy documents and political framework informing the public health initiative.

Hence, the research material for this article is the intervention report 'Public Health Project in Nursery Schools' (Dønnestad and Strandmyr, 2014), written by two project workers with background in sport sciences, Julie Dønnestad and Anette Strandmyr, performing an intervention in three nursery schools: Ammerudgrenda barnehage, Ammerudlia barnehage, and Grorud barnehage. The project is a part of 'The Groruddalen initiative' (2013) seeking to improve the general living conditions of this area of Oslo. This initiative encompasses several coinciding initiatives regarding environment and living conditions in the area, divided into four areas of action: (1) environmental friendly transportation; (2) the Alna River, greeneries, sport and cultural environment; (3) development of housing conditions and urban development; (4) childhood and adolescents, education, living conditions and inclusion.

The background of the intervention in question is the focus on social health inequity, and the writers state that: 'By promoting physical activity, a healthy diet and health enhancing habits among the children and their parents, we increase the probability of a lasting change of lifestyle in each family. Thus, we can contribute to the accomplishment of Bydel Grorud's long term vision of reduced social health inequity' (Dønnestad and Strandmyr, 2014: 7-8). This goal is then divided into two sub-goals: ensuring that the children's (1) diet and (2) physical activity are in accordance with the Norwegian Health Authorities recommendations (Dønnestad and Strandmyr, 2014). Different measures are taken in order to meet the state recommendation for diet: general guidelines for food served in the nursery schools have been established, the grocer contacted to improve the nutritional quality of the groceries offered, and the staff received training in nutrition. In addition, there were parent meetings with information about nutrition, regulations for packed lunches and birthday celebration were established, as well instructive brochures and pamphlets sent home with the children. (Dønnestad and Strandmyr, 2014). As for physical activity, a total of 95 children aged 3-6 years old were equipped with ActiGraph GT1M accelerometers – a biometrical device measuring the intensity of activity, with the goal of spending 60 minutes each day in an intensity zone defined as 'moderate' or 'intensive'. The first measurement was implemented in October 2013, and the second in December 2013, in order to perform a 'quality assurance that the activities we employed had the desired effect' (Dønnestad and Strandmyr, 2014: 11). The first measurement had shown that an average of 16% of the children reached a moderate to intensive level of activity in October, against 28% in December (Dønnestad and Strandmyr, 2014: 15). As this is a pilot project, it is designed to create general guidelines that can apply to all nursery schools, and the report can be seen as an application for further funding. Thus, the project can be seen as both a 'knowledge based intervention' operating with clearly defined aims of action and as a research project investigating what is needed and how to implement changes.

Results and discussion

Health inequity as practical problem

Ravetz defines a practical problem as 'a statement of a purpose to be achieved, whose means are to be established as the conclusion of an argument, with a plan for its accomplishment. In some respects, such practical problems are the most difficult to solve, as well as including some of the most important tasks facing our society, these coming under the general category of 'welfare'' (Ravetz 1996: 339). Such statements are necessarily ideologically based, and therefore the first part of this paper will be concerned with the political and ideological background referred to in the intervention report. The intervention finds legitimacy in a whitepaper on public health from 2003, entitled 'Prescription for a Healthier Norway' (Norwegian Ministry of Health and Care Services, 2002-2003). This whitepaper acknowledges diet and physical activity as the important areas of focus, particularly with regard to children and the young. This is seen as an important aspect of 'preventing more in order to repair less'. Furthermore, it focuses on social health inequity as an important political and societal issue, and emphasizes the necessity of 'considering initiatives directed towards geographical areas with particular problems concerning health and living conditions, with particular focus on Oslo, where there is a higher degree of social health in equity' (Norwegian Ministry of Health and Care Services, 2003). Interestingly, it notes that the health status of people living alone is the major cause of the national increase in health inequity (*ibid.*), and that there are more notable differences 'between groups of immigrants than between the particular groups of immigrants and the ethnically Norwegian population' (*ibid.*). Public health is explicitly acknowledged as a complex problem: 'Although it is important to emphasize the individual's options and actual responsibility – particularly with regard to habits – social health inequity will primarily be a political and societal matter. When differences are clearly visible in social structures, it is not first and foremost caused by individual's conscious choice of lifestyle' (Norwegian Ministry of Health and Care Services, 2003). Structural and individual preconditions of health are identified as intertwined, and, as such, not isolated aspects that can be reduced to manageable individual goals. This tension is further emphasized by a specific definition of health employed in the whitepaper, stated in explicit contrast to WHO's all-encompassing health definition which is described as 'conflating 'good health' with a 'good life'': health is not defined as a purpose in itself, but as a means of living a good life (Norwegian Ministry of Health and Care Services, 2003; WHO, 2010).

Taking a step down the ladder of political structure, the local public health strategy plan (Oslo Municipality, 2011) for the area of Grorud illustrates the complexity of health with a Venn diagram where living conditions (defined as social status), environment, lifestyle and health services overlap and surround the centre: health. Notably, health here becomes the purpose to be achieved rather than the means of a good life, something which illuminates the challenge of defining the purpose of practical problems. More importantly for this paper, the diagram models health as a complex size particularly with regard to causality and intertwining factors of influence within which lifestyle is one of several aspects.

Moving from the level of local politics to the specific context of nursery schools, the intervention report refers to the National Plan for Nursery Schools' Contents and Obligations (Norwegian Ministry of Education and Research, 2012), emphasizing 'the municipalities' responsibilities for promoting health and evening out social differences' (Dønnestad and Strandmyr, 2014: 7). A closer look at the official document, however, makes visible the complexity of child care and a range of potential value conflicts. 'Seeing care and bildung, play, everyday activities and learning as interconnected' is defined a central and distinctive feature of the Norwegian nursery school (Norwegian Ministry of Education and Research, 2012). This is already a complex task, with a range of potential implications. Under the headline of 'Body, Movement and Health', the goals range from the acquirement of 'knowledge about the human body and an understanding of the importance of good habits and a healthy diet' to gaining an 'understanding about and respect for one's own and other peoples body and that everybody is different' and ensuring that 'the culture of body and movement reflects the diversity in the children's cultural foundation ...' (Norwegian Ministry of Education and Research, 2012). Notable here, is the tension between the normative, yet undefined 'good habits' and the explicit respect for diversity.

In the policy documents informing this intervention, the tension between a notion of 'good' health behaviour and respect for diversity as core values is fundamental when it comes to the ethical considerations of public health intervention: 'Influencing other people is connected to great ethical challenges ... (and one must) respect different value choices ... In the opposite case, there's a danger that initiatives will appear as unwanted interference towards those who do not wish or do not know' (Norwegian Ministry of Health and Care Services, 2002-2003). The ethical challenges in turn relates to the knowledge base informing such initiatives. In a circular from the Norwegian Ministry of Children and Equality Circular (2007), this is explicitly stated: 'The reason why knowledge based initiatives are emphasized, is the consideration for the child. Children and youngsters in general, and specifically those who are already challenged, should not be exposed to initiatives which we do not know will have an effect, or which at worst can cause harm' (Norwegian Ministry of Children and Equality Circular, 2007). Furthermore, it states: 'Prevention work should as far as possible be based in available knowledge about what works and what does not work depending on the specific problem area and the specific target group' (Circular, 2007). Detectable here is the tension between the knowledge base, which by its nature must be complex, and the demand for a concrete result based in 'what works'.

The transition from purpose to function

In the above, social health equity as well as intervention in nursery schools have been established as complex practical problems. From the mere description of the intervention, it seems clear that there is a discrepancy between the purpose of tackling social inequity and the function of ensuring that the diet and level of physical activity in the nursery schools in are in accordance with the Norwegian Health Authorities recommendations. The obvious indicator of this is that the complexity of Health is done away with. Notwithstanding, the conclusion of the project report is that it has been successful, and that what is needed is more time and effort in order to complete the task (Dønnestad and Strandmyr, 2014). A prudent question, then, is, how can this discrepancy happen?

Ravetz observes that: '(T)he (practitioner) tries to argue (perhaps with sincerity) as if he were a scientist, establishing his conclusions on supposedly known and irrefutable facts' (Ravetz 1996: 349). In this intervention, the 'facts' most explicitly informing the practice are the state recommendations for diet and physical activity, to the extent that, in the course of the report, the term 'recommendations' is substituted with 'criterion' (Dønnestad and Strandmyr 2014). At the same time, the intervention report refers to a body of scientific knowledge, both as justification for the measures taken, and, as background for the choices that are made, and in the use of accelerometers. Thus, the project can be classified both a scientific research endeavor, in its measurements and mapping, and a practical public health endeavor. Acknowledging this duality, opens for a range of questions with regard to the knowledge base for this intervention, particularly when it comes to the relationship between the descriptive and the normative nature of the reference research. Ravetz claims that 'Although the body of doctrine in question may be nearly vacuous, an incompetent rationalization of a craft practice, it can still perform several functions in the internal workings of the field of practice' (Ravetz 1996: 350). The use of accelerometers illustrates the function of a fact: Is the ActiGraph a device of inquiry or is it designated to establish a norm of behaviour or practice? Evaluating the internal quality of this intervention would mean a closer look at the scientific base of the project.

As for the external quality, Ravetz sees the problem definition as crucial to the implementation of a function: 'The framing and solution of practical problems is at risk of encountering a multiplicity of pitfalls, so that the purposes served can turn out to be quite different from those intended .... (If) the inquiry avoids 'theory' and becomes 'empirical', it can encounter the pitfall of simplifying its objects of inquiry to homogenous populations defined by classes of simple data; then the complexity and contrariness of the situation, which created the problem in the first place, is lost from view' (Ravetz 1996: 355).

Introducing a simple set of parameters to fill a function as a solution to a practical problem carries with it an inherent danger, according to Ravetz, because it introduces a norm to which the citizen must adhere: 'Any citizen will be at risk of being victimized, with no possibility of redress, because of some deviance (real or erroneously imputed) from some expert's imagined 'norm'' (Ravetz 1996: 353). Solving the problem of social health inequity by ensuring that the children's, and their families', diets are in accordance with the state recommendations illustrates the pitfall of simplifications, particularly because the project workers do not appear to discriminate between the institutional changes in menus and diet guidelines and individual changes in behaviour within the families: 'According to the goals, we did not only want to influence the children's diet in the nursery schools, but also at home. As we have not had the opportunity to observe or document this, we cannot establish to which degree we have succeeded in this area. ... However, we suspect that many (parents) have improved the contents of the children's packed lunch without changing the diet at home. One important cause of this, is that the turnout on the parent-meetings was limited, and many (parents) have not learned sufficiently about nutrition and its importance to make the desired changes' (Dønnestad and Strandmyr, 2014: 19, my italics). Some prudent question might be: By whom are the changes desired? And from which initial state? There is no mention in the report, nor in the initial status report, of the diet habits of the families in question, other than a suspicion, so it is unclear whether these changes are desired on the grounds of an imagined norm, (and an imagined deviance from this norm) or necessitated by a diet that would or could lead to 'poor health' if continued. Furthermore, the official recommendations are based on a traditional Norwegian diet, which may or may not be in accordance with the values and norms of the people being subjected to the intervention. This holds particular relevance in the diverse community of Groruddalen Bydel. The only common denominator that the intervention report provides for this group of people, is the observation that 'there is a consistently higher number of people with bad health in Bydel Grorud as compared to the rest of Oslo'. The indirect definition of the population as a 'risk group', adds to Ravetz' observation that the establishing of a norm places the citizen in a position with no possibility of redress: What are the possibilities of rational resistance or conflicting values? In order to tackle the complexities of the real and messy world of a practical problem, Ravetz concludes that: 'What is required is an appreciation of the variety in the nature of the problems encountered, the difference in their methods, the criteria of adequacy of their solution, and in their characteristic pitfalls; and an awareness that there exist some important problems, of any of these sorts, which are incapable of a solution under any possible circumstances' (Ravetz, 1996: 356). Obviously, this is an ideal state of practice, which cannot reasonably be expected to permeate any intervention, but it may offer a prism for evaluation of such projects.

Conclusion

The concept of fitness for purpose makes it possible to untangle some of the structural pitfalls a project is at risk of encountering when not making the distinction between fitness for function and fitness for purpose. As such, a project may be evaluated as successful in technical terms even if it fails to encompass the complexity of the practical purposes of the social issue, in other words, it may be evaluated in terms of internal quality without adhering to any principle of external quality. As a result, a solution may potentially be counterproductive in terms of dealing with the issue at hand, in this case public health inequity, by not taking into account potential unforeseen consequences. While one cannot generalize on the basis of a single case study, an analysis of this intervention may illuminate the necessity of reflexivity on behalf of the practitioners or experts carrying out an intervention in order to avoid reducing a complex problem to a mere instrumental issue.

References

Dønnestad, J. and Strandmyr, A. (2014). Public health project in nursery schools. 'Folkehelseprosjekt i barnehager". Groruddalssatsingen, Bydel Grorud, Norway. Available at: http://tinyurl.com/nvfa44v.

Funtowicz, S.O. and Ravetz, J.R. (1991). A new scientific methodology for global environmental issues. In: Costanza, R (ed.) Ecological economics: the science and management of sustainability. Columbia University Press, New York, USA, pp. 137-152.

Funtowicz, S.O. and Ravetz, J.R. (1992), The good, the true and the post-modern. Futures 24: 963-976.

Norwegian Ministry of Children and Equality Circular Q-16-07 (2007). Preventive efforts for children and youth. 'Forebyggende innsats for barn og unge'. Available at: http://tinyurl.com/pq6afx4.

Norwegian Ministry of Education and Research (2012). White paper: national plan for nursery schools' contents and obligations'. 'Rammeplan for barnehagens innhold og oppgaver'. Available at: http://tinyurl.com/pyg35u6.

Norwegian Ministry of Health and Care Services (2002-2003). White paper: stortingsmelding no 16, prescription for a healthier Norway. 'Resept for et sunnere Norge'. Available at: http://tinyurl.com/qhudtw3.

Oslo Municipality (2011). Bydel Grorud plan document, Public health strategy. 'Folkehelsestrategi'. Available at: http://tinyurl.com/ktr8dky.

Ravetz, J.R. (1996 (1971)). Scientific knowledge and its social problems. Transaction Publishers, Piscataway Township, USA, 449 pp.

World Health Organization (WHO) (2010). Global recommendations on physical activity for health. Available at: http://tinyurl.com/chkzkvw.

Changing attitudes to genetic modification crops? Reviewing 10 years of playing a Democs card game on genetic modification crops with MSc students

D.M. Bruce
Edinethics Ltd., 11/6 Dundonald Street, Edinburgh, EH3 6RZ, United Kingdom; info@edinethics.co.uk

Abstract

Most studies on public attitudes to GM crops are limited to the short period of a funded study or consultation. The Democs card game has a special role amongst the tools used in public engagement on biotechnologies, not only in its style of grassroots interaction with diverse publics in their own local situations, but also in that once the game is created, it can go on being used indefinitely. It is thus possible to assess its results against the passage of time. Democs games have also proved valuable in a pedagogical context. This paper reviews the qualitative and quantitative results from playing a Democs card game GM crops with Edinburgh University MSc students on a sustainable development course, from 2007 to 2014. Each year 30-40 students played the game in groups of 6-8 people per game. This has provided a unique set of comparative data with graduate students of mixed science and social science backgrounds and from many different countries, but situated in a similar pedagogical context. The game's policy position votes and cluster statements are compared and analysed, to identify which issues have been consistently important and which more spasmodic, and looking for any trends in positivity/ negativity towards GM crop technologies. These findings are then compared with the original data obtained in 2003 from playing this Democs game with general public when it was created and used as part of the national UK 'GM Nation?' public consultation. It emerges that the widespread assumption of predominantly negative attitudes towards GM crops is not a fair judgement, even among people concerned enough about environmental, economic and social sustainability to study it at graduate level.

Keywords: GM crops, policy, democs, public attitudes, time

The Democs game concept

What is Democs?

Deliberative meetings of citizens (Democs; www.edinethics.co.uk) is a novel form of lay participation on complex technical issues, devised by Perry Walker, then of the New Economics Foundation (www.neweconomics.org) in 2001-2. The original game on embryonic stem cells was created with funding from the Wellcome Trust. Since then Democs games have now been created and developed for a wide range of issues from cloning to climate change, and in numerous European languages as part of the EC Decide and Fund projects. Games are normally provided free, to maximise their use, and are available in boxed form, or downloadable for the user to print out. The primary aim is to get otherwise unengaged citizens to engage with the issue. Games have been used effectively in a wide variety of contexts including government consultations, EC research projects, rare disease awareness raising, with homeless people, in school and university teaching. They have been played in homes, schools, churches, pubs, restaurants, public halls, science festivals and science centres (www.playdecide.eu).

Democs is not strictly a game you play to win, but a semi-structured group discussion for 6-8 people, in which the 'expert' is in the cards, which provide the necessary information and the basis around which discussion is then stimulated. A game can be played by any group of people (usually over about 16 years) anywhere, with no or minimal cost, without needing facilitation. It needs someone to get a group of people together, to send off for a game or else download and print it, to read the instructions, to be prepared to act as 'dealer' during the game, and finally to post off the results afterwards. A game typically takes about 1½-2 hours.

How a Democs game works

Series of cards are dealt out, read aloud, and discussed by the group in three rounds:
1. 6-8 Story cards. Imaginary narrative contexts, based on real life situations or ones reasonably envisageable. The aim is to introduce the technology through people who are involved in it, or affected by it, often presenting an ethical dilemma. Each player has a card and reads it aloud to the whole group when their turn comes, so that the group hears all the stories.
2. 32 Information cards. Essential information to cover the basic science involved, its potential applications and the regulatory context, aimed at being fair and objective.
3. 36 Issue cards. A range of questions, perceptions, issues and opinions to explore the ethical and social implications of the technology. Cards may present differing or even directly opposed views, to cover a fair range of opinions and attitudes known or likely to exist.

For the Information and Issue card rounds, each player is dealt a hand of 6-7 cards and chooses two which they read aloud to the group when their turn comes, and say why they chose them, then lay the two in the centre of the table, building up a set of selected cards.

The group uses the cards to discuss and gradually form their own views, agreeing or diverging. Towards the end of the game the group is asked to cluster the cards into subjects, which form the basis of into perhaps 3-4 statements or opinions by the group, made on 'cluster cards', aiming at consensus where possible, but admitting different views if necessary – it's entirely up to the group. Lastly a sheet of policy options is handed round for each person to vote on individually. In the case of the GM crops game these are 4 different broad policies for the technology. In other games like human enhancement and synthetic biology, the votes are on the acceptability or otherwise of particular applications. Thus the outputs are a set of cluster statements, recording which cards were used, and policy votes. In games voting on different applications, players are also invited to state in their own words their reasons for their votes.

The game is what each particular group of people makes of it. Participants are encouraged to contribute their own ideas and insights, even to write their own cards, or produce alternative policy positions to vote on, if they want to. People often do. The aim is that the expert input is in the cards, and does not need a technical specialist to be present, or a facilitator. It just needs that someone plays the role of dealer, having read the instructions, to explain the different phases of the game and to keep to time. Occasionally it may help to have someone present familiar with the material under discussion, for example with a subject like synthetic biology which is both unfamiliar and technically hard to grasp, but normally this is unnecessary.

The approach is a compromise between providing enough information for people to discuss meaningfully without prior knowledge, and an inevitable degree framing by the content of the cards. This puts a considerable onus on the writers of a game, and the peer review process, to achieve the necessary rigour and balance of both science and ethics, with as little personal bias as possible. But the same is true of the stimulus material for any method of discursive form of engagement like focus groups, citizens jury expert inputs, or in the framing of questionnaires.

Whilst, Democs games are aimed at lay people who are not expert and may know little about the subject in question, nonetheless, it has also found a considerable role in pedagogical contexts. One example is where ethics is being taught to science students, who are well versed in the science but not necessarily in the ethical and social questions explored in issue and story cards.

The role of Democs games amongst public engagement tools

The technique falls between focus groups and opinion polling in the type of information generated and numbers of people who can engage. It can provide considered views from a much wider range of people than are usually reached by focus group methods and in far more depth than from tick-box opinion polling. Most public engagement activities sample representative populations but access relatively few people. Democs aims at a much broader degree of lay participation, starting with people in the contexts where they normally get together – friends in a pub, neighbours, a church or community group, a group of students, and so on.

By its nature Democs seeks to engage people in the natural circumstances where they might gather together. It does not impose the strict methodology of having a socially balanced sample of people for each group, although this has occasionally been done. This poses the dilemma of sampling. While Democs cannot be sociologically rigorous and does not set out to be, it uses a method of sampling closer to real life situations and often involving existing groups of people, compared with a carefully socially balanced group which has been artificially constructed from strangers. The latter is made to reflect as fairly as possible the general population, but it probably would never actually occur in any real life situation, and in that sense is less typical.

The validity of the results from Democs games depends what questions are being asked of it, but useful analyses can be made provided one recognises its limitations. A degree of diversity can eventually be achieved if enough games can be played among enough varied sectors of the community, locations, age-groups, etc.

Democs game on the genetic modification of crops

Assessing genetic modification Democs game results over time

Most public engagement exercises on technological issues are limited to the short period of a funded study or consultation. The Democs card game has a special role amongst the tools used in public engagement on biotechnologies in that, once the game is created, it can go on being used indefinitely. It is thus possible to assess its results against the passage of time. This paper reviews the qualitative and quantitative results from playing a Democs card game on GM crops, which has been played every year from 2005 to 2014 with students on an Edinburgh University MSc course module on sustainable development. The students were mostly under 30 and of very mixed nationalities. Most years had 30-40 students, typically comprising some from Scotland and the rest of the UK, many from western and eastern continental Europe, a few from North and South America, Asia (especially), Africa (less) and the Australia's. Over the ten years, 54 such games were played with 325 players, in groups of 6-8 people per game. This has provided a unique set of comparative data with graduate students of mixed science and social science backgrounds and countries, but situated in a similar pedagogical context.

The game was originally written for the 2003 UK 'GM Nation?' national public consultation, as a tool for use in Tier 2 and 3 engagement exercises (PDSB, 2003). The content of the cards was required to be based on the official government 'toolkit' booklet of stimulus material, which attempted a balance by setting out 'for' and 'against' statements for about 60 questions. The Democs cards were necessarily shorter and sought to be more nuanced, in that some questions are more complex and conditional than a simple for-or-against choice. Feedback was received from 35 games, played with 220 participants (New

Economics Foundation, 2003). Changes have periodically been made to some cards, to correct out of date information or reflect more recent developments, but most have stayed the same.

The classes lasted 4 hours, sitting at tables cafe-style, beginning with a 45 minute lecture on GM crops and a short period of clarification questions. After a coffee break, students would play the Democs game for 90 minutes in groups, with feedback from all the groups at the end. The introductory lecture covered the basic GM concept contrasted with selective breeding; what crops, traits and locations are used and future applications, the claims and criticisms made, and an overview of the ethical and social issues presented from eight different value perspectives.

Analysis of policy voting

First we consider the cumulative votes averaged over the 10 years in Table 1. The highest votes were to disagree that GM crops should never be grown in the UK (53%) (only 8% supported this option) and to agree that GM food crops should not be grown until further evidence is obtained on side effects (50%). If one adds to the latter the 36% who 'could live with' this option, 86% agree or could live with not growing without this further evidence. Only 14% thought that GM food crops could be grown already, but 41% could live with this, 55% in all. Non-food products showed 25% support and 21% against but 45% could live with them.

The picture is thus quite nuanced. Less than 10% agree with a position completely opposed to GM food crops being grown, and over half the total participants disagree with this position. Over time the percent agreeing remained consistently low, only in one year (2008, 23%) did it exceed 15% (Table 2). This indicates that fundamental opposition to GM food has been consistently low in this group of students over ten years.

One does not see a correspondingly high support for growing GM food crops now, however (option 1). This averaged only 14% in agreement, whereas almost 40% disagreed. But the agreement figure

Table 1. All policy votes 2005-2014.[1]

Policy	p1 yes	p2 conditional	p3 non-food	p4 never	(p5)	(p6)
Agree	45	162	82	27	48	11
Live with	134	117	147	74	22	2
Disagree	125	27	68	172	8	0
Don't know	19	17	24	45	8	3
Agree	13.8%	49.8%	25.2%	8.3%	14.8%	3.4%
Live with	41.2%	36.0%	45.2%	22.8%	6.8%	0.6%
Disagree	38.5%	8.3%	20.9%	52.9%	2.5%	0.0%
Don't know	5.8%	5.2%	7.4%	13.8%	2.5%	0.9%
Agree + live with	55.1%	85.8%	70.5%	31.1%	21.5%	4.0%

[1] p1: policy 1: GM crops should now be grown commercially in the UK as long as sufficient safeguards are provided and maintained; p2: policy 2: GM crops should not be grown commercially in the UK until further evidence is obtained about possible side effects on human health and the environment; p3: policy 3: GM crops should not be grown for normal food, but might be grown for non-food applications such as medical products or renewable energy; p4: policy 4: GM crops should never be grown commercially in the UK under any foreseeable circumstances; p5: policy 5: players' alternative voting option 1; p6: policy 6: players' alternative voting option 2.

Table 2. Annual policy votes 2005-2014 – agree and disagree.[1]

Year	Games	Players	Agree				Disagree			
			p1	p2	p3	p4	p1	p2	p3	p4
2005	8	45	4%	44%	33%	9%	47%	0%	27%	47%
2006	5	26	23%	62%	27%	0%	19%	4%	19%	73%
2007	5	28	4%	25%	18%	4%	57%	14%	29%	75%
2008	5	30	20%	43%	7%	23%	53%	27%	13%	30%
2009	6	40	3%	33%	20%	15%	55%	13%	15%	33%
2010	7	40	30%	38%	35%	8%	38%	10%	13%	58%
2011	6	36	11%	67%	31%	0%	17%	6%	31%	56%
2012	2	12	8%	58%	17%	0%	25%	0%	0%	92%
2013	5	33	15%	64%	42%	12%	36%	9%	15%	39%
2014	5	35	20%	74%	11%	6%	26%	0%	34%	63%

fluctuates strikingly between years, alternating between 3-4% and 20-30% in the years from 2005-10. Moreover the fact that 55% could agree or live with growing GM food crops now, suggests again that basic opposition is not predominant.

The dominant policy with these students over this period is not to grow GM food until further evidence of side effects has been obtained, with 86% agreeing or could live with this. As with option 1, year on year variation in agreement with option 2 fluctuates markedly, sometimes 20-40% sometimes 55-75%. It is also interesting to note that 41% could also 'live with' growing GM food crops now. In several years the percent agreeing with option 2 is quite close to those who could live with option 1 (2005, 2007, 2009, 2011, 2012). This suggests that option 2 is the preferred option but the conditionality is not held very strongly.

The most striking overall finding of the variation over the ten years is that there are few clear trends against time in any of the votes. The fluctuation in agreeing to grow GM food crops now has been noted already. Percent disagreement with option 4 (never growing GM food crops) varies between the low thirties or the mid-seventies (the high 2012 figure is from a much smaller sample than other years). The pattern of voting with growing now does not correlate with corresponding agreement with never growing GM food crops. In 2007 only 4% agreed with either option, and in 2008 20% agreed with growing now and 23% with never growing. But in 2007 only 25% agreed with the conditional policy 2, but 43% in 2008.

The option of voting to 'live with' a policy has been used in Democs games from the beginning. It represents a conscious attempt to encourage players not only to think about what is their individual position, but also to work towards positions that found consensus with the group as a whole, as it could be in wider society. This option was used slightly more times (36%) than either agree (24%) or disagree (30%) voting categories. But this 'live with' category also fluctuates markedly between years. In 2008 and 2009, 47 and 38% were prepared, respectively, to live with never growing GM food crops but only 8% in 2006 and 2012.

Option 3 was an attempt to distinguish the specific issue of food from the issues of genetic modification, by asking if growing GM crops for non-food applications like medical products or renewable energy. At

25%, agreement with this fell between the never and the conditional options 2 and 4. The disagreement figure is comparable but it is not clear if this represents disagreement with non-food applications, or disagreeing that food applications should not be grown. Again the 'could live with' category is high (45%). The fifth and sixth options reflect that in some games, players wished to write their own policy position, but since each of these is unique to that game, no comparisons can be made.

Conclusions

On average, the students attending this course have preferred the policy that GM crops should not be grown without obtaining further evidence about possible side effects to human health or the environment. Whereas few agreed with growing GM food crops now and almost 40% disagreed, the largest single vote (53%) was those opposed to the option that GM crops should never be grown. The quite large percent (41%) who could live with GM crops being grown now suggests that for many people the conditional option 2 may not be held very strongly, if they are also prepared to 'live with' GM food crops being grown already.

When these games were played during the 'GM Nation?' public consultation in 2003, these policy option votes were not made, because players were asked instead to fill in the government's official questionnaire privately. In further research, a comparison will be made between the cluster card subjects between the two sets of games; but the most popular 2003 clusters – expressing uncertainty or the need for more time or more information (New Economics Foundation, 2003) – would be broadly in line with the preference for option 2 among the 2005-2014 student games. This suggests that rather than presenting GM food as an all or nothing option, in which consumers have no choice either way, a more sensible policy would be a slow, progressive approach using a pilot scale concept, to introduce a restricted range of clearly identified products, to allow people who wish to do so to try GM products, while at the same time doing long term testing for health and environmental effects.

These student data do not support the perception of a general opposition to GM food crops. Among people concerned enough about environmental, economic and social sustainability to study it at graduate level, there is cautious approach but one prepared to use the technology, rather than being merely sceptical towards it. The absence of any year on year trend in the voting pattern suggests that some years had a fairly negative view of growing GM food crops, and some years a more positive one. Rather than being mostly for or against GM food, it suggests that most find it quite difficult to make up their minds and have rather nuanced attitudes, reflected in some of the high 'could live with it' votes. The Democs card game has a special role amongst public engagement tools in that it can go on being used indefinitely, and can provide some indications of the development of attitudes against time. To make such temporal comparisons requires a fairly consistent group of people, and is then limited to that group. In this case, it has not shown clear trends of growing opposition or approval with time amongst these students, but a more complex response in which cautious and conditional adoption is the predominant theme, while some are more definitely for or against.

References

New Economics Foundation (2003). Democs and the GM Nation Debate, New Economics Foundation, London, UK.
Public Debate Steering Board (PDSB) (2003). GM Nation?, The Findings of the Public Debate. Department of Trade and Industry, London, UK.

Consumer preferences for welfare pork – is the interest for tenderloin greater than for minced meat?

S. Denver, P. Sandøe and T. Christensen*
Institute of Food and Resource Economics, University of Copenhagen, Rolighedsvej 25, 1958 Frederiksberg, Copenhagen, Denmark; sd@ifro.ku.dk

Abstract

Denmark has among the highest levels of organic consumption in the world when looking at the overall market shares. Nevertheless, the market shares for organic meat and other types of welfare friendly meat are low and animal welfare friendly production systems remain a niche. The main Danish animal welfare organisation, Dyrenes Beskyttelse, owns a label that approves premium levels of animal welfare similar to organic standards for pork, beef and poultry meat productions. Also, meats from production systems that guarantee medium levels of animal welfare are available in the Danish cool counters but these products are not approved by Dyrenes Beskyttelse. Instead, medium level animal welfare pork is marketed under privately owned super market labels. Other countries seem to be successful in using a different strategy where national animal welfare associations approve medium as well as premium levels of animal welfare (Christensen *et al.*, 2014; Heerwagen *et al.*, 2014). The purpose of this paper is to analyse whether there is a potential for increasing the market shares in Denmark for pork associated with medium levels of animal welfare without compromising the markets shares for premium animal welfare pork. Moreover, we investigate whether the interest in welfare meat depends on whether it is considered as a luxury meal or an every-day meal. The analysis is based on a large web-based survey that involves attitudinal questions related to pork production and consumption. Furthermore, a choice experiment is included where the respondents' preferences for pork originating from animal welfare friendly production systems are elicited. More specifically, the respondents' willingness to pay (WTP) for improved indoor space for slaughter pigs and for guaranteeing that sows are loose in longer periods of their production cycles were estimated. Random utility theory (McFadden, 1974) was used to estimate respondents' WTP for medium versus premium levels of animal welfare. Our results indicate that consumers are interested in animal welfare, and that moderate price premiums for medium levelled animal welfare products might attract consumers who would otherwise be inclined to purchase standard products. Labels guaranteeing medium levels of animal welfare may therefore indeed play an important role in promoting higher animal welfare standards. Consumers seem to be more willing to pay a higher price premium for animal welfare in luxury pork such as tenderloin than for every-day pork such as minced meat – but measured in percentages, the WTP are more alike.

Keywords: loose sows, consumer behaviour, food labelling, choice experiment, willingness to pay

Introduction

Denmark produces around 29 million pigs per year (Danish Agriculture & Food Council, 2014). This is a business where the main focus is on efficiency and meat quality whereas concern for animal welfare to a large extent has been limited to ensuring compliance with legal minimum requirements. This has created a concern among many Danish citizens as to whether the welfare of pigs is taken sufficiently into consideration in the animal welfare legislation and in the production decisions.

Almost 90% of the Danish pork production is exported with a value of 4.2 billion € (Danish Agriculture & Food Council, 2014). Hence, the contribution to the Danish economy is considerable and concerns about the competiveness of farmers operating under stricter local animal welfare legislations than their competitors and complexities in obtaining international agreements on animal welfare legislation have led the authorities in Denmark and other European countries to look for other ways to increase animal welfare than through legislation. Consequently, there has been an increased interest in market-based improvements in animal welfare (Christensen *et al.*, 2012; Heerwagen *et al.*, 2014) where demand for special products with focus on animal lead to standards exceeding the legal minimal requirements.

However, consumers do not constitute a homogenous group that can be addressed using a single type of instrument – such as a single label indicating a certain level of increased animal welfare. Several studies point to the existence of three distinct consumer segments: the first is very interested in animal welfare issues; the second does not show particular interest in animal welfare; while the third segment is somewhere in between encompassing consumers who are interested in animal welfare to some extent but who also focus on other attributes connected with gustatory quality and healthiness. This pattern has been identified around Europe in studies on consumer attitudes to farm animal welfare in general (Vanhonacker and Verbeke, 2014), to beef (Grunert and Vialli, 2001), and to pork (Meuwissen and Van der Lans, 2005; Mørkbak *et al.*, 2010). In particular, Vanhonacker and Verbeke (2014) link consumer interest in animal welfare with price sensitivity. They find a group of consumers who are dedicated to a range of products promising high animal welfare standards and who find that price is of secondary importance, while consumers in other segments who are also inclined to purchase these products are more sensitive to high price premiums. These consumer studies indicate that there is a market not for only bulk standard products at one end of the scale and premium standards (i.e. the highest standard being marketed) such as free-range and organic animal products at the other end of the scale but also for products with medium levels of animal welfare characterized by welfare standards that are set above the legal minimum but below the premium.

In the present paper we use Danish consumers as a case for studying preferences for different levels of improvement in animal welfare in the pig production. Our results shed light on the potentials for increasing the market shares for medium levels of animal welfare pork without compromising the markets for organic pork and other varieties of pork offering premium levels of animal welfare.

Animal welfare friendly pork in Denmark compared with other European Countries

In Denmark, the main animal protection organisation, Dyrenes Beskyttelse, certifies only pork from systems where sows are kept outdoors and finishing pigs have outdoor access, both in an organic and a non-organic variety each with market shares of about 2% of Danish pork consumption. The only other real alternatives to standard pork products are offered by Danish supermarkets who have introduced their own private labels for pork (Antonius, Den Goe Gris, Bornholmergrisen) with standards defined by the owner. There are similar standards for the three different labels guaranteeing levels of animal welfare which can be viewed as slight improvements compared to standard production. The total market share of these products is around 12%. The standard, medium and premium pork with respect to animal welfare differ with respect to the period where sows are confined and with respect to the space allowed to slaughter pigs.

- Sows in herds delivering pork to standard production are kept according to the national legislation. Sows should be kept in group housing from 4 weeks after mating to 1 week before farrowing. The indoor space requirement for slaughter pig is a minimum of 0.65 m^2 per pig.
- Sows in herds delivering pork to the supermarkets' private labels are kept in group housing already from weaning of their piglets which is approximately four weeks after farrowing until one week before farrowing and the slaughter pigs have minimum 0.85 m^2 each (slightly improved conditions compared to legislation).

- Sows in the outdoor systems are never confined; and the slaughter pigs are guaranteed a minimum of 1.3 m^2 per slaughter pig, respectively (considerably improved conditions compared to legislation).

While Dyrenes Beskyttelse exclusively approves of the premium animal welfare labels, a different strategy seems to be used in other countries such as the United Kingdom, the Netherlands, and Germany. In these countries, medium levels of animal welfare, including improved indoor conditions, can also be certified as animal welfare improvements by animal welfare associations. This approach provides the consumers with choices of increased animal welfare that are certified by animal welfare associations in the respective countries. For example, the Freedom Food labels in the United Kingdom set up by the Royal Society for the Prevention of Cruelty to Animals in 1994 operate with four categories; certified indoor, outdoor bred, outdoor reared and free-range. Freedom Food certifies almost one third of the pork production in the United Kingdom. In the Netherlands, the brand Beter Leven has operated since 2007 with a three stars animal welfare label certified by the Dutch Society of the Protection of Animals. The first star for pork production guarantees improved indoor conditions, the second star guarantees, among other things, access to outdoors areas and the third star guarantees animal welfare condition equivalent to organic standards. It is estimated that 40% of the pork sold in the Netherlands is labelled with at least 1 Beter Leven star (Christensen *et al.*, 2014).

Material

The analysis is based on a large web-based survey distributed to the market research institute Userneeds' online Danish consumer panel in June 2013. The questionnaire includes a number of questions related to pork consumption as well as questions concerning perception of pork production and animal welfare. In addition, we tested consumers' preferences for specific pork attributes in a choice experiment where respondents were asked to trade-off between different attributes. In particular, the following attributes were included: (1) housing conditions for sows; (2) indoor space for slaughter pigs; (3) the price of pork. In order to uncover possible differences in preferences for luxury and every day products we included both pork tenderloin and minced pork in the choice experiment. The attributes and levels are shown in Table 1.

The software Ngene 1.0 was used to design choice sets with two hypothetical types of meat (meat A and meat B) described by different combinations of the levels (as described in Scarpa and Rose, 2008). In order to mimic a real shopping situation the respondent could also choose none of these. A satisfactory design efficiency was obtained with nine choice sets which was also considered to be a feasible task for the respondents. Differences in preferences for minced pork and pork tenderloin were tested using a split-sample approach. Representativeness with respect to gender, age, place of residence, and education were pursued according to Statistics Denmark.

Table 1. Attributes and levels included in the choice experiment

Attributes	Levels
Housing conditions for sows	Confined except in the gestation period (current legislation)
	Confined around mating but loose during gestation, farrowing, and lactation
	No confinement
Indoor space for slaughter pigs	0.65 m^2 per pig (current legislation)
	0.85 m^2 per pig (30% more indoor space)
	1.3 m^2 per pig (100% more indoor space)
Price (DKK/500 g)	4 different prices for minced pork and for tenderloin

Method

Data from the choice experiment were analysed in a random utility framework (McFadden, 1974) and used to estimate respondents' willingness to pay (WTP) for medium versus premium levels of animal welfare. Each respondent is assumed in each choice (n=1, ..., n) to choose the alternative with the highest utility from the choice set (j=meat A, meat B, none of these). A multinomial logit approach (Train, 2003) was used where the utility associated with alternative *j* is described by:

$$U_j = V_j + \varepsilon_j = \alpha_j + \beta'_1 \times \text{housing}'_j + \beta'_2 \times \text{space}'_j + \gamma \times \text{price}_j + \varepsilon_j, \varepsilon_j \sim \text{iid}$$

Where $\beta'_1 \times \text{housing}'_j$ and $\beta'_2 \times \text{space}'_j$ are vectors representing the utility from reducing the periods in which sows are confined and increased space requirements for slaughter pigs, respectively. The term, $\gamma \times \text{price}_j$, captures the marginal utility of the price. A straight forward and robust way of expressing the marginal valuation of an attribute is to use WTP, which is the negative of the relation between the marginal valuation of an attribute and the price coefficient ($-\beta/\gamma$).

Results

Around 800 respondents completed the questionnaire which corresponds to a response rate of approximately 15%. The following analysis includes 348 respondents in split A and 446 respondents in split B. In both splits elderly people, respondents with higher educations and women were found to be overrepresented compared to the Danish population (Statistics Denmark, 2013).

Around 15% of the respondents claimed that they often or always buy products pork with premium labels as organic pork or Friland. At the same time, a widespread ignorance about which brands they frequently buy was found and there was a general unawareness of the animal welfare requirements behind different labels for pork. Only one third of the respondents were aware of the fact that confinement of sows is not allowed housing in the productions of organic and Friland pork. On the other hand, the majority of the respondents thought that sows were required to be loose under the supermarket labels. The latter is only partly true as the private labels guarantee that sows are loose for a longer period than required by the legislation but not as long as guaranteed by organic or Friland.

More than half of the respondents considered it as very positive that sows are never confined, approximate one fourth considered it to have small positive impact while 10% did not care. Concerning the ranking of the different welfare attributes, we found that the respondents valued freedom of movement for sows as less important than more space for pigs in the barn, access to outdoor areas, supply of straw, short transport time to slaughterhouses and the possibility to socialize with other pigs. On the other hand, the respondents considered freedom from confinement for sows as more important than organic production and avoiding tail docking and castration. Hence, housing of sows without confinement is not the attribute with the highest priority among the respondents, but it is considered as a positive attribute and is generally ranked in the middle of the field when being compared with other conditions for the pigs. It was also considered of medium importance when compared with other quality attributes of pork.

When asked directly whether they would pay more for pork from production systems where sows are permanently free from confinement, around one quarter of the respondents stated that they certainly would pay more. Almost half of the respondents stated that they might want to pay while one out of four respondents was not willing to pay anything. The estimated WTP from the choice experiment support these findings. The respondents were willing to pay a statistically significant price premium for improving animal welfare both by providing slaughter pigs with more space and allowing sows freedom to move around. In line with data obtained from the attitudinal questions, the average respondent is willing to pay more for additional space for slaughter pigs than for avoiding confinement of the sows. Interestingly though, we identified a stepwise increase in the amount that the respondents were willing to pay as the level of animal welfare increase. Similarly, the respondents stated a higher WTP for permanent freedom of movement than for production systems where sows are confined around mating. This indicates that consumers are able to distinguish between different levels of welfare improvements and that increased welfare is associated with increased stated WTP.

Similar rankings of attributes were found in split A and split B. However, the results indicate that respondents who evaluated the pork tenderloin would be willing to pay a higher amount of money for animal welfare improvements in absolute terms than the respondents who evaluated the minced pork. As tenderloin is considered a luxury piece of pork compared to minced pork which is considered an every-day type of meat, we calculated the price premiums in percentages and thereby taking differences in the market prices of the two types of product into account. Interestingly, when the WTP a price premium is formulated as percentage of the market prices, the difference in WTP diminishes.

Conclusion and discussion

Our study indicates that Danish consumers find it positive that sows are kept in systems without confinement but that they consider outdoor areas and additional indoor space for slaughter pigs as more important welfare parameters. Furthermore, consumers seem to be able to distinguish between standard, medium, and premium levels of animal welfare in the amount they are willing to pay for these attributes.

There is a considerable fear within the organic movements that animal products offering medium levels of animal welfare will induce consumers to buy these products instead of organic products. We cannot eliminate that fear but our results indicate that there is a distinct segment of consumers which can be attracted to a product offering a medium level of animal welfare (as freedom for confinement for sow in extended periods or more space to slaughter pigs). As our results are based on hypothetical behaviour, there is a future task in determining the real cut-off price that induces consumers to compare the medium levelled products with premium products rather than with standard products.

In light of the experiences with certifying medium levels of animal welfare products from other countries as well as our own findings, it is likely that there is a market opportunity in Denmark for pork labels guaranteeing medium levels of animal welfare at moderate price premiums. Additionally, our findings suggest that there is a larger interest in animal welfare friendly tenderloin than in animal welfare friendly minced pork which is important to include in further market considerations.

References

Christensen, T., Denver, S., Hansen, H.O., Lassen, J. and Sandøe, P. (2014). Dyrevelfærdsmærker – sammenligning af erfaringer fra seks EU-lande. Commissioned Report 2014/10, University of Copenhagen, Copenhagen, Denmark.

Christensen, T., Lawrence, A., Lund, M., Stott, A. and Sandøe, P. (2012). What can economists do to help improve animal welfare? Special issue Animal Welfare 21(S1): 1-10.

Danish Agriculture & Food Council (2014). Statistics 2013 pig meat. Danish Agriculture & Food Council, Copenhagen, Denmark.

Grunert, K.G. and Valli, C. (2001). Designer-made meat and dairy products: 640 consumer-led product development. Livestock Production Science 72: 83-98.

Heerwagen, L.R., Mørkbak, M.R., Denver, S., Sandøe, P. and Christensen, T. (2014). The role of quality labels in market-driven animal welfare. Journal of Agricultural & Environmental Ethics 28: 67-84.

McFadden, D. (1974). Conditional logit analysis of qualitative choice behaviour. In: Zarembka, P. (ed.) (1974) Frontiers in Econometrics. Academic, New York, USA, pp. 105-142.

Meuwissen, M.P.M. and Van der Lans, I.A. (2005). Trade-offs between consumer concerns: an application for pork supply chains. Food Economics – Acta Agriculturae Scandinavica, Section C2(1): 27-34.

Mørkbak, M.R., Christensen, T. and Gyrd-Hansen, D. (2010). Choke price bias in choice experiments. Environmental & Resource Economics 45(4): 537-551.

Scarpa, R. and Rose, J.M. (2008). Design efficiency for non-market valuation with choice modelling: how to measure it, what to report and why. The Australian Journal of Agricultural and Resource Economics 52: 253-282.

Statistics Denmark (2013). Statistical Yearbook 2013. Danmark Statistik, Copenhagen, Denmark.

Train, K. (2003). Discrete choice with simulation. Cambridge University Press, Cambridge, UK.

Vanhonacker, F. and Verbeke, W. (2014). Public and consumer policies for higher welfare food products: challenges and opportunities. Journal of Agricultural and Environmental Ethics 27(1): 153-171.

Section 9. Consumer and citizen concerns

'Ag-gag' laws increase negative perceptions of farm animal welfare and decrease trust in farmers

J.A. Robbins, D.M. Weary and M.A.G. von Keyserlingk*
Animal Welfare Program, Faculty of Land and Food Systems, University of British Columbia, 2357 Main Mall, Vancouver, BC V6T 1Z4, Canada; jesserobbins.ubc@gmail.com

Abstract

In response to undercover videos of poor farm animal welfare, livestock industry groups in the USA and Australia have lobbied for the introduction of laws designed to limit the ability of such investigations to occur. These 'Ag-gag' laws have garnered a great deal of media attention and public debate. This online study assessed support of USA participants (n=715) for these laws and experimentally tested whether awareness of these laws has any unintended reputational consequences. Opposition to these laws was high (74% opposed) and when participants became aware of them (as a result of participating in our survey) their perception of animal welfare on farms became more negative, as did their willingness to trust farmers as sources about animal welfare. Exposure to the 'Ag-Gag' laws also negatively affected the perceptions of participants regarding issues unrelated to animal welfare (e.g. environmental stewardship, food safety, worker rights, etc.). These results suggest that policies perceived as limiting transparency are perceived negatively and undermine public trust in agriculture.

Keywords: agricultural policy, source credibility, legislation, transparency

Background

Undercover videos documenting animal abuse and/or poor living conditions are an increasingly utilized tactic for raising awareness of farm animal welfare. Visual imagery of animal abuse has been shown to elicit strong emotional responses (Nabi, 1998; Westbury and Neumann, 2008), leading to negative attitudes towards livestock production (Scudder and Bishop-Mills, 2009). In response, livestock industry groups in the USA and elsewhere have introduced legislation that seeks to restrict the potential for these investigations to occur. Exact legislative language varies by state, but most seek to restrict: (1) willful deception agricultural employers; (2) taking or possessing photographs, video or audio recordings without consent; (3) delayed reporting of animal abuse. Since 2005, more than 16 USA states have introduced some version of this type of legislation (in the USA frequently referred to as 'Ag-gag' legislation) – of which five have successfully passed this legislation into law (e.g. Idaho, Iowa, Utah, Missouri and Arkansas) (HSUS, 2014). Despite much media attention, there has been no scientific evaluation of the impact of these laws on public perception.

Objective and methods

In addition to gauging overall support/opposition to 'Ag-gag' legislation, this study sought to test whether awareness of 'Ag-gag' legislation reduces public trust in farmers and fosters negative perceptions of farm animal welfare.

USA participants were recruited via Amazon's Mechanical Turk crowdsourcing service (Buhrmester *et al.*, 2011). After consenting, subjects were randomly assigned to either the treatment or control condition. In the treatment condition, subjects received basic information about 'Ag-gag' laws accompanied by common arguments used by specific organizations that are generally in favor and opposed to these types of laws. The term 'Ag-gag' was purposefully avoided due to its negative connotation. Next, participants were asked to indicate their level of support for the proposed legislation, followed by a previously validated 12-item inventory (Frewer *et al.*, 1996) assessing their level of trust in farmers as sources of credible information regarding farm animal care and welfare. Lastly, subjects answered a series of questions about controversial farm issues including, but not limited to animal welfare.

In the control group, subjects received information unrelated to 'Ag-gag' legislation, then responded to the same outcome measures previously described (with the exception of indicating support/opposition to legislation).

Preliminary findings

Across all subjects (n=715), trust in farmers was low and perceptions of modern farming were negative. In the treatment condition (n=324), 64% of participants were opposed to 'Ag-gag' legislation. Less than 10% of these participants reported previously awareness of 'Ag-gag' legislation. Participants exposed to information about 'Ag-gag' laws were less likely to agree that farmers are trustworthy sources of information (24 vs 64%), farmers provide factual information about animal well-being (24 vs 64%), farm animals have good lives (20 vs 40%) and farmers do a good job of protecting the environment (28 vs 46%) ($P<0.0001$ for all comparisons).

Preliminary analysis of qualitative responses suggests participants were sympathetic to mandatory-reporting language included in some versions of 'Ag-gag' legislation such as California Assembly Bill 343 which, if enacted, would require, 'any person, who willfully or knowingly photographs, records, or videotapes animal cruelty to provide a copy of the photograph, recording, or videotape to local law enforcement within 48 hours of taking the photograph, recording, or videotape...' Such requirements are not uncommon in professions dealing with vulnerable populations (e.g. elderly, children), but to our knowledge have never been extended to those working with animals.

Critics of 'Ag-gag' legislation argue successful prosecution necessitates demonstrating a pattern of abuse, and mandatory-reporting requirements would greatly restrict their ability to compile sufficient evidence (Carlson, 2013). Greater attention to the ethics of undercover investigations of animals abuse is greatly needed as the moral line between doing and allowing harm to animals can become blurred (Rollin, 2013).

Preliminary results of this study reveal that 'Ag-gag' laws can have negative reputational consequences. More generally, our initial results suggest industry initiatives perceived as limiting transparency and information flow are likely to be perceived negatively. We did not evaluate potential impacts of 'Ag-gag' legislation on the perceived credibility of its opponents. Previous work suggests whistleblowers are often viewed more favorably following media crises involving animal abuse (Tiplady *et al.*, 2013). Future research should examine whether industry initiatives aimed at increasing transparency (e.g. farm tours, live video streaming) can improve public perceptions, and perhaps serve to inoculate against future public relations crises.

Acknowledgments

This research is paid for through the generous donations made to the University of British Columbia's Animal Welfare Program. We also wish to thank the many volunteers who participated in this study as well as the support and guidance of Dr. Becca Franks.

References

Buhrmester, M., Kwang, T., and Gosling, S.D. (2011). Amazon's mechanical turk a new source of inexpensive, yet high-quality, data? Perspectives on Psychological Science 6(1): 3-5.
Carlson, C. (2013). How state ag-gag laws could stop animal-cruelty whistleblowers. The Atlantic. 25 March 2013.
Frewer, L.J., Howard, C., Hedderley, D. and Shepherd, R. (1996). What determines trust in information about food-related risks? Underlying psychological constructs. Risk Analysis 16(4): 473-486.
Nabi, R.L. (1998). The effect of disgust-eliciting visuals on attitudes toward animal experimentation. Communication Quarterly 46: 472-484.
Rollin, B.E. (2013). An ethicist's commentary on the use of undercover cameras. The Canadian Veterinary Journal 54(7): 629-631.
Scudder, J.N. and Bishop-Mills, C. (2009). The credibility of shock advocacy: animal rights attack messages. Public Relations Review 35(2): 162-164.
The Humane Society of the United States (HSUS) (2014). Anti-whistleblower bills hide factory-farming abuses from the public. Available at: http://tinyurl.com/crehxaf.
Tiplady, C.M., Walsh, D.A.B and Phillips, C.J.C. (2013) Public response to media coverage of animal cruelty. Journal of Agricultural and Environmental Ethics 26(4): 869-885.
Westbury, H., Neumann, D. (2008). Empathy-related responses to moving film stimuli depicting human and non-human animal targets in negative circumstances. Biological psychology 78(1): 66-74.

Contextuality of pig welfare – a study comparing public perception in three European countries

C. Thorslund, J. Lassen and P. Sandøe*
University of Copenhagen, Department of Food and Resource Economics, Rolighedsvej 25, 1958 Frederiksberg, Copenhagen, Denmark; ceth@ifro.ku.dk

Abstract

Existing research suggests that attitudes to pig welfare are perspectival: they depend on whether the welfare is viewed from the perspective of a politically engaged citizen or from the perspective of a consumer looking at options for the next family meal. Furthermore, this context-dependence may vary with cultural differences. Results from 12 semi-structured focus group interviews, each with 5-8 participants, conducted between November 2013 and February 2014 in Denmark, England and Sweden are reported. The interviews were thematically structured around two contexts, each presumed to have a distinctive system of values: everyday life and production. The results of the interviews demonstrate that it can be meaningful for an individual to perceive pig welfare differently depending on the logic and values of the context. Also, a preliminary analysis of the focus groups indicates that there are significant differences between the three countries, particularly over who is seen as responsible for animal welfare. The English focus groups seemed to favour the idea that animal welfare should be market-driven through consumer-choice. The Danish groups took the general view that animal welfare is also a political issue, and that responsibility for it is shared between parliament and the consumer. The Swedish groups displayed a tendency to assume that parliament is responsible for ensuring welfare.

Keywords: focus group interviews, Europe, attitudes and values, market-driven animal welfare

Introduction

The welfare of farm animals, including pigs, has been a topic of public discussion in many European countries for many years, and quite a few studies have identified animal welfare as an important aspect of the way the public views pig production (Krystallis *et al.*, 2009; Lassen *et al.*, 2006). Arguably the growing public interest in animal welfare during the twentieth century was the 'most outstanding ethical development' to affect modern day agriculture (Hurnich, 1995). Concerns about animal welfare are mirrored in current EU legislation (e.g. council Directive 2008/120/EC defining minimum standards for housing of pigs), large integrated EU projects (e.g. Q-PorkChains) and retailers' own introduction of welfare standards.

Since the legally defined minimum standards still fall some distance short of what wide segments of the public consider good pig welfare, there is room for initiatives to support labelled pork products based on 'welfare friendly' production methods being sold at a premium. However, if this is to be successful, and improve animal welfare in the pig industry, the higher level of welfare secured will need to be in line with what people generally think is essential. Importantly, the growing focus on the marketization of welfare is not only an expression of ideological cultivation of the market. It is also a response to the globalization of food markets, associated difficulties with local animal welfare legislation, and a lack of political will to strengthen common rules across borders (Sandøe and Christiansen, 2008, Chapter 5).

However, there is reason to believe that perceptions of pig welfare depend on whether welfare is viewed from the perspective of the politically engaged citizen or the perspective of the consumer trying to make the most of the household budget (Korzen and Lassen, 2010; Lassen *et al.*, 2006). In the latter perspective the context is everyday life and the animal is mainly seen as meat; in the former, the context is related to production, and the animal is seen as a living being with its own moral standing. Inevitably these contexts affect attitudes to the issues raised by animal welfare. Furthermore, this context-dependent view of animal welfare may vary across the cultural contexts found both intra- and internationally.

Using results from an ongoing research project on public perceptions of animal welfare in relation to pigs, this paper will discuss: (1) how people perceive pig welfare in an everyday and production context; (2) how perceptions differ in three European countries; and (3) who, in the various contexts, is seen as responsible for the welfare of the animals. The countries investigated are Denmark, Sweden and England. They were selected because they have high rates of pig meat consumption, and because they display differences in the public perception of market-driven animal welfare and legislatively driven improvements (Spoolder *et al.*, 2011).

Data production

We organized 12 focus group interviews with regular meat eaters in Denmark, Sweden and England in late 2013 and early 2014. Four group interviews were carried out in each country, spread evenly across rural and urban areas. Participants were recruited by a professional recruiting company using participant panels and a random telephone number generator in specific locations. The groups were organized heterogeneously to capture a range of ages (spanning 18-70), incomes and educational levels, and genders and urbanization. As such the groups were expected to show variation in the perceptions and arguments put forward by the participants.

The interviews followed a funnel-shaped, semi-structured interview guide, opening with broad discussions of food in general and everyday meals, moving on to discussions of pork and meat quality, and eventually focusing on the living conditions of animals in the intensive systems of pig production – and so eventually closing in on issues of animal welfare. The participants occupied two contexts during the interview: (1) the context of everyday life, introduced by asking the informants what they had for dinner yesterday and why; and (2) the production context, introduced by reminding the participants that meat comes from animal production systems.

The interviews were taped and subsequently transcribed before being coded using QSR NVivo 10. The analysis followed a two-step procedure owing its design to Coffey and Atkinson (1996): first, the interviews were coded into themes with different levels of detail (primary, secondary and tertiary). Second, the arguments within each theme were identified and characterized using a simple argumentation analysis based on the ideas of Stephen Toulmin (2003). In what follows, some of the results of this analysis are presented.

People eat meat, not animals: perceptions in context

In industrialized economies consumers are generally not inclined to conceptualize the food they eat as being of animal origin: people eat 'meat', not 'animals' (cf. Vialles, 1994). Further, shopping, cooking and eating have to be easy and practical (Holm and Kildevang, 1996), which in many cases, as a result of giving priority to other product qualities, allows limited attention to be given to animal welfare.

In keeping with the results of other studies (Korzen and Lassen, 2010; Lassen *et al.*, 2006), the discussions we observed couched in the everyday context rarely referred to the welfare of the living animal as an important issue, unless this was explicitly connected with the material quality of the meat. When we changed the context by introducing production as the framework, immaterial qualities such as animal welfare became central. Farm animal welfare was then either seen as a factor contributing to quality or as an explicit ethical commitment.

An example from the study nicely demonstrating the different perceptions and arguments is the case of outdoor access for pigs. When discussing outdoor access in the everyday context, participants focused on the possibility of a better muscular state, leading to leaner meat with better taste and less fat. Outdoor access for the sake of the animal was not an issue. When, on the other hand, participants were asked for their opinions of outdoor access in the production context, they showed little concern about the meat. The focus was now on the benefits for the animal, as a means to higher welfare and respect for integrity and naturalness.

In the following we will concentrate on the production context, where, as we say, discussions of animal welfare are more prominent.

Animal welfare matters

In discussions of pig production animal welfare was related to normative questions of justice and rights. The animal's freedom to exhibit natural behaviour, i.e. to live in a natural way and to satisfy natural desires, was also often brought in as a relevant concern. All of this is in line with previous research (Schröder and McEachern, 2004; Skarstad *et al.*, 2007; Te Velde *et al.*, 2002). One method of promoting welfare that was expressed was human contact and care, which overall was agreed to be an important way of ensuring that the basic needs of the animals are met. As such the care was part of surveillance, and the argument was that the caretaker could ensure a supply of food and water, and respond quickly, when needed, to sickness and injuries in the livestock. But other effects of human contact were debated, and there was significant variation in the evaluation of human contact. One argument, put by several in favour of increased human contact, was the possibility of individual focus on the welfare of each animal as seen through the eyes of the caretaker. It was argued that there was a contrast here with machines, which represented industrialization and mass production in the negative sense in which this conflicts with high levels of animal welfare.

Human contact going beyond basic care and surveillance by the farmer was rarely touched on in Denmark and Sweden, but it arose more often in England. Where the Danish and Swedish groups mainly defined human contact in relation to the farmer, the English groups showed greater variation in their conception of the role of the human and the value of the contact. First, the term 'human' was often taken to mean any human, not just a farmer or caretaker. Second, the benefits of contact were often attributed to the human more than the pig: contact was regarded as something worth having for the sake of the person – which could be a layperson – more than for the sake of the animal. Even when participants were prompted to discuss increased contact with a caretaker, the focus remained mainly on the benefits to the person involved with the livestock, rather than animals. Thirdly, in the English groups many argued that increased human contact is a benefit to society because, for instance, it informs the public, and provides entertainment for visitors and psychological benefits of bonding to animals.

Thus overall the English groups seemed to place more importance on increased human contact than the other two nationalities. When questions were asked about the value and importance of another welfare initiative, namely enrichment for the pigs (e.g. through access to chains, straw or pieces of wood to interact with), the Scandinavian groups were more eager to express an interest. By contrast, discussions in England were few and somewhat limited: here enrichment seemed not to be regarded as very important. This need not have shown a lack of interest in the welfare of the pigs. It could instead be explained by another finding in the study: that in England, unlike the two other countries, outdoor access was taken for granted, and that it was simply assumed that, by being outdoors, the pigs already had their enrichment needs met.

In general the English interviewees were under the impression that most or all pigs in their domestic industry are kept outside with sheds, or indoors with outdoor access. This is in contrast with the Danish and Swedish groups, who expected all or most pigs to be reared inside. When the English groups were asked to discuss outdoor access as an animal welfare initiative, the focus was not on the positive aspects of such access, but more often on negative perceptions, such as being 'left in muddy fields'. Accordingly the discussions here often reflected concerns about bad weather, and some would even argue that the animals should always be kept indoors instead, to protect them from rain and cold.

Who is responsible?

We know from previous research that citizens' perceptions of the best, or most appropriate, way to improve standards of animal welfare – by legislation, on the one hand, and market forces, on the other – vary across the EU. In Sweden upward pressure comes mainly from new legislation and compliance with existing legislation, whereas in England the improvements are sought through the development of, and compliance with, private standards (Spoolder *et al.*, 2011). Denmark is expected to reflect both types of pressure.

Reflections on who has responsibility for improving or securing high animal welfare were more or less explicit throughout the interviews. In the everyday context, people engaged in discussions about the supply and demand of meat and other end products, focusing especially on their own role as consumers. 'They produce, what we ask them for' was the underlying attitude; and the logic of the argument was that while consumers continue to buy low-welfare products, farmers and the food industry cannot be expected to prioritize expensive aspects of higher animal welfare. It was agreed, then, that people should engage through political activity or through 'ethical' consumption, focussing on better and more expensive meat, and connecting higher quality with price.

The same discussions took place in the production context, but here the focus on responsibility expanded from the consumer as a single person to consumers as a group of potentially political actors who can affect politics and law through democratic channels. This could also be seen as a shift from the consumer perspective to the citizen perspective; it included a greater focus on the roles of different stakeholders. Responsibility was lifted from the shoulders of the individual and given to the masses and industry.

Although this was a general view throughout the interviews, there were still some differences between countries. In England, many focus group participants pointed persistently to the market as a driver of improved animal welfare, stating that laws and regulation were an important minimum, but that the freedom to choose as a consumer was just as important. This was argued as a case of respecting the less well-off, who could not afford to choose higher quality (with that higher quality conceived in terms of higher cost and better animal welfare). The implementation of stricter laws to raise animal welfare would create higher costs for the private consumer. In Sweden the emphasis was on legislation first and foremost, where control is an important further requirement. The argument put forth by many was that if change is required in the industry, it has to be through legislation, and the role of the consumer is thus more political, since he or she should seek to influence those in political power using democratic rights. In Denmark the discussion was somewhat similar to the Swedish, focussing on a societal perspective – but there were also reflections on consumption habits and the way in which market-driven change plays a role.

These international differences could reflect conflicting views of citizenship and individual freedom. At one end of the spectrum there is a libertarian view of responsibility according to which it is up to consumers to make their own choices as long as there is a product label. At the other end there is the republican idea of freedom, where the key issue is taking part in a shared or public responsibility, and here responsibility is located in institutions and law, not personal choice (Halkier, 2010).

Interestingly, there also seemed to be ongoing discourse, across the nations, offering an alternative path of change in consumer habits: lowering consumption. The idea was that people can buy more expensive meat, but at the same time cut their need for meat. Two reasons were given: (1) the health benefits of reduced meat intake; and especially; (2) the reduced expense of lower levels of consumption. The argument which then followed was this: that a general reduction in meat demand would take some of the pressure off the industry and reduce mass production which is seen by some as the cause of low animal welfare. In a sense, this is a different form of consumer responsibility, focussing not only on the products being bought but also on the willed moderation of consumption as a means for change.

Conclusion

The present study raises several important issues. We shall highlight three.

Firstly, we have shown that it can be meaningful for an individual to perceive pig welfare differently depending on the context. What might look like inconsistencies in people's perceptions can be understood when a contextual analysis, setting the arguments in two contexts, is applied. This is in line with previous findings (e.g. Korzen and Lassen, 2010). Animal welfare as a theme is present in both contexts, but the role it plays differs. This is especially interesting as it represents a novel perspective on the well-known discussion of why it is that consumers' behaviour fails to match declared attitudes. That discussion often implicitly defines behaviour as a variable, contrary to attitudes which are seen as anchored. By stressing the importance of context, and by emphasizing the consumer/citizen dualism that follows, we take one step further into understanding the complexity of attitudes and values.

Secondly, we have found international differences in people's assumptions, and therefore in what topics of discussion they find relevant. When it was assumed that animals have outdoor access, discussion about providing more space and more enrichment took a different turn and seemed less vivid than when participants expected the pigs to be kept indoors at all times or even tied up. It is our impression that the explanation of these international differences has to do with lack of knowledge and, linked to this, different expectations. Arguably the discussions could have been different if the focus group participants in England had known that only few pigs actually have outdoor access.

Thirdly, we found international differences over who is seen as responsible for ensuring higher farm animal welfare. This suggests that a one-size-fits-all route to the improvement of farm animal welfare across the EU may be problematic. This is especially important where cross-national attempts to raise welfare standards through market demand are concerned. If the market is to be used as a lever for improving animal welfare in the Danish – or any other – pig industry, the initiatives taken will need to be broadly in line with what people generally think is important. For countries like Sweden with seemingly low interest in market-driven welfare it could be fruitful to look further into the values of the consumers. Arguments for buying higher welfare quality products, combined with an underlying discourse stating that a reduced consumption level is perhaps more and more socially acceptable, or even desirable, could point to a path of market-driven welfare alongside legislative structures.

References

Coffey, A. and Atkinson, P. (1996). Making sense of qualitative data: complementary research strategies. SAGE Publications, Thousand Oaks, USA, 206 pp.

Holm, L. and Kildevang, H. (1996). Consumers' views on food quality. A qualitative interview study. Appetite 27: 1-14.

Halkier, B. (2010). Consumption challenged: food ini medialised everyday lives. Ashgate, Farnham, UK.

Hurnich, J. (1995). Poultry welfare. In: Hunton, P. (ed.) Poultry production. Elsevier, Amsterdam, pp. 561-578.

Korzen, S. and Lassen, J. (2010). Meat in context. On the relation between perceptions and contexts. Appetite 54(2): 274-281.

Krystallis, A., De Barcellos, M. D., Kügler, J. O., Verbeke, W. and Grunert, K. G. (2009). Attitudes of European citizens towards pig production systems. Livestock Science 126(1-3): 46-56.

Lassen, J., Sandøe, P. and Forkman, B. (2006). Happy pigs are dirty! – conflicting perspectives on animal welfare. Livestock Science 103(3): 221-230.

Sandøe, P., and Christiansen, S.B. (2008). Ethics of animal use. Blackwell, oxford, UK.

Schröder, M.J.A. and McEachern, M.G. (2004). Consumer value conflicts surrounding ethical food purchase decisions : a focus on animal welfare. International Journal of Consumer Studies 28: 168-177.

Skarstad, G.Å., Terragni, L. and Torjusen, H. (2007). Animal welfare according to Norwegian consumers and producers: definitions and implications. International Journal of Sociology 15(3): 73-90.

Spoolder, H., Bokma, M. H., Harvey, D., Keeling, L.J., Majewsky, E., De Roest, K. and Schmid, O. (2011). EconWelfare, findings, conclusions and recommendations concerning effective policy instruments in the route towards higher animal welfare in the EU.

Te Velde, H., Aarts, N. and Van Woerkum, C. (2002). Dealing with ambivalence: farmers' and consumers' perceptions of animal welfare in livestock breeding. Journal of Agricultural and Environmental Ethics 15: 203-219.

Toulmin, S. (2003). The uses of argument. Cambridge University Press, Cambridge, UK.

Vialles, N. (1994). Animal to edible. Cambridge University Press, Cambridge, UK.

Knowing one's food – making food a public issue

S. Meisch
Junior Research Group: 'Ethics of Science in the Research for Sustainable Development', International Centre for Ethics in the Sciences and Humanities, University of Tuebingen, Wilhelmstr. 19, 72074 Tuebingen, Germany; simon.meisch@uni-tuebingen.de

Abstract

There is a lively debate on consumers' responsibility and the role of social institutions (e.g. market, state, common goods) in bringing about sustainable development. Subject to the institutional setting chosen, the claim for knowing one's food acquires a different meaning, e.g. better information for consumers that help to take different purchase decisions (certification) or participating in the food process itself (community-supported agriculture). Regardless of contestation on appropriate institutional settings to bring about sustainable food production and consumption, all approaches seem to agree (in one way or another) on the relevance of public debate on both issues. By referring to the social philosophers Michael Sandel and Alan Gewirth, the paper advances two claims. First, it is necessary to lead a substantial public debate on the aims and the social organisation of food production and consumption. Second, public debate can be guided by principles that have a universal moral claim and are still open for different ways of life. The paper starts with an overview of debates on different food regimes. It will be argued for a substantial public debate as well as moral principles that might guide these debates. Making food an issue of public debate would also answer the question why knowing our food is important.

Keywords: sustainable food, social theory, social ethics, governance, Alan Gewirth

Know your food! But why?

The claim 'Know your food!' has two perspectives: one addressing a bundle of social problems and potential solutions and another looking for ways of solving these problems and achieving solutions. The first perspective deals with the challenge that present industrial agriculture and food production are mostly on a non-sustainable path (Meisch, 2014a). They negatively affect humans in all parts of the production chain, greatly harm animals, destroy nature and the climate and overuse natural resources. Here, knowing one's food means both being aware of those negative effects as well as of the justice-based concept of sustainable development as this critique's normative basis: sustainable agriculture and food production should enable a good life of everybody. Its social organisation and state of technology aim for this objective and do not limit environment's ability to provide for everybody's good life. In this context, knowing one's food means acknowledging the problem and contributing to the solution. The second perspective looks at ways and means to reach the solution of a sustainable agriculture and food production. It deals with governance mechanisms that are supposed to achieve this aim. Depending on the mechanism chosen, knowing one's food means different things. In a market model, based on competition, consumers need information in order to buy more sustainable products and thereby to drive others out of the markets. Therefore, eco-labels play a crucial role as they are meant to mark morally correct products. In a common goods model, based on cooperation, people would for instance organise in community-supported agriculture. In this context, knowing one's food means knowledge about what to plant out and when to sow or to harvest, but also how to use one's food. Other examples with different governance mechanisms such as deliberation, negotiation or hierarchy (*ibid.*) would be possible.

The paper claims that knowing one's food is an issue that needs public and scientific debate on moral norms and social aims as well as on institutional rules that together ought to bring about sustainable agriculture and food production. It suggests an ethical approach (Alan Gewirth's Principle of Generic

Consistency (PGC)) in order to argue for a conception of sustainable development. Based on this approach, an outline of a more sustainable agriculture and food production will be sketched. Finally, the paper focuses on governance as a social science concept. It shows that conceptual problems on both the political and scientific level resemble and what help the PGC might be.

'Knowing one's food' as a public issue

It is contested that food related issues are indeed a public concern. Often, it is argued that what one eats and what food one buys is everyone's own choice based on individual preferences. Others and the state have to stay out of this private sphere. The German debate on the Veggie Day is an illustrative example of this belief (*ibid.*). In its 2013 election manifesto, the German Green Party suggested a greater variety of vegetarian and vegan menus and a Veggie Day in all public canteens. This measure was meant to inform consumers on the health, social and ecological effects of meat consumption. Resulting media and political outcry was enormous. Members of other parties called the suggestion an example of 'an ideological lifestyle dictatorship' or 'a disenfranchisement of citizens'.

It cannot be shown here that the critique of the Veggie Day rests on libertarian beliefs and that it seems particularly inconsistent with regard to the highly state regulated policy field of agriculture and food production (cf. *ibid.*). Neither does the paper provide a critique of the market model (cf. e.g. Crouch, 2011). However, it claims that if libertarianism ought to be the normative basis and the market the institution to deal with sustainable agriculture and food production, this has to be a matter of public debate in the first place – before pushing food related issues out of the public realm. The market model's preferability with regard to alternatives need to be justified and cannot be assumed implicitly. While the paper acknowledges the need for and the relevance of markets, it rejects libertarian notions of unrestrained markets and their implications on production and consumption. It shares Sandels (2012: 14) 'critique that these notions have (...)' drained public discourse of moral and civic energy, and contributed to the technocratic, managerial politics that afflicts many societies today. A debate about the moral limits of markets would enable us to decide, as a society, where markets serve the public good and where they don't belong. It would also invigorate our politics, by welcoming competing notions of the good life into the public square'.

A public discourse is needed on how social order can be organised that everybody (now and in the future) can live a good life. In this context, markets besides other governance models can play a role. Their limits are given by moral and social rules. It remains challenging to find ethical norms that allow for universal rules and still respect different notions of the good life. Gewirth's PGC will be introduced as an ethical approach to generate duties and rights but also to legitimise and limit social orders.

Conceptualising sustainable development: principle of Generic Consistency

Any conception of (and eventually any political programme based on) sustainable development has to refer to an ethical approach in order to determine which moral, legal and political obligations emerge from the concept of sustainable development. Within the conceptual core of sustainable development, there are several building blocks (e.g. inter- and intragenerational justice, direct and indirect claims for justice, human-environment interactions) that need to be specified in terms of obligations and corresponding rights and duties (unpublished data).

This paper suggests the PGC as ethical approach to conceptualise sustainable development. This approach regards human rights as based on a universalistic moral understanding of human dignity and aims to specify which aspects of human life are indispensable parts of human dignity and therefore create direct obligations to other humans and the state (Claassen and Düwell, 2013). Gewirth defines capabilities that are morally relevant and whose protection and support are morally, politically and legally obligatory. His approach claims that it can respect the plurality of various ways of living while at the same time it can identify morally relevant capabilities necessary to live a life of dignity.

Principle of generic consistency

The PGC is introduced as a supreme moral principle (Gewirth, 1978: 129-199) that reads as follows: 'Act in accord with the generic rights of your recipients as well as of yourself' (*ibid.*: 135). The approach focuses on human agency. Generic rights are rights to the necessary conditions an agent needs for purposeful action (*ibid.*: 64). These rights are freedom and well-being. The right of freedom consists of non-coercion and the ability to act according to one's own choice, while the right of well-being encompasses those general abilities and conditions that are necessary for an agent to reach the purpose of his/her action (*ibid.*: 64). Both rights are to be respected, protected and supported by other agents and by political institutions.

With regard to safeguarding an agent's capacity to act Gewirth distinguishes positive and negative rights. Negative rights mean the freedom from interference by others, while positive rights imply legitimate claims to those goods necessary to realise prospective agency. Gewirth distinguishes three kinds of goods an agent requires to achieve the purpose of his/her action (*ibid.*: 53-58): basic, nonsubtractive and additive goods. Basic goods are the generally necessary preconditions for all purposive action and '(...) comprise certain physical and psychological dispositions ranging from life and physical integrity (including such of their means as food, clothing, and shelter) to mental equilibrium and a feeling of confidence as to the general possibility of attaining one's goals' (*ibid.*: 54). Nonsubtractive goods are goods an agent also values and which he/she needs to maintain his/her level of purpose-fulfilment. Gewirth introduces 'level of purpose-fulfilment' as a reference point. It is the status quo of what an agent already has acquired (Steigleder, 1999: 53): 'Since he regards each of the particular purposes for which he acts as good, he regards as good in each case an increase in his level of purpose-fulfilment whereby he achieves the goal for which he acts' (Gewirth, 1978: 53). Therefore, an agent would be harmed if he/she loses (some of) these abilities in the course of an action. While the loss of a basic good would inflict basic harm on an agent, the loss of a nonsubtractive good would cause a specific harm (*ibid.*: 230). Additive goods are acquired in acting for a purpose and raise an agent's level of purpose-fulfilment. They cannot be basic or nonsubtractive. For instance, an agent struggling for life does not acquire additive goods as his/her actions deal either with maintaining the necessary preconditions of his/her agency (basic good) or with retaining what he/she already has (nonsubtractive good) (*ibid.*: 56). With reference to the PGC, it becomes possible to prioritise goods that deserve moral protection: 'If the normative idea is that we should be enabled to act as prospective agents, then the hierarchy follows from the relative necessity of each good for acting as a prospective agent' (Claassen and Düwell, 2013: 502). Therefore, the PGC allows to prioritise some goods over others and to deal with conflicts of duties of some agents to respect their recipient's generic rights:

> To structure rights to well-being and freedom, Gewirth doesn't propose a fixed hierarchy of rights but instead presents some criteria that allow to determine which capabilities in moral conflicts are more important than others. One of these is the 'degree of necessity for action' (Gewirth 1978: 343). On the basis of this criterion, the protection of an agent's freedom and basic well-being takes priority (at least prima facie) over nonsubtractive and additive well-being. In each specific case, there will be debate and disagreement about these criteria. (...) The PGC functions as the organizing principle for determining the scope and the relative importance of the basic capabilities that are forming the basis of rights and entitlements (Claassen and Düwell, 2013: 503).

As the generic rights are to be respected, protected and supported by other agents and by political institutions, the PGC also legitimises protective and supportive state action. In this regard, Gewirth distinguishes direct and indirect applications of the PGC (Gewirth, 1978: 200). The direct application refers to individual actions. An agent acts freely without interference by others (negative rights) or legitimately claims those goods required to realise prospective agency (positive rights). It might also imply that an agent fulfils his/her obligations with regard to other agent's generic rights. The indirect application refers to the state and its institutions insofar as they protect the negative rights of individual agents and promote the positive rights of developing basic capabilities and basic goods. With this, Gewirth distinguishes a static and dynamic justification of social rules. The first focuses on the protection and restoration of every person's equal possession of generic rights. The second acknowledges a disproportional inequality in person's ability to exercise their generic rights and aims to remove this inequality (*ibid.*: 292).

Ethical guidance for more sustainable food regimes

By enfolding the PGC in food related, morally relevant circumstances, it becomes possible to reach moral judgements on desirable courses of action with regard to agriculture and food production. As this paper chose a conceptual focus, it does not deal with a particular food issue that could be discussed here with regard to the PGC. The issue of food as such would be too wide. What is aimed for here is to outline in what respect the PGC might give ethical guidance for more sustainable approach to food.

One can safely argue that every form of producing and consuming food is legitimate that respects the generic rights of freedom and well-being of all persons affected. Food itself is a basic good and therefore a necessary precondition for engaging in purposive action. That is also true for attributes of the 'physical and social environment, ranging from habitability of the land and natural resources to some modicum of order and peace' (Gewirth, 1978: 212). Already with these general statements, it becomes obvious that there are rights to food, habitable land and sufficient resources. To the degree that inequalities keep persons from exercising their rights, social organisation has to remove these unequal conditions and protect the rights. That entails access to sufficient food as well as preventing agriculture and food production from violating generic rights e.g. by depleting and polluting environmental media (e.g. water, air or soil) or by feeding animals with food that (in the presence of hunger) could otherwise be used for human consumption. The PGC would also prevent slave-like labour as it can be found in many German slaughterhouses that exploit poverty in South-Eastern Europe (Meisch, 2014a). While the generic right of freedom has to be respected, forms of self-enslavement are regarded with a critical eye as they might result from a desperate need for basic goods. In this case, 'the PGC requires that there be institutional arrangements that help to assure for all persons the rights of freedom and well-being and hence minimize the circumstances that lead some persons to try to surrender one or both of these rights of action by inflicting basic harms on themselves' (Gewirth, 1978: 266/7).

Much more might be said on how the PGC informs the search for more sustainable food regimes. Solution approaches would have to pay attention to specific issues of global and intergenerational justice and the status of non-human animals in the process. This relates to individual rights and duties as well as social organisation and coordination.

Governance – blind in one eye

Taking the PGC as a starting point, governance research and policy recommendations look for governance models that respect the PGC and achieve sustainable agriculture and food production most appropriately. At this point, no governance model can be excluded *prima facie*. Most likely, there will be no single solution but many – in line with people's generic rights of freedom and well-being. Politics and social research will have to deal with polycentric governance (Ostrom, 2014) and the interaction of different governance models (Mayntz, 2009). However, it seems that some governance tools will lose relevance. These tools are policies designed to implement and enforce a specific governance mechanism. Eco-labels are tools aiming at competition that is the mechanism in the market model. By providing information on a sustainable product, it is hoped that this product gets a competitive advantage. However, as non-sustainable ways of agriculture and food production will vanish because they violate the PGC it will not be necessary any more to mark sustainable products.

The paper argued that problem analysis (including actor and power constellations) and normative reflection (based on the PGC, as it was suggested here) (have to) provide the basis for choosing an appropriate governance model. Several social scientists pointed out that the governance model itself has a blind spot when it comes to policy problems and normative aims (Mayntz, 2010; Offe, 2008; Streeck, 2009). This has epistemological reasons as the governance approach from its beginnings has been interested in modes of social coordination (Mayntz, 2009). However, it would be wrong to regard the approach instrumentally: a neutral or value-free tool for politics. The governance approach itself has a 'common goods bias'. That does not mean that governance always serves the common good. However, for genealogical reasons, the approach assumes that it does advance the common good. Mayntz (2010) argued that this might be the case but that it cannot be simply assumed. Others stressed that this bias has consequences as every governance model incorporates normative assumptions and value judgements that need to be reflected and discussed (Immergut, 2011; Offe, 2008; Streeck, 2009). For instance, the market model has a different notion of how a just society would look like than the common goods model. Here, these objections meet Sandel's critique. Streeck (2009: 12-14) criticised social sciences for their lack of engagement with normative issues as well as their naïve belief that a just social order can be reached by governance mechanisms that resemble the invisible hand and make the state as democratically controlled guarantee of a just social order indispensable.

Therefore, reasoning on governance in politics and social science requires as precondition an analysis of the social problem and the normative aims. On this basis, appropriate governance mechanisms and instruments can be chosen (Meisch, 2014b). Here, both above-mentioned aspects of knowing one's food meet one another: first, knowing about the present situation of agriculture and food production and possible aims for a more sustainable development and second, knowing as a possible governance tool (e.g. labelling, knowledge on growing and processing). While the first reflects on desirable social aims and orders, the second deals with ways and means to achieve these aims. Both need an ethical approach (such as the PGC) that informs critique of the status quo, provides guidance in the search for individual and social rules and allows choosing or rejecting possible coordination mechanisms.

Conclusion

In social ethics and social theory, a tendency in politics as well as the social sciences is sharply criticised. By taking on the governance perspective, it ignores analysis of political problems, power relations and normative aims. This results partly from the epistemological interest of the governance approach and does not delegitimise the approach *per se*. However, it leads to problematic results when underlying normative questions remain unaddressed and the preferability of a particular governance model together with its normative assumptions is taken for granted, yet without substantive ethical and social debate on implicit political aims.

Building on this critique, the paper argued that knowing one's food can be the basis for substantive discourse on sustainable food regimes. Choosing a governance approach, implicitly or explicitly entails a decision for a social order. As this choice has to be taken responsibly, public debate on normative and evaluative aims of this order is needed before choosing a governance mechanism to regulate a policy field such as food.

Food related issues pose global and local, national and transnational challenges alike that very often cannot easily be separated from each other. The challenge would be to find a conception of sustainable development that guides public debate on food, has a universal claim but still allows for context-sensitivity. The paper argued that public debate is necessary and suggested Alan Gewirth's PGC as ethical guidance of this discourse. It was outlined that this approach allows making claims with regard to food on an individual and social level. Most importantly, Gewirth argues for the foundation of a just political order and provides a principle to deal with issues related to food. Further research needs to elaborate how both can form the basis for more sustainable food regimes and corresponding governance approaches.

References

Claassen, R. and Düwell, M. (2013). The foundations of capability theory: comparing Nussbaum and Gewirth. In: Musschenga; A.W. and Heeger, F.R. (eds.) Ethical theory and moral practice 16, pp. 493-510.

Crouch, C. (2011). The strange non-death of neo-liberalism. John Wiley & Sons, Hoboken, USA.

Gewirth, A. (1978). Reason and morality. University of Chicago Press, Chicago, USA.

Immergut, E. (2011). Democratic theory and policy analysis: four models of 'policy, politics and choice'. Der moderne staat 1: 69-86.

Mayntz, R. (2009). New challenges to governance theory (1998). In: Mayntz, R. (ed.) Über governance. Institutionen und Prozesse politischer Regelung. Campus Verlag, Frankfurt, Germany, pp. 13-27.

Mayntz, R. (2010). Governance im modernen Staat. In: Benz, A. and Dose, N. (eds.) Governance – Regieren in komplexen Regelsystemen. Eine Einführung. Verlag für Sozialwissen-schaften, Wiesbaden, Germany, pp. 37-48.

Meisch, S. (2014a). Bewusster Fleischkonsum – wie kann Politik gestaltend eingreifen? In: Voget-Kleschin, L., Bossert, L. and Ott, K. (eds.). Nachhaltige Lebensstile. Metropolis, Marburg, Germany, pp. 411-436.

Meisch, S. (2014b). The need for a value-reflexive governance of water in the Anthropocene. In: Bhaduri, A., Bogardi, J., Leentvaar, J. and Marx, S. (eds.) The global water system in the Anthropocene. Springer, Cham, Switzerland, pp. 427-437.

Offe, C. (2008). Governance – „empty signifier' oder sozialwissenschaftliches Forschungs-programm? Politische Vierteljahresschrift, Sondervelt 41: 61-76.

Ostrom, E. (2014). Beyond markets and states: polycentric governance of complex economic systems. In: Sabetti, F. and Aligica, P.D. (eds.). Choice, rules and collective action. The ostroms on the study of institutions and governance, ECPR Press, Cholcester, UK, pp. 167-209.

Sandel, M. (2012). What money can't buy. The moral limits of markets. Farrar, Straus and Giroux, New York, USA.

Steigleder, K. (1999). Grundlagen der normativen Ethik. Der Ansatz von Alan Gewirth. Alber Verlag, Freiburg im Breisgau, Germany.

Streeck, W. (2009). Von der gesteuerten Demokratie zum selbststeuernden Kapitalismus. Die Sozialwissenschaften in der Liberalisierung. WestEnd 6: 13-33.

Employing Nussbaum's eighth capability to reason sustainable food production?

L. Voget-Kleschin
Institute of Botany and Landscape Ecology, Greifswald University, Soldmannstr. 15, 17487 Greifswald, Germany; voget-kleschin@philsem.uni-kiel.de

Abstract

The capability approach argues that in evaluating human quality of life, we should focus on what human beings are able to do and to be, that is, capabilities. In this line of argument, Martha Nussbaum has proposed a list of central human capabilities aiming to ground political justice (Nussbaum 2007). The eighth capability on the list reads 'being able to live with concern for and in relation to animals, plants and the world of nature' (*ibid.*, 77). The paper explores in how far it is possible to draw on this capability to reason claims for environmental protection, focusing specifically on claims for (more) sustainable food production.

Keywords: preconditions of justice, encountering food production, sustainable food consumption

Introduction

The capability approach argues that in evaluating human quality of life, we should focus on what human beings are able to do and to be, that is, capabilities. In this line of argument, Martha Nussbaum has proposed a list of central human capabilities aiming to ground political justice (Nussbaum, 2007). The paper aims at exploring in how far it is possible to draw on the capability approach to reason claims for environmental protection generally and (more) sustainable food production specifically.

One way to do so lies in substantiating how realizing any capability on this list requires certain conducive environmental preconditions. Thus, by way of example, because climate change features negative health effects it negatively impacts on capabilities such as being able not to die prematurely (the capability 1: life, Nussbaum, 2007: 77-78) or to have good health (capability 2: bodily health, *ibid.*). Furthermore, one could argue that because climate change will result in sea level rises and an increase in extreme weather events it will force people to migrate and thus negatively impact on capabilities such as being able to have attachments to things and people outside ourselves (capability 5: emotions, *ibid.*), to have the social basis of self-respect (capability 7: affiliation, *ibid.*) and being able to participate effectively in political choices that govern one's life (capability 10: control over one's environment, *ibid.*) (similarly Holland, 2012: 151-156). In this line of argument, Breena Holland points out that because certain environmental preconditions are necessary to realize any and all capabilities on the list, environmental protection constitutes a precondition of social justice (Holland, 2014). Similarly, in the last few years an increasing amount of work focuses on possible links between sustainability and the capability approach (e.g. Schultz *et al.*, 2013; Voget-Kleschin, 2013a), that is, on elaborating in how far allowing all contemporary and future human beings to live a good human life as conceptualized in the capability approach requires maintaining certain aspects of nature into the future.

In contrast to these works' emphasis on the importance of certain environmental conditions for the entirety of capabilities on Nussbaum's list, I will initially focus on one of the capabilities only (for a similar focus on one specific capability cf. e.g. Nussbaum, 2000, Chapter 3 on the role of religion, which is part of capability 3: sense, imagination and thought).

Nussbaum defines the eighth capability on her list, titled 'other species', as 'being able to live with concern for and in relation to animals, plants and the world of nature' (Nussbaum, 2007: 77). To begin with, I explore in how far it is possible to draw on this capability to reason claims for environmental protection, focusing specifically on claims for (more) sustainable food production. In doing so, I disclose a second issue: Nussbaum argues that the capabilities on the list need to be specified in an abstract and general way so as to respect reasonable pluralism pace Rawls. In discussing the potentials and limits of drawing on the eighth capability to reason claims for (more) sustainable food production, I carve out in how far this general vagueness of the list affords or rather aggravates the feasibility of drawing on the list to develop concrete, action guiding conclusions.

Nussbaums capability list: why vagueness

Nussbaum emphasizes that her capability list should not be understood as a comprehensive notion of the good pace (Rawls, 1988) but rather as potential object of an overlapping consensus (Nussbaum, 2007: 70):

> What is asked of (citizens) is that they endorse the basic ideas of the Capabilities Approach for political purposes only, not as a comprehensive guide to life, and that they view them as operative within a distinctive domain, namely that of the political (Nussbaum, 2011: 90, emphasis in the original).

Nussbaum argues that her list constitutes a plausible candidate for an overlapping consensus because it allows for respecting pluralism in six ways (cf. Nussbaum, 2007: 78-80; Nussbaum, 2000: 105). One of these is that the list is a list of capabilities rather than (so called, cf. *ibid.*) functionings. That is, the list does not prescribe that human beings actually realize the items on the list but demands that they should be able to realize them if they so choose. I will come back to this aspect in the next but one section. Another aspect indicates that the items on the list are formulated somewhat abstract and general so as to leave room for the activities of specification by deliberation. In the subsequent section I will discuss the implications of this latter feature.

Direct action guidance

What kind of environment allows us to 'live with concern for and in relation to animals, plants and the world of nature' (Nussbaum, 2007: 77)? Or rather: can we imagine any kind of environment that actually precludes us from doing so? We might answer this question by pointing to completely artificial environments that do not features any other species at all – such as a space-capsule, a hospital's isolation ward or perhaps the environment of people who spend most of their life in the virtual spheres of the internet. Meanwhile, even such a second life adherent needs to eat as well as get out every now and then to buy food or other daily necessities. In consuming food, every individual necessarily encounters other (animal- and plant) species, namely in terms of eating them. Furthermore, even assuming the above-mentioned second life adherent lives in one of today's megacities, she will probably encounter quite a few different species on her way to the supermarket. After all, cities generally feature an abundance of species. This demonstrates that even environments we conceive as manmade and artificial throughout do not actually preclude us from realizing Nussbaum's eighth capability.

More to the point, let us look at a second example, focusing on agrarian landscapes from a nature conservation point of view: in comparison to a diversified crop rotation, a high abundance of oilseed rape features comparatively lower species diversity. Nevertheless, a lot of people enjoy the bright yellowness of flowering oilseed rape as a beautiful expression of spring. They might even favor oilseed rape above other, less striking crops. Accordingly, one could say that for these people increasing the abundance of oilseed rape in crop rotation might actually increase their ability to 'live with concern for and in relation to (...) the world of nature' – even though from a nature conservation point of view it negatively impacts on a region's biodiversity. Thus an agrarian landscape that qualifies as 'more valuable' from a nature conservation perspective does not necessarily offer better possibilities to realize Nussbaum's eighth capability.

These examples show that because capability eight is defined in an abstract and general way, it does not allow distinguishing between an environmental quality that allows us to 'live with concern for and in relation to animals, plants and the world of nature' (Nussbaum, 2007: 77) and one that does not do so. The eighth capability is too unspecific to allow for directly inferring specific requirements regarding land management and food production.

In what follows, I will discuss two implications of this insight. The first regards how capability eight can nevertheless be fruitfully employed for reasoning rather specific claims regarding food production – though by a more indirect argument. The second transfers the argument that because capability eight is abstract and general it does not allow for direct action guidance to the other capabilities on Nussbaum's list.

Indirect action guidance

Though the dedicated nerd/computer freak in my first example might actually encounter quite a few natural entities on her way to the next supermarket as well as in consuming her groceries, this does not necessarily mean that she will pay any attention to them, allowing for a conscious encounter. Thus living 'with concern for and in relation to animals, plants and the world of nature' (Nussbaum, 2007: 77) does not only necessitate the presence of animals, plants and the world of nature. Furthermore, it requires that people consciously perceive this presence. Therefore, besides asking what kind of environmental conditions advance or obstruct realizing capability eight in terms of providing the physical preconditions for this capability, we can also ask what kind of conditions advance or obstruct the psychological disposition required for the realization of this capability. Relating this question to the field of food production allows pointing out some ways in which an argument drawing on capability eighth can be employed to reason certain specific claims regarding food production.

I propose that providing possibilities for individuals to actually encounter food production constitutes an important possibility for individuals to develop concern for and relations towards the world of nature. Such encounters can for example be realized in buying food on famers markets or participating in community supported agriculture or urban gardening. Such direct contact with food producers and food production can enhance capability eight because of at least two reasons: first, every human has to eat and most food derives from animals and plants. Therefore, food consumption constitutes one of the most common points of contact between human beings and the world of nature. This is not to say that in consuming food human beings necessarily consciously focus on this contact with nature. Meanwhile, because food consumption is both a common, everyday activity and because physically incorporating food constitutes quite an intimate act, food seems a good starting point for individuals to develop concern for and relations towards nature.

Second, producers aiming at more sustainable production methods often incur higher production costs and thus have to sell their products for a higher price. In selling their products they thus depend on potential customers ready to pay a certain surplus to remunerate their higher expenses. Accordingly, possibilities for more sustainable food production inter alia depend on consumer's readiness to consume products produced in a more sustainable manner (*ibid.*).

Finally, it is important to note that what has been said so far does only mean that providing possibilities for individuals to actually encounter food production constitutes an important way for realizing capability eight, that is, allowing individuals to develop concern for and relations towards the world of nature. To assume that experiencing such encounters increases individuals concern for sustainability and sustainable food production and consumption respectively constitutes a further argumentative step (Voget-Kleschin, 2014). In offering possibilities for such encounters one can thus hope but not guarantee that individuals develop concern for and relations to animals, plants and the world of nature in ways that change their attitude towards food (consumption) and thus concurrently advances possibilities for more sustainable food production.

Employing the capability approach in reasoning justice-based claims to a certain environmental quality

In discussing direct action guidance, I have argued that capability eight is too unspecific to allow for directly inferring specific requirements regarding the environmental quality necessary to realize this capability. Furthermore, in the introduction I have argued that any capability on Nussbaum's list requires certain conducive environmental preconditions (cf. Holland, 2014). Finally, in introducing Nussbaum's list, I have pointed out that all capabilities on the list are formulated in an abstract and general manner so as to allow for specification by deliberation.

Against this background, I put forward the thesis that actually neither any of the other capabilities on its own nor the capability list as a whole are specific enough to delineate what kind of environment we need to maintain so that all capabilities can be realized. The problem consists in that while there are clear links between capabilities on the one hand and benefits human beings derive from animals, plants and the world of nature on the other, these links are indeterminate. By way of example, it is clear that if pollution of environmental media such as water diminishes contemporary and future humans' capabilities of life and bodily health, contemporary humanity should delimit water pollution. However, this does not allow distinguishing what kind or level of pollution is tolerable rather than objectionable. Once again, if climate change leads to an increase of extreme weather events again threatening human capabilities (see above), it is clear that contemporary humanity should delimit greenhouse gass (GHG)-emissions. However, this does not imply to what degree GHG-emissions need to be reduced.

Meanwhile, many environmental threats are not caused deliberately but rather constitute negative side-effects of actions with other legitimate aims, such as actions aiming at an extension of certain capabilities for people who are not yet able to realize them to a sufficient degree (e.g. development of/ in developing countries). Hence, it does not suffice to argue for an unconditioned minimization of environmental threats.

Furthermore, there is a second reason why it is not possible to employ the capability list in exactly delineating the environmental quality necessary to allow for the realization of all capabilities. This is that the relation between capabilities on the one hand and environmental quality on the other can be depicted as a many-one-correspondence. This term is explained by Sen in reference to the relation between capabilities and the commodity bundle necessary to allow for capabilities:

> ...often commodity requirements may not be at all derivable from a specified set of capabilities, since the relation between commodity bundles and capability bundles may quite plausibly be a many-one-correspondence, with the same capabilities being achievable by more than one particular bundle of goods and services (for example, different combinations of food and health services may produce the same level of nutrition) (Sen, 1997: 513-514).

A similar many-once-correspondence holds in regard to the relation between capabilities and environmental quality. That is, we can imagine different combinations of say biodiversity, climatic conditions, and other natural entities and processes that allow for a similar level of capabilities to be realized.

Now, it is important to note that neither Nussbaum nor Holland claim that the capability list was specific enough to exactly delineate how much of any social or environmental preconditions needs to be preserved to allow for sufficient capabilities for all. Rather, both emphasize that delineating such thresholds should be object of a political process. Meanwhile, this does not mean that we can say nothing in regard to what level of environmental quality we should maintain for reasons of justice as understood by the capability approach. First, even though the list does not allow inferring which aspects of nature need to be preserved, the insight that all capabilities on the list presuppose certain environmental qualities emphasizes nature's multifunctionality in regard to human wellbeing.

Second, so far I have argued: (1) that our natural environment offers multiple benefits necessary for the realization of different capabilities and is thus multifunctional; (2) that we are not able to infer an exact demarcation between those aspects of nature that need to be preserved and those which we could (theoretically) do without.

If we combined this insights with the precautionary principle, we would arrive at a strong argument in favor of 'if in doubt about the necessity to preserve a certain natural entity or process, better err on the side of caution'.

Before closing, I would like to briefly point to two caveats. First, it has to be noted that such an argument rests on the acceptance of the pp. Second, I do not hold the claim to 'better err on the side of caution' to justify conceiving of all of nature as in need of conservation. However, in my view the argument suffices to reason a shift in the burden of proof: as long as there is no sufficiently sound argument that it is otherwise, we should act on the assumption that we need to preserve a certain natural entity or process.

Finally, in 'sustainability terminology' the above argument can be read as a reasoning of a conception of strong sustainability, that is, a conception of sustainability that assumes that natural capital cannot be substituted for by other forms of capital. Therefore, according to this argument, to qualify as more sustainable, food production needs to comply with guidelines for strong sustainability (Voget-Kleschin, 2013b).

Conclusions

In this paper, I have argued that because Nussbaum defines the capabilities on her list in an abstract and general way, the list is too unspecific to reason a precise amount or degree of environmental conservation. Meanwhile, I have subsequently demonstrated that it is possible to refer to capability eight to argue in favor of providing locations and opportunities to actually encounter food producers and food production. Thereby, one can hope but not guarantee that these concerns and relations result in individual's acceptance and support of (more) sustainable food production.

Second, I have argued that if we: (1) acknowledge that our natural environment is multifunctional; (2) accept that we are not able to delineate the exact amount of nature that needs to be preserved; and (3) approve of the precautionary principle, we arrive at an argument in favor of strong sustainability.

Acknowledgement

I would like to thank Simon Meisch for discussion and advice and two anonymous reviewers for their comments and suggestions regarding this paper.

References

Holland, B. (2012). Environment as a meta-capality: why a dignified human life requires a stable climate system. In: Thompson, A. and Bendik-Keymer, J. (eds.). Ethical adaptation to climate change. Human virtues of the future. MIT Press, Cambridge, USA.

Holland, B. (2014): Allocating the earth. A distributional framework for protecting capabilities in environmental law and policy. Oxford University Press, Oxford, UK.

Nussbaum, M.C. (2000). Women and human development. The capabilities approach. Cambridge University Press, Cambridge, USA.

Nussbaum, M.C. (2007). Frontiers of justice. Disability, nationality, species membership. The Belknapp Press, Cambridge, USA.

Nussbaum, M.C. (2011). Creating capabilities. The human development approach. Belknapp Press, Cambridge, USA.

Rawls, J. (1988). The priority of right and ideas of the good. Philosophy & Public Affairs 17(4):251-276. Available at: http://tinyurl.com/ol349px.

Schultz, E., Christen, M., Voget-Kleschin, L. and Burger, P. (2013). A sustainability-fitting interpretation of the capability approach: integrating the natural dimension by employing feedback loops. Journal of Human Development and Capabilities 14(1): 115-133. DOI: http://dx.doi.org/10.1080/19452829.2012.747489.

Sen, A. (1997). Resources, values, and development. Harvard University Press, Cambridge, USA.

Voget-Kleschin, L. (2013a). Employing the capability approach in conceptualizing sustainable development. Journal of Human Development and Capabilities 14(4): 483-502. DOI: http://dx.doi.org/110.1080/19452829.2013.827635.

Voget-Kleschin, L. (2013b). Large-scale land acquisition: evaluating its environmental aspects against the background of strong sustainability. Journal of Agricultural and Environmental Ethics 26(6): 1105-1126 DOI: http://dx.doi.org/10.1007/s10806-013-9448-9.

Voget-Kleschin, L. (2014). Reasoning qve. Journal of Agricultural and Environmental Ethics DOI: http://dx.doi.org/10.1007/s10806-014-9503-1.

What about the calves? How society perceives dairy farming

I. Christoph-Schulz, P. Salamon and D. Weible*
Thünen-Institute of Market Analysis, Bundesallee 50, 38116 Braunschweig, Germany; inken.christoph@ti.bund.de

Abstract

Animal husbandry has a prominent role in discussions about ethics in food production. This may be compounded by the increasing criticism towards the present system for agriculture and food production in media and public discussions during the last years. Since most consumer studies focus on production systems regarding pig and poultry, this paper examines society's perspective regarding dairy farming and addresses if it is ethically correct separating the calves from their dam of origin. The study aims to give orientation to farmers as well as policy-makers about the societal point of view. Six focus groups are carried out in three German cities to capture a variety of opinions and concerns among the population: perception and expectation of the participants regarding dairy farming in general and the separation of the calves from their mothers are particularly discussed. Results show that participants' perception is differentiated. It is well distinguished between big farms that are described as industries, where the cows have insufficient space, and small farms of part-time farmers. In the latter case it is assumed that the cows have more space. It is explicitly criticised that the cows get more and more concentrated feed to enhance the milk yield. With respect to the calves, some participants harshly criticized that the calves are separated after birth from their mothers. It is argued that cows will lose their motivation for lactation if the calves are separated. It is also assumed that milk quantity as well as quality is affected. Participants assume that today's cows are not able to suckle the calves due to the intensive breeding. The knowledge about rearing calves is rather low and participants are uncertain about common practices.

Keywords: society, dairy farming, perception, expectations, calves

Introduction

In Germany, modern agri-food production has developed into a highly mechanized sector, a change critically perceived by society in general although farmers have tried to capture societal concerns by different measures i.e. by improving animal health or reducing emissions. Reasons for the on-going societal criticism may be based on the fact that people's view on agriculture is quite strongly formed by childhoods' picture book stories and later intensified by messages transported via product marketing while continuous changes taking place in the sector are neither reflected in the media nor perceived by society. Small farms like in picture books seem to be preferred whereas the problems of these small farms are, in general, unknown (Salamon *et al.*, 2014).

One spot light of criticism is on modern animal husbandry. With respect to pigs and poultry there are numerous studies dealing with the expectations of society and/or consumers (e.g. Andersen, 2011; Krystallis *et al.*, 2009; Weible *et al.*, 2013). In contrast, expectations concerning rearing of cattle and dairy farming are only seldom tackled. This is astonishing as dairy production is an important agricultural sector. Milk has a significant input for the food industry, processed into numerous different products, which are in high favor of consumers and have an increasing demand. In Germany, per head consumption of fresh milk products reached 84.8 kg and of cheese 23.7 kg in 2012 (BMELV, 2013). Compared to the total turnover of the food industry milk and dairy products account for the second highest share (BVE, 2013).

Due to the limited knowledge relevant questions which would allow to capture societies' concerns more precisely are:

- How is dairy farming in Germany perceived in general?
- Which are important topics and main points of criticism?
- What does people expect by a good practice in dairy farming, what adjustments do they want to see?
- Do people relate individual perception and individual buying behaviour?

Not all of these relevant questions can be answered within the presented study. Aim of this study is to capture the scope of the societal perceptions concerning the current quality of dairy farming in detail. What do people have in mind if talking about dairy farming? What do they think about the feeding, the stable, the way cows are treated? Based on those results, expectations towards a good dairy farming will be extracted. How should dairy farming be adjusted in future if they could decide? What aspects are highly relevant to them and what aspects are of less importance? Results should allow to induce in the long-run improvements in the current systems of dairy farming. Farmers and decision makers will be given an orientation with respect to major topics of the current criticism.

Background

Perception of the society of current agriculture and especially of modern animal husbandry systems has been analyzed in various national and cross-national studies; however, depending on the focus of the respective studies, in some cases people were addressed as consumers as well. In one group of studies, the main focus is on animal welfare and on animal protection. In those cases people are mostly asked in their property as citizens. In general, the population of most member states of the EU indicates a preference for species-appropriate animal husbandry. Respondents state that they would be willing to pay more for food out of species-appropriate animal husbandry, but they also demand a better labeling policy to characterize these specific food items (Eurobarometer, 2005). These people placed the importance of animal welfare at 7.8 on a scale of one to ten (Germany: 8.1). Although about 60% of respondents believe that animal welfare has been improved in the last 10 years 75% of questioned people requested further improvements in this area (Eurobarometer, 2007). When compared to other types of animal husbandry acceptance of dairy cattle and sheep husbandry was higher (Harper and Henson, 2001).

In Germany, responsible handling of animals and production factors as well as production of food items with a high product quality are important with respect to agriculture (TNS Emnid, 2012). Some studies found that species-appropriate animal husbandry has the same (Zander and Hamm, 2009) or a higher importance (SGS Fresenius, 2011) for consumers than regional origin. Deficits in current animal husbandry are attributed to space limitations (Kayser, 2012). Many citizens notice a connection between available space for animals and other critical items in husbandry, like drug applications of antibiotics or increased numbers of behavioural disordered animals. Species-appropriate animal husbandry with sufficient space is seen as a solution for those problems (Weible *et al.*, 2013).

In a second group of studies the focus is on the quality of food; here, people are regarded more as consumers. In this context animal welfare and ethical problems are seldom mentioned (Harper and Henson, 2001). Mostly, criteria like food safety, food quality and health are discussed more deeply. Hence, if adjacent, citizens are specifically questioned with respect to animal welfare, they will express strong concerns against modern animal husbandry and they will mention their concerns as reasons for changes in their consumption pattern. To conclude: modern animal husbandry induces negative feelings in general which will not (yet) be reflected in purchase decisions or are superposed by other impressions (Von Alvensleben, 2002).

With regard to dairy production, Boogard *et al.* (2011) show that Dutch citizen prefer advantages due to a modern and hygienic system while at the same time they request traditional and nature-orientated farms. Perception and evaluation of modern dairy farming appears to be ambivalent. Neither is it fully accepted nor completely rejected. Therefore authors conclude that a medium acceptance of modern dairy cow husbandry exists in the Dutch population combined with a cautious attitude against new modern developments. Hellberg-Bahr *et al.* (2012) depict that a preference for high standards in animal welfare and environmental protection can be found in Germany. About half of the respondents prefer ecological produced milk or milk produced on grassland; however, stated preferences are not revealed in buying decisions. Zander *et al.* (2013) point to the fact that the view on agriculture of the German society is quite divers, however, in its basic attitude – to some extent – negative. A quite critical opinion concerning modern agriculture can be found within those groups which depicted a wider knowledge on agriculture.

Method

To get an idea of perceptions and expectations towards dairy farming an exploratative method was used: focus groups. Focus groups do not provide any representative results, but allow to capture the range of different opinions existing in the society or in part hereof. Typically, discussions focus on one distinct topic. Focus groups are organised as moderated discussions of a small group of participants (in our case: citizens). An interview guide structures the discussion and ensures that the points of interest are discussed. Focus groups are carried out for eliciting respondents' perceptions, attitudes and opinions (Wilson, 1997: 209) and take advantage of group interactions. Participants are confronted with other participants' opinions, attitudes or perceptions, and may have to justify the own opinion, attitude or perception.

In total six focus groups were carried out in Hannover, Dresden and Munich with six to eleven participants in December 2013. Although most of the participants directly lived in those cities, others are commuting from rural areas close by the cities into the bigger to take part in the discussions. Approximately 30% to 40% of the participants grew up in rural areas. Cities were chosen as locations for the focus groups for practical reasons: market research agencies hardly provide enough participants in rural areas. Past experience showed that even in smaller cities participation rates are low. In addition, there was also a rationale related to contents to concentrate on urban regions because the main objective was to capture the outsiders' view on dairy farming. Even though focus groups are not representative, a limited size of discussions are sufficient. It is also not possible to reflect the German society in total by focus group discussions. For this reason, a quantitative study will be carried out based on the findings of the focus groups (see conclusion). In the current study no new aspects were raised by the participants after the fourth discussion group. The interview guide was discussed in advance with an expert on dairy cows. The interview guide dealt, inter alia, with participants' perception of dairy farming in general as well as on conventional and organic dairy farming, the pros and cons of these two types of dairy farming and the participants' expectations regarding dairy farming.

Participants were recruited by an online panel of a market research agency. Quotas concerning age (20-70 years old), gender (33.3-66.6%), and employment (not more than 67%) were specified in advance in order to achieve heterogeneous groups. Three groups consisted of participants buying mainly (more than 50%) organic food; the other three groups consisted of citizens buying mainly conventional food. The study is part of a larger project which deals with the perceived additional benefits of organic dairy products. In this respect also the question of calve rearing became relevant: it was hypothesized that consumers of organic food are more sensitive concerning this topic. Today's conditions of animal breeding in general or in dairy farming in particular could affect individual's decision to consume or avoid the consumption of animal products. Therefore, focus group participants not consuming dairy products were not excluded from the discussions. People with a professional background concerning agriculture, food or market research were not recruited. Discussions had a length of about 90 minutes and were documented by audio and video. Afterwards, they were verbatim transcribed and analysed using open coding, a process to analyse the transcripts for a first understanding of underlying concepts and dimensions. Transcripts are interpreted line by line or even word by word (Lueger, 2009). Respondents' perception, assessment and expectation towards dairy farming in general and the separation of the calves from their mothers in particular were discussed. The topic was not announced in advance to avoid that participants prepare themselves for discussion. Even if questions were raised directly to the moderator they were not answered during the discussions and no additional information was given.

Results

Respondents perception of dairy farming

Discussants had quite different ideas about the quality of dairy farming conditions whereas the perceptions of the participants were strongly influenced by the herd size. They emphasized the need to distinguish between 'smallholder part-time farms'[3] and 'agrarian factories' before talking about their perceptions in more detail. In the latter, cows were seen as objects whereas the cows on part-time farms were better treated. Also vacation memories were used to describe the current farming practice: one respondent said, that she goes to Garmisch (south of Germany) for holidays for many years where the cows are kept outside.

Most discussants perceived dairy production as intensively mechanized and farms as economic enterprises where animals could be found instead of machinery. One discussant described 'big stables' where cows were tended by an 'assembly line'. Expressions like 'manufactory', 'industry' and 'aseptic' were often used. Some were convinced that cows were kept in boxes with a minimum amount of space to move. Hence, other participants reported about free-stall barns and mentioned that stables without walls (so called cold stalls) would have a good air ventilation. Whether cows were kept on meadows or not depended on the number of dairy cattle kept and on the farm size which was estimated to range between 50 and 1000 animals. For large farms this would be impracticable due to the herd size.

With respect to fodder, most participants mentioned that it would be primarily concentrate, but that also grass, hey and straw would be fed. Several discussants stated that concentrate would contain feeding stuff additives like vitamins and minerals. Also in this respect a strong relationship to land availability was presumed as the following statements proves: 'If a farmer has five meadows then the critter will grow up quite easily. If he has maize they will get maize. But if he has no land then he will need to buy concentrate and then he himself does not know what it's content is'. In this context prophylactic application of medical products, especially antibiotics, was addressed. Respondents criticized that fodder of animals would be enriched by pharmaceuticals prophylactically so that animals would not infect each other when being kept together in small stables. And also the performance-enhancing impact of antibiotics was seen as a reason. Hence, there was no unified opinion why animals would be treated with antibiotics. A good portion of discussants speculated that the constriction in barns would enable easy infection and therefore animals were in need of high quantities of pharmaceuticals. In contrast, others presumed that cows kept in stables would need less pharmaceuticals, as cows kept outside would be more easily infected with pathogenic germs.

No clear picture was found with respect to the rearing of calves. Participants harshly criticized the separation (weaning) of the calf from its mother. Especially they expressed their view that weaning would

[3] 3 Citations were translated from German into English

interfere with the calves nature and, thus, both animals would be harmed and more, the cows would 'lose their motivation' for lactation if the calves were separated. It was not directly stated whether the weaning was regarded as too early or not. It was further discussed if milk's quality and quantity would be better if calves could stay with their mothers due to reduced stress hormones. Other respondents were sure that nowadays the cows are not able to suckle their calves anymore due to the intensive breeding. It was also mentioned that the treats would be too sensitive and therefore milking too painful for the cows. Participants were also worried about the possibility that calves could be kicked to death by their mothers.

Respondents expectations towards dairy farming

Asked concerning the expectations in a good dairy farming the participants stated by themselves they would like to see that the dairy cows would have more space and the possibility of free-ranging on meadows. Expressions like 'free-range', 'a green meadow' or 'free-range, no boxes' characterize those expectations. It was not captured whether the participant had any knowledge about dairy farming or conditions in husbandry of dairy cows. More free movements are not the only reason to request of free-range. Additionally, as also important for the participants to see interactions between animals. Here, communication should not be restricted with animals next to them but should be enabled with their comrades farer away. A further aspect for grazing was to allow cows to choose what they prefer, and not what is provided generally.

Some participants preferred an open barn where the cow could decide to move outside or to stay in the barn. With respect to fodder the discussants had the notion it should be 'fresh' and 'without chemical additives'. However, the idea of additives like vitamins and minerals were not rejected in general, but chemical additives were refused. Here it was also discussed to what extent concentrates should be used, whether it would be good to reach the boundary of possible milk yields to the limit. Important to the participants were also regular visits of the veterinary and controls including fodder controls. It was requested to check for the wellbeing of the cows and whether they would be fed properly.

Within the discussion about expectations concerning dairy farming an intensive tendency of pathetic fallacy was noticed. Participants often draw conclusions based on human needs. Cows would require 'alternations' and should not have 'always the same temperature'. One discussant argued that she knows on her own 'how nice the sun would be' and she would imagine that the cow would feel similar. It was discussed that the animals would be unable to tolerate stress and that they would be stressed 'by the same things than humans'. One participants summarized his wishes as follows: 'a cow is just another human'.

Hence, another participant wondered whether the fulfillment of all those expectations would not have consequences for the sector as more staff and modern techniques would be required. He asked himself whether this would not induce a price explosion and a decline in demand.

Conclusion

By qualitative means the dairy farming in Germany was perceived as quite technical and often not species appropriate. This results still have to be underpinned by quantitative methods; however a similar research conducted for pig husbandry (Weible *et al.*, 2013) could show that outcomes of qualitative and quantitative method were very much inline. Main areas of concerns in dairy farming were, among others, the lack of space and the prophylactic use of medical products. Main areas of desirable improvements – based on the expectations of the focus group discussants with respect to good dairy farming – are more space per animals if possible with free-range on meadows and regular monitoring carried out by vets. This latter item in particular is more or less easy to be implemented by agricultural associations or policy makers. It will be very important to communicate these aspects credibly, and to be open about the frequency of controls as well as about the sanctions in case of misconduct. This study focused on dairy cattle. Therefore, results cannot be directly compared with findings of other studies regarding other animals' husbandries. It just could be imagined that points of criticism are less severe compared to pig husbandry. This impression is based on findings for societal perceptions towards pig production.

A proactive communication strategy on cows' needs could be very helpful too. This has to be underpinned by further research. Further research is also necessary for the question about society's expectations towards calves rearing. Here, no clear picture was found so far.

The presented qualitative research delivers a differentiated picture on perceptions and expectations regarding dairy farming. This would not have been provided with solely literature review or a standardized survey. But, of course, results are not representative. Therefore, an online-survey based on the outcome of this study will be carried out in 2015.

References

Andersen, L. (2011). Animal welfare and eggs – cheap talk or money on the counter? Journal of Agricultural Economics 62: 565-584.

Bundesministerium für Ernährung, Landwirtschaft und Verbraucherschutz (BMELV) (2013). Zeitreihe Pro Kopf-Verbrauch der EU-Mitgliedstaaten von Milch-erzeugnissen. Available at: http://tinyurl.com/l6oeuov.

Boogaard, B.K., Bock, B.B., Oosting, S.J., Wiskerke, J.S.C. and van der Zijpp, A.J. (2011). Social acceptance of dairy farming: The ambivalence between the two faces of modernity. Journal of Agricultural and Environmental Ethics 24(3): 259-282.

Bundesverband der Ernährungswirtschaft (BVE) (2013). Die Ernährungsindustrie in Zahlen 2013. Available at: http://tinyurl.com/os3ej3r.

Eurobarometer (2005). Attitudes of consumers towards the welfare of farmed animals, Eurobarometer Spezial 229. Available at: http://tinyurl.com/ehyjb.

Eurobarometer (2007). Attitudes of EU citizens towards animal welfare, Eurobaro-meter Special 270. Available at: http://tinyurl.com/ogkgtwd.

Harper, G. and Henson, S. (2001). consumer concerns about animal welfare and the impact on food choice. Final report. EU FAIR Project CT98-3678.

Hellberg-Bahr, A., Steffen, N. and Spiller, A. (2012). Marketingpotentiale für Weidemilch. In: Hambrusch, J., Hoffmann, C., Kantelhardt, J. and Oedl-Wieser, T. (eds.) Jahrbuch der Österreichischen Gesellschaft für Agrarökonomie. Facultas Verlag, Wien, Austria, pp.3-12.

Kayser, M., Schlieker, K. and Spiller, A. (2012). Die Wahrnehmung des Begriffs „Massentierhaltung' aus Sicht der Gesellschaft. Berichte über Landwirtschaft 90(3): 417-428.

Krystallis, A., De Barcellos, M.D., Kügler, J.O., Verbeke, W. and Grunert, K.G. (2009). Attitudes of European citizens towards pig production systems. Livestock Science 126: 46-56.

Lueger, M. (2009). Grounded theory. In: Buber, R. and Holzmüller, H.H. (eds.) Qualitative Marktforschung. Gabler, Wiesbaden, Germany, pp.189-206.

Salamon, P., Bürgelt, D. and Christoph-Schulz, I. (2014). Societal expectation towards structural change in agriculture. How can the sector cope with it? In: Rickert, U. and Schiefer, G. (eds.) Proceedings in system dynamics and innovation in food networks. International Center for Food Chain and Network Research, 17-21 February 2014, Innsbruck-Igls, Austria.

SGS Fresenius (2011). Verbraucherstudie 2011 'Lebensmittelqualität & Verbrauchermacht'.

TNS Emnid (2012). Das Image der deutschen Landwirtschaft: Ergebnisse einer Repräsentativbefragung. März 2012. Available at: http://tinyurl.com/ldely4c.

Von Alvensleben, R. (2002). Neue Wege in der Tierhaltung: Verbraucheransichten und – einsichten. In: Kuratorium für Technik und Bauwesen in der Landwirt-schaft e.V. (Hrsg.): Neue Wege in der Tierhaltung. Darmstadt: KTBL-Schrift 408.

Weible, D., Christoph-Schulz, I. and Salamon, P. (2013). Does the society perceive its own responsibility for modern pig production? In: Röcklinsberg, H. and Sandin, P. (eds.). The ethics of consumption: the citizen, the market, and the law; conference proceedings. Wageningen: Academic Publ, pp. 386-394.

Wilson, V. (1997). Focus Groups: A useful qualitative method for educational research? British Educational Research Journal 23(2): 209-224.

Zander, K. and Hamm, U. (2009). Farmer Consumer Partnerships: Information search and decision making – the case of ethical values of organic production. CORE Organic Project Series Report, 51p. Witzenhausen. Available at: http://orgprints.org/15199.

Zander, K., Isermeyer, F., Bürgelt, D., Christoph-Schulz, I., Salamon, P. and Weible, D. (2013). Erwartungen der Gesellschaft an die Landwirtschaft. Münster: Stiftung Westfälische Landschaft.

Section 10. Food cultures in comparison

Beyond access: integrating food security and food sovereignty models for justice

E. Murdock and S. Noll*
Michigan State University, Department of Philosophy, 503 S. Kedzie Hall, East Lansing, MI 48824, USA; murdoc23@msu.edu; nollsama@msu.edu

Abstract

This paper situates itself within the conversations about food security and sovereignty. Food security requires that 'all people at all times have physical and economic access to sufficient, safe and nutritious food to meet their dietary needs and food preferences for an active and healthy life' (FAO 1996). Food sovereignty focuses on the right of people and countries to determine their agricultural policies and food cultures, a critique that comes from populations that have been historically oppressed and marginalised. The growing literature on food justice is moving away from models that focus purely on making communities food secure to models that incorporate a wider range of issues beyond trade policies and commodity distribution. In order to understand this trend within the food justice literature, this paper argues that food security is based on a particular model of justice, distributive justice, which limits the sovereignty and autonomy of communities as food producers and consumers. The authors argue that focusing on a purely distributive model of justice, as instantiated by food security models, is an incomplete method for obtaining deeper and more significant forms of justice. For example, replacing salmon with a variety of commoditised foodstuffs can lead to the breakdown of cultural traditions and the increase of food born diseases. So, while in terms of access these communities are food secure, the replacement of traditional foods and the cultural methods for obtaining these foods has led to various other injustices. As such, the authors argue that food security is a necessary part of food justice and food sovereignty, but food security on its own and as the primary model for alleviating hunger is insufficient for creating sustainable communities and limiting harms. This paper relies on an expanded conception of justice common to Indigenous cosmologies, which disallow for exclusionary and divisive categories and practices.

Keywords: agricultural philosophy, indigenous philosophy, environmental ethics, applied ethics, political philosophy

Introduction

For over half a century, combatting worldwide hunger and malnutrition has been a priority for global development programs. In fact, access to food is often considered a condition that must be established before various other issues are addressed, such as myriad environmental, healthcare, and political goals (Schanbacher, 2010). Today, a wide range of non-government organisations, charities, and government agencies run programs aimed at bringing about greater access to food or 'food security'. This term is often used by these organisations to signify various development efforts that share the common goal of eliminating hunger. However, previous strategies that predominantly utilise market-based mechanisms have recently been critiqued (Lyson, 2004; Morales, 2011; Pimbert, 2008), sparking the development of an alternative model called 'food sovereignty' (Schanbacher, 2010). Food sovereignty is comprised of a diverse array of social movements (such as small-scale, landless farmers, indigenous rights movements, etc.) that critique the theories at the heart of previous food security efforts, the policies informed by this model, and the effects of market-based initiatives. Indeed, in the current literature, there appears to be a conflict between food security and food sovereignty models as mechanisms for alleviating global hunger. This paper situates itself within these conversations, exploring what motivates the current conflict in the literature in an attempt to break down the dichotomy between food security and food sovereignty. While we draw upon case studies, it should be noted here that the paper is largely theoretical in scope.

Specifically, this paper will explore the ways in which the conflicts between food security and food sovereignty are not simply about food or market-based mechanisms, but fundamentally concern competing conceptions of justice. We argue that efforts to alleviate hunger that predominantly accept limited concepts of justice, such as the distributive justice model, may cause unintended harms in various contexts. For example, focusing on food security alone might lead to harmful judgments and assumptions about certain communities' preference for nutritionally 'valueless' food, without paying attention to the ways in which certain cultural narratives and traditions have important relationships to and histories of food that not only ground their cultural identities and worth, but also act as a touchstone of cultural remembrance and healing. Conversely, solely focusing on the end goal of making communities food secure might be inadequate in providing communities with the autonomy necessary to make them sovereign food producers and consumers, an idea that is especially important in historically long oppressed and discounted populations. However, adopting a more holistic concept of justice as the foundation for food programs could help dissolve the tensions between food sovereignty and food security. When viewed through the lens of a more expansive conception of justice, food security becomes an integral part of food sovereignty, complicating the rigid distinction between food security and food sovereignty. Adopting one of the many Indigenous models of justice makes it impossible to see these two concerns as separate. However, before presenting this argument, we first give a brief definition of our key terms: food security and food sovereignty.

Food security and food sovereignty

Today, many multinational organisations and government agencies (such as the World Bank, World Trade Organization (WTO), and International Monetary Fund (IMF)) use the term 'food security' to signify the large-scale project of eliminating malnutrition and hunger worldwide (Schanbacher, 2010). While 'food security' was coined after reconstruction efforts post World War II and was originally understood to focus on the national level or state level, more recently this term has come to apply to individuals, as well. According to the Food Agricultural Organization (FAO), a nation is food secure when 'all people, at all times, have physical and economic access to sufficient, safe and nutritious food to meet their dietary needs and food preferences for an active and healthy life' (FAO, 1996). Within this definition, lack of food security is largely understood as an 'access' issue and thus encompasses a wide range of social goals aimed at increasing access or removing distribution barriers, such as the eradication of poverty, emergency access to food staples, public and private investment in 'developing' countries, and the creation of stable environments through the eradication of war, terrorism, and other disruptions. While these goals are laudable, several scholars (Pimbert, 2008; Schanbacher 2010; Wittman *et al.*, 2010) and local movements (such as 'La Via Campesina' and those represented in the Declaration of Nyeleni (Nyéléni, 2007)) have critiqued food security efforts because such projects were historically built upon the idea that 'economic growth, via market mechanisms, provides the most suitable solution for curbing poverty and achieving food security' (Schanbacher, 2010). For example, both Pimbert (2008) and Morales (2011) argue that, while we live in an era where industrial agriculture produces record quantities of food, the modern food system, built on market-based distribution, is associated with increasing food insecurity, social exclusion, malnutrition, and deepening poverty, especially in the developing world.

Due to these critiques, an alternative method of addressing hunger is steadily gaining support: specifically, food sovereignty is the right of people and countries to define their agricultural policy without the above negative effects. It organizes food production and processing 'according to the needs of local communities, giving priority to production to local consumption' (Schanbacher, 2010). Definitions of food sovereignty place a wide-range of other issues above trade policies and simple commodity distribution, such as sustainability, equal land access, ecological impact, protection of heritage breeds and seeds, and the creation of a more 'just' food system. It is concerned with the idea that in many ways food and eating (and all that that entails) is an integral part of human identity and self-actualization.

Differing conceptions of justice

Food security and food sovereignty rely on different conceptions and models of justice. Food security with its focus on access and distribution of certain material goods represents a distributive model of justice. Distributive justice locates justice in the process of equitable distribution of goods to all people. While this model captures a particular type of injustice, namely material inequality, distributive justice alone is not sufficient for guaranteeing the just treatment of individuals and communities. For example, imagine situations where citizens are given identical and equal shares of the corn, but despite this material equality some citizens are still viewed as lesser than other citizens. In this case, distributively, justice has been served, but injustices still abound at different levels. Political philosophy as a literature has been moving away from purely distributive accounts of justice because it recognizes the inadequacy of this single model alone. Nancy Fraser makes the distinction between socioeconomic injustice that is generally remedied by redistribution and cultural/symbolic injustice that should be remedied by transformative methods, which aim to not only change bad or unjust outcomes, but also the structures that produce and reproduce them (Fraser, 1997). While Fraser asserts that these differing types of injustice interact and influence each other in complicated ways, they still capture distinct aspects of justice such that examining one without the other is problematic.

Food security could also be understood through the lens of a capabilities model of justice that sees the equal access to an acceptable quality of food as a basic capability, which needs to be met in order for other higher order capabilities to be achieved. However, as we see from our exploration of the limits of distributive justice, a basic capability is the groundwork for other capabilities or just a piece of the project for justice. Similarly, focusing solely on distributive aspects even within the service of other forms of liberation (ultimately) can ignore procedural concerns. Procedural models of justice are concerned with 'the fairness of who gets to participate, and to what extent, in the decision-making processes used to allocate risks and goods' (Whyte, 2011). Models that focus purely on distributive models might be enacting other harms by excluding members of communities from the procedural and decision-making processes that let them define and realise what justice means for them. For example, making a community food secure might still involve negative colonial attitudes of excluding previously and currently oppressed populations from the formation of sustainable and culturally relevant local food systems and outcomes.

Food sovereignty as a theory and project expands itself beyond the purely distributive model and could be argued, as we do here, to be a model that encompasses and entails food security. We are not arguing that food security is irrelevant; rather we are arguing that food security is a necessary, but insufficient requirement for justice. Food sovereignty focuses more on the larger structures and procedures that problematically create injustice and one of those injustices is food insecurity. In terms of solutions, then, food sovereignty has to be sensitive and aware of the different models of justice and the different ways in which harms can be perpetuated.

A multifaceted understanding of the different ways in which harms are perpetuated and the ways in which solutions are formed through different models of justice means that we can no longer talk about food security and food sovereignty as if they are mutually exclusive categories. We can explore the real world implications of these separations between categories by seeing how solutions offered may suffer or indeed instantiate more harms through attempting to maintain these separations. The next section of this paper uses a case study to illustrate how food security may be inadequate to address various food related harms in particular contexts.

A case study: the Columbia River salmon contamination

Amerindian tribes living around the Columbia River have cultivated a special relationship with the salmon of the river since the beginning of their history (Lambert, 2008). In fact, the land surrounding Columbia River's Celilo Falls was a large trading area, where as many as 5,000 people would gather to trade the fish. Historically, tribes of the northwest continually struggled to maintain fishing rights in the face of powerful political interests, such as timber, farming, mining, and hydroelectric power (Lambert 2008). Today, tribal people of this region continue to use traditional fishing methods to harvest salmon from the Columbia River for a variety of purposes, such as commercial, ceremonial, and subsistence practices. According to the Tribal Fish Commission, 'Salmon is important and necessary for physical health and for spiritual well being' (CRITFC, 2011). However, the waters of the Columbia River are now contaminated and dams impede the natural migration patterns of the salmon. Tribal members have noticed that the fish are sick and/or deformed, with curved spines and unusual tumours. An Environmental Protection Agency (EPA) study found concentrations of mercury, Polychlorinated Biphenyls (PCBs), and Dichlorodiphenyldichloroethylene (DDE) in salmon and concluded that a person eating an average of 48 meals a month from the contaminated catch will be 50 times more likely to develop cancer. Despite these findings and a push to replace fresh salmon with canned fish and other processed foods, tribal members still practice traditional methods of harvest and consume the fish.

From a food security standpoint, agricultural production in the northwest region of the USA could adequately address the loss of food reserves due to salmon becoming unsafe to eat. In fact, according to Hormel and Norgaard (2009), the strategy of replacing traditional foodstuffs with processed food, such as canned fish, is the current strategy being implemented to address the contamination of fish in the Columbia River. This potential band-aid fits FAO's definition of food security, as the tribes could be given access to 'safe and nutritious food' that would meet their dietary needs (FAO, 1996). Thus, if we focus only on distributive justice or providing access to an adequate amount of 'food,' then the problem appears to be solved in an acceptable manner.

However, as Hormel and Norgaard (2009) argue, providing communities with processed foods while destroying local food traditions produces a multiplicity of harms. This argument can be extended to models of food security, in general. Specifically, we argue that models based on distributive concepts of justice miss key ethical components of the contexts that they are applied to. Here it is important for tribal members to have access to food that is safe to eat, but the salmon contamination issue is not simply a matter of distribution or food access. Humans and salmon have a long-standing relationship that is of both cultural and spiritual significance that cannot be displaced and, indeed, could be disrupted by the implementation of possible solutions that utilize food security models. Efforts to rectify the situation based on more holistic conceptions of justice, such as those that form the basis for food sovereignty movements, would be more sensitive to the wide array of justice issues (such as ecological, social, cultural, health, etc.) that tribal members are facing in this situation. Proposed solutions made by the tribes themselves include various strategies to mitigate the effects of the contamination for the better health of all involved, including the river basin ecosystem, the salmon, and the people (as well as the commercial viability of the fisheries). This is an excellent example of hunger mitigation efforts founded on a more robust theory of justice. We can now examine more holistic models of justice based on the cultural heritage and cosmologies of Indigenous peoples, which begin from a place of relationality that makes these rigid distinctions between ideas like food security and food justice more difficult to create/maintain.

Indigenous models of justice

Generally, Indigenous models of justice focus on the aspect of relationality of all things. In this way, the multiplicity of Indigenous models of justice offer a more holistic view of the cosmos such that justice involves right relations established and maintained amongst and between the cosmos. As Barbara Wall states, 'Western philosophical concepts generally reflect a discursive process of thinking that results in a dichotomous way of conceiving the world – for example, universal/particular, one/many – emphasizing distinctness rather than interrelatedness across concepts of meaning' (Wall, 2001). While there are various approaches in Western thought, this general penchant toward differentiation and distinctness has already been observed in the separation of food security and food sovereignty as well as the differing conceptions of justice outlined above. In contrast, the Navajo people's cosmology entails a conception of justice 'that is relationship-centered, not based on a political authority beyond an individual' (Wall, 2001). For example, taking the Navajo conception of justice (or a Western concept that does not reflect the discursive process, such as one created using poststructuralist theory), we can reinterpret how harm and repair is articulated. Since injustice is conceived of as any relations that are not in balance with the entire cosmos, it is easier to understand how food security divorced from food sovereignty, as a process oriented toward justice, is insufficient.

In terms of the Columbia River case study, divorcing the outcome of the salmon contamination from the context through which it arose, which encompasses myriad harms, is yet another form of injustice. Not understanding the centrality and cosmological significance of salmon to the Indigenous peoples in that area means that the proposed solution is not only inadequate, but also offensive. If we rely solely on a Western, distributive model of justice in assessing this situation, we would not only not foresee this harm, but we would also not be able to articulate how and why there might be more to be done.

Using this more relational model of justice informed by the rich history and practices of Indigenous peoples (or another model not grounded in distribution, such as a model coming out of care ethics), we find fewer tendencies toward the distinctiveness that leads to unjust outcomes and unforeseen harms being carried out in the name of 'justice'. By interrogating the relationship between food security and food sovereignty, we have determined that food security is an important aspect of what it means to practice food sovereignty, food justice and overall justice, or right relations within and among the entire cosmos. This integration allows us to come up with more sustainable, less harmful solutions to the complex harms we face regarding food and food justice in our world.

Conclusion

As the case study illustrates, conflicts between food security and food sovereignty are not simply about food or market-based mechanisms, but concern competing conceptions of justice. Efforts to alleviate hunger that accept limited concepts of justice may cause unintended harms. Focusing on the goal of making communities food secure might be inadequate for providing communities with the autonomy necessary to make them sovereign food producers and consumers. When viewed through the lens of more expansive conceptions of justice, such as Indigenous forms of justice, food security becomes an integral part of food sovereignty, complicating the rigid distinction between food security and food sovereignty. Hence, food security is a necessary, but insufficient requirement for justice. In terms of solutions, then, food sovereignty has to be sensitive and aware of the different models of justice and the different ways in which harms can be perpetuated.

References

Fraser, N. (1997). From redistribution to recognition? Dilemmas of justice in a postsocialist age. In: Fraser, N. (ed.) Justice interruptus: critical reflections on the postsocialist condition. Routledge, New York, USA.

Hormel, L. and Norgaard, K. (2009). Bringing the salmon home! Karuk challenges to capitalist incorporation, Critical Sociology 35(3): 343-366.

Lambert, L. (2008). Salmon and contamination in the Columbia river. In: Enduring legacies: native case studies. Evergreen State College, Washington, USA. Available at: http://tinyurl.com/q7ke2uj.

Lyson, T. (2004). Civic agriculture: reconnecting farm, food, and community. Tufts University Press, Medford, USA.

Morales, A. (2011). Growing food and justice: dismantling racism through sustainable food systems. In: Alkon, A.H. and Agyeman, J. (eds) Cultivating food justice: race, class, and sustainability. MIT Press, Cambridge, USA.

Columbia River Inter-Tribal Fish Commission (CRITFC) (2011). Columbia River treaty tribes. Available at: http://tinyurl.com/pbcpn5h.

Food and Agriculture Organization of the United Nations (FAO) (1996). Rome declaration on world food security. World food summit: 13-17 November 1996, Rome, Italy. Available at: http://tinyurl.com/q4e77hb.

Nyéléni (2007) Synthesis report. Nyéléni 2007 – forum for food sovereignty. 23-27 February 2007. Sélingué, Mali

Pimbert, M. (2008). Towards food sovereignty: reclaiming autonomous food systems. International Institute for Environment and Development, London, UK.

Schanbacher, W. (2010). The politics of food: the global conflict between food security and food sovereignty. Praeger Security Press, Oxford, UK.

Wall, B.E. (2001). Navajo conceptions of justice in the peacemaker court. Journal of Social Philosophy 32(4): 532-46.

Whyte, K.P. (2011). The recognition dimensions of environmental justice in Indian Country. Environmental Justice 4(4): 199-205.

Wittman, H., Desmarais, A. and Wiebe, N. (2010). The origins and potential of food sovereignty. In: Wittman, H., Desmarais, A. and Wiebe, N. (eds) Food sovereignty: reconnecting food, nature, and community. Fernwood Publishing, Oakland, USA.

Sustainable dairy cattle production in Southern Brazil: a proposal for engaging consumers and producers to develop local policies and practices

R. Anthony[1], J.A. Fregonesi[2] and A. De Paula Vieira[3]*
[1]Department of Philosophy, University of Alaska Anchorage, 3211 Providence Drive, Anchorage, AK 99508, USA; [2]Department of Animal Science, Universidade Estadual de Londrina, Rod. Celso Garcia Cid, Campus Universitário, CEP: 86057-970 Londrina, Brazil; [3]School of Communication and Business, Universidade Positivo, Rua Prof. Pedro Viriato Parigot de Souza, 5.300, CEP: 81280-330 Curitiba, Brazil; rxanthony@uaa.alaska.edu

Abstract

Internationally, consumers want dairy products that are produced ethically, sustainably, endorsed by reputable food safety assurance schemes, and promote cattle health and welfare. This demand challenges regulators and the industry to develop socially and scientifically informed policies and practices that transcend mere support for economic expansion of the industry. Today, as Brazil, begins to regulate the sector towards sustainability, greater attention should be paid to the quality preferences of local communities for good governance, social justice (including animal welfare) and long-term environmental resilience. Sustainable management practices, when adopted, should be informed by international research results (e.g. animal welfare) and standards from export markets, international organizations and dairy corporations. However, peculiar Brazilian farming realities, local science and values should complement these international perspectives and need to be documented scientifically. Awareness of local stakeholders' perspectives and commitments to native science are key in constructing a sustainable Brazilian dairy industry that is consonant with both national and international expectations. Through stakeholder analysis and a systemic view of the milk chain, we show the importance of understanding how consumers and producers of Southern Brazil integrate science, ethics and amoral preferences in choosing which type of dairy systems and technologies to support. We use sustainability assessment of food and agriculture systems (SAFA) indicators, namely, good governance, environmental integrity, economic resilience and social well-being, for our investigation. SAFA is an international reference for assessing trade-offs and synergies between different central components of sustainability. In addressing Brazil's ethically based sustainability climate, scientifically, we also consider: (1) existing local policies and the effectiveness of available channels by which producers and consumers communicate their preferences to policymakers and industry actors; (2) applied research for development of preventative strategies to maintain the sustainability of dairy systems; and (3) assessment of technologies that benchmark and streamline sustainability criteria on-farm. By studying evolving consumer preferences in Southern Brazil for value added dairy products we intend to: (1) identify research and policy questions that can close the gap between local and international demands; and (2) create a sustainability framework for the longevity and resilience of the Brazilian dairy sector in support of local and global food security.

Keywords: sustainable agriculture, dairy production, stakeholder engagement, ethical assessment

Introduction

This research has just been approved for funding by Coordenação de Aperfeiçoamento de Pessoal de Nível Superior (CAPES), Brazil. Investigation will commence in January 2015 and will be completed by January 2018. Below is a brief account of what will be investigated. Consumers around the world are increasingly demanding food products of high quality that are produced sustainably (Anthony, 2012a; Bovay and Sumner, 2013; FAO, 2006; Lagerkvist and Hess, 2011). However, consumer conception of sustainability might differ from that of producers (Thompson and Nardone, 1999). Amidst this backdrop, policymakers and those producing food commercially are challenged to develop: (1) policies and practices that are scientifically informed (Anthony, 2012b; Wilsdon and Willis, 2004); and (2) innovative sustainable systems that address consumers' social and environmental concerns beyond economic ones (Thompson, 2010). Development strategies should not merely be reactive to market or other external factors, but involve a holistic understanding of the system and also reflect important values of the production sector (Borner and Menz, 2005). Food related policymakers in Brazil, like their counterparts throughout the world are currently wrestling with how to design appropriate food and agricultural policies to meet consumer demands and to promote economic expansion for the industries. Government interventions and voluntary industry initiatives continue to emphasize the growth potential of the Brazilian dairy sector and efficiency-enhancing goals for the industry. However, Brazilian food consumers, especially those who are not merely living by subsistence standards or in rural poverty, seem to be equally motivated by ethically based food preferences. This interdisciplinary research project focuses on stakeholder analysis (Pinstrup-Andersen and Watson II, 2011) and a holistic and systemic view of the food system through a participatory research model (Anthony, 2012b). It aims to provide vital information to inform the food and agricultural policymaking process and to bridge the chasm between both consumers and producers by enhancing the communication of preferences between producers and consumers of dairy products in order to promote a more vibrant and sustainable dairy industry in Brazil (Croney and Anthony, 2011). Dairy production is highly valuable to the Brazilian public and economy and is integral to the food and nutritional security (e.g. dairy systems provide protein, energy for newborns and are vital sources of lipids, calcium and potassium) of many across the country. In Brazil, milk is one of the most important agricultural products. Brazil produces 32,304,421,000 l of milk/year and 22,803,519 cows are milked/year, with an average production of 1,417 l/cow/year (Instituto Brasileiro De Geografia E Estatística, 2013). Currently, the Brazilian government, through public policies for the dairy sector, plans to increase production by 40% over the next 10 years to meet both food security needs and to pursue economic opportunity. They intend to do this in a sustainable and internationally competitive manner (e.g. 'More livestock plan', MAPA, 2014). According to the government's strategic plan, achieving this goal of enhanced production and food security will involve ensuring that producers have access to cutting-edge technologies such as genetics, balanced nutrition, disease control, treatment of diseases, and access to specialized information. Capital to implement these practices is also a necessary ingredient. Any expansion of milk production must not only be accompanied by technical solutions that increase productivity and efficiency, but it should also be informed by the perspectives of different actors engaged in the dairy chain so that it will receive wide public acceptance and support. For the dairy industry in Brazil to grow in a sustainable manner, it is necessary to know what producers and consumers are thinking and how their preferences impact the structures of governance, environmental integrity, economic resilience and social well-being. Incorporating scientific and ethical information about the preferences and attitudes of producers and consumers of milk can mitigate market uncertainty. These preferences serve as catalyst for the actions of food processors and retailers. This project will: (1) analyze, document, compare and contrast the perspectives of Southern Brazilian dairy consumers and farmers towards sustainability concepts applied to dairy systems of high and low external inputs; and (2) further map and contextualize a governable, environmental friendly, economically resilient and socially responsible sustainability framework for the dairy production chain in Southern Brazil. The outcomes obtained will facilitate future studies on the validation of local indicators in a continuous manner. It will also serve to inform the evaluation of the sustainability of other sectors of the dairy chain, such as processing plants and retailers.

Methods

This research investigates the information needs (scientific, ethical and/or practical) of the dairy industry in southern Brazil. Understanding the information needs of the dairy industry in Brazil has not been done before in the manner articulated above. Conducting this research is vital for the sustainability for the Brazilian dairy chain as it closes existing knowledge gaps and helps to advance producer sovereignty in a time of rapid technological and economic change. Doing so will also assist the industry, farmers and policymakers as they formulate and implement critical public policies for this sector, both nationally and internationally. The four central methodological steps for this project include:

1. Analyzing and documenting the attitudes of consumers of milk from southern Brazil on sustainability concepts applied to dairy systems of high and low external inputs.
2. Analyzing and documenting the attitudes of milk producers in southern Brazil on sustainability concepts applied to dairy systems of high and low external inputs.
3 Comparing and contrasting the perspectives of consumers and producers towards sustainability concepts applied to dairy systems in high and low external inputs, in order to identify constraints.
4. Launching a governable, environmental friendly, economically resilient and socially responsible sustainability framework for the dairy production chain for southern Brazil.

Briefly, this project surveys (through phone interviews and follow up focus groups) producers and consumers' attitudes towards sustainability concepts applied to dairy production using the sustainability assessment of food and agriculture systems (SAFA) framework (FAO, 2013) as well as other dimensions of sustainability not accounted for by SAFA (for example, ecological diversity and integrity and scientific and economic relationships to sustainability), but yet important to local stakeholders and the research team. Our target sample are consumers and farmers in Rio Grande do Sul, Santa Catarina and Parana, important dairy states in southern Brazil. Our representative sample will be according to the distribution data provided by Instituto Brasileiro De Geografia E Estatística (IBGE) (2011; 2013). We will analyze, document and compare the attitudes of Brazilian dairy consumers and farmers towards sustainability concepts applied to dairy systems of high and low external inputs in Brazil. The SAFA model was chosen as a starting-point because it is international, holistic in focus and promotes multi-way and synergistic relationships in an open environment. Our survey questions will be divided according to central dimensions of sustainability, namely: (1) good governance (transparency, stakeholder dialogue, sustainability management plan, etc.); (2) environmental integrity (ecosystem diversity, species diversity, genetic diversity, use of materials, energy use, reduction and elimination of waste, animal health, freedom from stress etc.); (3) economic resilience (profitability, production stability, stability of supply, market stability, liquidity, risk management, food security, etc.); and (4) social well-being (quality of life, skills development, responsible buyers, suppliers' rights, employment relations, forced labor, child labor, freedom of association and the right to bargaining, non-discrimination, gender equality, support to vulnerable people, workplace safety and health provision, public health, etc.). In step 3, the survey and focus group data collected and documented in steps 1 and 2 will be compared and contrasted to identify the values that motivate decision-making between the two groups in terms of the categories specified in the SAFA framework identified above and other relevant evaluative decision-making criteria. The framework for values or ethics-based thinking involves consideration of two questions: (1) On what values and preferences consumers and producers base their preferences and motivations? (2) How are these values and preferences applied to specific situations by consumers and producers when they interact with the dairy value chain? In step 4, we synthesize our findings in the previous steps regarding unique Brazilian values and expectations regarding their dairy system to help farmers, scientists, policy makers, industry agents, and consumers begin to consider core pillars for the future of sustainable dairy production in the country. We anticipate conversations to be centered around: (1) the relative merits of current production practices with respect to animal welfare, sustainability indexes and how best to support producers who are open to producing milk sustainably and absorbing higher costs involved in order to ensure that sustainability requirements are met; (2) necessary ingredients to foster change in production practices to accommodate processing plants' greater interest in sustainability; (3) understanding cost structures and challenges for processing plants, the retail sector and consumers; (4) when is it acceptable to sacrifice sustainability criteria for lower production costs in a less than perfect world; (5) the importance of participating in certification programs to continuously improve the sustainability of the production; and assisting in the identification of who should participate in future actions.

Expected results

Since the research is yet to begin, we do not have any results to share at this time. However, we envision the following:

- The successful transitions to more sustainable dairy production systems does not only depend on the capacity of the various actors in the food system to integrate their divergent interests in a balanced manner and to rethink existing models, but also on their willingness to embark on a creative learning process that incorporates scientific research and values aware public policies in the emergence of sustainable innovations.
- It will be necessary for initiatives aimed at sustainability to be sensitive to the complex nature of milk production in Brazil and to target solutions at both: (1) low external input that depend almost entirely of local resources, human and animal labour; and (2) high external input, namely those which are intensive, highly mechanized and highly dependent on fossil fuels and human-made agricultural chemicals.
- Brazilian consumers may value being informed about the trustworthiness of the sources of food purchased by them, and the type of technology that was used to produce them and their intended and unintended consequences. They may also prioritize the credibility of sources like policymakers and the scientific community to navigate the maze of moral choices involved from farm to glass and plate.
- In contrast, producers may be concerned with efficiency and risk minimization. They may be interested in knowing whether and to what extent consumers, so as to reduce their risk and cost, will actually support their practices and initiatives. Producers may also seek help in examining the social and ethical aspects of how they should produce food in order to support healthier eating habits and more efficient resource use and income.

Conclusions

By identifying the preferences of consumers and dairy farmers from southern Brazil (which accounts for 33% of national milk production in a homogeneous area in terms of climate, soil, management practices and level of professionalism in the industry), we intend to assist all those interested in the milk production in the region to appreciate how social preferences, science, technology, moral and amoral values can be used to identify the future needs of the dairy sector. Ultimately, this investigation will help to promote a more sustainable dairy chain, for the region, for Brazil and for the world.

Acknowledgements

We would like to thank CAPES for providing funding for this multi-year, interdisciplinary and inter-institutional project. We would also like to thank the researchers and students from UEL, Universidade Positivo and UAA who will be involved with this project.

References

Anthony, R. (2012a). Building a sustainable future for animal agriculture: an environmental virtue ethic of care approach within the philosophy of technology. Journal of Agricultural and Environmental Ethics 25(2): 123-144.

Anthony, R. (2012b). Taming the unruly side of ethics: overcoming challenges of a bottom-up approach to ethics in the areas of climate change and food security. Journal of Agricultural and Environmental Ethics 25 (6): 813-841.

Borner, A. and Menz, W. (2005). Bioethical controversies and public advice: the production of ethical expertise and its role in the substantiation of political decision-making, In: Maasen, S. and Weingart, P. (eds.) Democratization of expertise: exploring novel forms of scientific advice in political decision-making. Springer, Dordrecht, pp. 21-40.

Bovay, J. and Sumner, D.A. (2013). Voter and consumer evaluation of restrictions on farm animal management practices. Paper presented at the Annual Meeting of the Agricultural and Applied Economics Association, 4-6 August 2013, Washington, USA.

Croney, C. and Anthony, R. (2011). Ruminating conscientiously: scientific and socio-ethical challenges for the US dairy industry. Journal of Dairy Science 94(2): 539-546.

Food and Agriculture Organization of the United Nations (FAO). (2013). Sustainability assessment of food and agricultural system: guidelines version 3.0. Available at: http://tinyurl.com/pl5jqqs.

Food and Agriculture Organization of the United Nations (FAO) (2006). Livestock's long shadow: environmental issues and options. Available at: http://tinyurl.com/7yzdoy.

Instituto Brasileiro De Geografia E Estatística (IBGE) (2013). Diretoria de Pesquisas, Coordenação de Agropecuária, Pesquisa da Pecuária Municipal. Rio de Janeiro, Brazil,

Instituto Brasileiro De Geografia E Estatística (IBGE) (2011). Sinopse do Senso Demográfico de 2010. Rio de Janeiro, Brazil.

Lagerkvist, C.J. and Hess, S. (2011). A meta-analysis of consumer willingness to pay for farm animal welfare. European Review of Agricultural Economics 38(4): 55-78.

Ministério da Agricultura, Pecuária e Abastecimento (MAPA) (2014). Plano Mais Pecuária. 'More livestock plan'.

Pinstrup-Andersen, P. and Watson II, D. (2011). Food policy for developing countries: the role of government is global, national and local food systems. Cornell University Press, Ithaca, USA.

Thompson, P.B. (2010). The agrarian vision: sustainability and environmental ethics. The University Press of Kentucky, Lexington, USA.

Thompson, P.B. and Nardone, A. (1999). Sustainable livestock production: methodological and ethical challenges. Livestock Production Science 61: 111-119.

Wilsdon, J and Willis, R. (2004). See-through science: why public engagement needs to move more upstream. Demos, London, UK.

Teaching interdisciplinary ethics as interactive process: advancing the Hohenheim-Tübingen model

M.D. Núñez Burbano de Lara[1*], *S. Lemke*[1] *and T. Potthast*[2]
[1] *Universität Hohenheim, Institute for Social Sciences in Agriculture, 70593 Stuttgart, Germany;* [2] *Eberhard Karls Universität Tübingen, International Centre for Ethics in the Sciences and Humanities (IZEW), Wilhelmstrasse. 19, 72074 Tübingen, Germany; dnunez@uni-hohenheim.de*

Abstract

The release of the International Assessment of Agricultural Knowledge, Science and Technology for Development (IAASTD) led the student initiative Food Revitalisation and Eco-Gastronomic Society of Hohenheim (FRESH) at the University of Hohenheim to start a participatory process that resulted in the establishment of the seminar 'Ethics of Food and Nutrition Security' in 2010 as a collaborative effort with the Chair of Gender and Nutrition at the University of Hohenheim (UHOH) and the International Centre for Ethics in the Sciences and Humanities (IZEW) at the University of Tübingen. This seminar follows the overall goal to equip young scientists with a firm foundation in interdisciplinary ethics-comprising basic knowledge of both ethical theories and tools to identify and analyse implications of food systems design and governance at the local and global levels – while developing their skills for practical application. We assess the development of the seminar since its inception, with particular focus on the introduction of a dynamic group work project in 2013/14 based on the ethical matrix originally suggested by Ben Mepham. The specificity of the new format was to work with the matrix in a culturally and ethically diverse student group, fostering interactive discussion among students, combined with input from external academic experts. Further, taking on different actors' perspectives demanded students' active and early engagement in the search for and critical examination of empirical evidence. These adaptations served the aim to strengthen the connection of the seminar's two thematic blocks 'Ethical theory and argumentation' and 'Practicing ethical analysis and argumentation'. The assessment of group work and a review of students' evaluations indicate that we succeeded to provide students more effectively with a tool to integrate empirical, normative and evaluative knowledge. In order to proceed from ethical analysis to judgment, however, additional philosophical tools would need to be employed more intensely in future seminar rounds. Further recommendations include a stronger contextualisation of guest lectures, interactive selection of the actors considered in group work, and more involvement of non-academic experts. Institutional challenges are the financial limits to full co-teaching of UHOH and IZEW staff and the fluctuation of engaged FRESH members to sustain the participatory development of the specific format of this non-standard seminar.

Keywords: ethical skills, ethical matrix, participatory curriculum development

Envisioning and shaping the seminar 'Ethics of Food and Nutrition Security'

In April 2008, the International Assessment of Agricultural Knowledge, Science and Technology for Development (IAASTD) concluded that '(b)usiness as usual is no longer an option' (McIntyre *et al.*, 2009: 3) and set off a serious debate on the legitimacy and long-term sustainability of the currently prevailing agricultural productivity paradigm. This paradigm assumes that food and nutrition security depends above all on the further development and diffusion of both green revolution technologies (e.g. synthetic fertilizers and pesticides) and more recent innovations (e.g. improved seed varieties and genetically modified organisms). According to the IAASTD, however, economically, environmentally, and socially resilient food systems first and foremost depend on our re-cognition of agriculture's multidimensionality and multifunctionality and of small-scale farmers' primary role therein.

This *re*-cognition demands from young agricultural scientists and practitioners, among others, knowledge of and skills for ethical reflection and argumentation. Understanding that this in turn requires an appropriate university training, the student initiative Food Revitalisation and Eco-Gastronomic Society of Hohenheim (FRESH) started in 2008 an energetic participatory process that resulted in the establishment of the seminar 'Ethics of Food and Nutrition Security' (EFNS) as a collaborative effort with the Chair of Gender and Nutrition at the University of Hohenheim (UHOH) and the International Centre for Ethics in the Sciences and Humanities (IZEW) at the University of Tübingen (Georgiadis *et al.*, 2010). The German Commission for the United Nations Educational Scientific and Cultural Organization (UNESCO) acknowledged the potential long-term impact of the project and awarded FRESH as part of the United Nations Decade of Education for Sustainable Development (Universität Hohenheim Presse und Öffentlichkeitsarbeit, 2011).

The EFNS seminar was offered in English to international students at the UHOH Faculty of Agriculture for the first time during the winter semester (WS) 2010/11 as an elective, unblocked module with 6 ECTS credit points. Since its creation, EFNS aimed to provide students with space, knowledge, and skills to identify, reflect, and discuss ethically relevant food and nutrition security issues. To this aim, EFNS was structured to resemble a seminar rather than a lecture series. Moreover, enrolment of students was conditioned upon previous submission of a letter of interest and limited to 15-20 participants per term (Georgiadis *et al.*, 2010).

In all four cycles completed so far, the EFNS seminar encompassed two thematic blocks. The first, 'Ethical theory and argumentation', was imparted by IZEW staff with the support of a facilitator from the Chair of Gender and Nutrition at UHOH. This block comprised an average of 17 teaching periods of 45 minutes and introduced basic knowledge of moral philosophy, basic concepts for ethical argumentation (Dietrich, 2008), as well as the ethical matrix as a tool for ethical analysis (e.g. Mepham, 2000). During the second block, 'Practicing ethical analysis and argumentation', students practiced the use of these concepts and tools in the context of controversial food and nutrition security issues, integrated into discussions with academic and non-academic guests, and moderated by the UHOH facilitator. During WS 2010/11, the seminar also served as a platform for the development of an innovative model for empirical analysis of ethical competence (Dietrich *et al.*, 2012; Lutz *et al.*, 2011).

Developing the 'Ethics of Food and Nutrition Security' seminar as an interdisciplinary and interactive process

The distinctiveness of the seminar lies in its participatory approach at two levels. First, the inception of the seminar involved FRESH members, other students and researchers from the UHOH Faculty of Agriculture, external agricultural practitioners, civil society representatives, and, importantly, IZEW staff (Georgiadis *et al.*, 2010). At a second level, EFNS has undergone an interactive review at the end of each cycle based on three evaluation modes: (1) a detailed survey, completed individually by students, comprising questions on structural and didactic issues as well as on student's perception of their own ethical competence, permitting numerically coded and explanatory answers; (2) a plenary feedback round; and (3) the assessment of students' performance by IZEW and UHOH staff, according to the seminar structure in the respective term. Deliberations on possible modifications of the seminar structure and contents were based on these evaluation modes and involved UHOH and IZEW staff, as well as FRESH members.

Modifications of the seminar undertaken in winter semester 2013/14

Students commented overall positively after the three first completed seminar cycles (Table 1). But they also recurrently emphasized the need for a stronger link between the seminar's two thematic blocks, frequently recommending to ensure the guidance of an ethicist in all seminar sessions, i.e. also in those sessions involving guest speakers. While guest speakers were experts in their respective discipline and therefore able to back their individual positions in an eloquent manner, they seldom explicitly and systematically identified and analysed underlying ethical questions. Students' evaluation of discussions and critical reflection in class, both in terms of quantity and quality, also point to this need (Table 1).

Until WS 2012/13, students' major task for successfully completing the seminar consisted of an individual journal exercise. Students were asked at minimum to revisit their experience in exploring the ethical dimensions of the food and nutrition security issues addressed in class. Ideally, entries would be made regularly and include: (1) the use of the tools for ethical deliberation introduced during the first thematic block on the discussed issues; (2) the use of these tools on other food and nutrition security topics of interest; (3) a critical reflection on the appropriateness of the tools; and (4) students' self-reflection of their views and competences. Despite students estimating their ethical competence to have improved through the seminar (Table 1), the need to link more strongly the two seminar blocks was also reflected in their journal writing performance: entries were too often limited to a descriptive revision of sessions and lacked a deeper ethical reflection (cf. Dietrich *et al.*, 2012: 512). Additionally, only few students devoted the time or discipline needed to make regular notes in their journals, having

Table 1. Student's yearly evaluation of the 'Ethics of Food and Nutrition Security' seminar.[1]

Exemplary survey questions	Arithmetical mean for the winter term			
	2010/11 (n=17)	2011/12 (n=17)	2012/13 (n=16)	2013/14 (n=14)
How was the balance between ethical theory and the exemplary guest lecturers?[a]	2.08	2.62	2.24	2.12
How was the balance between ethical theory and group work?[a]	–	–	–	2.00
How did you like the teaching methods?[b]				
ethical theory	1.57	1.69	1.53	1.59
guest lectures	2.29	1.69	1.94	2.34
journal exercise	–	–	4.41	–
group work	–	–	–	1.59
individual reflection paper	–	–	–	1.94
Was discussion sufficient in terms of				
quality and quantity?[c]	2.08	2.13	–	–
quantity?[c]	–	–	2.29	1.75
quality?[c]	–	–	2.06	1.56
Was critical reflection sufficient in terms of				
quality and quantity?[c]	2.69	2.69	–	–
quantity?[c]	–	–	2.18	2.12
quality?[c]	–	–	2.24	1.76
Do you feel more capable				
to see, articulate, and analyse ethical dimensions of issues of food and nutrition?[d]	2.21	1.89	1.47	1.41
to relate your studies to the ethical questions they raise?[d]	1.93	2.00	1.65	1.29
to solve real-life ethical dilemmas/to identify hidden ethical aspects?[d]	2.43	2.22	2.12	1.53

[1] Answers to these questions were coded numerically within a range of 1 to 5 as follows: [a] 1='balanced', 5='imbalanced'; [b] 1='very much', 5='not at all'; [c] 1='sufficient', 5='insufficient' and; [d] 1='yes, a lot', 5='not at all'. '–' stands for those terms in which a particular question was not part of the survey.

to complete all entries at once as the submission deadline approached. This represented a heavy work load and in part explains students' overall dislike of this teaching method (Table 1). Some students, however, regretted this circumstance as they belatedly realised the opportunity for self-reflection that the journal exercise had offered.

Based on the evaluations, in WS 2013/14 we sought to continue the EFNS seminar while at the same time enhancing the link of the 'Ethical theory and argumentation' with the 'Practicing ethical analysis and argumentation' thematic blocks to improve students' practical skills for identifying and analysing ethical issues. One possible response to students' feedback and unsatisfactory journal performance would indeed have been to ensure the presence of an ethicist throughout the seminar. Yet, due to the high time demands the seminar placed on IZEW staff commuting between Tübingen and Hohenheim and the limited financial budget, this proved difficult. A promising and, within the given institutional constraints, plausible modification was to adjust the requirements for successfully completing the seminar. Thus, the journal exercise was replaced by a dynamic group work assignment based on the ethical matrix (cf. Mepham, 2000) and a short individual reflection paper (Table 2).

The group work assignment aimed at leading students to work regularly and early in the term and at fostering both individual ethical reflection and group discussion. Furthermore, having students complete an ethical matrix was thought to serve as basis for a subsequent and more in-depth ethical analysis of the food and nutrition security issue at stake. The reflection paper sought to give students individually the opportunity to critically reflect on the structure and the contents of the seminar in relation to their own learning process.

Group work for building the ethical matrix

At the beginning of WS 2013/14, students were asked to join one of five groups consisting of three to four members. Each group was assigned to search for empirical data on the impact of all three of the following issues: (1) livestock consumption; (2) agroecology; and (3) biofuel production, on the enjoyment of well-being, autonomy and justice by one of the following stakeholder groups: (1) small-scale producers; (2) large-scale producers; (3) consumers; (4) farm animals; and (5) the environment. This investigation was carried out in parallel to the seminar's first thematic block 'ethical theory and argumentation'. The information gathered through literature review was then discussed with an external academic expert in each of the chosen topics at the beginning of the seminar's second thematic block 'Practicing ethical analysis and argumentation'. Selection of guest speakers was guided to the greatest possible extent by the speaker's previous experience in the field of ethics. During the last three sessions of the seminar, each group was asked to present their findings on their assigned row (i.e. stakeholder

Table 2. Yearly requirements for successful completion of the 'Ethics of Food and Nutrition Security' seminar.

		Winter term			
		2010/11	2011/12	2012/13	2013/14
Weight (in %)	Attendance and participation	30	30	30	10
	Journal exercise	50	50	50	–
	Group work	–	–	–	40
	Individual reflection paper	–	–	–	20
	Oral examination	20	20	20	30
	Total	100	100	100	100

group) of the ethical matrix. Combining all individual presentations resulted in three complete matrices that served as the basis for plenary debates about the underlying ethical issues that had become explicit. Taking on different stakeholder's perspectives demanded students' continuous search for and critical examination of empirical evidence and prevented repetitive discussions.

Achievements in winter term 2013/14

The group work was evaluated by students as being useful to learn ethical theory in connection to specific food and nutrition security topics (Table 1). The interactive and self-organised group discussions supported a broader thinking and revealed different dimensions of ethical concern. Especially the connection between empirical findings and the evaluative and normative dimensions could be practiced extensively, which is in turn necessary for moving towards the so-called mixed judgements in application-oriented ethics. Viewed from the ethicists' side, an important improvement compared to previous seminar cycles appeared in the depth of individual ethical reflections and the group outcomes.

The intense work with the matrix, however, came with a price: the step from: (1) using the ethical matrix to identify ethical points to consider towards; (2) a judgment concluding from and balancing the different cells of the matrix often remained to be made. Furthermore, students seemed to face difficulties in making use of their newly acquired knowledge on argumentation theory, not only but also due to time constraints. On a more general level, the ethical matrix as a consequentialist approach is not easy to combine with deontological approaches based on dignity, human rights and the right to adequate food, food sovereignty, among others. These aspects can of course be incorporated into several matrix cells, but then the weighing and prioritisation issues come up again. The two-pronged, in-depth use of the matrix as a tool for inter- and transdisciplinary ethical analysis on the one hand and concurrently as groundwork for judgement formation on the other, remains a philosophical and didactic challenge. Hence, argumentation theory and models for ethical judgment and deliberative decision-making will have to complement the matrix approach in future seminar cycles in more depth. With regard to deliberation, the structure of the seminar could be more transdisciplinary, i.e. more actively including the perspectives of non-academics experts, e.g. from civil society organisations and social movements, and transforming their insights into an ethical argument in the sense of moral philosophy.

Conclusions and structural challenges

Linking ethical theory with a dynamic group work on the ethical matrix proved effective for teaching interdisciplinary ethics as an interactive process. Elaborating relevant issues by using the matrix was confirmed as a feasible introductory tool to a culturally and ethically diverse class. Employing additional methods for ethical analysis remains nonetheless necessary, especially regarding the complex process of weighing the partial results of the individual matrix cells. It is both a philosophical and didactic challenge within application-oriented ethics to move between 'scilla' of simplistic formal decision algorithms and 'charybdis' of individualistic decisionist judgements.

Moving forward will not be achievable without an even more time-demanding participation of both ethics and agricultural sciences staff in all seminar sessions. Also, continuity from the student initiative FRESH is at stake due to the insecure recruitment of new members who are keen on getting actively involved in this project. These institutional and financial challenges demand the continuation of the participatory seminar process with the aim of moving ethics beyond a 'nice to have addition' in agricultural sciences education.

Acknowledgments

The authors would like to acknowledge Anne C. Bellows, Julia Dietrich, and FRESH members for their groundwork in establishing the module. We further thank Falk Kullen, María Isabel Matute Girón, and Katharine Cresswell-Riol for their support in summarizing the survey results.

References

Dietrich, J. (2008). Ungewissheit in der ethischen Urteilsbildung – ein Überblick. In: Benk, A. and Thomas, P. (eds.) Negativität und Orientierung. Könighausen und Neumann, Würzburg, Germany, pp. 65-77.

Dietrich, J., Lutz, R., Hilscher, M., Manoharan, D., Matute Giron, M.I., Mauser, J., Schweizer, S. and Bellows, A.C. (2012). The ethical matrix as an instrument for teaching and evaluation. In: Potthast, T. and Meisch, S. (eds.) Climate change and sustainable development: ethical perspectives on land use and food production. Wageningen Academic Publishers, Wageningen, the Netherlands, 526 pp.

Georgiadis, P., Schumacher, J., Hilscher, M., Manoharan, D., Birkenberg, A., Schweizer, S., Idel, A., Hudson-Wiedenmann, A., Herren, H.R., Gottwald, F.T, Fadani, A., Bellows, A.C., Kruse, M. and Zeller, M. (2010). Food ethics: a new and necessary academic approach to improve food and nutrition security. In: Tielkes, E. (ed.) Tropentag 2010. International research on food security, natural resource management and rural development. World food system – a contribution from Europe. Book of abstracts. 14-16 September 2010. Zürich, Switzerland, pp. 510-511.

Lutz, R., Dietrich, J., Voget Kleschin, L., Hilscher, M., Manoharan, D., Schweizer, S. and Bellows, A.C. (2011). Ethics of food and nutrition security oder: kann ethische Theorie die ethische Wahrnehmungskompetenz steigern? Eine Pilotstudie zur empirischen Unterrichtsforschung. Zeitschrift für Didaktik der Philosophie und Ethik (ZDPE) 3: 156-166.

McIntyre, B.D., Herren, H.R., Wakhungu, J. and Watson, R.T (eds.) (2009). Agriculture at a crossroads. International assessment of agricultural knowledge, science and technology for development: synthesis report. Island Press, Washington D.C., USA, 97 pp.

Mepham, B. (2000). A framework for the ethical analysis of novel foods: the ethical matrix. Journal of Agricultural and Environmental Ethics 12: 165-176.

Universität Hohenheim Presse und Offentlichkeitsarbeit (2011). Student group 'FRESH' receives distinction from UN. Universität Hohenheim, Stuttgart, Germany. Available at: http://tinyurl.com/m79x2yd.

Section 11. Food for the future

Entomophagy – why should it bug you? The ethics of insect production for food and feed

M. Gjerris[1], C. Gamborg[1] and H. Röcklinsberg[2]*
[1]Department of Food and Resource Economics, University of Copenhagen, Rolighedsvej 25, 1958 Frederiksberg, Copenhagen, Denmark; [2]Department of Animal Environment and Health, Swedish University of Agricultural Sciences, P.O. Box 7068, 750 07 Uppsala, Sweden; mgj@ifro.ku.dk

Abstract

Humans have, as far as the archeological records show, always eaten insects, reaching over 2000 edible species today. Given a growing global human population and high pressure on resources, interest in insects as a nutrious protein source for human food (entomophagy) and novel protein source for animal feed is developing. Compared to most other sources of animal protein insects are considered to be an environmentally low-impact source of nutrients. In a Western context the search for sustainable food and feed products has therefore lead to a growing interest in the area. However, as insects score low on the socio-zoological scale, but high on the 'disgust'-scale, mainly based on culturally transmitted information, utilizing insects in the food sector will probably face consumer resistance. The use of insects as a major human food and feed source is thought to present two major challenges: how to turn insects into safe, tasty socially acceptable feed and food and how to cheaply yet sustainably produce enough insects? The focus in this paper is on the animal ethics aspects of the second challenge. While invertebrates make up the majority of animal species, their welfare is overlooked compared to vertebrates. Absence of proof is, however, not proof of absence, cf. growing recognition of cognitive and emotional abilities in fish. Thus the welfare of different insects within different 'production' systems needs to be taken into account and – in parallel to fish – the welfare capacities of each species need to be studied to accomplish this. But the growing interest in entomophagy also evokes other ethical issues: Is there such a thing as 'insect integrity', and if so, are we ethically obliged to protect it? Does it make sense to ethically speak of insect commodification and its pros and cons for ecosystems, animals and humans? In this thematic survey of welfare and ethical issues we build both on the at present scarce literature of insect welfare and ethics and the ever-increasing literature on ethics of animal production.

Keywords: animal ethics, welfare, integrity, mini-livestock, sustainability.

The challenge of animal protein

There is little doubt that livestock production substantially affects its surroundings. FAO talks about 'a long shadow' cast over the environment, nature and landscapes on a global scale – as well as in specific countries (Steinfeld *et al.*, 2006). If we take it as a premise that production of animal protein will continue for the foreseeable future, the question is whether meat can be produced in a way that is sustainable in relation to the land use, nature, the environment and socio-economical aspects?

Livestock production – in competition with other types of food production, or use of the land for producing feed or energy is on a global scale directly or indirectly (through the impact on climate, air, soil and water) affecting the environment (Ilea, 2009). Agriculture is responsible for a total of 70% of global freshwater consumption, which contributes to regionally putting scarce water resources under pressure. ⅔ of all arable land is already used for animal production and, livestock production is known to be contributing to deforestation, changes in savannas, drainage of wetlands and desertification, hereby affecting ecosystem services (Norris *et al.*, 2010). The contribution of the livestock sector to the anthropogenic GHG emissions is estimated very differently ranging from 18% by FAO (Steinfeld *et al.*, 2006) to more than 50% by World Watch Institute (Goodland and Anhang, 2009). Even though estimates vary and there are huge differences between animal species, there is very little doubt that the production of animal proteins already has a substantial effect on the planetary ecosystems and as consumption rates and demands are expected to rise, this challenge is only growing.

It is in this light one must see the 'mini-livestock' strategy: insects as a potential source for production of scale of protein, either for direct human consumption, or indirectly in recomposed foods; as well as a protein source into feedstock mixtures (FAO, 2013).

Eating bugs is old news

Although there has been a recent surge in the interest in entomophagy understood as the dietary consumption of insects by any organism (but here focused on humans) in Europe and the USA, there is nothing new about entomophagy (Gahukar, 2011). In fact insects probably form one of the earliest ingredients in human diet among hunters and gatherers along with edible plants. Humans have, as far as the archeological records show, always eaten insects (McGrew, 2014). Insects are listed as 'permissible' food in the Old Testament, and throughout recorded history, insects have been on the menu: the Greeks consumed locusts and beetle larvae along with the Romans – even Aristotle reflected on how to harvest cicadas.

Today most cultures around the world including more than 3,000 ethnic groups in 130 countries have some of the 2,136 edible insects as part of their diet (Jongema, 2012). About 80% of the world population eats insects intentionally (Srivastava,, 2009). The edible insects mainly consist of beetle, ants, bees and wasps as well as butterflies, moths, grasshoppers and crickets. Most of these are eaten in the larval and pupal stages, but also adult insects are used fried, roasted, cooked as stand-alone dishes or added to stews, soups and rice. Entomophagy is most common throughout tropical and subtropical countries linked not only to the abundance and variety of species but as well as to poverty, food security, food choice, and food culture including taste (Gahukar, 2011).

Insects are, however, not consumed by many in Europe and North America. One speculated reason, besides the 'ick' factor, is because of the history of farming and livestock production which set insects aside as a pest; something to eradicate. In the past hundred and fifty years in Europe and North America, attempts have been made to promote human insect consumption (e.g. Packard, 1877) observing the habits of other cultures. Insects have been endorsed at various times for various reasons: around the turn of the 19th century as a new cheap food source and as possible supply when Europe faced food shortages during World War I (Cruz and Dossey, 2013). Maybe this is why the misconception exists that insects are only consumed in times of hunger; as 'emergency' food or food for the poor.

So why the fuzz?

Eating insects – supplementing or replacing food and food ingredients from (vertebrate) livestock – purportedly carry many benefits: (1) nutritional value – high on protein, vitamins and essential nutrients and low on fat and carbohydrates (Okaraonye and Ikewuchi, 2008); (2) disease – with insects there are low risk of transmission of zoonotic diseases, although allergies are known to occur (comparable to allergies to crustaceans (Cruz and Dossey, 2013)); (3) abundance – insects are an abundant food source throughout the world; (4) efficiency – because they are cold blooded, insects have a much higher feed conversion ratio: where cattle use 8 kg of feed to produce 1 kg of body weight gain, insects on average use less than 2 kg (Vogel, 2010). Other advantages are that insects can be grown in areas not amenable to e.g. cattle such as drought ridden areas, and insects can feed on bio-waste (eg. food waste); (5) environmental sustainability – farming and harvesting of insects require much less input such as water

(Van Huis, 2013), fuel and are less land dependent when compared to livestock production or crops; and have much less GHG emissions compared to livestock such as e.g. pigs – thus helping to mitigate climate change. In all these ways insects may be seen as a solution to some of challenges of increased demand for animal protein in the face of climate change (Oonincx *et al.*, 2010). Also in terms of food safety, things are changing. For example, just recently in 2014 the Belgian Federal Agency for the Safety of the Food Chain (FASFC, 2014) approved ten insect species for consumption, and in the Netherland there are 15 industries breeding insects for food (De Goede *et al.*, 2013).

For insect protein to provide a more sustainable, stable and efficient solution to these challenges it will, however, be necessary to move from the currently prevailing hunter-gatherer methods and scattered, unsystematic attempts at production of scale into semi-cultivation as is already taking place many places in the world or into full scale intensive production systems (FAO, 2013). Thus, if it is to be possible to produce sufficient amounts of 'mini-livestock' to meet rising demands of animal protein and to contribute to food security, livelihoods, climate change mitigation and environmental protection in a more sustainable way than current livestock production, evidently there are many scientific and technical challenges to solve.

There are, however, also a number of important ethical issues related to such a challenge – and part of these – the animal ethical issues – will be the focus of this paper. When looking at the latest, informative FAO (2013) factsheet on edible insects, the ethical considerations are mentioned as issues that 'should be considered in the protection of insect populations in natural environments'. They clearly are important issues, but have their sole focus on the effects of the production directly for humans or indirectly through concerns for the aspects of the natural environment that are of importance for humans: respecting local people, protecting food chains, conserving 'beneficial' (to humans) insect populations and the avoidance of releasing non-endemic (domesticated) insect species into natural environments. Thus, it is striking how – compared to other livestock production systems – animal welfare and animal ethics issues of mini-livestock production are left out of the considerations.

Ethical challenges

While invertebrates make up the majority of animal species, welfare issues related to them have been overlooked compared to vertebrates (Horvath *et al.*, 2013). As long as there is only little knowledge about the capacity of insects to experience better or worse welfare, we propose that absence of proof should not be misunderstood as proof of absence. Further: the welfare of different insects within different production systems needs to be taken into account and – in parallel to e.g. fish – the welfare capacities of each species need to be studied to accomplish this (Erens *et al.*, 2012; De Goede *et al.*, 2013). Finally: besides welfare issues, there are a range of ethical questions related to human use and consumption of animals that could be relevant with regard to insects, such as commodification, naturalness and integrity.

Animal welfare

There are more than 1 million species of insects but very little research into the potential capacity of insects to experience welfare. However, recent studies show nociception and the capacity to integrate information into complex decisions are present in some invertebrates (e.g. honey bees and spiders) (Elwood, 2011). Studies on shore crabs show they adopt their behaviour in order to avoid an electric shock, and referring to several studies performed to proof the inability of crustaceans to experience pain, the authors came to the opposite conclusion, calling for humane care (Magee and Elwood, 2013). Transferring this to insects, the high number of species poses quite a challenge in itself as also recent studies into fish have shown the need for species specific welfare research (Braithwaite *et al.*, 2013). Taken together this indicates insect welfare research need to be species-specific to recognize specific emotional and cognitive abilities and specific behavioural needs. This is important to make informed decisions regarding insect welfare and even for more applied reasons such as enabling legislators to formulate accurate regulations; farmers to ensure basic knowledge and compliance with legislation; and enabling the general public, i.e. citizens and potential consumers to take both curiosity and potential welfare concerns into account: 'What did this cricket experience before it died?' Needless to say, such knowledge will take time and resources to acquire. Doing so is in itself an ethical issue (Röcklinsberg, 2014). So far the pronounced scientific uncertainty in the area makes the discussions on what constitutes ethically justifiable ways of utilizing insects a challenge (Erens *et al.*, 2012).

To assess the welfare of a specific kind of insect within a production system one does not only need knowledge of the insect in question in its natural environment, but to study it in the context of a range of production systems to see where the animal fares better – given that it is clear if/how it makes sense to talk about 'better'. Such comparisons might be performed under different conditions: insects being hunted/gathered in the wild; the ones already being semi-cultivated and those already under domestication in insect mass-rearing systems. Furthermore, based on the experiences with the welfare issue regarding prevalent livestock species such as cattle and pigs, and – more recently – fish, it seems fair to envision that different welfare paradigms will clash and come up with different evaluations (Yeates *et al.*, 2011). It is in this context also relevant to reflect upon whether insect welfare can simply fit into the existing welfare paradigms present in the debate (Dawkins, 2006) on 'vertebrate' farm animal welfare, like the five freedoms, or if they somehow have to be adjusted to accommodate these new beings in the welfare discussion (De Goede *et al.*, 2013)?

Finally: insects have so far not been included in any welfare legislation of EU. As non-vertebrate species (e.g. octopus) and vertebrates at the foetal stage are included in the Directive on animal research (2010/63/EU) the next step might be to include crustaceans and insects *if* solid scientific research will show insect capacities of sentience and ability to experience individual welfare. If so, we foresee a development of growing public concern and demand for legislation for insects, based on the fundamental ethical principle of making a difference only where there is a relevant difference.

Other animal ethical issues

What challenges are perceived relevant in relation to a certain opportunity is always dependent on the perspective, values and desires of the evaluator. In general, from a conventional anthropocentric perspective the main challenges when facing a world of growing needs and dwindling resources, diminishing eco-system services and climate change might lie in how to 'change the world to fulfill our needs', looking into technical solutions on how to increase resource efficiency and developing new resources to replace those depleted. From an ecocentric perspective, one would rather call for ideas to 'change our behaviour to fit the world', leading to a need to define 'fit' in a plausible manner and for ways of thinking that could inspire people to change their life-style. Here, we focus on 'somewhere in between' – the animal ethical issues – if insects are to be utilized within the intensive food production system.

Animal food production raises other animal ethical issues than welfare. Notions such as death, integrity, rights and naturalness that lie beyond the typical animal welfare discussions (Yeates *et al.*, 2011) are nonetheless both part of the public discourse on animal production and has been subject of academic discussion. Issues like whether the integrity of a cicada will be affected in an insect mass-rearing system, whether the killing of ants will be considered similar to the killing of a chicken and how commodification of yet another part of nature will influence us as humans will thus be part of an ethical evaluation of the inclusion of insects into the feed and food production. Further there are issues around breeding goals and perhaps the use of biotechnologies such as genetic modification that are relevant to cover and that may or may not involve animal welfare issues as well.

None of these are new issues: they have been analyzed and discussed with regard to farm and research animals for many years and more recently in relation to fishery and aquaculture. Nonetheless it will be necessary to reconsider them based on both the available and hopefully growing scientific knowledge about the emotional and cognitive capabilities of these animals, but also including considerations on the underlying reasons for the emotional reaction of many Westerners to include insects in their diet and even phenomenological analysis of the possibility to experience empathy towards insects as a basis for including insects into the moral community.

Conclusion

As demands for animal protein for food and feed increases, insect mass-rearing systems are increasingly being considered a realistic supplement to livestock production. It has been heralded as the world's largest unexploited biological resource and seemingly provides new opportunities as negative effects on e.g. regional livelihoods through climate change materialize and concern for animal welfare within livestock production is growing. 'Mini-livestock' farming is envisioned and advocated to meet demands of nutrition, food safety, feed efficiency transformation and sustainability. So far, except for a few exclusive production sites, it is only wild caught insects that are eaten by humans. This, however, is a widespread and established culinary practice in many regions of the world, but, social and cultural barriers to actually eat insects is extremely pronounced in some regions, most notably the Western world.

In this paper we have addressed animal ethical aspects of large scale insect production. Some of these relate to welfare issues: an area where there is limited research on the cognitive and emotional capacities of the many relevant insect species. Very little is known about what welfare demands or natural needs the 2,136 edible species of insects actually have. As seen recently with fish the conception that an insect is just an insect will not do: each species could have different capabilities and needs that have to be studied, before anything substantial can be said about the welfare of the individual species (Erens *et al.*, 2012). Today animal welfare legislation covers all sentient beings, and it therefore important to establish which species belong to this category, independently of whether they are vertebrates or not – even more so, if 'new' species are to be intensively housed and used as the production related welfare issues within intensive livestock production shows.

Finally, a range of ethical questions that are independent of the welfare capacities of the animal but rather concerns our responsibility and view on human-animal-relation, the role of empathy in animal ethics, the question of animal rights in relation to insects and integrity of the individual, are also essential to analyze to present a comprehensive ethical evaluation of the plans for insect mass-rearing systems. The paper thus points to a range of ethical issues we have a chance to address before beginning large utilization and farming of a new group of beings based on the experiences we have with ethical issues related to those that we have already included into such systems.

References

Belgian Federal Agency for the Safety of the Food Chain (FASFC) (2014). Food safety aspects of insects intended for human consumption. SciCom dossier 2014/04; SHC dossier N° 9160. Available at http://www.health.belgium.be.

Braithwaite, V.A., Huntingford, F. and van den Bos, R. (2013). Variation in emotion and cognition among fishes. Journal of Agricultural and Environment Ethics 26: 7-23.

Cruz, M.S. and Dossey, A. (2013). Insects for human consumption. In: Morales-Ramos, J.A., Rojas, M.G. and Shapiro-Ilan, D.I. (eds.): Mass production of beneficial organisms invertebrates and entomopathogens. Elsevier, Amsterdam, the Netherlands, pp. 617-651.

Dawkins MS (2006). A user's guide to animal welfare science. Trends in Ecology & Evolution 21: 77-82.

De Goede, D.M., Erens, J., Kapsomenou, E. and Peters, M. (2013). Large scale insect rearing an animal welfare. In: Röcklinsberg, H. and Sandin, P. (eds.) The ethics of consumption. The citizen, the market and the law. Wageningen Academic Publishers, Wageningen, the Netherlands, pp. 236-242

Elwood, R.W. (2011). Pain and suffering in invertebrates? ILAR Journal 5(22): 175-184.

Erens, J., Van Es, S., Haverkort, F., Kapsomenou, E. and Luijben, A. (2012). A bug's life: large-scale insect rearing in relation to animal welfare: project 1052. Wageningen: Wageningen University, Wageningen, the Netherlands.

Food and Agricultural Organisation of the United Nations (FAO) (2013). The contribution of insects to food security, livelihoods and the environment. Food and Agricultural Organisation of the United Nations, Rome, Italy.

Gahukar, R.T. (2011). Entomophagy and human food security. International Journal of Tropical Insect Science 31: 129-144

Goodland, R. and Anhang, J. (2009). Livestock and climate change. What if the key actors in climate change are... cows, pigs, and chickens. World Watch Institute, Washington D.C., USA.

Horvath, K., Angeletti, D., Nascetti, G. and Carere, C. (2013). Invertebrate welfare: an overlooked issue. Annali dell'Istituto Superiore di Sanità 49: 9-17.

Ilea, R.C. (2009). Intensive livestock farming: global trends, increased environmental concerns, and ethical solutions. Journal of Agricultural and Environment Ethics 22: 153-167.

Jongema, Y. (2012). List of edible insects of the world (April 4, 2012). Wageningen Wageningen University, Wageningen, the Netherlands.

Magee, B. and Elwood, R.W. (2013). Shock avoidance by discrimination learning in the shore crab (Carcinus maenas) is consistent with a key criterion for pain. The Journal of Experimental Biology 216: 353-358.

McGrew, W.C. (2014). The 'other faunivory' revisited. Insectivory in human and non-human primates and the evolution of human diet. Journal of Human Evolution 71: 4-11

Norris, K., Potts, S.G. and Mortimer, S.R. (2010). Ecosystem services and food production. Issues in environmental science and technology 30: 52-69

Okaraonye, C.C. and Ikewuchi, J.C. (2008). Rhynchophorus phoenicis (F) larva meal: nutritional value and health implications. Journal of Biological Science 8: 1221-1225.

Oonincx, D.G., Van Itterbeeck, J., Heetkamp, M.J., Van den Brand, H., Van Loon, J.J. and Van Huis, A. (2010). An exploration on greenhouse gas and ammonia production by insect species suitable for animal or human consumption. PLoS One 5, DOI: http://dx.doi.org/10.1371/journal.pone.0014445.

Packard, A.J. (1877). Half hours with insects. Estes and Lauriat, Boston, USA.

Röcklinsberg, H. (2014). Fish consumption: choices in the intersection of public concern, fish welfare, food security, human health and climate change. Journal of Agricultural and Environment Ethics DOI: http://dx.doi.org/10.1007/s10806-014-9506-y.

Srivastava, S.K., Babu, N., Pandey, H. (2009). Traditional insect bioprospecting – as human food and medicine. Indian Journal of Traditional Knowledge 8: 485-494.

Steinfeld, H., Gerber, P., Wassenaar, T., Castel, V., Rosales, M. and De Haan, C. (2006). Livestock's long shadow – environmental issues and options. Food and Agricultural Organisation of the United Nations, Rome, Italy.

Van Huis, A. (2013). Potential of insects as food and feed in assuring food security. Annual Review of Entomology 58: 563-583.

Vogel, G. (2010) For more protein – filet of cricket. Science 327: 811.

Yeates, J., Röcklinsberg, H. and Gjerris, M. (2011). Science in regulation: a critique of policymaking based solely on animal welfare science. Animal Welfare 20: 423-432.

Food visions revisited: organic and health foods

V. Beusmann[*] *and S. Stirn*
Research Centre for Biotechnology, Society and the Environment (FSP BIOGUM), University of Hamburg, Ohnhorststr. 18, 22609 Hamburg, Germany; volker.beusmann@uni-hamburg.de

Abstract

Visions about the future of food are contested and they change over time. We take a look at former visions and proposals for governance: What has changed in the meantime and why? Induced by the BSE-crisis in 2001, the German government announced a turn in agricultural policy and set the target of a 20% share of the organic sector in 10 years. As boundary trajectories we identified and focused on two major paths for the future: Genetic Engineering (GE), directed towards functional food and organic food. In two projects we analysed both routes and performed in depth empirical investigations of a similar design. Here we look at the state of the organic sector and functional food in Germany today to identify deviations from former expectations, try to explain the reasons, and ask, what can be learned for the future.

Keywords: sustainable food production and consumption, genetically modified plants, governmental policies, market processes

Introduction

Contemporary food production and consumption cannot be regarded as sustainable and, in the face of a growing population and increasing land-use conflicts, the problems will likely become more serious (Reisch *et al.*, 2013). Therefore, visions for future foods have to be in line with the criteria of a sustainable development (SD). We follow the Brundlandt-definition that 'sustainable development meets the needs of the present without compromising the ability of future generations to meet their own needs'. We regard this as the overarching vision. Fifteen years ago two different pathways have been controversially discussed in Germany, genetic engineering vs organic agriculture. We analysed both routes in projects financed by different Ministries.

Case 1: organic farming and food

Our project 'How to extend organic farming in Germany – preconditions, strategies, implications, political options' was financially supported by the Federal Ministry of Consumer Protection, Food and Agriculture (BMVEL) (Kuhnert *et al.*, 2004). We started from observing two gaps: first, the importance of widely varying perceptions and interpretations of the present situation, and second, of new means and modes of steering in a democracy, augmenting the role of the state – traditionally bound to setting rules and incentives – by communicative instruments of persuasion in the sense of convincing actors.

The research design comprised discourse analysis, empirical investigations like scenario and strategy workshops, 20 focus-group discussions with consumers and farmers (organic and conventional), 44 interviews with actors of two selected value chains and feedback by experts and stakeholders. Kuhnert introduced the concept of the Ministry as social marketer, incorporating problem perceptions and needs of relevant groups and using tools of marketing to pursue desired aims of society. Crucial questions asked included the following: What are the goals of actors? Does organic agriculture (OA) help to solve their problems? How could the state help to overcome hindrances?

The 'Bovine Spongiform Encephalopathies (BSE)' – crisis in Germany triggered the announcement of the 'Agrarwende': BSE was diagnosed as a consequence of industrialization of agriculture. The minister shifted emphasis from producers to consumers, opened discourses for critical consumer and environmental NGOs and announced the goal of raising the share of OA to 20% within 10 years. A federal program to promote OA was established, a federal label for OA introduced, and conditions to support OA in the second pillar of the European Agricultural Policy were improved.

The role of OA for sustainable development (SD) is discussed controversially among stakeholders, ranging from a small market segment to a guiding vision for the whole sector. In the scientific debate, ecological merits of OA are widely acknowledged; however, overall sustainability is disputed. In our empirical analysis hindrances for the conversion to OA could be identified in all elements of the value chains and on the side of consumers. Some are typical for niche markets, others specific for OA. These prevent actors from conversion: for farmers with successful conventional farming systems an incentive to conversion is missing, rather risks are associated with such a decision.

We suggested three general strategies with three specific ones each. A plan to extend OA cannot be based on a single strategy; rather a balanced mix of supply-, demand- and communicative strategies has to be chosen to address several issues simultaneously. This should be pursued in a participatory process of composing and fine-tuning the mix in order to establish an action plan, which has to be evaluated to facilitate learning processes in policy and self-organization of actors.

What has happened since the publication and dissemination of the study? Between 2000 and 2013 the number of organic farms increased from 12,740 to 23,271 and the acreage from 546,023 ha to 1,060,669 ha, i.e. with average annual growth rates of 4.7% and 5.2%, respectively. However, growth rates have slowed down to around 2% at the end of the period. The share of OA in total agriculture in 2013 reached 8.2% of the number of farms and 6.5% of the farmland. The share of organic products in the German food market in 2013 is only 3.7% (7.55 billion €), reflecting different consumer baskets.

Germany is by far the biggest market for OF in Europe and second behind the USA in the world (Federal Ministry of Food and Agriculture, 2014). An analysis of imports reveals, that between 2 and 95% of the products, which could have been produced in Germany, were imported in 2009/10 (Schaack *et al.*, 2011). Countries include neighbours, esp. for fresh milk, but increasingly more distant countries all over the world, e.g. for fodder. As demand outgrew internal supply, environmental advantages in the production process and value added accrue to exporting countries, and the idea of regional supply generating income and employment as well as reducing transport costs and green-house gas (GHG) emissions could only partially be realized within Germany.

The access to organic products has been improved for consumers by additional market channels, esp. conventional food chains and discount shops. The chains now control the biggest market share of organic food (OF) for a selected number of goods, often under the retailers' own brand name. The Janus face of diverging interests is obvious: farmers face pressure on producer prices. If these lower prices are partly passed on to consumers, new customers can be attracted. On the cost of production side farmers experienced increasing land and land-rental prices in areas with intense livestock production and with bioenergy production, a huge extension induced by very favourable policies. Therefore, OA-farmers find themselves in the sandwich position of being squeezed – partly out of the market – by higher land-rental prices, when contracts have to be renewed and lower product prices due to concentration among retailers and growing imports from countries with lower production costs. The bioenergy policy was recently ended, so this source of competition will not expand further.

Re-conversion of organic to conventional farming and farm cessation of organic farms have been analysed by Kuhnert *et al.* (2013). The average annual rate of farm cessation between 2003 and 2010 was 1.4%, i.e. half the rate of conventional farms, indicating competitive pressure, but at a lower level. Farmers working part-time, on smaller farms and above 65 years were especially affected, economic reasons were most important. 60% of the land went to conventional farmers. The same groups revealed higher probability for re-conversion. In addition farmers older than 50 years returned to conventional farming to reduce their load of labour. As a rule, there was no single reason for conversion, rather a bundle of internal and external factors. A coherent, reliable and lasting political support frame for OA, increased research and productivity as well as extension before conversion from conventional to organic are proposed in order to avoid wrong expectations.

Turning to consumers, Eisinger-Watzl *et al.* (2015) showed, that German buyers of OF consume more fruits and vegetables and less meat than non-buyers and thus follow much closer the recommendations of the German Society for Nutrition. Moreover, they live healthier with regard to smoking, physical activities and body weight. Regional labels recently were ranked higher than organic labels. The large food retailers offer both. An analysis for the metropolitan area of Hamburg revealed that organic stakeholders regard themselves as regional promoters and do not see necessities to invest more in regional labelling (Kuhnert *et al.*, 2011). Rapid growth led to problems of securing quality standards and limiting fraud. This led to CEC-proposals for a completely new EU-legislation for OA, integrating OA-control into the general public food control. However, OA-stakeholders and scientists advocated in favour of stepwise learning including improved self-control of the sector itself, which sets higher standards for their private brands and has a long history of experience (Sanders *et al.*, 2013). For a more comprehensive and detailed view on trends and issues of sustainable food consumption in general we refer to Reisch *et al.* (2013), for a critical review on OA from inside to Yussefi-Menzler and Zerger (2014).

Including their results we conclude that many issues originally raised in connection with OA like environmental and health aspects, animal welfare, food culture, seasonal and regional products, meat consumption and increases in vegetarian and vegan life styles have moved from the fringe more to the centre of the public debate, not always linked one to one to OA, and that OA increasingly experiences market forces like competition, concentration and globalization the more it moves from the niche to the centre of society, despite and partly due to the fact, that demand outgrows German supply.

Case 2: health foods based on transgenic rapeseeds

Our second case study refers to the subject of functional foods in the context of a healthy and sustainable diet. The aim of our project 'Socioeconomic implications of health foods based on transgenic rapeseeds', financed by the Ministry of Education and Research, was to analyse the respective expectations, hopes, and fears of researchers and developers as well as consumers with regard to the development of genetically engineered (gm) functional foods (Boysen and Beusmann, 2004).

Functional foods are defined differently, but in general, these foods exhibit an additional health benefit besides their basic nutritional functions (Doyon and Labrecque, 2008). One class of ingredients extensively studied are the so-called polyunsaturated fatty acids (PUFAs), which are essential to human development and health, but cannot be produced by mammals. Within this group, omega-3 and omega-6 fatty acids are distinguished according to the respective position of the double bonds. Especially, the intake of omega-3 fatty acids is thought to be too low in western societies. This results in an unbalanced ratio of omega-6 to omega-3 fatty acids which is thought to be one reason for the increase in lifestyle diseases like cardiovascular diseases or cancer. In addition, omega 3 fatty acids are also essential for the early development of the human eye and brain (Ruxton *et al.*, 2004).

Since the primary source for the long chain omega-3 fatty acids are marine algae and phytoplankton, which are then consumed by fish, several national nutrition societies recommend the regular consumption of fatty fish at least twice a week. Due to the overfishing of certain fish species, the sustainability of this recommendation was questioned. Concurrently, alternative sources of omega-3 fatty acids were sought. One option was to enable the production of these fatty acids within oil plants with the help of genetic engineering (Heinz, 2006; Riuz-Lopez *et al.*, 2014). In addition to the presumably more sustainable production the developers hoped for a greater acceptance of genetically modified foods when health benefits were offered to the consumers.

Within our subproject we performed semi-structured interviews with leading scientists involved in the project as well as focus groups with consumers. The outcome of the expert interviews were, not surprisingly, in line with the project's approach to allow for a sustainable production of large quantities of omega-3 fatty acids in of gm plants. The second most raised argument relates to the expectation that the low acceptance of gm foods in Europe could be changed by offering products with direct health benefits to consumers. This was contrasted by the results of the focus groups with consumers: in guided questionnaires the WTP was compared between a conventional cholesterol-lowering margarine ('becelpro.active') and the hypothetical margarine made from the gm omega-3 plant oil. As we expected, the share of those willing to buy decreased substantially. In contrast to the expectations of the developers, the key message is: despite direct benefits to the consumer the share of rejecters of gene technology remains relatively high.

Looking back from today not much has changed at the first glance: there is still no gm food product with direct health benefits available in Europe. The general acceptance of gm foods is even declining in Europe (Gaskell *et al.*, 2011). Nevertheless, Europe is not free of food products which have been produced with the help of genetic engineering: enzymes, vitamins, and food additives produced in gm microorganisms are used in or added to processed foods and gm plants are used as animal feeding stuffs. Most consumers are unaware of this fact since these food products will do not have to be labelled.

In the USA, the situation is quite different: several gm plants with altered product quality have been approved for marketing, including high oleic soybeans (ISAAA, 2014). Soybeans with high concentrations of the omega-3 fatty acid stearidonic acid (SDA) have been developed and are near approval (Monsanto, 2014). Monsanto already concluded a contract with DSM, a global player in food and dietary supplements, for commercializing the gm omega-3 oil. They have also applied for approval for import and processing of the gm soybeans to the EU (GMO Compass, 2014).

This is in line with the growing market for functional food in Europe: since 2006, functional foods might be advertised as exhibiting a health benefit, when the respective health claim has been approved by the EFSA. One of the few organic substances which gained approvals were the omega-3 fatty acids: health claims related to cardio vascular benefits as well as to early eye and brain development in infants have been approved and are widely used today (EU Register of Nutrition and Health Claims for approved DHA/EPA claims: http://ec.europa.eu/nuhclaims). The omega-3 market amounted to 5.57 billion € in 2010 (Daniells, 2011) and is rapidly increasing due to the increasing popularity of omega-3 enhanced functional foods and supplements.

Most of the omega-3 for human consumption comes from fish. In a recent report, British researchers raised the question whether declining wild fish supplies will be able to meet healthy eating recommendations in the future. They showed that, globally, fish supplies have been falling by nearly a third during the last 40 years. Only rapid growth in fish farming has shielded consumers from the consequences of overfishing and human population increase. Half of our seafood now comes from farms. They asked for a 'Blue Revolution' in fish farming to sustainable production methods (Thurstan and Roberts, 2014). This seems to be necessary not only regarding possible pollution of surrounding ecosystems but also regarding sustainable fish feed. Due to increasing pressure on fish oil supplies farmed fish are being fed more and more vegetable oil. This leads to lower omega-3 levels in farmed salmon (Naylor *et al.*, 2009). Alternatives to conventional plant oils are necessary: first trials are performed with e.g. seaweed enrichment in fish feed (Wilke *et al.*, 2014) but also the oil from gm plants is taken into account.

Then, we have come full circle: the genetically modified functional food will not be consumed directly but via animal products which have been raised on gm feedstuff (Thompson, 2014): field trials with gm plants will be performed in Great Britain and hopes are steered to help fish farming become more sustainable: 'If everything goes smoothly, the plants could produce omega-3 fatty acids for mainstream fish farm use in the next 10 years'. However, this would increase the competition for land used for food, feed, fibre and fuel.

Conclusions

The case studies chosen stand for different visions and concepts: OA is directed at the 'health' of a production system, allowing abstaining from using certain inputs like mineral fertilizer and pesticides. Functional foods are directed at improving health by adding specific ingredients. These are extremes in a range of options to improve health. Both routes can be justified and they offer specific potentials. They are used as well to raise public awareness and funds for research and development. The development of both paths revealed substantial learning processes with expected and unintended effects requiring continuous readjustments. For OA expanding from the niche meant to be more exposed to mechanisms of the conventional economic mainstream. In the case of functional food high in omega-3 fatty acids the approval of respective health claims might shift overuse of marine resources to higher pressure on land resources. Therefore, regular reassessments are necessary for learning from unintended effects, which cannot completely be avoided by ex-ante evaluations of their contributions to the overarching vision of sustainability. However, focussing on such boundary trajectories attracts attention away from the diversity of other options in between. Orientation on best cases and their dissemination as well as changes in behaviour like less meat consumption and reducing waste are equally important steps towards sustainability. Requested are changes in awareness and behaviour, in rules of the game and in technology in a balanced mix.

References

Boysen, M. and Beusmann, V. (2004). Gesunde Lebensmittel aus transgener Rapssaat in der Sicht von Entwicklern und Verbrauchern: BMBF-Leitprojekt: Ernährung – moderne Verfahren zur Lebens-mittelerzeugung'. Napus 2000 – Teilprojekt 6: 'Sozioökonomische Implikationen'. Universität Hannover, Hannover, Germany.

Daniells, S. (2011). The varied sources of omega-3. Nutraingredients special edition: omega-3. Available at: http://tinyurl.com/nhnbkae.

Doyon, M. and Labrecque, J.A. (2008). Functional foods: a conceptual definition. British Food Journal 110(11): 1133-1149.

Eisinger-Watzl, M., Wittig, F., Heuer, T. and Hoffmann, I. (2015). Customers purchasing oganic food – do they live healthier? Results of the German National Nutrition Survey II. European Journal of Nutrition Food Safety 5(1): 59-71.

Federal Ministry of Food and Agriculture (2014). Ökologischer Landbau in Deutschland. Available at: http://tinyurl.com/ozga9kl.

Gaskell, G., Allansdottir, A., Allum, N. (2011). The 2010 Eurobarometer on the life sciences. Nature Biotechnology 29(2): 113-114.

GMO Compass (2014). GMO database mon87769. Available at: http://tinyurl.com/oaozgm4.

Heinz, E. (2006). First breakthroughs in sustainable production of 'oceanic fatty acids'. European Journal of Lipid Science and Technology 108(1): editorial.

International Service for the Acquisition of Agri-biotech Applications (ISAAA) (2014) GM Approval database. Available at: http://tinyurl.com/oqemu5t.

Kuhnert, H., Feindt, P.H. and Beusmann, V. (2004). Ausweitung des ökologischen Landbaus in Deutschland – Voraussetzungen, Strategien, Implikationen, politische Optionen. Münster-Hiltrup: Landwirtschaftsverlag. English summary available at: http://tinyurl.com/p2head5.

Kuhnert, H., Behrens, G., Hamm, U., Müller H., Nieberg, H., Sanders, J. and Strohm, R. (2013). Dauerhafte Ausweitung des ökologischen Landbaus in Deutschland: Analyse der Ausstiege von Betrieben und Entwicklung eines Konzepts zur nachhaltigen Vermeidung. Braunschweig, Hamburg and Witzenhausen, Germany. Available at: http://tinyurl.com/nkxk7mq.

Kuhnert, H., Behrens G. and Beusmann, V. (2011): Strukturdaten Hamburger Ökomarkt. BIOGUM research paper nr. 24 (long version) and nr. 25 (executive summary). University of Hamburg, Hamburg, Germany.

Monsanto (2014). Stearidonic acid (SDA) omega-3 soybeans. Currently in phase IV of Monsanto's R&D pipeline. Available at: http://tinyurl.com/p2zwrfv.

Naylor, R.L., Ronald W. Hardy, R.W., Bureauc, D.P., Chiua, A., Elliott, M., Farrelle, A.P., Forstere, I.,Gatlin, D.M., Goldburgh, R.J., Hua, K. and Nichols, P.D. (2009). Feeding aquaculture in an era of finite resources. PNAS 106(36): 15103-15110.

Reisch. L., Eberle, U. and Lorek, S. (2013): Sustainable food consumption: an overview of contemporary issues and policies. Sustainability: Science, Practice, & Policy 9(2): 7-25. Available at: http://sspp.proquest.com.

Ruiz-Lopez, N., Haslam, R.P., Napier, J.A. and Sayanova, O. (2014). Successful high-level accumulation of fish oil omega-3 long-chain polyunsaturated fatty acids in a transgenic oilseed crop. The Plant Journal 77: 198-208.

Ruxton, C.H.S., Reed, S.C., Simpson, M.J.A. and Millington, K.J. (2004). The health benefits of omega-3 polyunsaturated fatty acids: a review of the evidence. Journal of Human Nutrition and Dietetics 17(5): 449-459.

Sanders, J., Zander, K., Padel, S., Vieweger, A., Stolze, M., Huber, B., Schmid, O., Nocentini, L., Devot, A., Clement, T., Brèche, O., Polakova J. and Keensleyside, C. (2013): Evaluation of the EU legislation on organic farming. Available at: http://tinyurl.com/o658ct2.

Schaack, D., Rampold, C., Willer, H., Rippin, M. and Von Koerber, H. (2011). Analyse der Entwicklung des ausländischen Angebots bei Bioprodukten mit Relevanz für den deutschen Biomarkt. Agrarmarkt Informations-Gesellschaft mbH, Bonn, Germany. Available at: http://tinyurl.com/o9l6z8x.

Thompson, H. (2014). Fish oil could (one day) come from plants. Smithsonian.com. Available at: http://tinyurl.com/nednbfj.

Thurstan, R.H. and Roberts, C.M. (2014). The past and future of fish consumption: can supplies meet healthy eating recommendations? Marine Pollution Bulletin 89: 5-11.

Wilke, T., Faulkner, S., Murphy, L., Kealy, L., Kraan, S. and Brouns, F. (2014) Seaweed enrichment of feed supplied to farm raisedAtlantic salmon (Salmosalar) is associated with higher totalfatty acid and LC n-3 PUFA concentrations in fish flesh. European Journal of Lipid Science and Technology. Available at: http://dx.doi.org/10.1002/ejlt.201400166.

Yussefi-Menzler, M. and Zerger, U. (2014). Ökologischer Landbau – Weichen für die Zukunft stellen. Der kritische Agrarbericht. Agrarbündnis e.V, Munchen, Germany.

Sustainability and food in Margaret Atwoods 'MaddAddam' trilogy

M. Rockoff
Junior Research Group 'Ethics of Science in the Research for Sustainable Development', International Centre for Ethics in the Sciences and Humanities, University of Tuebingen, Wilhelmstraße 19, 72070 Tübingen, Germany; marcus.rockoff@izew.uni-tuebingen.de

Abstract

In Margaret Atwood's science fiction trilogy 'MaddAddam', we read about food innovations such as ChickieNobs that are genetically modified and extremely fast growing chicken parts. They are not only presented as more efficient than conventional meat products but also as superior with regard to problematic side-effects caused by intensive husbandry, e.g. harming animal welfare or increasing greenhouse gas emissions. ChickieNobs are part of a widespread belief that research and innovation in biotechnology are the most promising way to figure out humankind's challenges such as overpopulation, environmental crisis or climate change. As counter-image, the God's Gardeners are presented, an environmentalist sect, which tries to minimize their ecological footprint. They meet their demand for food and other resources through self-sufficiency as they are living in small communities and imposing on themselves a very rigid code of conduct: „no consumption of animal protein except in emergencies, no corporate products unless in recycled form, strict limitations on the use of water'. This very different approach imposes partly very strict limitations on the member's personal freedom. Focussing on the topics of food production and food consumption, the paper wants to follow the discussion about and aesthetical configuration of future and sustainable food depicted in Atwood's novels. The discussion turns on the question whether sustainable food should be achieved by means of an economically driven and science-based approach or should it rest on a dramatic change of individual behaviour as represented by the Gardeners.

Keywords: human nature, genetic engineering, consumerism, fiction, humanities

Introduction

Margaret Atwood's popular science fiction trilogy 'MaddAddam' – consisting of three separate volumes 'Oryx and Crake' (2003)', 'The Year of the Flood' (2009) and 'MaddAddam' (2013) – is mainly known for its imaginative portrait of a dystopian and near-future society. It depicts a world of severe environmental damage. Climate change, ozone depletion, decline of natural resources, food scarcity and overpopulation had already rendered Earth an almost inhabitable place when a lethal human-made virus caused a massive die-off of the human race.

The character Crake who invented the lethal virus illuminates the underpinned diagnosis, why the earth became such a hard place to live in. The main problem – in his view – is that humans, despite diminishing resources, have not reduced their population growth like other species: 'the less we eat, the more we fuck' (Atwood, 2003: 120). Consumerism led humanity to exploit and pollute nature in order to fulfil individual demands and needs. In doing so, they destroyed their own livelihood and set themselves onto an inevitable path of human extinction. Crake aims to accelerate this predicted development with his virus.

His diagnosis is echoed by the God's Gardeners, an environmentalist sect more in focus in the second volume. They also believe that humanity is doomed because of its inability to live in harmony with nature. They share Crake's conviction that a generic error in human nature, which in the end lead to materialism and consumerism, is responsible for the upcoming disaster.

'Food' seems to be the most important marker for the on-going problems in this fictional world. It serves as an indicator for its current status. The reader is made to recognise the current predicaments of this pre-apocalyptic era not by various depictions of an increasingly damaged nature or by the stories of a character or narrator. Instead, the reader merely has to deduce the problems of this fictional world by recognizing the production, consumption and appreciation of food. In the three novels, many interesting 'food innovations' are introduced: 'soy-sausage dogs' (soy-based hotdogs); 'SoyOBoyburgers' (soy-based hamburgers); the genetically modified 'Happicuppa' coffee beans; 'SoYummie Icecream' in flavours like chocolate soy and roasted-danelion green-tea soy; 'SecretBurgers' with meat patties of which „no one knew what sort of animal protein was actually in them' (Atwood, 2009: 40); 'ChickieNobs Bucket O'Nubbins' – genetically modified chicken meat and also the so called 'Pignoons'. The latter are pigs originally invented and used to cultivate human tissue for organ transplantation. Later, however, when meat scarcity became widespread, they were also used for food purposes. All these food products indicate responses to a generic food scarcity from which the fictional society suffers. The lack of conventionally produced food led to an optimisation of livestock with the help of genetic engineering. The substitution of meat and dietary products by soy became mandatory. The reader observes a society already forced to find technological solutions for an ever growing shortage of animal protein.

Atwood caricatures the search for technological remedies with, for example, her depiction of the so-called SecretBurger fast food franchise. Several SecretBurger branches are involved in different criminal activities such as running a human corpse disposal business by 'harvesting organs for transplant, then running the gutted carcasses through the SecretBurgers grinders' (*ibid.*) and serving the resulting meat patties to their customers. The 'so few supply-side costs' (*ibid.*) enabled SecretBurgers to provide 'real' meat in an easy to consume form for an unbeatable price, and thus contributed to the success of its business model. Another dimension of this satirical cannibalism can be seen in the critism of standardised food supply by industry. Today's industry provides the possibility to consume clean and often controlled food products at an extremely low price. But these often-processed products enforced the problem that consumers do not know their food. This trend decouples not only our imagination of food and meat production from its socio-ecological context. The provision of easy to access and cheap nutrition also contributes to a widespread ignorance about social effects and possible environmental damage of these practices. We forget to valuate food (Meisch, 2014).

Atwood's thematic choice of food and food scarcity as an indicator for on-going environmental problems, reflects the causal relationship between a growing world population and an increased global production of food, above all: meat. Already in today's non-fictional world, a meat-producing industry, which raises, keeps, and slaughters nonhuman animals, is one of the top drivers of resource depletion (Bossert and Boscardin, in press). Atwood's trilogy sketches in only a few passages an already devastated landscape that may become reality in the non-fictional world within the next years. The reader can register how 'a change in ocean temperature and salinity caused by the great downpour of fresh waters did harm to some Species' (Atwood, 2009: 108) or how 'the weather had become so strange and could no longer be predicted – too much rain or not enough, too much wind, too much heat – and the crops were suffering' (Atwood, 2003: 118).

Without being stated explicitly, it becomes obvious what Atwood's novel tries to highlight: humanity's food and especially meat consumption and production is causing climate change and environmental pollution. Moreover, climate change will have a negative impact on agricultural productivity that food security is high at stake. I argue that we have to understand food consumption and food production in Atwood's trilogy as a proxy for humanity's dealing with nature in general. It stands in line with other pressure we put on Earth systems such as surging consumption of fossil fuels and rare metals, deforestation or a perpetuation of the automotive industry (Assadourian, 2010: 4).

Dealing with the situation

In summary, it can be noted that production, consumption and appreciation of meat in the novel is used to criticise humanity's unsustainable behaviour. At the same time consumerism is presented as cause of environmental pollution and the pressure on the ecosystems. The fictional society already experiences the adverse consequences of its own behaviour and tries to find technological solutions to maintain its standard of living and protect the remaining functionality of the ecosystems. In this context, the mentioned 'food innovations' are relevant.

This critique is even put forward by outlining an anthropological hypothesis within the novels Protagonists like Crake or the Gardeners are convinced that the ecological crisis is not preventable because there are some serious 'flaws in humanity's biological make-up' (Bergthaller, 2010: 729). They believe that humans are unsustainable by nature because 'to stay human is to break a limitation' (Atwood, 2003: 269). Today's scholars like the bioecologist William Rees (2010: 4) support that conviction. He also assume that human biological traits lead to an unstoppable expansion into all accessible habitats and to an unstoppable consumption of all available resources until a negative feedback will finally stop us. In line with this Crake and the Gardeners seem to argue for a bio-evolutionary precondition, which influences our behaviour and leads to a constant desire for growth. But, in contrast to other species, humanity's technological capacities make it possible that we do not 'limit reproduction in the face of dwindling resources' (Atwood 2003: 120). Instead humans started for colonizing more and more ecosystems until they occupied the entire planet. Now the fictional society in the novels seems to struggle with the negative feedbacks of the earth systems caused by bio-evolutionary and later sociocultural patterns of their behaviour. With this anthropological hypothesis the novels contrast many scholars in the sustainability discourses, which agree that dissipative consumerism is not a viable path to deal with the earth's ecosystems but which are still convinced that the current unsustainable behaviour is only a cultural paradigm which has 'to be re-engineered in cultures of sustainability' (Assadourian, 2010: 113).

Technological fix for nonhuman nature

By accepting the anthropological hypothesis, we find three approaches to achieve sustainability presented and criticised within the novels. The first can be understood as a search for a technological fix for nonhuman nature. In order to do so researcher invented the so-called ChickieNob. It is 'a large bulblike object that seemed to be covered with stippled whitish-yellow skin. Out of it came twenty thick fleshy tubes, and at the end of each tube another bulb was growing' (Atwood, 2003: 202). ChickieNobs are genetically modified and extremely fast growing chicken parts with a 'mouth opening at the top (to...) dump the nutrients in there. No eyes or beak or anything, they don't need those' (*ibid.*).

Taking this description into account they cannot be seen as proper non-human animals. They are rather just meat with rudimentary brain functions, needed for digestion, assimilation, and growth. This bio-technologically supported meat production made it possible to provide sufficiently cheap broilers to meet the requirements of the growing population, to reduce the environmental harming emissions, and to silence 'the animal-welfare freaks (...) because (...these) thing(s) feel no pain' (*ibid.*). ChickieNobs can be interpreted as an artifical life form that consists only of 'useable' and edible body parts. This former nonhuman species is genetically altered to meet current eating habits and may be seen as 'an outgrowth of modern supermarket meat presentation, in which all traces of the living creature have been minimized to increase the separation between humans and their once-alive food' (Galbreath, 2010: 3). Further, its artificial origin delineates not only the border between natural and unnatural but also the boundaries between nonhuman animal and vegetable. No wonder that, despite the intended benefits, some consumers perceive ChickieNobs as ambiguous. The first reaction of Jimmy, one of the main protagonists in the books, when he saw one of these creations in the laboratory was pure disgust: 'This is horrible', said Jimmy. 'The thing was a nightmare. It was like an animal-protein tuber' (Atwood, 2003: 202). In the novels, ChickieNobs are marketed as greasy and fried chicken nuggets in fast food chains with a 'bland tofu like consistency and their inoffensive flavour', which even let a mild capon taste like game (*ibid.* 292). It becomes obvious that ChickieNobs – in line with the SecretBurger motif – are a critique on consumer habits. Jimmy forgets his initial repulsion in favour of cheap fast food. Therefore, the ChickieNobs 'can serve as a metaphor for a humanity (...) that only takes in nutrients and expels waste in a blind, sensory-deprived fashion' (Galbreath, 2010: 4).

In addition to this, I assume that Atwood with the ChickieNobs also criticises current developments in the poultry industry. ChickieNobs are not only presented as more efficient than 'the most (...) low-light, high-density chicken farming operation so far devised' (Atwood, 2003: 203) but also as superior regarding the problematic side-effects caused in the farming conditions like harming the animal's welfare and increasing greenhouse gas emissions. Today's poultry industry argues that breeding supported by molecular biology and genetic engineering represents a promising approach that enables to breed not only more profitable but also more environmental friendly (Twine, 2010: 115-160). In doing so. they understand sustainable development just as a technical or mangerial challenge. With such a approach, the search for solutions towards a more sustainable world mainly trys to improve the efficency of the business model and implicitly bypass more fundamental approaches that aim to alter consumption pattern or reduce the overall animal consumption.

Technological fix for human nature

But these scientific efforts to achieve sustainable food and to stop environmental pollution by means of an economically driven and science-based strategy, including its underpinning belief in progress, are viewed critically within the novels. Crake's scientific project called Paradice is here in focus. It can be seen as the second approach mentioned in the novels. His scientific answer, on the one hand, is in line with today's widespread belief that research and innovation, especially in biotechnology, is the most promising way to address the current challenges of humankind. On the other hand, his disappointment about humanity's inability to live environmentally-friendly as well as its on-going egoism and consumerism convinced him to find the ultimate fix. So he decided to invent a special virus that would kill nearly all humans on earth – something that, he holds, will happen sooner or later anyway. Specifically, he creates a new nearly human, transgenic species called the Crakers. With the help of genetic engineering, this post-human species got a new biological make-up that enhanced its chances to survive in the polluted environment. For instance, the carcinogenic effects of the sun do not affect the Crakers. But most importantly, their digestion and social behaviour is designed to be extremely environmental-friendly. They do not farm or hunt other nonhuman animals and just 'graze on green grass, leaves, and the shoots of plants' (Ingersoll, 2009: 116). Furthermore, they live peacefully in small social groups and have the opportunity to recycle their own excrements, and keep away enemies with the help of their special urine.

The invention of Crakes' new transgenic species follows the same logic as the food innovations. Science and technology are considered to solve environmental problems. In this line of argument, the Crakers seem to be the end of the line. Why, so the argument goes, fight against pollution, food scarcity and resource depletion by only fixing food production and food consumption? The better strategy seems to be to fix all this by building up humanity from scratch. Now Crake's slogan 'to stay human is to break a limitation' (Atwood, 2003: 354) gains another dimension. At first glance the statement serves as a description to reveal the problematic and true nature of humanity. But taking the Craker project into account, it seems that the suggestion can also be understood as a description of his scientific challenge:

modifying the genetic makeup of humanity and mixing it with characteristics of nonhuman animals is not only a transgression of current natural boundaries. It is also a successful scientific experiment that overcomes human nature by reinstalling a lost sense for limitations. Crake's intervention clearly aims to bring the ability to self-restrain and to live in harmony with nature back to human nature again. The Crakers question our anthropocentricity because their post-human nature figuratively represents a return to the wild, which gets them closer to animals and nature than to human beings (Rockoff, 2014: 264-266). But the question still remains, whether or not these highly sustainable Crakers are still humans?

Sustainability by building up a meaningful context

There is another story told in Atwoods trilogy, too. Especially in 'The Year of the Flood', the daily life as well as the beliefs of Gardeners become central. They represent the third approach to achieve sustainability. Their ideology is presented as a mixture of Deep Ecology and Christian dogmatism. As already stated, they share similar convictions like Crake regarding the origin of humanity's unsustainable behaviour but add to this a religious dimension. According to their leader, the 'Fall of Men' to non-sustainability began at the point when men separated from primates: 'then they fell from vegetarianism into meat-eating. Then they fell from instinct into reason, and thus into technology; from simple signals into complex grammar, and thus into humanity; from firelessness into fire, and thence into weaponry; and from seasonal mating into an incessant sexual twitching (...)' (Atwood, 2009: 224). This rush through the (cultural) evolution of humanity is mixed up with a re-interpretation of central topics in Christian theology. In this sense, the Fall of Men in the Garden of Eden is the fall from the original behaviour of our evolutionary ancestors, which were more or less able to live in harmony with the surrounding environment. Therefore, humanity has disobeyed God's 'commandment to live the Animal life in all simplicity' (*ibid.*: 63). As Atwood writes: 'In our efforts to rise above ourselves we have indeed fallen far, and are falling farther still (...) Ours is a fall into greed (...). God's commandment to 'replenish the Earth' did not mean we should fill it to overflowing with ourselves, thus wiping out everything else' (*ibid.*).

In line with their central beliefs, the Gardeners try to minimise their ecological footprint. They meet their demand of food and other resources by self-sufficiency, living in small communities, and by imposing a very rigid code of conduct upon themselves: „no consumption of animal protein except in emergencies, no corporate products unless in recycled form, strict limitations on the use of water' (Bergthaller, 2010: 739). This kind of food production and appreciation differs extremely from the already discussed industrial food production with all its substitutes, in particular from the ChickieNobs. Especially the gardening tries to establish a connection between food consumption and food production, which is already lost for most parts of the fictional society. It tries to recall the fact that food does not only stem from factories, but from nature; that it is sometimes difficult to generate. Modern food industry with its cheap and easily accessible products tends to veil this connection. Regarding contemporary discussions about sustainable consumption, with all its claims of organic food, local food or the understanding of a consumption 'that restores natural capital and the resilience of ecosystem processes' (Maniates, 2014: 204), the lifestyle of the Gardeners goes even beyond these proclaimed duties. For them, abstinence and simplicity are central. These two topics are unpopular in today's sustainability discourses and are therefore more or less eschewed. But according to the Gardener's belief, these values generate the only appropriate behaviour pattern that might be able to tame human greed.

In sharing Crake's conviction of a generic error in human nature, which leads to unsustainable behaviour, the Gardeners follow a different approach to achieve sustainability. While Crake follows a science-based strategy that finally leads to the abolition of humanity and the advent of the Crakers, the Gardeners rely on a dramatic change of individual behaviour, which implies a radical change of personal (food) consumption. It needs to be stressed again, however, that this reanimation of human self-restraint is not powered by the idea that 'there is a stable and knowable natural order within human beings have their proper place' (Bergthaller, 2010: 731) and that cultural developments made humans unsustainable in the first place. The ideology of the Gardeners outlines that these behavioural patterns 'that lead to environmental destruction are not in any way 'unnatural'' (Bergthaller, 2010: 741) and were already installed in the brains of our ancestors (Atwood, 2003: 305). To solve the sustainability challenge, the Gardeners have to create complex religious practices to tame humanity's unsustainable nature. Every encountering with animals, food and natural resources is integrated in symbolic orders and ritualised activities to provide meaning to this transformation of behaviour towards abstinence and simplicity. Since all technical possibilities to remake humankind's biological nature to make human beings more sustainable would eventually abandon their very nature, a cultural change is seen as necessary to serve this purpose. But by creating a meaningful context – like a religion or an rigid ideology – the discussed flaws of human nature should be taken into account.

Conclusion

Margaret Atwood's works of fiction must be seen as a straightforward contribution of the arts and humanities to the challenges of making the world a more sustainable place. In addressing possible disastrous consequences of today's food production and consumption patterns, Atwood, firstly, raises the readers' awareness of the environmental impact of consumer choices. Furthermore, she connects this topic with a complex reflection on the key features of human nature. Atwood outlines the conviction that the biological traits of humanity decrease the possibilities of becoming a more sustainable species. In her novels she investigates different approaches. Modifying nonhuman nature by means of technology to achieve sustainability, however, is viewed critically. As the novels suggest, it is not enough to invent 'sustainable' future food only. To take actions against the on-going consequences of current depletion, humanity needs to change itself. One possibility seems to be a biotechnological fix for human nature towards more sustainable bodies and pattern of behaviours. Although in the novel this experiment was successful, Atwood raises questions whether this species can be still seen as fully human. The other possibility is to accept humanity's unsustainable nature and try to tame human greed by creating a meaningful cultural context that creates incentives for an abstinent and simple lifestyle without wasteful consumerism. Yet, such a strategy of a cultural reform is not promising either, because 'no matter how much the Gardeners deprived themselves of proper food and clothing and even proper showers (...) and felt more high and mighty and virtuous than everyone else, it wouldn't really change anything' (Atwood, 2008: 136) – because this strict belief in abstinence and simplicity never became mainstream.

Acknowledgements

I would like to thank Robert Ranisch, Simon Meisch and Johannes Lundershausen for providing me valuable comments and substantial feedback for this manuscript.

References

Assadourian, E. (2010). The rise and fall of consumer cultures. In: Starke, L., Mastny, L. (eds.) State of the World 2010, W.W. Norton, New York, USA, pp. 3-22.

Atwood, M. (2003). Oryx and Crake, Doubleday, New York, USA.

Atwood, M. (2009). The Year of the Flood, Bloomsbury, Hachette, UK.

Atwood, M. (2013). MaddAddam, Bloomsbury, Hachette, UK.

Bergthaller, H. (2010). Housebreaking the human animal. humanism and the problem of sustainability in Margaret Atwood's Oryx and crake and The year of the flood. English Studies 91(7): 728-743.

Bossert, L. and Boscardin, L. (in press). Sustainable development and nonhuman animals: why anthropocentric concepts of sustainability are outdated and need to be extended. In: Meisch, S., Bossert, L., Lundershausen, J. and Rockoff, M. (eds.) Ethics of science in the research for sustainable development. Nomos, Baden Baden Germany.

Galbreath, M. (2010). A consuming read. The ethics of food in Margaret Atwood's Oryx and crake. Florida Gulf Coast University's 2nd International Humanities and Sustainability Conference. Fort Myers. 7-9 October 2010. Available at: http://tinyurl.com/qfmp6ox.

Ingersoll, E.G. (2009). Survival in Margaret Atwood's novel Oryx and crake. In: Bloom, H. (ed.) Bloom's modern critical views: Margaret Atwood (new edition). Bloom's Literary Criticism, New York, USA, pp. 111-126.

Maniates, M. (2014). Sustainable consumption – three paradoxes. GAIA – Ecological Perspectives for Science and Society 23(3): 201-208.

Meisch, S. (2014). Nachhaltige Entwicklung, Gesundheit und Fleischkonsum. In: Voget-Kleschin, L., Bossert, L., Ott, K. (eds.) Nachhaltige Lebensstile. Welchen Beitrag kann ein bewusster Fleischkonsum zu mehr Naturschutz, Klimaschutz und Gesundheit leisten? Metropolis, Marburg, Germany, pp. 112-132.

Rees, W. (2010). What's blocking sustainability? Human nature, cognition, and denial. Sustainability: Science, Practice & Policy 6(2): 13-25.

Rockoff, M. (2014). Literature. In: Ranisch, R. and Sorgner, S.L. (eds.) Post- and transhumanism. An introduction. Peter Lang, New York, USA, pp. 251-270.

Twine, R. (2010): Animals as biotechnology. Ethics, sustainability and critical animal studies. Earthscan, London, UK.

Bioeconomy and the future of food – ethical questions

B. Kröber and T. Potthast*
International Centre for Ethics in the Sciences and Humanities (IZEW), University of Tübingen, Wilhelmstraße 19, 72074 Tübingen, Germany; birgit.kroeber@uni-tuebingen.de

Abstract

There is farming for food, biofuels, biomaterials, chemicals and biorefinery. All this, and more, is being lumped together under the new term 'bioeconomy' or 'bio-based economy'. In 2012, the European Commission launched a new strategy 'Innovating for sustainable growth: a bioeconomy for Europe'. It involves a radical change in biomass production, consumption and recycling, also opening up new and diversified markets in food and bio-based products. Bioeconomy takes place in the name of sustainable development. However, there are great discrepancies between adequate conditions of a bio-based economy and the current situation. On the one hand, we are dreaming of a 100% sustainable world in which fossil resources are not needed anymore, while everybody values nature and its services, resulting in a very efficient commerce with energy and natural resources. On the other hand, there are conflicting economic, political and societal interests. Concerning biomass production, there are many trade-offs and problems: (1) monocropping; (2) the competition of land use; (3) changing cultivation methods; (4) environmental problems; and (5) societal acceptance. The majority of the bioeconomic literature implies that genetic modifications are safe, healthy and socially acceptable, despite the problems and the already known negative consequences. Another problem is that the demand for biomass, e.g. for fuels in Germany, exceeds possible domestic production by far. Most of the globally traded biomass is produced under questionable conditions with respect to ethical and societal aspects. Hence the approach of bioeconomy is in need of a detailed critical ethical analysis. Our conceptual and ethical analysis of bioeconomy strategies will address the following issues: the meaning of sustainability in bioeconomy is opaque and the problems are rarely addressed. Sustainability in the sense of efficiency, sufficiency and resilience is not really dealt with. Bioeconomy does not yet address issues of insecurity, especially regarding long-time consequences on food production, food sovereignty, environmental conservation, human health, and loss of biodiversity. This is due, among others, to the employment of a business-as-usual model of high-input, high-tech agriculture and to not reflecting on recent ethical debates on the values of biodiversity beyond monetary terms. Re-thinking a bioeconomy in these respects will remain a major theme with regard to a sustainable future of food.

Keywords: sustainability, land-use conflicts, values of biodiversity

Introduction: global challenges and 'bioeconomy'

The world is facing grand challenges, and we shall initially sketch this well-known 'big picture' narrative (cf. Rockström *et al.*, 2009). There is an ever-growing increase of demands for food, energy, and water, among others. To date, more than seven billion people live on our planet. By 2050 more than nine billion people are estimated – and they will consume food and water like 12 billion people as compared with current per capita rates (Alexandratos und Bruinsma, 2012). It is of utmost importance to understand that not only – and with regard to global issues maybe even not mainly – the increasing number of people makes up the high rate of increase of future demands, but – also – changing lifestyles and hence the size of the ecological footprint per capita. Not only food, energy, and water demand and supply are at stake. People will consume textiles, mobility devices, materials for housing and a lot of other industrial products which all stress ecological systems especially when supply is based on biomass. The anthropogenic acceleration of climate change adds to the challenges of the 21st century, since the conditions of the production of food, feed and other biomass will be much more complicated. The pressure on the environment will be fostering the decrease of biodiversity and soil degradation, losing invaluable resources for food and human health. The fossil resources are at their limits and will shrink in usability. Hence, new methods are developed to produce things on the basis of renewable materials. This is where 'bioeconomy' comes in: 'Bioeconomy is the knowledge-based production and use of biological resources to provide products, processes and services in all economic sectors within the frame of a sustainable economic system' (Bioökonomierat, 2014a). Although this broad understanding reaches far beyond agriculture, aquaculture and forestry, exactly these domains are the most important so far. They form the starting point as they have already proved to be capable of producing goods from renewable sources, thus supporting the shift to a post-fossil economy. Following the definition, 'bioeconomic' is not the same as organic, is not natural, is not per se (un)sustainable, and it is not only biogas and bioenergy. The term 'bioeconomy' or 'bio-based economy' is multi-faceted and sometimes misleading. It is important to know what this term includes, what it does to the food sector, and what it means for the overall mission of sustainability.

In this paper, we will give an overview of what bioeconomy stands for, it's goals, and the problems or trade-offs to be anticipated from the perspective of ethics. We will analyse different strategy papers and the existing critical discussion. In this connection, some interesting questions are being recognised, which are mainly not directly addressed within the policy documents. We point to the history of bioeconomy and its global challenges and show the desiderate for linking up with ethical issues of the precautionary principle, the principle of responsibility, and inter-generational justice.

Goals, strategies and the bioeconomy claim

Bioeconomy or bio-based economy was initiated around 1998 by, among others, Enriquez-Cabot and Martinez addressing the commercial use of biotechnological and genomic knowledge for industrial purposes (Enriquez, 1998). In the following years it developed, aiming at new forms of cooperation between different industrial sectors. Initially, industry realised that fossil resources would run out in the near future, at least as an economically solid basis, and subsequently, there is the need for new resources in order to preserve the capabilities for being innovative and competitive in the future. The hope was that a bio-based economy would generate solutions to cope with the future and its challenges and at the same time create economic growth and, hence, also profit. In the beginning, bioeconomy was focussed on the substitution of fossil resources. The goal was to make products from regenerative materials in the same quality as products from fossil resources. To date, it has been repeatedly stressed that bioeconomy strategies focus on sustainability, whole value chains, and a broad spectrum of production (Blade, 2012; European Commission, 2012; High-Level Group, 2009; OECD, 2009; Popov, 2011; United Kingdom, 2012). When looking at bioeconomy strategies of the OECD, Germany, Russia, Ireland, the United Kingdom and Canada, it is noticeable that almost all of them focus on the grand challenges, on which they will act. Most strategies face food, feed, fibre, flower and fun. India and Argentina focus on biomedical engineering as well (Albrecht and Ettling, 2014). Strategies include the following goals: reducing CO_2 emissions, recycling, cascading use, sustainability, low ecological footprint of production and the food sector (Blade, 2012; European Commission, 2012; OECD, 2009; Popov 2011). Furthermore, the strategies talk about global impacts and cooperation as well as of regional adaptive bioeconomies. The main claim of bioeconomy is the total conversion from fossil resources to natural resources. This sounds very sustainable at first, but the question is whether this holds as a general assumption and, more specifically, how to reach this overall goal.

German bioeconomy strategy: it has been shown that this strategy, concerning its contents, is very close to the others because Germany was one of the first countries which worked on it (cf. Gottwald and Krätzer, 2014). The German Biooeconomy Council started its work in 2008. In 2009 the OECD published its bioeconomy policy agenda. The OECD policy (OECD, 2009) for a bio-based economy is very close to the German documents. The EU started in 2007 with 'En Route to the Knowledge-Based Bio-Economy'. In 2012, the European Commission launched a new strategy 'Innovating for sustainable growth: a bioeconomy for Europe' (European Commission, 2012). It involves a radical change in biomass production, consumption and recycling, also opening up new and diversified markets in food and bio-based products. However, Germany published its own policy in 2013, but its research strategy on bioeconomy already in 2010 (BMBF, 2010).

The German strategy has been and still is developed by a brain trust, called 'Bioökonomierat' (Bioeconomy Council). Its members are scientists and businesspeople, especially from the biotechnology sector (www.biooekonomierat.de). The main normative statements are food first, sustainability and acting in accordance with society, connected with the programmatic goal of a growing economy. The Bioeconomy Council promotes a sustainable intensification of agriculture, conservation of biodiversity, the use of new technologies in agriculture (i.e. precision farming, genetic engineering, new breeding techniques), the improvement of soil fertility, the improvement of water use management, and resource efficiency.

Trade-offs and conflicts in bioeconomy

An analysis of the Bioeconomy Council shows some trade-offs, conflicts und weaknesses of the German strategy. Agriculture is seen as the most important primary producer, and the general model is a continuation of high-input (fertilisers, pesticides), high-tech agriculture. Concerning biomass production, a couple of trade-offs emerge: (1) monocropping; (2) the competition of land use; (3) changing cultivation methods; (4) environmental and ecological pollution; and (5) societal acceptance.

Monocropping

Monocropping is connected with abundant cultivation of maize and canola. Especially for biogas, maize is the most important plant. At the same time, it is accompanied by a decreasing species diversity. Reduction of species diversity in monocropping (of maize and crops in general) leads to a changing landscape, a strong enhancement of diseases and pests. It also results in loss of natural habitats (cf. Albrecht *et al.*, 2012). This could mean an intensified use of herbicides, changing nutrients in soil, water and air, which will have direct consequences on human health and human food security. In addition, the problem of soil degradation results, among others, from monocropping.

Competition of land use

Competitions concerning the use of land result from conflicting interests between land use for energy cropping, environmental conservation areas, food and feed production (Albrecht *et al.*, 2012: 19). Due to the lack of land resources, prices for agricultural products have continued to rise extremely since 2005 (Bioökonomierat, 2014c). People in poor countries are not able to pay these high prices for food or feed, contributing to problems of undernourishment (Alexandratos and Bruinsma, 2012).

Changing cultivation methods

Changing cultivation methods include intensification, genetic engineering and high-tech precision farming. Environmental problems should be decreased with precision farming (cf. Godfray *et al.*, 2010). The majority of the bioeconomic literature implies that genetic modifications are safe, healthy and socially acceptable, which, however, has been severely questioned (ENSSR, 2013).

Environmental and ecological pollution

Environmental and ecological pollution comes along with different reasons, i.e. land use change from grassland or forests into arable land or conversion of peat to increase cultivable land. Greenhouse gas emissions will increase and stress the atmosphere, which influences human health. The demand for biomass in Germany easily exceeds the domestic production (37% import). Most of the imported biomass is produced under questionable conditions with respect to ethical and societal aspects (Leopoldina, 2013).

Societal acceptance

Societal acceptance is very important to retain political added value – if one wants to express it in economic terms – and to develop democratic deliberation. However, if food prices will rise due to bioeconomy shifting land use towards other products, society will not accept these products and the whole strategy; see e.g. the implementation of 'biofuel' E10 in Europe: most users still avoid E10 although it is even cheaper but viewed as problematic. Furthermore, aesthetic shifts in landscapes due to maize monocropping ('Vermaisung' in German) shall be taken into account. Notwithstanding, the global dimensions of food production in competition with biomass production are much more serious.

What effects do these conflicts and problems have on the food sector? The latter is put under massive pressure, e.g. concerning the following aspects:

- Monocropping can further strengthen large seed companies, especially if they market their genetically modified plants internationally. The global diversity of plant varieties is thus diminished even further.
- The competition for land use is additionally fuelled by the subsidies policy of the EU. According to this policy, the cultivation of energy plants is subsidised in a better way than the cultivation of food plants, which can result in ever more areas being used for the cultivation of biomass for fibre and energy utilisation.
- A further aspect is the application of genetic engineering as a contested issue in several respects.
- Already today, Germany has to import more than a third of its biomass from other countries, predominantly from less developed countries. There, the conflict between cash crops and plants cultivated for their own national requirements is also increased and thus threatens the local food production.
- Bioeconomy is not yet aware enough of this fundamental issue of distributive justice of food, food production and competing land use, which goes far beyond acceptance, since it addresses general ethical questions of (conditions of) acceptability.

General ethical perspectives on the sustainability of bioeconomy: an agenda

Albrecht *et al.* (2012), Gottwald and Krätzer (2014) and Unmüßig (2014) have already highlighted essential points of criticism, which have also found some consideration above. To start with a few remarks concerning the line of argument of the German Bioeconomy Council: It is noticeable that the term 'sustainability' occurs frequently. Sometimes, it only refers to the durability of systems. However, if a Sustainable Development is pointed to, as is the case in this strategy, it should always stand in the context of the debate which was initiated in Rio de Janeiro in 1992. Since then, specific theories as well as operationalisation approaches of sustainability have become established. As milestones, the integrative concept of sustainable development by Kopfmüller *et al.* (2001) and the theory of strong sustainability by Ott and Döring (2008) shall be mentioned here. In the bioeconomy strategies, the concept of sustainability is not addressed in detail. It is only stated without further consequences that ecological risks, such as the loss of biodiversity, shall not proceed further. Analysing bioeconomy under aspects of sustainable development cannot and may not be reduced to cascading use of biomass (Essel and Carus, 2014) as sustainable form of productivity. A thorough argumentation of what is understood as sustainable is required. In this context, a discussion on the principle of responsibility and the precautionary principle has to be initiated. We can detect a clear argumentation only to the effect that different technologies require funding in order to function more efficiently and intensively in the environment. Furthermore, the proponents of bioeconomy advocate the reshaping of international treaties, such as the redesign of the Convention on Biological Diversity (Bioökonomierat, 2014b: 27). It is not described to whose advantage or disadvantage it is to be reorganised. All in all, a critical examination of technological procedures within the sciences and within society is not (yet) up for debate in the bioeconomy community.

To date, mainstream bioeconomy is not aiming to make societies aware of new patterns of consumption and food utilisation. For this purpose, intensive communication would be required. However, if one looks at the research strategy of the Federal Government of Germany, there is only a very small number of research projects within social-scientific or ethical disciplines.

With reference to the food sector, one gets the impression that the main point is the increase of modern plant varieties in order to provide for the more than 800 million under-nourished people in a better way and to supply other, especially poor, regions of the world with certain vitamins. This involves the utilisation of transgenic breeding as well. Dangers, lack of knowledge, uncertainties concerning human health or ecological risks are completely ignored here. For a sustainable bioeconomy, awareness of global responsibilities (of producers, retailers, consumers as well as policymakers), based on ethical principles of autonomy, nonmaleficence, beneficence and justice would be key objectives and could also serve as sustainability indicators. However, for example, if agriculturists and farmers can only use certain seeds, all these principles are severely limited to begin with.

In order to apply ethical analysis and deliberation to bioeconomic processes within food production, a number of questions must be specified. To that end, Table 1 gives an initial overview. Subsequently, the people affected in each case have to be identified and the consequences must be evaluated. However, for this purpose, far too few elaborations from bioeconomy are available at present.

The 'Biooekonomierat' and other strategic bodies ultimately aim at the standard goal of economic growth, but not at sustainability taken seriously; in that sense bioeconomy is not innovative and was dubbed as being ' a dead end' (Unmüßig 2014). In this vein, it is also seen as the ultimate tool to completely monetarise life on earth, reduce life and nature to sheer economic resources – and thus devaluate life from an ethical point of view (Gottwald and Krätzer 2014). We would not go so far as to completely dismiss the whole idea of bioeconomy. But in its current form, bioeconomy simply follows the pathways of unsustainable standard western economic models and goals. This means, among others, the employment of a business-as-usual model of high-input, high-tech agriculture and not reflecting on

Table 1. Points to consider for a sustainability analysis of bioeconomy.[1]

Measures in bioeconomy for food production	Measures in bioeconomy for food utilisation
Application of chemicals, fertilisers	Handling of
Application of genetic engineering and related biotech-breeding techniques	• waste, residues • surplus production
Implementation of the 'Food First' strategy	• communication (throw-away society, new patterns of consumption…)
Inclusion of biodiversity protection	• morally-based food choices (animal welfare, vegetarianism, veganism)

[1] Examples, not exhaustive; of course these issues are not exclusive for bioeconomy.

recent ethical debates on the values of biodiversity beyond monetary terms. One perspective could be to focus much more on small-scale dimensions, cascade-use, employment of existing local or regional agricultural diversity, and soil-protection. Re-thinking a bioeconomy in these respects will remain a major theme with regard to a sustainable future of food.

References

Albrecht, K. and Ettling, S. (2014). Bioeconomy strategies across the globe. Rural 21(3): 10-13.

Albrecht, S., Gottschick, M., Schorling, M. and Stirn, S. (2012). Bioökonomie am Scheideweg. Industrialisierung von Biomasse oder nachhaltige Produktion? GAIA 21(1): 33-37.

Alexandratos, N. and Bruinsma, J. (2012). World agriculture towards 2030/2050. The 2012 revision. ESA Working Paper (FAO) No. 12-03. Available at: http://tinyurl.com/pn4y8rc.

Bioökonomierat (2014a). What is bioeconomy? German Bioökonomierat. Available at: http://tinyurl.com/nbkg547.

Bioökonomierat (2014b). Positionen und Strategien des Bioökonomierates. German Bioökonomierat. Available at: http://tinyurl.com/nbkg547.

Bioökonomierat (2014c). Landwirtschaft in Deutschland – ihre Rolle für die Wettbewerbsfähigkeit der Bioökonomie. German Bioökonomierat. Available at: http://tinyurl.com/lg64nfy.

Blade, S. (2012). Building the bioeconomy: Alberta integrates energy, forestry and agriculture. Available at: http://tinyurl.com/pog33hu.

Bundesministerium für Bildung und Forschung (BMBF) (2010). Nationale Forschungsstrategie BioÖkonomie 2030. Unser Weg zu einer bio-basierten Wirtschaft. Bonn, Germany.

Enriquez, J. (1998). Genomics and the world's economy. Science 281(5379): 925-926. DOI: http://dx.doi.org/10.1126/science.281.5379.925.

European Network of Scientists for Social and Environmental Responsibility (ENSSR) (2013). No scientific consensus on GMO safety. Available at: http://tinyurl.com/krss2wr.

Essel, R., Carus, M. (2014). Increasing resource efficiency by cascading use of biomass. Rural 21(3): 28-29.

European Commission (2012): Innovating for sustainable growth: a bioeconomy for Europe. Available at: http://tinyurl.com/mj6hgb7.

Godfray, H.C.J., Beddington, J.R., Crute, I.R., Haddad, L., Lawrence, D., Muir, J.F., Pretty, J., Robinson, Thomas, S.M., Toulmin, C. (2010). Food security: the challenge of feeding 9 billion people. Science 327(5967): 812-818. DOI: http://dx.doi.org/10.1126/science.1185383.

Gottwald, F.T. and Krätzer, A. (2014): Irrweg Bioökonomie. Kritik an einem totalitären Ansatz. Suhrkamp, Berlin, Germany.

High-Level Group on Green Enterprise (2009): Developing the green economy in Ireland. Available at: http://tinyurl.com/phlxk7s.

Kopfmüller, J., Brandl, V., Jörissen, J., Paetau, M., Banse, G., Coenen, R., Grunwald, A. (2001). Nachhaltige Entwicklung integrativ betrachtet. Konstitutive Elemente, Regeln, Indikatoren. Edition sigma, Berlin, Germany.

Leopoldina – Nationale Akademie der Wissenschaften (2013). Bioenergie – Möglichkeiten und Grenzen (with executive summary and recommendations in English). Halle, Germany. Available at: http://tinyurl.com/d7l3x5w.

Organisation for Economic Co-operation and Development (OECD) (2009). The bioeconomy to 2030. designing a policy agenda. Main findings and policy conclusions. Available at: http://tinyurl.com/leyc5p8.

Ott, K., Döring, R. (2008): Theorie und Praxis starker Nachhaltigkeit, 2nd edition. Metropolis, Marburg, Germany.

Popov, V. (2011). Prospects of bioeconomy in the Russian federation. Available at: http://tinyurl.com/mr8jh8h.

Rockström, J., Steffen, W., Noone, K., Persson, A., Chapin III, F.S., Lambin, E.F., Lenton, T.M., Scheffer, M., Folke, C., Schellnhuber, H.J., Nykvist, B., De Wit, C.A., Hughes, T., Van der Leeuw, S., Rodhe, H., Sörlin, S., Snyder, P.K., Costanza, R., Svedin, U., Falkenmark, M., Karlberg, L. Corell, R.W., Fabry, V.J., Hansen, J., Walker, B., Liverman, D., Richardson, K., Crutzen, P. and Foley, J.A. (2009). A safe operating space for humanity. Nature 461(7263): 472-475.

United Kingdom – Department for Transport, Department of Energy and Climate change, Department for Environment, Food and Rural Affairs (2012). UK Bioenergy Strategy.

Unmüßig, B. (2014). Bioeconomy – a dead end. Rural 21(03): 34-35.

Moralizing food technology

C. Coff
University College Metropol, Nutrition and Health, Pustervig 8, 1126 Copenhagen, Denmark; coff@phmetropol.dk

Abstract

Food technologies are common on many levels in society and used by both food professionals and consumers. Food technologies are not neutral. They inform and shape the behaviour of people. This paper presents a theoretical framework for analysing the mediating role of food technology and its influence on food ethics. Post-phenomenology and the idea of a technologically mediated morality are central theoretical approaches. Four elements are included in the analytical framework: perception, interpretation, intentionality, and mediated morality. The framework is applied to two cases; food safety and the mechanical tomato harvester.

Keywords: story-telling, perception, interpretation, intentionality, mediated morality

Introduction: food technology

Food technologies transform foodstuff through processing. It is very much the purpose of food technologies to turn the inedible into something edible. But there is more to food technologies than just processing food. They also transform and 'process' our relationships to food. This paper points to some of the wider consequences of this transformation as regards the attitude and behaviour of people. Since only limited research so far has looked into how food technologies shape human actions and relations, there is good reason to further examine this. Inspired by the philosophy of technology this paper proposes a framework for the analysis of the impact of food technology on human behaviour and human relations. The question raised here is how food technologies influence human perception and behaviour.

First, a clarification is needed of what should be understood by the term food technology. Food technology can be distinguished from agricultural technology in that the later belongs to the field of production of crops and animals on farms. In agriculture the focus is on producing 'raw materials' and not on processing. Food technologies on the other hand involve the processing, preserving, and packaging of 'raw materials' that mainly originate from agricultural production.

Technology is often perceived as mixture of technique and science. The ancient Greek word 'technê' can be translated into craft or craftsmanship. Technê was for the Greeks an ability to make or perform something. The combination of craftsmanship and science gives technology its unique character. Science adds something new to traditional craftsmanship. It qualifies and refines craftsmanship and it introduces new utilizations and practices that could not have been realized without science. The production of high fructose corn syrup (HFCS) by an enzymatic process that converts glucose (starch) found in corn into fructose is an example of a new food technology. HFCS was first produced in the laboratory in 1957 and came into industrial production in the 1960s. HFCS is a sweetener and is a cheap replacement for sugar.

The case of HFCS highlights the importance of scientific knowledge for the development of food technologies. Contemporary food technology can thus be understood as an application and integration of food science in food processing and preservation practices. This understanding of food technology excludes for instance traditional bread-making, as it is not based on science. On the other hand, modern industrialized bread-making definitely bears the signs of technology, as it is highly mechanised and based on science. Food technology is not only used for industrialized purposes; some modern household appliances like the refrigerator, the freezer, the electric oven, the coffee machine, the microwave oven, food processors, the bread machine, or more recently the sousvide devices also bear the signs of food technology.

Since many people enjoy communicating about food, food production and consumption is deeply intertwined with communication technologies. These cannot be considered as food technologies in themselves but likewise they are important for the kind of attitudes and behaviours that can be found in relation to food.

Story-telling around food technology

Both in professional and private spheres technology has become an integrated part of modern people's lives. Humans constantly make use of technology; for transportation, during work, for leisure activities, and also for preparing food. Technology affects and changes human lives. Indeed, it is the purpose of technology to change human lives. The American artist and performer Laurie Anderson once described technology as the campfire around which modern human beings tell their stories. But what does this analogy imply? In what sense do we tell stories around technology?

People gather around campfires because they provide light when it is dark, warmth when it is cold, and constant protection from dangerous animals. They are also a good place to cook, to share a meal, and to enjoy company. They are thus an ideal place for story-telling.

Technology can provide some of the features of the campfire. It can enlighten us. It can provide us with some level of comfort and security. We also use technology, industrially and in households, to prepare food. However, just as campfires themselves do not tell stories, nor do technologies. Stories are told around campfires and around technologies because both attract people and invite company. It is in their capacity as meeting places that campfires and technologies have acquired a central position for story-telling: stories only come to life when there is someone to tell them and someone to listen to them.

Anderson did not go on to discuss the differences between telling stories around a campfire and around technologies. But it is not hard to imagine that it makes a difference whether stories are told around a television or a campfire. Or when stories are told around pre-cooked convenience food or around homemade food in household kitchens. Or when stories evolve around shopping in an on-line food store or around shopping in the local grocery.

Man is a story-telling animal, and story-telling is a way of understanding and giving meaning to life experiences. Stories connect people and create community because through them ideas and experiences are shared. Stories have the capacity to change the perspective of the listener and reader so as to allow for new interpretations and understandings of themselves, of others, and of the world. In the end this can lead to behaviours and actions that would have been different had the story not been told, or told differently. From this narrative view the quest is to explain the implications of stories being told around food technology.

Technology and mediated perception

To examine the questions raised above the approach adopted here is to consider how food technology shapes perception and informs understanding. The following describes the theoretical approach adopted to examine the mediating role of food technology.

Looking through an optical lens changes perception. The picture perceived through a lens is different from the picture perceived without a lens. Likewise, our perception of geography and landscapes is different whether we have a car or a bike at our disposal. And the world does not look the same whether you have a loaded gun in your hand or not. These examples illustrate why it can be said that technology changes our perception of the world. Whenever technology is involved, it imposes new views on the world and thus changes experiences and actions.

Technology mediates perception. For instance, perception is changed by smartphone apps dealing with nutrition/health or animal welfare. By informing citizens and consumers apps (co-)shape the experiences of what can be considered healthy food and what can be considered animal-friendly farming practices.

The technological mediation of perception has been explored and described by Don Ihde (1990) in his post-phenomenological variant of phenomenology. Post-phenomenology analyses technology as a constitutive part of human beings' life-world. In the post-phenomenological perspective human beings and things constitute each other (Verbeek, 2011). In a technological world human experience cannot be considered as something that stands on its own but rather as something framed by and mediated by technology.

The technological mediation of perception has further implications as it also influences interpretation and understanding. Perception is dealt with in phenomenology, and interpretation/understanding in hermeneutics. In daily language the distinction between the two might not always be clear. However, perception is generally associated with becoming aware of or knowing things by means of the senses. Interpretation on the other hand is associated with explaining something. Explanations put experiences into context and deduce meanings. The purpose of interpretation is to create an understanding of the issue at hand.

To give a concrete example, testing for pathogenic microbes at a certain point in a food chain is an activity that involves both scientific knowledge and specific techniques for identifying microorganisms that are otherwise invisible to the eye. For instance, the presence of *Bacillus cereus* can be revealed on a cereus selective agar. A microscope can be used to confirm the presence of the bacteria. Perceptions are mediated by the specific qualities of the agar and the magnifying ability of the microscope. Identification of the pathogenic microbes in the sample would hardly be possible without these techniques.

Interpretation takes it a step further by asking what it means when a certain number of a specific microorganism can be detected. The meaning of the perceived result can only be deduced by putting the result into the context of established scientific knowledge produced in the field of food science and especially food safety. By doing so the combined effort of perception and interpretation provides a scientific answer to the question of whether the food in question should be considered safe or unsafe for human consumption. Ihde calls this process of interpretation a hermeneutic relation.

Even a simple technology like the thermometer creates representations of reality: the temperature is translated into numbers on a scale. The numbers cannot be understood without a knowledge of the scale. For anyone interested in food safety the interpretation of a given number on the temperature scale means considering if the measured and perceived temperature complies with standards for food safety. More complicated technologies like apps for smartphones are also perceived by the senses but the interpretation of the data/information provided is much more complex and often more ambiguous than in the case of the thermometer.

The human-technological intertwinement does not end here. With the term technological intentionality Ihde indicates that technologies are not neutral, but are often designed to represent reality in a specific manner with a specific focus. The test for pathogenic microbes in the food chain is designed to focus and thus represent reality from the perspective of food safety. Technology thus directs human intention towards some issues and away from others.

Technologically mediated morality

Ihde's analysis of technology is gathered around the structure of experience, whereas Bruno Latour's focal point is on action. In Latour's view technology provides a framework that shapes the scope of possible actions. Technologies 'want' people to do things in specific ways. Corresponding to the idea of technological intentionality Latour speaks of a script that prescribes how users are to act when employing technologies. Testing for pathogenic microbes in a food chain or measuring a temperature aim to keep up with hygienic food safety standards and thus to avoid food-borne diseases; they prompt specific actions by the people involved in the food chain.

Verbeek (2011) uses the term mediated morality about the way that technologies embody morality. Latour's famous speed bumps that urge car drivers to slow down and thereby force them to show care for more vulnerable road-users I here interpret as an example of mediated morality. This almost self-evident example of a technology that embodies morality can also be seen as a kind of delegation. Latour (1992: 154) describes delegation as the 'transformation of a major effort into a minor one'. Although Latour focus on how technology transforms demanding tasks into minor tasks – like entering a house is made a lot easier by the use of doors – the idea of delegation can also be applied to ethics. In the moral sphere delegation of the speed bumps can be understood as the transformation of the major effort of the car driver to show care for other vulnerable road-users to a minor effort by the fact that not slowing down the car may cause serious harm to both car and driver. Speed bumps make the decision to slow down much easier. This kind of delegation comes close to some forms of nudging that incite people to act in ways that are considered morally right. However, it is far from always that the ethical dimension of technology can be so easily discerned.

Often, when some sort of technology is involved the ethical judgment of what to do or not to do will be located somewhere between the human-technological association. Technology offers or prompts certain perceptions of the world, which in turn favour certain interpretations that inform moral judgements and decisions made by the individual.

It is easy to imagine that the above-mentioned smartphone app on nutrition/health or animal welfare could inform moral decision-making concerning the individual's diet. To paraphrase Latour (1999): you are different with a diet app in your hand; the diet app is different with you holding it. Human and non-human entities are in this sense intertwined and folded into one another. Hence, technologies are not just a neutral ensemble of tools and instruments that can be separated from ethical considerations.

From this understanding of the role of technology it is not far-fetched to state that technologies have moral implications in the sense that they exert an influence on human action and interaction. To put it succinctly, morality and ethics are here understood as human interactions with oneself, with others, or with the natural environment that have to do with understandings of right/wrong and good/bad. For example, using a diet app may indeed have implications for one's perception of what is 'good' and 'bad' food and hence influence one's diet, which in turn may influence one's social relations. To sum up the theoretical approach presented here it can be stated that food technologies: (1) co-shape the perception of reality; (2) frame interpretation and the construction of meaning; (3) co-shape intentionality by directing human attention towards some issues and away from others; (4) mediate morality and contribute to ethical ideas of right/wrong and good/bad.

The moralizing character of the mechanical tomato harvester

What implications does this theoretical approach have for the moral dimensions of food technology? To answer this, let us look into the case of the mechanical tomato harvester. Although the case, strictly speaking, has to do with agricultural technology it resembles in many ways that of food technologies, as the quality of tomatoes is affected. In the article 'Do artifacts have politics?' Langdon Winner discusses the social and political consequences of technologies (Winner, 1980). His analysis of the mechanical tomato harvester is on a macro sociological level but it serves here as a suitable entry to discuss the moralizing character of food technologies.

To illustrate the intertwinement of the technological and the social Winner describes how the mechanical tomato harvester developed by researchers at the University of California in the late 1940s has had a number of effects; handpicking and farm workers have been replaced by machines, harvesting costs have been reduced, the number of farms has declined because the machines are costly and only few can afford them, yields have increased, farms have become bigger, and the tomatoes have become less tasty as new hardier and sturdier varieties were developed to fit the new harvesting method. For Winner, this process was not the result of a plot, but 'a social process in which scientific knowledge, technical invention, and corporate profit reinforce each other in deeply entrenched patterns that bear the unmistakable stamp of political and economic power' (Winner, 1980: 126).

The example describes the industrialization of agriculture that has made the separation of food production and food consumption possible on a large scale. A number of other technologies were also needed in that process. For instance, the introduction of artificial cold, the 'cold chain', was paramount to efficient and industrialized mass production and the distribution of food (Truninger, 2013: 84). The main conclusions Winner draws are that: (1) the introduction of new technologies tends to favour certain developments and delimit others; (2) technology is the symbol and embodiment of a specific social order; (3) the chosen technologies become 'strongly fixed in material equipment, economic investment, and social habit, (and) the original flexibility vanishes for all practical purposes once the initial commitments are made' (Winner, 1980: 127-28); and (4) technology establishes a pattern of power and authority: certain kinds of technology are strongly linked to particular institutionalized patterns of power and authority (Winner, 1980: 134).

Winner's analysis operates on the macro social level and reveals some of the implications of the technological development for social structures, whereas the post-phenomenological/moral mediation approach focuses on personal experience and thereby accentuates the individual involvement. Common to both, however, is the preoccupation with the moral dimensions of technology.

The mechanical tomato harvester can be analysed from different actor perspectives using the post-phenomenological and moral mediation approach. The consumer is a suitable actor to be considered in deeper detail. How can we hypothetically imagine the outcome of the introduction of the mechanical tomato harvester from the consumer perspective? In general, consumers will not have any personal experience of the new mechanical tomato harvester due to the separation of production and consumption in industrialized societies. Only the end-product, the tomatoes, is visible to the consumers, whereas the production history and production practices are not (Coff, 2006).

However, in this case also the product has been altered, which means that consumer perception is likely to be altered too. The pure sensuous perception of the new varieties, which in this case is a culinary perception, differs from the perception of the old tomato varieties, as the new ones are less tasty yet hardier compared to the old varieties.

From this first banal statement on the perception we can move to the next step in the analysis and look at how this change in taste is interpreted. How is meaning constructed around the new taste of the new tomato variety? Consumers who have tasted the old varieties may consider it as a decline in food quality. It can be assumed that such consumers will be less satisfied with the new variety compared to the old ones. Consumers who do not know the taste of the old varieties have nothing to compare with. Still it must be assumed that their interpretation of the new tomato varieties will rank these new tomatoes low on a culinary scale compared to other foodstuffs.

The lower ranking of tomatoes on the culinary scale can be expected to have consequences for consumers' intentionality. Attention will be directed away from fresh tomatoes and towards other foodstuffs. Multiple scenarios can be imagined. One is that canned tomatoes would be preferred over fresh tomatoes. Another would be that other kinds of vegetables would be preferred. A third possibility is that fresh tomatoes would be substituted by completely other kinds of foods such as bread or meat. A fourth possibility is that consumers get used to the new variety and accept the lower culinary quality without changing eating habits. Cheapness of the new varieties and the development of new adjusted recipes would support the fourth possibility.

Consumers construct stories around the industrialization of agriculture and food processing. Consumers with a culinary interest realize that modern tomato varieties are often less tasty than old ones. However, they seldom know why. The diverted effects of the mechanical tomato harvester, like the decrease in tomato taste and texture, mean that new stories are told around tomatoes. Some of these stories are about how to incorporate tomatoes meaningfully into the consumer context.

By changing perception and interpretation, the intentionality and morality of consumers also change. As vegetables, tomatoes are in general considered healthy, but the technologically induced change in taste and price influences thoughts 'around' good/bad and right/wrong. This may in practice lead to a change in eating habits, which may for instance have individual and public health related consequences.

The coupling of industrial food technologies with tasteless products and undesirable production practices has been documented by many food writers in the last decade. Today, however, breeding efforts are made to make tomatoes both hardier and tastier. Whether such efforts can change the public perception of large-scale food production remains to be seen.

This final case illustrating the consumer perspective on food technology should be seen as a first effort to apply the proposed theoretical framework to an analysis of the moral mediation of food technology. The case is brief, and I hope to have the opportunity to develop it further in another context.

References

Coff, C. (2006). The taste for ethics. In search for food ethics. Springer, New York, USA.

Ihde, D. (1990). Technology and the life world. Indiana University Press, Bloomington, USA.

Latour, B. (1992). Where are the missing masses? The sociology of a few mundane artifacts. In: Bijker, W.E. and Law, J. (eds.). Shaping technology/building society: studies in sociotechnical change. MIT Press, Cambridge, USA, pp. 225-258.

Latour, B. (1999). Pandora's hope. Harvard University Press, Cambridge, USA.

Truninger, M. (2013). The historical development of industrial and domestic food technologies. In: Murcott, A., Belasco, W. and Jackson, P. (eds.) The handbook of food research. Bloomsbury, London, UK.

Verbeek, P.P. (2011). Moralizing technology. Understanding and designing the morality of things. The University of Chicago Press, Chicago, USA.

Winner, L (1980). Do artifacts have politics? Daedalus 109(1): 121-136.

Section 12. Limits of food knowledge

What does it mean to 'know' your food? Philosophical reflections on consumer responsibility

M. Huth
Messerli Research Institute, Veterinärplatz 1, 1210 Vienna, Austria; martin.huth@vetmeduni.ac.at

Abstract

Responsibility and knowledge are closely linked together. But what kind of knowledge is required for responsible action? This question will be treated using the example of meat production. In animal ethics as well as in public debates we find the assessment of current collective behaviour as irrational, weak-willed or even as schizophrenic. While a lot of people treat pets as family members, livestock breeding reduces animals with similar (socio-cognitive) abilities like our pets to mere production units. Alleged maltreatment of pets or e.g. zoo animals leads occasionally to public outcries whereas the comparable treatment of meat providers is more or less hidden or hushed up. The ones who work in the front line of meat production are possibly harmed by their activity – they experience the reifying current treatment potentially as emotional toll. Against the background of inalienable universal moral principles the assessment of current conditions as inconsistent or 'schizophrenic' is at least understandable. But if one finds herself trapped in moral demands that she cannot or does not really want to fulfill one might have to ask for different accounts to responsibility. Relationalism offers an alternative. The embedding in societal structures and practices that predetermine collective conduct is taken into account. Practical reason is pre-determined by the significances in our lifeworld. Thus, this reason is not the application of abstract norms but a practical understanding as one finds exemplified in Aristotle's *phronesis*. It requires a socialization and habituation constituting practical knowledge. Mere factual knowledge or knowledge about principles is not sufficient to orient consumption behaviour. But as a lot of people find themselves highly alienated from the processes of food, especially meat production, it is to be seen as political agenda to open access to these practices. This is a required step to have an opportunity to deal with the already processed products ethically responsible.

Keywords: practical knowledge, relationalism, visibility, moral judgment

Introduction

Embedded responsibility

In this text the conditions of individual responsibility in food consumption will be shown with the example of meat. This will be done against the background of the assumption of an embedding of practical reason in societal and political structures and practices. Practical reason is not provided by an 'a priori' autonomy independent of any context we are living in (like in Kant's approach to morality). On the contrary, moral convictions and significances are acquired by socialization within a respective lifeworld (cf. Diamond, 2012; Held, 2007). A principal witness for the validity of this account is the famous US-American philosopher Judith Butler: '(...) (W)e are (...) constituted in cultural norms that precede and exceed us, given over to a set of cultural norms and a field of power that condition us fundamentally' (Butler, 2004: 45). Thus, moral obligations (towards humans, animals, and environment) are considered as relational. Relationalism is an account to ethics and especially animal ethics that was developed e.g. by Alice Crary (2010), Cora Diamond (2012) or Jeff McMahan (2005). The crucial point of this approach is the moral relevance of societal significances. Considerable moral entities are not abstracted from our relations to them. It turns out to be relative without being arbitrary – because people live in a shared world in which deliberation about moral agendas is possible and necessary.

But moral claims oftentimes appear as absolute without alternative and abstracted from the relations individuals are living in.

Animal ethics, environmental ethics or food ethics are the related fields of academic research. Within these realms and also in public one faces the fact that this topic is a true battleground of argumentation. One side is uttering demands according to moral principle like the equal consideration of beings with equal (or similar) characteristics. Here, the protagonists claim to speak in the name of reason or at least from neutral grounds (cf. Singer, 2011: 90f). That leads on the other side to counter-reactions referring to traditions and branding the opponents as radical. But this division happens even within the individual. Apparent reasonable arguments for an ethically legitimate meat consumption or vegetarianism/ veganism clash with traditions or habits that seem to be stronger than reason – feelings of guilt and a self-assessment as will-weak and irrational might arise. But how can one then support a responsible consumption behaviour or at least a proper reflection about moral obligations?

Eating the products of violence!? The apparent gap between knowledge and conduct

In current debates in animal ethics an interesting phenomenon is visible: the (polemic) assessment of usual collective behaviour in meat consumption as (kind of) schizophrenic. A lot of people criticized harshly the killing of a giraffe and of four lions due to space problems in Copenhagen's zoo in 2014. The media even reported a 'public hysteria' with an online petition for closing the zoo and murder threat for the manager. (Der Spiegel, 2014) But could we then assume that all of them lead a vegetarian or vegan life so that they would also protest against such a treatment in pigs, cows or sheep? The answer is almost certainly: no. As the title of Melanie Joy's most famous book 'Why we love dogs, eat pigs and wear cows' (Joy, 2011) already indicates: we live in clear inconsistencies regarding the treatment of animals. We use animals as resources which have similar (socio-cognitive) capabilities like e.g. dogs (Benz-Schwarzburg, 2012). We keep these dogs as pets and respect them not at least because of these capabilities. We try to save living beings denoted as wonderful creatures (the giraffe and the lions) or as family members (the dogs). We eat industrially raised and killed living beings denoted as production units. They seem to be victims of reification and violence. Indeed, this seems to be schizophrenic – at a first and maybe superficial glance.

Another, heuristic example supports a different suspect than the currently powerful one of collective schizophrenia – the suspect of collective alienation. A friend of mine leads guided tours in an exhibition about Neanderthals. When talking about nutrition an approximately 5 year old child asked wondering: 'Why did they eat animals?' Only after a while my friend recognized that this child was not aware of the fact that eating meat equals eating animals. The little girl had so far obviously only access to meat as processed product. This shows the possibility to have no closer relation to animals that are not considered as pets or e.g. zoo animals – than grabbing packed meat in the shelf.

Responsibility as practical understanding

Within the debates of animal ethics as well as public debates there is the fundamental demand to end the mentioned inconsistencies in dealing with animals. Equals (in regard of socio-cognitive abilities or the ability to suffer) should be treated equally. (cf. Singer, 2011) A lot of people would theoretically put forward this request, too. But reality looks different. The realms of the human-pet-relation, of the human-zoo animal-relation or the human-farm animal-relation seem to be practically unconnected. Nevertheless we are to a certain extend aware of these inconsistencies; this leads obviously to feelings of guilt and powerlessness. We experience ourselves as overburdened selves, as clamped in between subjectively unchangeable lifeworld normalities (or even: norms) of meat consumption and the ethical demands already pointed out above (cf. Bandura, 1997). These ethical demands are basically of high persuasiveness not at least because auf the stringency of argumentation. According to relevant authors (e.g. Singer, 2011 or Joy, 2011), there is the underlying normative principle to treat equals equally, no matter in which circumstances. This is reasonable. If a practice is in an ongoing contradiction to reason, how could one then prevent herself from feeling guilty and powerless?

But to turn this question upside down: Why fail these principles to help people more efficiently to change their practice? Wittgenstein might have a first answer: '(For the crystalline purity of logic was, of course, not a result of investigation: it was a requirement.) The conflict becomes intolerable; the requirement is now in danger of becoming empty. We have got on to slippery ice where there is no friction and so in a certain sense the conditions are ideal, but also, just because of that, we are unable to walk. We want to walk: so we need friction. Back to the rough ground!' (Wittgenstein, 2007: §107) But how can one get friction when even the ground is not visible? How can people reflect about and change their consumption behaviour accordingly when knowledge is theoretical but often not practical or experienced (because of a far-going alienation of a lot of individuals)?

In contrast to that it is fruitful to look back to Aristotle's *phronimos* as 'native actor' (Waldenfels, 2006: 43) Having *phronesis* means: having a well-developed understanding derived from experiences in our lifeworld. Acting towards '(...) the person one should, to the extend one should, when one should, for the reason one should, and in the manner one should – this no longer for anyone, nor is it something easy, which explains why getting things right us a rare thing, a proper object to praise, and something fine' (Aristotle, 2002: II 1109 28ff.). The difficulty named by Aristotle is not a simply intellectual one. It refers to a kind of practical know how and a kind of sensitivity that is not provided by blunt facts or reasonable principles. On the contrary, moral behaviour and moral thinking are altogether acquired by habituation (what Aristotle calls *hexis*) or by rehearsal compared to learn to e.g. play guitar. No-one learns to play guitar by learning the line. While Aristotle's philosophical father, Plato, was convinced that factual (epistemic) knowledge suffices for practice, we now find a gap between knowledge and know how. Moreover, according to Aristotle, practical reason (*phronesis*) is not a clear, distinct and proved but a fuzzy kind of knowledge and thus different to theoretical reason (*episteme*). (Aristotle, 2002: I 1094 b12ff. and VI 1143 b18ff.) Adequate behaviour and an elaborated power of discrimination is to be taught and trained. People are socialized within a certain socio-cultural setting that establishes horizons of significances and habits. The human self-understanding is deeply determined by the world one is living in, also by the Others one is living with – and it rests on shared ways of conduct (habits). (cf. Heidegger, 2001: 54; Held, 2007; Wittgenstein, 2007, esp. §§1, 6, 9, 10, 12). Cora Diamond also states that even our wishes, dislikes, emotions, perceptions or inner pictures are determined by implicit normative orientations (Diamond, 2012: 219).

Regarding our topic, these orientations might partially rest on questionable grounds. As commonly known, livestock farming is not part of a usual everyday life. A majority lives outside a space of experience that gives the way free to a notion of the path from breeding procedures to the packed meat in the shelf. Traditions of animal keeping have changed radically especially in the last century (cf. Derrida, 2010: 50ff.). Human-animal interactions are often reduced to certain practices like pet keeping, animal-assisted therapy or zoo visits. This is the superficial practical paradigm of our being with animals. But for average consumers there is only hardly a relation to meat animals or to the current conditions of livestock breeding. There is no common experience of these animals since they do hardly occur in our lifeworld.

Supporting relationalism in animal ethics or food ethics – what kind of relation can one rely on, should one interpret or can one take as theoretical starting point? And how should one then assess responsibility assuming that it derives from or bases on shared habits in a society? Bodily (human) beings are not fully rational and autonomous beings that grow up and decide from outside common significances. Thus one has to conclude that responsibility is dosed. The US-American philosopher Judith Butler states that minding conditions as so-to-say responsible does not equal with a surrender of individual responsibility: 'Conditions do not act in the way that individual agents do, but no agent acts without them. They are presupposed in what we do, but it would be a mistake to personify them as if they acted in the place of us' (Butler, 2004: 11). It just opens the tasks '(...) to situate individual responsibility in light of its collective conditions' (*ibid*.: 15).

But this does not dismiss ourselves from the task to reflect critically on the conditions of responsibility. Of course there is not the possibility to step outside the moral infrastructure we are in. You cannot abstract yourself from having habits constituting your understanding of moral demands (cf. Held, 2007: 24).

Morality and politics

To come closer to a well-informed relational responsibility one does not have to force herself more efficiently (like in Kant's terminology of the force by the moral law qua categorical imperative) but has to go back to the social conditions of responsibility. The afore-mentioned five-year-old, unaware of what she was eating, is just an example for our alienation from what we eat. The powerlessness and the alienation from basic experiences can only be dissolved by enablement and empowerment regarding the possibility of these very experiences. The general hiddenness of meat production is an undemocratic structure of at least the current western societies. It prevents people from taking responsibility due to the fact that they do not see possibly regularly what they are dealing with, namely the process behind the processed product in the shelf. It is not seen, it is not visible that there is a basic vulnerability that might be regarded as anchor of moral consideration. Speaking for a majority, the mere information about meat production is available and displayed for instance in documentaries. But this is not knowledge at the level of *phronesis*, at the level of collective and individual habitual skills. This is something abstracted from the everyday routines and their normative implications we are living in.

Basically, people know that the killing and/or dying of a lived body, an animal, is not an event that one can handle like an everyday trifle. A bodily human being is inevitably affected by such a happening as philosophers like Derrida (2010: 31), Levinas (1992) or even writers like Tolstoy (1958: 1348) have shown. All of these authors would consent to the assessment that humans are open to the vulnerability or mortality of the Other. The own vulnerability serves as kind of a gateway for the demand originating from the other suffering (or even: dying without suffering) (cf. Huth, 2014: 64f.). This is not only found within the theoretical reflections by philosophers or men of letters but also in reports of experiences by persons involved in practices that cause animals' death. An impressive example is the following by an US-American slaughterhouse worker: 'The worst thing, worse than the physical danger, is the emotional toll (...). Pigs down on the kill floor have come up and nuzzled me like a puppy. Two minutes later I had to kill them – beat them to death with a pipe. I can't care' (Drillard, 2008: 391). The inability expressed in the phrase 'I can't care' is to be interpreted on the one hand as impossibility to take time to concentrate on what happens (e.g. to mourn), on the other hand as radical helplessness facing the injury and killing of another living being (Huth, 2014: 69). This affects an individual on a pre-reflexive level of existence. One cannot argue against being affected by violence against other living beings. But these experiences remain so to say reserved for slaughterhouse workers or e.g. laboratory staff. A heuristic example is the following quote by a worker in animal experimentation: 'They did not like having rats in clear cages because the animals could look at you' (Acampora, 2006: 100). The animals as living beings should remain invisible, they should remain mere examples without perceivable vulnerability – and within the respective structures (industrial slaughterhouses, laboratories,...) they are to be treated like mere production units. The ones who are in the front line seem to pay for the usual invisibility of animal killing. They fulfill the task of 'producing cheap meat' by applying practices that in our lifeworld would probably not considered as morally permissible. This might even lead to suffering from post-traumatic stress disorders, burnout syndrome or feel trapped in structures without the opportunity to change things. They lack self-efficacy regarding a proper practice of killing animals (if something like that is possible) which might cause psychological or psychosomatic disorders (cf. Bandura, 1997).

This example shows that vulnerability has to be visible and recognized to serve as basis for conscious ethical consideration. To use a formulation by Judith Butler: 'There are ways of distributing vulnerability, differential forms of allocation that make some populations more subjects to arbitrary violence than others' (Butler, 2004: xii). There are dominant forms of representation (*ibid*.: xviii) which are able '(...) to control the way in which people see, how they hear, what they see'. (*ibid*.: xx). Heuristically, we can look back to the afore-mentioned author Melanie Joy who claims that 'carnism' is a common view in western society which blocks empathy with some animals (cf. Joy, 2011). It pre-determines the allocation of vulnerability – but Judith Butler also says: 'Our capacity to feel and apprehend hangs in the balance' (Butler 2004, xxi). There is the possibility to make the invisibility visible, and that equals: to make use of the balance. This would open the possibility to know our food. One would be able to transfer her experiences in the mentioned conscious considerations. But prior to the experiences there is the political demand for access to precisely these experiences and the structures of animal keeping and killing – e.g. by opening the respective institutions obligatorily for visits and excursions. Responsibility for consumption bases on fundamental political demands to know (in the meaning of practical wisdom qua *phronesis*) what gives us lives and sustains us. Only then one would have the full opportunity to consume responsible and to develop self-determined habits.

Conclusion: towards a relational responsibility

Albert Bandura's concept of (lacking) self-efficacy shows that powerlessness in practice (or work) leads to an experience of being overburdened, to burnout symptoms, etc. (Bandura, 1997). This was displayed by the afore-mentioned examples of the slaughterhouse employee and of the involved in animal experiments. But this assessment is also suitable for the consumer as excluded from the processes in which food (here: flesh) is produced (here: raised) and processed. A lot of people assess themselves as incapable of acting rational or even as schizophrenic against the background of reasonable principles which should actually guide our conduct. But practical reason understood according to Aristotle's *phronesis* requires concrete involvement in the respective structures and practices, requires a being experienced and also habitual acting. Only this involvement enables individuals to reason and act conscientiously according to own opinions because of their being trained and because of a conscious relation to the things, entities, living beings they are acting with or they are treating. A relation to animals as meat providers would open a self to their vulnerability and would reveal the breeding and keeping conditions as well as slaughter methods in a less abstract way. Political decisions would be the condition to have access to these structures and practices – moral responsibility must be considered as embedded in societal and political structures and traditions. Thus one can speak of a conditioned autonomy.

Agents have to have access and thus relations to the (vulnerable) animals to take responsibility properly; otherwise the abstract deed of grabbing in the shelf remains strangely (morally) insignificant although one seems to know on a vague level 'very well' that meat production might be ethically questionable. Only then people are able to reflect and act properly, according to the very nature of practical reason and of our power of discrimination. If one can emancipate from current conditions of the hiddenness of food production one can take the relations as significant starting point of ethical decisions. Then one can overcome the (self-)assessment as irrational, weak-willed or even schizophrenic.

References

Acampora, R. (2006). Corpreal compassion. Animal ethics and philosophy of body. University of Pittsburgh Press, Pittsburgh, USA.
Aristotle (2002). Nicomachean ethics. Oxford University Press, Oxford, UK.
Bandura, A. (1997). Self-efficacy. The exercise of control. Freeman, New York, USA.
Benz-Schwarzburg, J. (2012). Verwandte im Geiste – Fremde im Recht. Soziokognitive Fähigkeiten bei Tieren und ihre Relevanz für Tierethik und Tierschutz. Harald Fischer, Erlangen, Germany.
Butler, J. (2004). Precarious life. The powers of mourning and violence. Verso, New York, USA.
Crary, A. (2010). Minding what already matters: a critique of moral individualism. Philosophical Topics 38(1): 17-49.
Derrida, J. (2010). Das Tier das ich also bin. Passagen, Wien, Austria.
Diamond, C. (2012). Menschen, Tier und Begriffe. Aufsätze zur Moralphilosophie. Suhrkamp, Frankfurt am Main, Germany.
Drillard, J. (2008). A slaughterhouse nightmare. Psychological harm suffered by slaughterhouse employees and the possibility to redress through legal reforms. Georgetown Journal of Poverty Law and Policy XV, 2: 391-408.
Heidegger, M. (2001 (1927)). Sein und Zeit. Max Niemeyer, Tübingen, Germany.
Held, K. (2007). Lebenswelt und politische Urteilskraft. In: Leghissa, G. and Staudigl, M. (eds.) Lebenswelt und Politik. Perspektiven der Phänomenologie nach Husserl. Königshausen und Neumann, Würzburg, Germany, pp. 17-29.
Huth, M. (2014). Ihr Tod geht uns an. Eine Phänomenologie des Sterbens von Tieren. Tierstudien 05/2014: 59-71.
Joy, M. (2011). Why we love dogs, eat pigs and wear cows. An introduction to carnism. Conari Press, Berkeley, USA.
Levinas, E. (1992). Jenseits des Seins oder anders als Sein geschieht. Alber, Freiburg im Breisgau, Germany.
McMahan, J. (2005). Our fellow creatures. The Journal of Ethics 9: 353-380.
Singer, P. (2011 (1979)). Practical ethics. Cambridge University Press, Cambridge, UK.
Der Spiegel (2014). Zoo Kopenhagen tötet jetzt vier Löwen. Available at: http://tinyurl.com/ndnz3rh.
Tolstoy, L. (1958). Krieg und Frieden. List, Munich, Germany.
Waldenfels, B. (2006). Schattenrisse der Moral. Suhrkamp, Frankfurt am Main, Germany.
Wittgenstein, L. (2007). Philosophische Untersuchungen. Suhrkamp, Frankfurt am Main, Germany.

Climate change and agri-cultural knowledge: Bangladesh through a mirror and magnifying glass

A. Blanchard and S. Bremer*
Centre for the Study of the Sciences and the Humanities, University of Bergen, Allegaten 34, 5020 Bergen, Norway; anne.blanchard@svt.uib.no

Abstract

The prospect of climate change threatens food security in the developing world, and demands that 'agri-cultures' re-examine the way they mobilise knowledge relative to climate and nature. Bangladesh is particularly vulnerable to climate change given its burgeoning population and its dependence on domestic agriculture. Only recently has Bangladesh achieved a food surplus, owing to both techno-scientific innovation from the green revolution, and local traditional knowledge. However, Bangladeshi agri-cultures are now being forced to face the limits of their knowledge relative to fundamental changes to their climate. Arguably, such pervasive changes defy a 'problem-solving' approach alone, characterised by targeted calibrations to existing technologies. Indeed, the idea of climate change can be deployed as a creative force to stimulate new thinking on the ways Bangladeshi agri-cultures mobilise knowledge of their climate. Mike Hulme introduces the metaphor of climate change as both a magnifying glass and a mirror; it at once demands examination of each of our human projects, and reflection on their objectives. This paper uses Hulme's metaphor, and places the climate knowledge of Bangladeshi agri-cultures under the magnifying glass, to for instance examine how climate science can be downscaled to a more relevant local scale. The paper also argues for deeper critical reflections on the underlying purpose and means of knowing the climate; that is, looking in the mirror. The paper forms a point of departure for a newly initiated research project in northeast Bangladesh. The TRACKS project, 'Transforming Climate Knowledge with and for Society', seeks the co-production of high-quality climate knowledge with local and scientific communities.

Keywords: climate knowledge, food security, Bangladesh

Climate change and food security: challenges for Bangladeshi agri-cultures

The prospect of climate change threatens food security in the developing world, with Bangladesh one of the most acute cases. Bangladesh is highly dependent on its domestic production of food grains like wheat and rice to feed its growing population, and agriculture is a critical source of income, directly employing about half the labour force (GoB, 2012). Rice accounts for 90% of the food production in Bangladesh, but it is sensitive to changing weather conditions (rainfall, temperature, humidity), as well as extreme events such as floods, droughts and cyclones (Alauddin and Sarker, 2014). This makes agriculture-dependent rural communities highly sensitive to climatic change, because their livelihood is tied to the distinct seasons and the weather they bring. For instance, Etzold *et al.* (2014) worked with communities in Kurigram district and found that most interviewees noticed the impact of rainfall variability on their own livelihoods, whether directly through their own food production or indirectly through higher food prices.

Climate change represents a challenge not only to the livelihoods of Bangladeshi agricultural communities, but also to their wider relationship with nature and the climate. With communities' livelihoods closely intertwined with seasonal rhythms, this influences their representations of climate, their ways of knowing the climate and acting accordingly, and their value systems; what can be considered their 'agri-culture'. For instance, Bengal folklore shows the importance of agriculture as part of social and cultural life: the 12th century poetry of Khana distils experience-based knowledge through rhymes and sayings. Farmers also have knowledge of the weather from their own experience and that of their family and friends, using natural signs to predict rainfall such as when birds nest or ants swarm. Since the 1970s, this relationship with nature and the climate has also been shaped by science and technology, through installing irrigation, planting high-yield crops, teaching climate science in schools, or using meteorological forecasts. However, Bangladeshi communities have started to experience changes to their weather that fundamentally challenge this sophisticated and long-standing body of traditional, local and scientific knowledge of climate. Communities face erratic temperatures and rainfall, with droughts and flash-floods that force them to abandon agricultural practices based on generations of experience. In the Kurigram study, communities complained that while they used to experience six seasons, these past years they have only observed three or four (Warner and Afifi, 2014).

This paper looks at how climate change presents a challenge to Bangladeshi rural agri-cultures, with a particular focus on how it demands fundamental transformation of their knowledge of the climate. It examines Bangladeshi agri-cultural knowledge using the metaphor of Mike Hulme (2009), who describes climate change as a magnifying glass for examining and adjusting our knowledge of climate, and as a mirror for reflecting on the very foundations of this knowledge and transforming it. Hulme's metaphor sees climate change as a 'creative force', and it that way contributes to give an impetus to Bangladeshi agri-cultures to challenge existing knowledge systems, and rethink the ways they come to know their food and climate. The paper is the point of departure for and early reflections of the TRACKS research project; 'Transforming climate knowledge with and for society: mobilising high quality knowledge on climate variability with communities in northeast Bangladesh'.

Climate change: a 'wicked problem' acting as a magnifying glass and a mirror

The framing or representation of climate change can be argued to determine the way we come to know about this phenomenon, and what counts as high quality and useful knowledge. In his book 'Why we disagree about climate change', Mike Hulme (2009) shows that in many different spheres, climate change remains represented as an environmental problem to be solved, with common parallels drawn to the hole in the ozone layer caused by a build-up of chlorofluorocarbons (CFCs) in the atmosphere. Such well-framed and scientifically knowable problems lend themselves to a suite of targeted regulatory and technical solutions. However, in this paper we join Hulme in arguing that such a problem framing is incongruent with the challenges presented by rapid and unpredictable changes to our weather and seasons, and will not help Bangladeshi agri-cultures transform their knowledge of climate.

Here we argue that climate change is better framed as a complex and pervading 'wicked problem' (Rittel and Weber, 1973). While a number of authors have written about the features that distinguish this category of wicked problems, Funtowicz and Ravetz (1993) present a useful typology that when applied to climate change, highlights that: (1) its mechanisms are complex, and therefore associated with significant uncertainty; (2) there is a lack of consensus on its definition and its most appropriate solution; and (3) the decision-making process is pressured by urgency and high stakes. By emphasising uncertainty as a barrier to scientifically knowing climate change, this framing opens up to socio-cultural and historical representations of climate change. We argue that such a framing is more useful for helping transform Bangladeshi agri-cultures, and their knowledge of climate, by recognising these communities' cultural relationship with climate. Understanding climate change as a wicked problem provides an impetus to Bangladeshi agri-cultures to think about how best to mobilise knowledge that accommodates the complexity of climate change, and also to question the social structures, institutions and norms that underpin this knowledge.

Hulme also departs from the idea of climate change as a wicked problem, describing it as a creative force to encourage new thinking about the relationship between humanity, nature and climate. He develops the metaphor of climate change as both a magnifying glass and a mirror: 'As a magnifier, climate change allows us to conduct examinations (...) of each of our human projects (...) As a mirror, climate change teaches us to attend more closely to what we really want to achieve for ourselves and for humanity' (Hulme, 2009: 363). We argue here that this metaphor presents two useful lenses through which Bangladeshi agricultural communities can interrogate the way they mobilise knowledge of climate.

As a magnifying glass, climate change gives agri-cultural communities a reason to rethink or readjust their current climate knowledge in order to adapt to the changing climate that they are experiencing; their new lived reality. By modifying the conditions for Bangladeshi agriculture, climate change provokes anomalies in the agricultural systems and cycles, threatening crop harvests; monsoons arrive too early or too late or periods of drought are longer. This forces communities to put their agricultural practices under the magnifying glass to evaluate their current capacity to adapt to their new climate, putting in place new local practices or technologies such as riverbank protection against erosion, or irrigation systems for the longer dry season. These local solutions are short-term initiatives; more like incremental tinkering with the existing knowledge and tools.

As a mirror, climate change encourages Bangladeshi communities to reflect on the very foundations of their agri-cultures and knowledge systems. As seen from the mirror, climate change is not an external problem, but an internal symptom of a 'sick society'. As argued by Rotmans (2006), many of our most pressing environmental issues are persistent in that they represent endemic failures that have crept into our existing socio-cultural systems, requiring fundamental changes to our society and its governance. The mirror reflects Bangladeshi agri-cultures' profound representations and values, as well as their previous actions and practices. Beyond local solutions to tackle climate change, Bangladeshi agri-cultural communities are encouraged to question the very foundations on which their responses are built, and challenge the ideas that pervade society, such as the predictability of their daily-lived reality.

Method: pilot survey in Northeast Bangladesh

This paper represents the point of departure and the early reflections of the TRACKS project: 'Transforming Climate Knowledge with and for Society: mobilising high quality knowledge on climate variability with communities in northeast Bangladesh'. TRACKS is a three-year research project that started in June 2014, and that was funded by the Norwegian Research Council. It is led by the authors' Centre at the University of Bergen, and has several partners in Bangladesh.

The study site of TRACKS is limited to the Sylhet Division in Northeast Bangladesh (Figure 1). This region has a varied topography and is home to many agri-cultural communities, with a strong tradition of tea and rice farming. The agri-cultural communities in Sylhet Division are particularly impacted by the weather, as they experience a unique early season rainfall that comes before the monsoon season, and which causes drastic floods that impact on the crops in a way that is not seen elsewhere in the country.

In this context, the objectives of the TRACKS project are to first gain a greater understanding of current weather variability in the Sylhet Division, and its impacts on the agri-cultural communities living there, to support adaptive action. Second, TRACKS aims to critically challenge the 'normal' climate science approach, by adopting a 'post-normal science' approach, and thus increase the capacity within communities of Sylhet Division to engage with different forms of climate knowledge. This is why the metaphor of climate change as both a magnifying glass and a mirror is central to TRACKS.

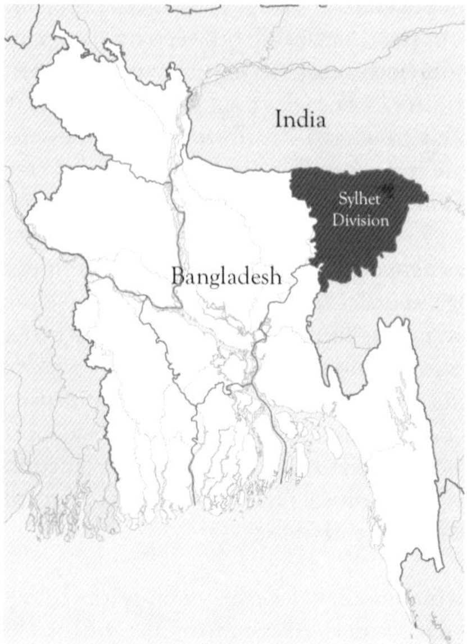

Figure 1. Map of Bangladesh.

The early reflections of this paper draw on a pilot survey undertaken in September 2014 in Bangladesh, and composed of: (1) Eleven narrative-based interviews of agri-cultural stakeholders in Sylhet Division, including rice farmers, a tea garden manager, local administration representatives, teachers, a boatman and a shop owner. These interviews were conducted to elicit stories of the weather and its impacts on the life of these stakeholders; and (2) Two workshops, one in the capital Dhaka, and one in Sylhet, with in total 60 people, principally government officers, scientists, and NGO representatives, to gather a list of key weather impacts in Sylhet Division.

This paper is not an empirical paper, and will not report in details the outputs of this pilot study. Rather, it will draw upon it to stimulate our early reflections on the ways climate knowledge can and ought to be mobilised by agri-cultural communities in Sylhet Division. We will structure our reflections according to the metaphor of climate change as both a magnifying glass and a mirror in the two following sections.

Putting agri-cultures' climate knowledge under the magnifying glass

The metaphor of climate change as a magnifying glass gives Bangladeshi agri-cultural communities a reason to readjust their current climate knowledge and find local solutions for enhancing their agri-cultural practices. We here discuss five ways climate knowledge can be readjusted with regard to agri-cultural practices in the Sylhet Division.

The first way to put climate knowledge under the magnifying glass is to think about how climate science that is done at a global scale (like global climate models (GCMs) – from the IPCC) could be in the future downscaled to the Sylhet Division, as there are currently no climate models available for this region. Having climate models that focus on the Sylhet Division with a higher resolution might allow a greater understanding of the region's climatic system, by taking into account its unique early season rainfall and its distinctive flat topography surrounded by the Himalayas. As one participant in the Sylhet workshop put forward: 'We should use advanced GCMs to analyse the impacts of climate change on local crop production'. Moreover, making global climate science more local also shows the importance of building models that correspond to natural boundaries rather than political ones. The Sylhet Division is surrounded by Indian borders, with the Himalayas playing an important role in the mechanisms of the early season rainfall. Despite India and Bangladesh having their own weather stations, climatic data is rarely shared between the two countries. Regional models that are attentive to the coherence and comprehensiveness of local mechanisms could thus help overcome these geopolitical boundaries.

A second way of readjusting agri-cultures' climate knowledge is to improve the quality of the climate science that is currently undertaken locally. With relatively little institutionalised climate research in Sylhet Division, NGOs have taken a leading role in providing local communities and decision-makers with knowledge on the climate and adaptation. However, this knowledge is not produced according to the usual norms of science; it is often not peer-reviewed, or transparent in its methods. Government officers, scientists, and NGO representatives at the Sylhet workshop noted a need to improve scientific capacity in the Sylhet Division, by for instance 'improving research facilities in local public and private universities, as well as providing funds and facilities (including the access to peer-reviewed articles) to local university students to complete Masters and PhDs in climate science', or by 'introducing long-term climate change projects on the Sylhet Division'.

Third, the stakeholders of the Sylhet workshop expressed a necessity for improving the education of agri-cultural communities relative to climatic issues (Alauddin and Sarker, 2014). Notions of climate change are taught at the primary school level, but it was argued that this effort could be pursued beyond that level to improve the scientific literacy of agri-cultural communities: 'We should inform those who are involved in agricultural activities like fisheries or livestock about climate change, and suggest procedures to help them tackle changes in the climate'. Some workshop stakeholders raised the importance of improved science communication, by for instance 'making the academic knowledge as close to the people as possible', such as through Union Information and Service Centres.

Fourth, Sylhet workshop stakeholders also argued that current climate knowledge would benefit from eliciting the tacit knowledge of agri-cultural communities, and compiling it as a coherent whole. Accordingly, future research should 'collect information from the Sylhet Division stakeholders' and 'rely more on stakeholder engagement'. Indeed, the narrative-based interviews of our pilot survey provided an overview of the richness of local knowledge about the changing climate and its impacts. For instance, two female high school teachers in Sunamganj Sadar observed that bombax trees were not growing as tall as compared to the previous ten years, and a rice farmer in Barlekha explained how the delayed monsoon season could negatively affect rice production.

Fifth, creating new institutions and processes for compiling and bridging the knowledge of agri-cultural communities would be a useful way to store this knowledge and make it available to the agri-cultural communities of the Sylhet Division. These new institutions could be a way of improving not only the dialogue among the agri-cultural communities, but also between these communities, climate scientists and decision-makers. As argued in the Sylhet workshop: 'We should make links between local institutions and research organisations'.

Putting agri-cultures in front of the mirror

Hulme's metaphor of climate change as a mirror opens up a number of possible reflections on the foundations of climate knowledge in Bangladeshi agri-cultures. Here we briefly discuss four such challenges to knowing the climate and climate change in rural Bangladesh.

First, the rapid and largely unpredictable changes to the weather experienced by Bangladeshi communities present a powerful challenge to the inductive knowledge of climate, based on long-accrued observations and experiences with the weather. Communities are forced to question their traditional knowledge, captured in the Bangla calendar and the Khana sayings for instance, that does not correspond with their lived reality anymore. This detachment from traditional knowledge not only breaks communities' link with traditions, but also their long-learned relationship with nature and climate. Communities must re-learn to live with a new climate, a process challenged by uncertainty. Uncertainty is also a major challenge for climate science and meteorology in Bangladesh, particularly in the Sylhet Division. Science of climate in the area must muddle forward in the face of uncertainties associated with incomplete science, conceptual uncertainties on the underlying processes causing the weather, and an inability to model and predict these processes. Science is forced to reflect on how it can work with these uncertainties.

Secondly, climate science in Bangladesh faces a challenge to what it can meaningfully say about climate at the local scale; can 'universal' science be legitimate for local and contextual challenges? Climate science has until now been largely concerned with GCMs, but these abstract models are based on spatial grids that are totally out of scale relative to the lived experience of communities. Not only does this offer challenges to the fine-scale calibration of models for local conditions, but also fundamental challenges to the legitimacy of climate science in a local place, when its tools, methods and cultures are unrecognisable to agri-cultures. Place-based science means recognising that universal science may not be 'fit-for-function' in a particular place.

Reintroducing climate science to its context also presents a third fundamental challenge, to the primacy of science as the most powerful source of knowledge of climate. Science has long assumed primacy from its claims to extricate itself from the messiness of context and provide objective, value-free and universal knowledge that is 'true' independent of place. Workshop participants discussed a professional arrogance among the Bangladeshi scientific community, particularly at the national scale. This has seen the paradox of a science-poor Bangladesh that has made itself dependent on science-based climate adaptation. However, by acknowledging uncertainties in climate science, the scientific community is acknowledging its limits and acquiescing to a role of 'knowledge partner' alongside local communities; thus nesting knowledge in local agri-cultures: 'We should include general people (...) for an appreciation of local cultural lifestyles'.

Accepting the shared legitimacy of traditional, local and scientific knowledge of climate presents a forth challenge. How can agri-cultures create an inclusive setting for sharing and bridging different forms of climate knowledge, while also recognising that not all knowledge is of an equal quality? Some forms of knowledge are fitter for the function of supporting the adaptation of agri-cultures to their changing climate, but how can we judge this quality? It is not sufficient to use scientific norms because they are designed specific to scientific approaches, so local and traditional knowledge will always perform poorly by these measures. An inclusive setting implies the inclusive negotiation of what counts as important climate knowledge, ways for collecting this knowledge, and evaluating its quality. The TRACKS project will give effect to this via a 'citizen science' approach, which will bring together scientists and communities to: (1) choose a portfolio of indicators; (2) design a monitoring regime; (3) share in the monitoring duties; and (4) analyse and evaluate the findings.

Conclusion

We have seen in this paper how engaging with climate change as both a magnifying glass and a mirror provides an impetus to Bangladeshi agri-cultures to focus on and deeply rethink their current climate knowledge. This metaphor presents climate change as a creative force for challenging existing knowledge systems, and forms the point of departure for the TRACKS project's efforts to transform climate knowledge with Bangladeshi agri-cultures.

References

Alauddin, M. and Sarker, A. (2014). Climate change and farm-level adaptation decisions and strategies in drought-prone and groudwater-depleted areas of Bangladesh: an empirical investigation. Ecological Economics 106: 204-213.

Etzold, B., Ahmed, A., Hassan, S. and Neelormi, S. (2014). Clouds gather in the sky, but no rain falls. Climate and Development 6(1): 18-27.

Funtowicz, S. and Ravetz, J. (1993). Science for the post-normal age. Futures 25(7): 739-755.

GoB. (2012). Bangladesh economic review. Ministry of Finance, Finance Division, Dhaka, Bangladesh.

Hulme, M. (2009). Why we disagree about climate change. Cambridge University Press, New York, USA.

Rittel, H. and Weber, M. (1973). Dilemmas in a general theory of planning. Policy Sciences 4(2): 155-169.

Rotmans, J. (2006). Tools for integrating sustainability assessment. The Integrated Assessment Journal 6(4): 35-57

Warner, K. and Afifi, T. (2014). Where the rain falls. Climate and Development 6(1): 1-17.

Ethics approach in food choices along the agri-food supply chain: a survey in Apulia region

F. Contò, M. Fiore and R. Dicecca*
Department of Economics, University of Foggia, Largo Papa Giovanni Paolo II, 71121 Foggia, Italy; mariantonietta.fiore@unifg.it

Abstract

Food issues as increased prices and prices volatility, resources scarcity and climate change, are more and more emerging in today's fast moving global marketplace. In this scenario, understanding the dynamics of the agri-food chain is a crucial driver in order to tackle rapid and fast-spreading challenges related to the food products and their knowledge. Moreover, an increasing distance between production and consumption determined the need of acquiring knowledge and others determinants on food products. Nevertheless it is important to underline the relation between food ethics and innovation process (new technologies to production or management practices) whether the latter helps and boosts the development of the agro-food supply chain. The purpose of the paper is to investigate the link between food and ethics under several perspectives in the supply chain in order to understand how ethical issues may be traced to the analysis the agro food supply chain and how innovative food processes assess potential effects on the agri-food stakeholders. Agricultural ethics was being envisioned as an interdisciplinary sub-discipline that nowadays should become a significant basis to analyze norms and values being implicitly assumed by agricultural researchers, practitioners and policy makers assessing ethical and societal values equal to the economic concerns of production and consumption. Several works and policies are on the role of ethics including health, quality, trust, environmental welfare aspects in influencing consumer behaviour. Making use of a qualitative survey administrated to main actors involved in a supply chain the ethics approach is analyzed in food choices along the agri-food supply, whether consumer rely on trust and labeling practices, or whether producer foster the participation in innovation process and spread knowledge along the chain. Results corroborate the importance to build a fair and healthy food system for people and the environment. The paper is structured as follows: after a literature review on food and ethics, the present situation is described, especially ethical aspects, of the supply chain of wine in Apulia region in Southern Italy. Then, methodology is presented; an exploratory survey is carried out on two main steps of a supply chain: producers and consumers. Findings highlight that consumers choices are affected by some ethical motivation and also that wine producers increasingly shifted attention on sustainable practices and innovative methods to meet ethical issues. Finally conclusions are drawn.

Keywords: wine chain, ethical analysis, quality, innovation process

Literature review

Furthermore, a plethora of works and policies are on the role of ethics and quality, health and nutrition, trust and food safety, environment and animal welfare in influencing consumer behaviour (Grunert, 2005; Krystallis *et al.*, 2012; OECD, 2008; Young *et al.*, 2010). In particular, ethics is examined in the framework of social sustainability in an innovative perspective in relation to agricultural practices and the food supply chain. Population growth and unsustainable practices, global inequalities and environmental degradation are more and more increasing and it is crucial to know which aspects need to be strengthened from a business ethics perspective and where appropriate interventions are desirable (Kalfagianni, 2014). Many food scandals all over the world get worse consumers' confidence in food safety along food chain; ethical implications stand out (Burkhardt, 2012; Olsen and Banati, 2013). So consumers around the world are becoming progressively concerned with ethical, environmental and health issues (Brugarolas *et al.*, 2001). Organic agriculture, integrated agriculture, fair trade, and ethical trade can be identified such as the main agrifood-related movements in the context of environmental and social ethics (Codron *et al.*, 2006; Cullen and Kingston, 2009). Organic purchasing motives are principally attributed to consumers' environmental and health consciousness and/or food safety and quality issues (Krystallis *et al.*, 2012). Then several authors highlight the ethical behaviours which provide benefits either to society in general, to specific groups, or individuals. Community-supported agriculture (Brown and Miller, 2008; Cone and Myhre, 2000; Hayden and Buck 2012; Nost, 2014), Solidarity Purchasing Group (Brunori *et al.*, 2012; Fonte, 2013), the so called 'alternative agro-food networks' as well as the well known fair trade initiatives or social co-operation (for a better understanding of alternative food network see among others Harris 2009; Higgins *et al.*, 2008; Tregear, 2011;) make a rich set of practices according to concepts of 'ethic, 'local', 'embeddedness', 'quality', and 'short food supply chains' (also recognized as 'ethic of care'). Hence this current debate about ethics and sustainability of production and consumption in food and supply chain could be seen as a new consensus to develop innovative ways and to support local agribusinesses.

The wine chain: the case of Apulia region in Southern Italy

Recent trends in preferences move towards innovative oenological products characterized by low content in sulphites and lower content in alcohol. Demand changes, new consumer preferences and global market challenges drive wine producers in recognizing in consumer expectations for quality, sustainability and safety a significant opportunity to differentiate their products and to face competition (Contò *et al.*, in press). Organic production, for example, represents a good starting point. Following the publication of the new (EU) Regulation N° 203/2012 for organic wine production, excellent export performances and consumption growth stand out more and more (Nomisma, 2014). In 2012, 8% of the Italian vineyard area was organic (compared to the world average of 4%). In terms of absolute value, Italy ranks third in Europe with just over 57 thousand hectares of organic vineyards (+8.6% compared to 2011 and +81% compared to 2003). At the regional level, Sicily is the leader (16,144 hectares), followed by Apulia (10,173 hectares) that has a formidable wine area (about 100,000 hectares (*ibid.*). The main grape varieties of Apulia region are *Primitivo* and *Negroamaro*. Currently the region is experiencing a significant changes period in the wine sector; after years of wine production based on the high yields per hectare of especially cutting wines, due to their coloration and texture, it is now trying to exploit the territory reaching quality levels for some wines mainly made from native grapes varieties. Apulia region only follows Sicily in the Italian region rank of producers and is also second in the rank for distribution of organic wine in 2012, with 8,365 acres of surface cultivated of organic grapes (+11.9% compared to 2009) (Contò *et al.*, in press). Furthermore, Apulia is a virtuous example of Italy's agro-food sector; indeed, it is the first region to introduce ethical standard labeling in agriculture. Apulia has its strength in young population working age that is 60% of the region (ISTAT, 2014), whose 10% engaged in agriculture. So we can say that ethical dilemmas about food are more rooted owing to the use of social networks by young people that are so more responsible and awareness of food issues and challenges. Ethical standard labeling in agriculture means recognition for the companies which respect workers following labor laws. Actually Apulia authority is running to update the regulations in order to take into account technological innovation too. Therefore Apulia is a pioneer in this important ethic path in food sector so essential for human survival. Correlatively, Apulia wineries provide strategies that pay attention consumer market by means the 'Prodotti di Qualità Puglia' brand that is a brand of EU collective quality with an indication of origin guarantying the quality and origin of the product. It is an 'umbrella recognized label', ensuring that consumers can discover all the excellences along the entire agri-food chain; a special attention is provided to biological sector to promote those who have chosen to undertake an 'ethical' production marked by the warranty and the transparency of agricultural production. The brand is strengthened by the financial support from the EU quality regime and recognized by the regional assembly. Furthermore, Apulia has 25 protected designation of origin (PDO) wines so assuring that the entire product must be traditionally and entirely manufactured (prepared, processed and produced) within the specific region and thus acquires unique properties. In this matter, tradition and old experience are assured; many wineries have implemented a number of changes in the production of Apulia wine, enhancing and bringing innovations, but without forgetting tradition and the territory authenticity so respecting the ethics values and issues.

Methodology

The exploratory survey is carried out in Apulia region on two main steps of the supply chain: producers and consumers. The work of investigation has been carried out on following basis, giving rise to the following actions:

- Analysis of the competitive strategies of wine-making companies of the local system, carried out through the implementation of a questionnaire prepared by means Survey Monkey software. A web based survey (cg. e.g. Gilinsky *et al.*, 2008) has been submitted to a target population of 280 wineries involved in 3 integrated projects of food (IPFs) chain in wine sector; 204 responses have been collected. IPFs are implemented thank to the measure 124 of rural development programme of Apulia region and are aimed at the development of new products, processes and technologies; so, the involved wineries are representative for the wineries target population and can provide suggestions and insights. The main answers are on rating Likert scale equal to 1-7 score (1: never; 2: very rarely; 3: rarely; 4: occasionally; 5: frequently; 6: very frequently; 7: ever). About 75% of wineries choices sustainable agricultural and winemaking practices (such as efficient use of water for irrigation operations and for production processes), 47% introduce new varieties but only 11% use innovative processes such as ITC and GIS; attention for product quality is related to 94% of wineries and customer orientation (preferred to competitor orientation) is chosen by 84% of them. The main results is that 87% carries out strategies for reaching the consumer value and 87% finds info by all chain actors.
- Data collection about the consumer profile seeking for quality and perceiving social and ethics values, carried out through a face to face survey during a special wine event: the data have been collected during the Starry glasses Apulia event, in Trani, which is a annual wine tasting event fostering the tourism of wine in regions whose production is aimed to invest mainly on high quality and sustainability of their products. The questionnaires were submitted to 300 respondents, but complete questionnaires were collected only in 258 cases. Several literature corroborate this our psychographic segmentation: the wine tourists (Alebaki and Iakovidou, 2011; Contò *et al.*, in press; Yuan *et al.*, 2005) that represent a large percentage of consumer of wine based on quality production. The sample is composed on average by males (55.5%), forty years old, non-single, about 50% employed in the public and private sector dependent and freelance professionals and about 43% of respondents have college or higher degree.

As regards results deriving from the first survey administrated to wineries, the radar graphics (Figure 1 and 2) show higher values related to the positive consensus (5-7 Likert values) about environment respect (79.26%), sustainable practises promotion (70.21%), eco-friendly partners (71.81%), water resources optimization (66.49%). Generally speaking wineries are facing new challenges of recent scenarios paving attention on ethics/sustainable issues for reaching the always more awareness consumer. On the other hand, consumers declare to have a sustainable behaviour; 79% do separate waste collections and 62% saves power consumptions even if only 20% buy eco-friendly product. Focusing on organic sector, for consumers consuming organic food products is associated with positive effect to the human health (76.9% of the maximum score) and then impact to a less extent on the environment (72.4%). Scarcity and/or lack of information from operators on the organic food processing is highlighted by the most of the sample (70.4%). Furthermore, analyzing the open questions on mean of quality, for consumers

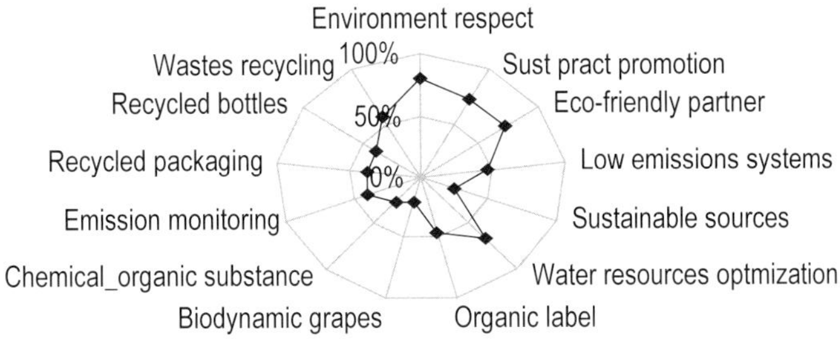

Figure 1. Radar graphics on results from wineries.

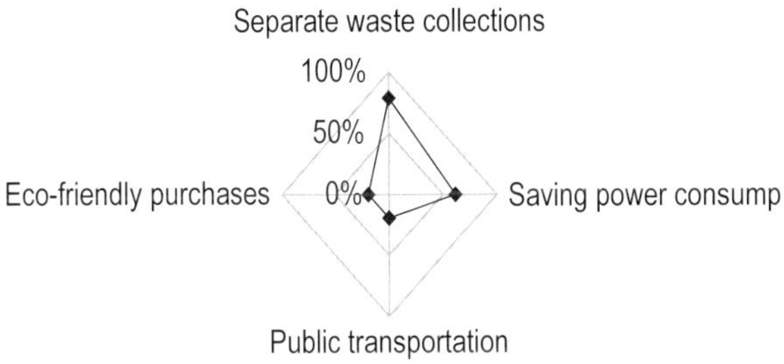

Figure 2. Radar graphics on results from consumers survey.

quality is typicality, origin, health, nutrition, safety corroborating the results from literature review (Grunert, 2005; Krystallis *et al.*, 2012; OECD, 2008, Young *et al.*, 2010). These aspects converge on concept of food ethics as conceived from consumers, producers and policy strategies in this new global scenario.

Findings

Results from the analysis seem matched with the desired supply chain emerging from theoretical literature. Looking under the consumer prospective, Zander and Hamm (2010) show the role of ethical attributes (among others 'regional production') to explain the new consumer attitudes and higher willingness to pay (WTP) for organic product. Moreover, Manning *et al.* (2006) in their review of ethical methodology and mechanisms for assisting in ethical decision, stress a very key component referring to trust. Ethics issues have to be traced along all the supply chain, so that according to Manning et other authors, it concerns policy makers and those manufacturing and supplying food organizations as well. Consumers need to trust that the supply chain stakeholders provide information which is accessible, accurate and affords reasoned choice when purchasing food products. They demand diversity and distinctiveness especially in wine supply chain, establishing a system of innovation which is able to adapt to a changing context (Brunori *et al.*, 2012). Conversely, when it comes to producers, ethic processes include caring about aspects of wine production that affect the natural environment, people, and welfare. Higgins *et al.* (2008) referring to those alternative agro-food networks, emphasize that producers should 'short circuiting' the supply chain characteristics by supporting consumers with a great deal of information both related to geography and quality attributes. That is easily translated for wineries that, particularly, via labeling, tackle the growing demand of local wine in ethical, environmentally sustainable and socially ways.

Conclusions

Food raises ethical issues, covering the entire process and its actors, starting from primary producers to final consumer and involving all aspects of the value chain. The aim of our study was to provide insights into the link between food and ethics under several perspectives in the wine supply chain. Ethical issues may be traced to the analysis the agro-food supply chain and how innovative food processes assess potential effects on the stakeholders. This paper seeks in the literature about the ethical motivations under both the consumer and wine-making companies lens in a south Italy region. Particularly the paper have focused on the ethic motivations that explain local consumer choices and the sustainability of the wine production processes. The findings highlighted that consumers choices are affected by some ethical motivation; they acknowledged that words as typicality, origin health, safety and nutrition are translated into quality products; they behave in a sustainable manner, focusing also their choices in the organic sector with a high (more than 70%) share of respondents who believe there are positive effects on human health and environment. The study highlighted also that wine producers increasingly shifted attention on sustainable practices and innovative methods to meet ethical issues. The brand 'Prodotti di Qualità Puglia' ensures wineries an innovative way to built trust between consumers that 'recognize' an added value in the local Apulia wine chain. In accordance with other studies (Manning *et al.*, 2006; Grunert, 2005), trust in local production and demand for health and safety, local identity and ethics upon the products, are perceived as crucial factors for consumers choices that are willing to pay even a higher price (Zander and Hamm 2010; Navin 2014). Wineries are trying to structure systems of innovation able to adapt to a changing context linking origin and quality, safety and environment for respecting the ethics issues; on the other hand, consumers definitely ask for quality and sustainable products, but emerged reluctance to change buying habits. This process may be related still to a weak information flow system along the supply chain, and in general between stakeholders, that bring mistrust in some of the respondents. Policy-makers could intervene stimulating a better matching between supply and demand side and providing transparency; wine sector operators could reduce scarcity and/or lack of information on the organic food. Some limitation of our research can be found, calling further investigations in this domain. Firstly, according to the survey, it is not clear whether labeling processing ensure the perfect information within the supply chain and for final users. Secondly the sample can be not quite representative and a comparison with another region or country could be very interesting. Furthermore, understanding if ethical approach is able to protect the local economy (even if you know that these products could be more expensive) can be crucial. Finally, politicians, producers and consumers, and rural social scientists should dialogue and renew their efforts to conceive ethics as a crucial collaborative domain (Thompson, 2014; Olsen and Banati, 2013; Burkhardt, 2012) and in an inter-disciplinary and trans-disciplinary manner.

References

Alebaki, M. and Iakovidou, O. (2011). Market segmentation in wine tourism: a comparison of approaches. Tourismos 6(1): 123-140.

Brown, C. and Miller, S. (2008). The impacts of local markets: a review of research on farmers markets and community supported agriculture (CSA). American Journal of Agricultural Economics 90(5): 1298-1302.

Brugarolas M. and Rivera L.M. (2001). Comportamiento del consumidor valenciano ante los productos ecológicos e integrados. Revista Española de Estudios Agrosociales y Pesqueros 192: 105-121.

Brunori, G., Rossi, A. and Guidi, F. (2012). On the new social relations around and beyond food. Analysing consumers' role and action in Gruppi di Acquisto Solidale (Solidarity Purchasing Groups). Sociologia Ruralis 52(1): 1-30.

Burkhardt, J. (2012). The ethics of food safety in the twenty-first century: who keeps the public good? The Philosophy of Food. University of California Press, Ewing, USA, pp. 140-160.

Codron, J.M., Siriex, L. and Reardon, T. (2006). Social and environmental attributes of food products in an emerging mass market: challenges of signaling and consumer perception, with European illustrations. Agriculture and Human Values 23(3): 283-297.
Cone, C.A. and Myhre, A. (2000). Community-supported agriculture: a sustainable alternative to industrial agriculture? Human organization 59(2): 187-197.
Contò, F., Antonazzo, A.P., Fiore, M. and La Sala, P. (in press). Assessing perceptions of wine tourists on organic wine. REAA. 1-2.
Cullen, F. and Kingston, H. (2009). Analysis of rural and urban consumer behaviour toward new food products using a food-related lifestyle instrument. Journal of Food service Business Research 12(1):18-41.
European Union (EU) (2013). Facts and figures on organic agriculture in the European Union, Unit Economic Analysis of EU Agriculture, October. Available at: http://tinyurl.com/ndc4j2d.
Fonte, M. (2013). Food consumption as social practice: solidarity purchasing groups in Rome, Italy. Journal of Rural Studies 32: 230-239.
Gilinsky, A., Santini, C., Lazzeretti, L. and Eyler, R. (2008). Desperately seeking serendipity: exploring the impact of country location on innovation in the wine industry. International Journal of Wine Business Research 20(4): 302-320.
Grunert, K.G. (2005). Food quality and safety: consumer perception and demand. European Review of Agricultural Economics 32: 369-391.
Harris, E.M. (2009). Neoliberal subjectivities or a politics of the possible? Reading for difference in alternative food networks. Area 41(1): 55-63.
Hayden, J. and Buck, D. (2012). Doing community supported agriculture: tactile space, affect and effects of membership. Geoforum 43(2): 332-341.
Higgins, V., Dibden, J. and Cocklin, C. (2008). Building alternative agri-food networks: certification, embeddedness and agri-environmental governance. Journal of Rural Studies 24(1): 15-27.
Italian National Institute of Statistics (ISTAT) (2014), Data on population census 2010. Available at: www.istat.it.
Kalfagianni, A. (2014). Addressing the global sustainability challenge: the potential and pitfalls of private governance from the perspective of human capabilities. Journal of Business Ethics. 122(2): 307-320.
Krystallis, A., Grunert, K.G., De Barcellos, M.D., Perrea, T., Verbeke, W. (2012). Consumer attitudes towards sustainability aspects of food production: insights from three continents. Journal of Marketing Management 28: 334-372.
Manning, L., Baines, R.N. and Chadd, S.A. (2006). Ethical modelling of the food supply chain. British Food Journal, 108(5): 358-370.
Navin, M.C. (2014). Local food and international ethics. Journal of Agricultural and Environmental Ethics 27: 349-368.
Nost, E. (2014). Scaling-up local foods: commodity practice in community supported agriculture (CSA). Journal of Rural Studies 34: 152-160.
Nomisma (2014). Nomisma newsletter. 04 April 2014. Available at: http://tinyurl.com/o8mv6l6.
Olsen, N.V. and Bánáti, D. (2013). Ethics in food safety management. In: Motarjemi, Y. and Lelieveld, H. (eds.) Food safety management: a practical guide for the food industry: Elsevier, Amsterdam, the Netherlands, pp. 1115-1125.
Organisation for Economic Co-operation and Development (OECD) (2008). environmentally responsible food choice. OECD Journal. General Papers 2: 125-162.
Thompson, P.B. (2014). Agricultural ethics: then and now. Agriculture and Human Values 32: 77-85.
Tregear, A. (2011). Progressing knowledge in alternative and local food networks: critical reflections and a research agenda. Journal of Rural Studies 27(4): 419-430.
Young, W., Hwang, K., McDonald, S. and Oates, C.J. (2010). Sustainable consumption: green consumer behaviour when purchasing products. Sustainable Development 18(1): 20-31.
Yuan, J.J. (2005). An analysis of wine festival attendees' motivations: a synergy of wine, travel and special events? Journal of Vacation Marketing 11(1): 41-58.
Zander, K. and Hamm, U. (2010). Consumer preferences for additional ethical attributes of organic food. Food quality and preference 21(5): 495-503.

Ethical concerns in Romanian beekeeping

C.B. Pocol[1] and L. Stan[2]*
[1]University of Agricultural Sciences and Veterinary Medicine of Cluj-Napoca, Faculty of Horticulture, 3-5 Manastur Street, 400372 Cluj-Napoca, Romania; [2]University of Agricultural Sciences and Veterinary Medicine of Cluj-Napoca, Faculty of Food Science and Technology, 64 Calea Floresti, 400509 Cluj-Napoca, Romania; laurastan@usamvcluj.ro

Abstract

The relationship between humans, bees and nature, is highly complex, with roots in ancient times. Honeybees contribute significantly to ensure food for humans both directly – through the bee products obtained, and indirectly – via crop pollination services. Because of pollination, bees play a crucial role in the proper functioning of terrestrial ecosystems. In the last 20 years a massive bee disappearance – colony collapse disorder (CCD) – has been recorded, which has yielded drastic repercussions on economy and nature. Therefore many governmental bodies have taken initiatives to sustain beekeeping. Romania is a country with a strong apicultural potential on the world market, as the 10th most important exporter in the world. In the last decade there has been a revival of beekeeping in Romania due to financial support from public policies. However, honey consumption in Romania remains low compared with other European countries. Therefore Romania exports about 75% of honey production. In this article an ethical matrix proposed by Mepham in 2006 is used to identify and perform a critical analysis of principles and factors in the Romanian beekeeping sector. The matrix is constructed by applying the *prima facie* principles of wellbeing, autonomy and fairness. The relevant interest groups taken into consideration are: honeybees, beekeepers, consumers, society and the environment. This matrix can be used to raise an awareness of apicultural ethical issues and to encourage ethical reflection.

Keywords: bee products, beekeepers, consumers, sustainability, fairness

The sustainability of beekeeping

Sustainability is an integrative concept based on three main pillars: economic, social and environmental. The sustainability of the Romanian beekeeping sector was previously evaluated by Pocol *et al.* (2012). The economic aspects identified by the authors in the first pillar are:

- production resources (named inputs – natural, material, human, financial);
- production results (named outputs – high productivity, superior quality and high value added bee products).

The link between the economic pillar and the social one is realised by food chain performance and satisfaction of food safety requirements. The direct outcome of hazard analysis critical control points (HACCP) implementation is the identification of contamination risks and the preparation of preventive and control measures to obtain safe food products for consumers. This outcome is reflected in a healthier lifestyle and in the valorisation of traditional products. Other components of the social pillar refer to the affordability for consumers and the knowledge of product quality criteria. The environmental pillar of sustainable beekeeping refers to management (good apicultural practices), eco-friendly packaging, the well-being of the colony and to pollination services.

Concerning this triple bottom line of sustainability, Norman and MacDonald (2004) concluded that it is more difficult to evaluate social and environmental performance than to measure financial results. It is well known that bees are essential pollinators, bringing substantial benefits to the environment, but it remains necessary to demonstrate the full costs and benefits that lead to economic profitability.

What ethical aspects are behind sustainable beekeeping?

The ethical matrix adapted from Mepham *et al.* (2006) is based on three principles: well-being, autonomy and justice. The first ethical matrix in beekeeping was developed by Heaf (2009) and takes into consideration four categories of stakeholders: bees, producers, consumers and biota. In addition to these interest groups, our ethical matrix (Table 1) includes society.

There is a close relationship between the components of the matrix, as presented in Figure 1. Each component of the matrix is discussed in detail in Figure 1.

Honeybees

The honey bee is not domesticated. Bees collect nectar and pollen for their own use; to create food for themselves and for future generations. But how does a bee benefit from having people around? Beekeepers provide the hive (living conditions) and treatments when needed (Chauzat *et al.*, 2014; Crane, 1999). When the queen suddenly dies, the bees or beekeepers replace it with a new queen. Improper climatic conditions negatively impact the life of the bees. In such instances, the beekeepers bring additional feed into the hive to ensure their survival. A beekeeper's actions are intended to secure the well-being of bees.

Table 1. A proposal for a beekeeping ethical matrix.

Respect for	Well-being	Autonomy	Fairness
Bees	Welfare, disease resistance Problems: colony collapse disorder	Behavioural freedom Problem: intensive pastoral beekeeping	Intrinsic value Access to qualitative food sources Maintain bee colonies
Producers	Satisfactory income and working conditions	Managerial freedom of action Access to information Good practices of manufacturing Problem: transhumance in organic beekeeping	Fair trade laws and practices Regulatory stakeholders: beekeepers' associations, governmental bodies, formal and informal networks Entrepreneurial ethics regarding authenticity and safety of bee products
Consumers	Safety and acceptability Nutritional quality of bee products	Informed choice of bee products-shopping guide Freedom of choice Opportunities to influence the production	Availability and affordability of bee products
Society (distributors, laboratories, legislative bodies, future generations)	Increased income from beekeeping and pollination	Freedom to manage resources for the benefit of society Financial programs for beekeepers and farmers	Equal access to the resources (information, bee products) Legislation
Environment	Conservation and protection.	Preservation of biodiversity by pollination.	Regional sustainability.

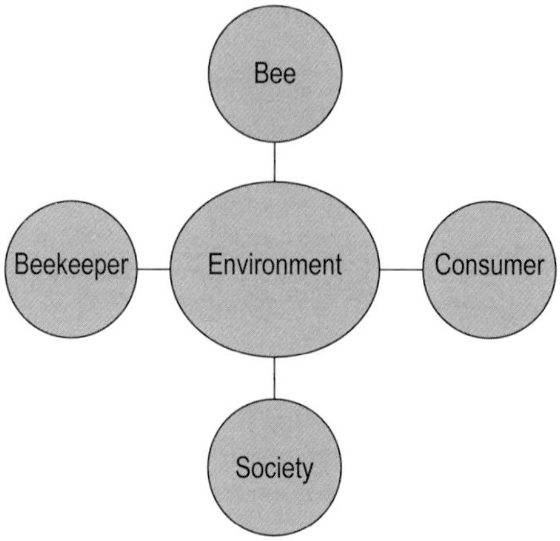

Figure1. Relationship between matrix components.

On the other hand, the evolution of society (industrialisation, new communication technologies, and the overuse of chemical products in agriculture) has led to CCD – a pressing global problem in the last 10 years (Dainat, 2012; Williams, 2010). Although bees have behavioural freedom in a 3 km radius, intensive migratory beekeeping negatively influences their welfare. The bees have a short life span (about 35 days) and intensive pastoral practices offer limited sources of feed (especially monofloral); in effect, bees are put on a diet. But if the period of migratory beekeeping is limited to a week in length, bees have access to various food sources that help to maintain colony health and to provide high-value bee products. Therefore beekeepers have the freedom to choose which beekeeping practice to apply for the best of bees and for society as a whole.

Producers

Most Romanian beekeepers practice beekeeping as a hobby, considering it a traditional, historical, cultural and recreational activity. A small percentage of them are professional (with more than 100 bee colonies), and in this case beekeeping is usually a family business. In both situations, beekeepers obtain decent revenues. In the last decade the number of Romanian beekeepers has increased; on the one hand due to the economical attractiveness of the sector and to European/governmental support, and on the other hand due to good working conditions: freedom in activity planning, the spiritual approach to life and a strong contact with natural cycles (Dumitrescu, 2012).

Heaf (2009) distinguished four possible moral stances concerning the managerial freedom of beekeepers' actions: dominator, steward, partner and participant. The series runs from the most anthropocentric (the dominator) to the most biocentric attitude (the participant), or from the most utilitarian (the steward) to the most respectful for the intrinsic value of bees (the participant). The author maintains that none of these stances is more important or ethically sound than any other.

Producers who are also members of beekeepers' associations have access to information regarding legislation, innovative technologies and practices, management and marketing of bee products. Therefore they have the freedom and ability to choose good practices. However in the case of transhumance in organic beekeeping, the risks of diseases increase significantly. The costs of treatments are higher than in conventional beekeeping, as are the production costs. According to Pocol *et al.*(2014), there remains a need for supplementary support for organic beekeepers in terms of creating added value for bee products.

Europe and the USA import honey from poor regions of developing countries. The beekeepers in these regions often don't have access to information. They also suffer from reduced possibilities for logistics and transportation, and a limited power of negotiation. In these conditions, the price obtained for such products is very low compared with the real price on the international market (Kevan *et al.*, 2007). Ethical issues arise from these situations. The solution could be the application of fair trade laws and practices that link beekeepers directly with buyers. In this case, the regulatory stakeholders would be the associations and cooperatives that support beekeepers to fight against poverty.

Consumers

Unfortunately, the economic results of beekeeping are strongly influenced by climatic conditions. During years with low production yields, beekeepers are tempted to cut the losses by product adulteration. In the case of honey adulteration, inverted sugar syrup is often used. This behaviour is detrimental to both the authenticity and safety of bee products, and by extension, to the consumer's health. Regrettably, there is a lack of population education about the quality and use of bee products. Romanian consumers are not aware of quality criteria, neither for the authenticity of bee products nor for the signs of its adulteration. The first sources of information for Romanian consumers are family members, as well asthose producers who are also providers of bee products (Kumar, 2011). This means that the information reaching the consumer is not necessarily correct. For example, crystallised honey is falsely considered by most consumers as adulterated honey, and this information is often passed among members of the family. In order to avoid confusion, beekeepers usually liquefy the crystallised honey by heating to dissolve the crystals. This processing step negatively affects the quality of honey by destroying the enzymes, vitamins, and formation of degradation compounds (e.g. hydroxymethylfurfural). Usually, consumers prefer to buy bee products from beekeepers rather than from the supermarket or from other sources.

Bee products cover the spectrum of all nutritional factors important for human well-being: carbohydrates (from honey and bee pollen), proteins and minerals (from bee pollen) and lipids (from pollen and royal jelly). Traditionally in Romania bee products are consumed during cold seasons, for prevention and treatment of specific medical conditions associated with winter. In this respect, it is even more important to consume authentic bee products of good quality. Due to this restricted seasonal use, Romania registers the lowest amount of honey consumption in Europe at about 400-600 g/year/person. This can be contrasted with 5 kg/year/person in Denmark, 1,5-2 kg/year/person in Germany and 1.5 kg/year/person Holland and Belgium (Dumitrescu, 2012). The most appreciated honey by Romanian consumers is the acacia variety, due to its clear colour, suave and delicate taste. The popular saying 'sweet as honey' fits perfectly for acacia honey. This variety is among the most expensive on the market, and is falsely considered by consumers to be of the best quality due to its neat and clear appearance and high price. However from a nutritional perspective, acacia honey is the poorest in minerals and proteins compared to honeydew. At the same time, acacia honey is the easiest to adulterate with sugar syrup, and consumers with little experience are easy to mislead. Consumers have the opportunity to influence production via their buying choices: they prefer monofloral honey and pollen. This attitude encourages the beekeepers to practice migratory beekeeping. As mentioned earlier, this is not the best option for bee health and for the environment, nor for the consumer.

Price is an important factor in the buying decision of bee products. For consumers with low income, supermarkets become the main source of honey purchases due to a convenient price. What is the reason to sell at such a low price? In many cases the product is imported, with a mixture of sugar syrups other than honey included. Therefore product content, as with nutrition and quality, can also be a problem of consumer education; many consumers are not accustomed to reading labels.

Society

The well-being of the whole society is influenced by beekeeping. Kevan *et al.* (2007), suggests that pollination is significant at regional, national and international levels. The elasticity of supply and demand is strongly influenced by pollination services. Therefore it is ethical that international bodies create a uniform methodology in order to assess regularly the contribution of pollination services to agricultural productivity. In Romania governmental bodies have taken initiatives to sustain beekeeping. The result of applying The National Beekeeping Programme was an increase of the number of bee colonies from 614,000 in 2000, to 1,248,298 in 2013 (Pocol *et al.*, 2014). Romanian governmental bodies gave their support for the development of regional Quality Control Laboratories of Bee products. This measure improved the export of bee products, especially in the EU.

Beekeeping practices and open classes for students help to increase the awareness among society members about natural life cycles and adaptation to nature. Training vulnerable groups could serve to maintain local employment and to aid in rural development.

Environment

Beekeeping's environmental benefits are reflected throughout the entire ecosystem: in landscape biodiversity, quality and diversity of bee products, and in healthy bees. There is a direct dependency between plants and bees. Many crops and wild plants also depend upon bee pollination for survival. Likewise, bees' health depends on the quality of food (nectar, pollen and water). Numerous factors have negatively affected this relationship: continuous pollution of the environment, use of genetically modified crops, and CCD have been the gravest factors in recent decades. The impact of losing honeybees is economically and environmentally immeasurable. Einstein said that if honeybees were to die, human kind would too die off within four years. Although many scientists have intensively studied the possible reasons for CCD (COLOSS Honeybee Research Association: http://www.coloss.org), a conclusive answer has not been found. Alternative options are being explored, such as Robobees (http://robobees.seas.harvard.edu).

Conclusion

Ethical concerns in beekeeping are presented at several levels: bees, producers, consumers, environment and society. More than 30% of the world's food depends on bees due to pollination services. The organization of life in a bee colony can serve as a model of sustainable development (e.g. division of labour, management of natural resources: feed, comb building from own secretions, natural materials used for hive insulation). The ethical focus of society should be to maintain the melliferous areas and to preserve the genetic heritage of bees, aspects that sustain local beekeeping. Therefore, at a societal level, beekeeping will have an increased economical impact if it is supported by apicultural policies, including performing services for the quality control of bee products. Bee products can be used as food or medicine, therefore their quality in the sense of authenticity is essential. An important condition to respect the ethical consumption of bee products is the quality of information, as well as the resources that reach consumers. This article has focused on the main ethical issues in beekeeping and is a starting point for further investigations.

Acknowledgments

Authors wish to thank Molly McDonough for the linguistic revision of the article.

References

Chauzat, M.P., Laurent, M., Saugeon, M., Hendrikx P. and Ribiere-Chabert, M. (2014) Epilobee report 2012-2013. A pan-European epidemiological study on honeybee colony losses. European Union Reference Laboratory for honeybee health (EURL). Available at: http://tinyurl.com/n9zgllu.

Crane, E. (1999). The world history of beekeeping and honey hunting. Routledge, New York, USA

Dainat, B., Van Engelsdorp, D. and Neumann, P. (2012). Colony collapse disorder in Europe. Environmental Microbiology Reports 4(1): 123-125.

Dumitrescu, M.J. (2012). Potențialul de export al României – Miere. Centrul Român pentru Promovarea Comerțului şi Investițiilor Străine. Bucuresti, Romania.

Heaf, D. (2009). Towards sustainable beekeeping. The Beekeepers Quarterly 91: 2-4.

Kevan, P.G., Eisikowitch, D., Kinuthia, W., Martin, P., Mussen, E.C., Partap, U., Taylor, O.R., Thomas, V.G., Thorp, R.W., Vergara, C.H. and Winter, K. (2007). High quality bee products are important to agriculture: why, and what needs to be done. Journal of Apicultural Research 46(1): 59-64.

Kumar, R., Lorenc, A., Robinson, N. and Blair, M. (2011). Parents and primary healthcare practitioners' perspectives on the safety of honey and other traditional paedriatic healthcare approaches. Child: Care, Health and Development 37(5): 734-743.

Mepham, B., Kaiser, M., Thorstensen, E., Tomkins, S. and Millar, K. (2006). Ethical matrix manual. LEI, the Hague, the Netherlands.

Norman, W. and MacDonald, C. (2004). Getting to the bottom of 'triple bottom line'. Business Ethics Quarterly 12: 243-262.

Pocol, C.B., Marghitas, L.A. and Popa, A.A. (2012). Evaluation of sustainability of the beekeeping sector in the North West Region of Romania. Journal of Food Agriculture and Environment 10:1132-1138.

Pocol, C.B., Moldovan-Teşelios, C. and Arion, F.H. (2014). Beekeepers' association: motivations and expectations. Bulletin of University of Agricultural Sciences and Veterinary Medicine Cluj-Napoca. Horticulture 71(1): 141-147.

Williams, G.R., Tarpy, D.R., Van Engelsdorp, D., Chauzat, M.P., Cox-Foster, D.L., Delaplane, K.S., Neumann, P., Pettis, J.S., Rogers, R.E.L. and Shutler, D. (2010). Colony collapse disorder in context. Bioessays 32: 845-846.

Section 13. Poster session

Edible but unsalable food: ethical and legal challenges in the EU

A. Arroyo Aparicio[1] and L. Escajedo San-Epifanio[2]*
[1]UNED. Commercial Law, Law Faculty, UNED, Calle Obispo Trejo, 2, 28040, Madrid, Spain;
[2]University of the Basque Country/EHU, Constitutional Law and History of Political Thought, Faculty of Social and Communication Sciences, Aptdo 644. Bilbao 48080, Spain; leire.escajedo@ehu.es

Abstract

In recent years the concern about food waste has grown in the EU. In accordance with official data, almost half of the world's food is thrown away. It has been said that if nothing is done, food waste in the EU is expected to rise to about 126 million tons by 2020. The European policy target is to reduce food waste by 50% by 2030 and institutions and actors in the food chain are considering different measures to achieve it. In this paper we analyze the ethical and legal aspects of some proposals that currently promote the donation of leftover food in ways that do not match the current EU and member state's legislations. In other words, we focus on some proposals that would involve changing the EU members' legislation and practices on food donation. Among others, we will reflect on the so called 'Food banking solution to food waste and hunger' proposal, which involves reducing legal barriers to food donation, following the model of the 'Good Samaritan laws' of other countries (like the USA) and giving supportive tax benefits to the potential donors. We think that some of the premises behind the above mentioned proposal are controversial and there are some relevant facts that have not been taken into account by its design.

Keywords: food donation, food waste, good samaritan law, food banking, EU food poverty

May food waste and food poverty be somehow reduced by just changing the legal framework? Ethical issues arising from the proposal for an EU Samaritan law

Up to half of the world food production is wasted (cf. e.g. FAO, 2011). According to the EU parliament the current wastage in EU27 is 179 kg per capita (89 million tons per annum) and if nothing changes it will get worse, with a major impact in environmental and economic terms as well as ethical implications. Model European parliaments (MEPs) have called on the commission to implement a coordinated strategy combining EU-wide and national measures to improve the efficiency of food consumption chains sector by sector and to tackle food wastage as a matter of urgency (EU parliament, 2011). When developing strategies for a more efficient food chain, we must take into account that the reasons for the waste are not always the same.

There is an amount of food waste, statistically still not totally clear, that is described as avoidable. Edible but not sellable food is presented as a subcategory of avoidable food waste. Widespread in some European countries, like Spain, we can find several projects – mainly local food banks and community larders – that aim to create new, more effective and efficient collection and distribution of surplus food that is not suitable for sale to the public, although it is still suitable for human consumption (Generalitat de Catalunya, 2013). They carefully design logistics circuits to bring the surplus food as close as possible to the final consumers (individuals or families). Such initiatives, we agree, clearly show that collaborative spaces can be found when trying to resolve food waste and urban food insecurity. However, it cannot be implied that this is a one-size-fits-all solution. This practice takes numerous forms and provides a variety of social, economic and environmental benefits. (Cotugna and Beebe, 2002).

We would like to reflect on some particular proposals that involve regulatory changes in order to promote the donation of food surpluses within the EU. Needless to say, first of all, the local food recovery processes we have mentioned above are lawful. We do not need to change the law to allow those local practices of preventing surplus foodstuffs from being dumped in the trash. The approach we take is focused on an EU level proposal that underlines the convenience of changing the EU food donation guidelines.

Following the documents of the European federation of food banks (FEBA), which brings together 256 food banks in 21 countries, we need to change our food donation law model. FEBA's proposal is based on the assumption that many potential food donors hesitate to donate their surplus food as they think that doing so will 'expose their enterprise to unmanageable and unnecessary risk of liability for negative consequences for the ultimate consumers of recovered food' ('Every crumb counts', FEBA, 2013). The FEBA has pointed out how some non EU-countries – and Italy in 2003 – have extended the so-called 'Good Samaritan laws' to food donations.

It is important to mention that those laws do not only address civil liability issues but also confer some tax advantages. We doubt the validity of the argument that elevated VAT taxes or absence of good Samaritan laws in the EU are 'major obstacles' for food manufacturers and retailers to donate surplus food to EU food banks, as 'Every crumb counts' document suggests. Therefore, we find it difficult to agree that a new regulation may solve all the involved issues. Furthermore, we would like to make some remarks on the comparative experience of the 'Good Samaritan laws'.

The good Samaritan food donation acts: a comparative perspective

Food waste is a tricky concept. It signifies that organic residues are generated by the handling, storage, transport, distribution, sale, preparation, cooking and/or serving of foods. It can be read in many ways. Of course, we guess there is an abundance of food not consumed and lost to human use by many actors. In developed countries, individuals and families who do not have sufficient access to food – maybe not all of them – can access free or reduced price food through emergency food facilities. These food facilities are among others food banks, soup kitchens, community larders or community kitchens; they have different names in different countries. Such food facilities receive food and/or cash donations. However, some actors maintain that the fear of liability is an obstacle to some potential food donors (Morenoff, 2002).

In 1964 Kitty Genovese was stabbed to death in the city of New York without passers-by providing any help. Some of them were afraid, but others admitted their fear of the responsibility for the consequences of their actions. For those who are legally compelled to provide emergency aid (e.g. police officers, firefighters or health care givers) there is a legal framework defining their eventual civil liability. There was not such kind of definition regarding the people that do not have any legal obligation to provide aid, and who act driven by compassion.

The public debate on this and other similar cases promoted the creation of many state laws supporting emergency aid. Due to their connection to the parable by Lucas (10:25) those laws were named 'Good Samaritan laws' (GS laws). However, instead of speaking about GS laws it may be more precise to speak about a legal concept of 'Good Samaritan' (Jaeck *et al.*, 2014). In fact, what those laws did was to define the legal status (the rights and obligations) of a person who – driven by compassion – provides emergency aid to another one.

At the beginning, food was not explicitly included in the GS laws. This changed when president Clinton signed into law the 'Federal Bill emerson good Samaritan food donation act'. This law protects the food donor and the food recipient agency against civil liability, excepting only gross negligence and/or intentional misconduct. Subsequently many states passed GS laws that provide liability protection to good faith food donors. Those laws protect good faith donors from civil and criminal liability, should the product later cause harm to its recipient. The scope of those laws is limited to food and grocery products that meet all quality and labeling standards imposed by federal, state and local laws and regulations, even though the food may not be 'readily marketable due to appearance, age, freshness, grade, size, surplus or other conditions' (Cohen, 2006).

Do we need good Samaritan food donation acts in Europe? The current debate

As has been mentioned previously, food surplus is not salable to the final consumer for several reasons and one of the ways to manage it may be using it for other purposes outside the market, such as for the support of the most deprived persons. However, the food recovered by non-profit organizations creates new risks related to company's reputation, for example in the case of products that are closer to the expiration date (Aiello *et al.*, 2014). But it is clear that also food donation can be an operation with positive consequences for companies. In this sense, it can contribute to ameliorate the company's reputation and to increment consumer fidelity and sales. Both of these statements are perfectly reconcilable. It is clear too, that tax concessions and other incentives can increase food donation by companies.

On the other hand, food banks and non-profit organizations are interested in obtaining a legal framework on the basis of 'GS laws'. This legal framework of American origin, as noted before, can be linked to some prior German and French judicial doctrine. But above all else, the Italian 'Federation of food banks' succeeded to impose a legal perspective and law N°. 155 was enacted on 16 July 2003. The Italian law applies only to the purposes of charity, free distribution of food to the most deprived persons, and the non-profit organizations are considered, within the limits of the service provided, to final consumers for the correct state of preservation, transportation, storage and use of foods.

The law 155/2003 is revolutionary in its simplicity. The law includes a single article though not without relevant consequences: according to the law 155/2003, non-profit organizations that retrieve food are not kept to the rules on food safety as an entrepreneur. This framework facilitates the donation of food to charities. In 2003 18620 meals were recovered and Italian food banks counted 654751 meals recovered in 2012.

It must be stated that considering non-profit organizations as final consumers is a very ground-breaking legislation and perspective. Indeed, the notion of consumer as it is known in a legal sense differs from the concept of consumer as used in other matters (e.g. 'Sociology or marketing'). In the legal sense, in a contractual way for instance, a precise definition of the 'consumer' is essential in order to delimit the persons entitled to extended legal protection regarding traders. The wider the definition of consumer, the more extensive the scope of consumer law provisions is and the less reason there is to speak about consumer law as a special regulation (Arroyo, 2013). This is clear so far, but also it's clear that the notion of consumer itself contains several specificities. The primary idea is that there is not only one notion of consumer. Linked to this first basic idea is the view that a distinction can be made between an abstract notion and different concrete notions (Bercovitz, 1987).

On this basis, the Italian use of the notion of consumer differs from these last scopes exposed. Indeed, in all concrete notions the functional element distinguishes the consumer from the professional. Conversely, in the Italian law, non-profit organizations that get food surplus and consumers get the same treatment.

Could a similar interpretation be concluded from the Spanish legal framework? According to the European consumer protection laws, consumer means any individual who, in contracts covered by the Directive (applicable in the specific case), is acting for purposes outside his trade, business, craft or profession. The expressed reference to natural persons must be highlighted as further discussed below.

Under Spanish law, there is a general law for the protection of consumers together with other special laws. The purpose of the general law is to establish the regime for the protection of consumers and users and the general scope of application is 'the relations between consumers or users and entrepreneurs' (Art. 2). For the purposes of this general law, consumers and users are natural or legal persons acting in a sphere that falls outside entrepreneurial or professional activity (Art. 3). In the opposite corner, for the purposes of this law, 'entrepreneurs are considered to be those natural or legal persons acting in the context of their entrepreneurial or professional activity, whether this be public or private' (Art. 4) and producers are considered to be 'goods manufacturers, service providers, intermediaries thereof, or importers of goods or services into the territory of the EU, as are persons presenting themselves as such by providing their name, trademark or other identifying feature along with the goods or services, whether on the container, wrapping or any other protective or presentational component' (Art. 5). Lastly, Article 7 gives the concept of supplier as 'entrepreneur that supplies or distributes products on the market, irrespective of the entitlement or contract by virtue of which such distribution is carried out'.

Article 3 of general law for the protection of consumers provides – for the purpose of the law- that legal persons and entities without legal personality acting non profit in a foreign field to a trade or business are also food consumers. For instance, from EU law perspective, 'Member States may decide to extend the application of the rules of the Directive 2011/82 to legal persons or to natural persons who are not consumers within the meaning of this Directive, such as non-governmental organizations, start-ups or small and medium-sized enterprises' (Directive 2011/83, whereas 13). It can therefore be stated that Spanish general law concerning consumer notion has an extended notion including not only natural but also legal persons.

In view of the above, it is the Italian law that it is necessary to bring up now. The purpose of the Italian law is directly related to the duty of care to be expected of the non-profit organizations concerned with food donations. In contrast to that specific legal framework, the Spanish law is a general one. This different scope of application undermines a totally identical interpretation. At least, there is the possibility under the Spanish law to consider non-profit organizations as consumers, in general. This consideration evidences, however, two problems. The first concerns the scope of the term 'legal persons' covered by the Spanish law. The second refers to the application of this law only to the relations between consumers and professionals.

Without a clear specific legal framework, the liability related to food donations could be considered under the general provisions (contractual or non-contractual liability) or rules on civil liability on defective products. Similarly, food donations are subjected to food safety and hygienic rules (178/2002 and 852/2004 European Regulations). In the absence of this specific legal framework, some guidelines for the safe handling of food are published.

Final reflections

Food insecurity is a systemic and severe problem that should not be seen as separated from the social exclusion and poverty in which many people barely survive. Whether the law is changed or not, that should not reinforce the idea that hunger can simply be alleviated through a single piece of legislation. Individuals and families without food security need means and abilities to become healthy and socially included (Riches, 2002).

In terms of civil liability, EU countries have rules that provide adequate protection to food donors and recipients. Some minor changes can be made, at best, to clarify those terms. Therefore, the final goal of the so-called 'Food banking solution' may not only seek to clarify liability issues, but also to provide fiscal advantages that for the moment exist only in Italy. Should the fiscal perspective be the key for those proposals that apparently tend to remove 'barriers to food donation', we think their advantages and drawbacks need to be examined in depth.

Acknowledgements

This research has been done at the Urban Elika research group, funded by the University of the Basque Country, University & Society Research Grant's Call, 2014-2016; besides, Leire Escajedo is member of the Research Team Grant of the Directorate on Scientific Policy of the Basque Country IT 743-13 'Multilevel constitutionalism and integration of diversity (political, legal-institutional and cultural-religious)'.

References

Aiello G., Enea M. and Muriana C. (2014). Economic benefits from food recovery at the retail stage: an application to Italian food chains. Waste Management Journal 34: 1306-1316.

Arroyo, A. (2013). Who is a 'consumer' on food law: some reflections on the notion of consumer and the EU food law. In: Röcklinsberg, H. and Sandin, P. (eds.) The ethics of consumption. The citizen, the market and the law. Wageningen Academic Publishers, Wageningen, the Netherlands, pp. 465-470.

Bercovitz, A. (1987). Estudios Jurídicos sobre protección de los consumidores, Tecnos, Madrid, Spain, 326 pp.

Cohen, J.A. (2006). Ten years of leftovers with many hungry still left over: a decade of donations under the bill emerson good samaritan food donation act. Seattle Journal for Social Justice 5: 455-496.

Cotugna, N. and Beebe, P.D. (2002) Food banking in the 21st century: much more than a canned handout. Journal of American Dietetic Association 102(10): 1386-1388.

European Federation of Food Banks (FEBA) (2013). Every crumb counts. Joint food wastage declaration.

European Union (EU) parliament (2011). Non legislative resolution on avoiding food wastage: strategies for a more efficient food chain in the EU (2011/2175 (INI)).

Food and Agriculture Organization of the United Nations (FAO) (2011). Global food losses and food waste. Extent, causes and prevention.

Generalitat de Catalunya (2013). Guia de prácticas correctas de higiene para el aprovechamiento seguro de la comida. Available at: http://tinyurl.com/mwly4br.

Jaeck, F., Cooke, P., Verstrepen, W., Henriksson, T., Schetter, P., Teixeira De Matos, J., Massiello, M., Beades Martin, I. (2014) Good Samaritan law across Europe. Available at: http://tinyurl.com/lxbubr8.

Morenoff, D.L. (2002). Lost food and liability: the good samaritan food donation law story. Food and Drug Law Journal 57: 107-132.

Riches, G. (2002). Food banks and food security: welfare reform, human rights and social policy. Lessons from Canada? Social Policy & Administration 36(6): 648-663.

Cattle husbandry without slaughtering: case studies from Europe and India

P. Meyer-Glitza
Humboldt University Berlin, Faculty of Life Sciences, Division of Resource Economics, Hannoversche Straße 27, 10115 Berlin, Germany, patmg@web.de

Abstract

Vegetarianism and veganism are becoming more and more common, also often for ethical reasons. As a lacto-vegetarian diet results in the slaughtering of non-productive milk cows, fattened male calves and bulls, the question arises: Are there ways to keep cattle without slaughtering, and with dairy products for lacto-vegetarians? On 5 farms without slaughtering, narrative and semi-structured interviews were conducted and analysed with qualitative methods of grounded theory and sequential micro analysis. In a cross-case comparison of the 5 case studies (4 from Europe, one from India) the ethics, their context and the given system of husbandry were examined. Along the case studies 5 rules of the care-system and its ethics are identified: (1) an universal (moral) status of the different animals for which they are taken or kept on the farm; (2) an unconditional care for the animals, without a need for the individual animal to be productive; (3) a lifetime of care as a promise not to (re)commodify the animals, but to care for them in age and dying – similar to homes for elderly or handicapped; (4) a familising of the cattle akin to pets, where the cattle are seen as individualities and are treated as part of the enlarged family, and (5) prevention of slaughtering by example and campaigning. The analysed husbandry of the agri-care-system are specified as 3 farm stiles: (1) the pure sanctuary without any kind of animal products being sold; (2) the agricultural sanctuary where innovative dung-products and products fertilized with dung are sold (oxen may also be used for draft); and (3) vegetarian milk production with cattle breeding. Along with Sanctuaries there are cattle-farms which do not slaughter their cattle and which have the potential to supply lacto-vegetarians with milk products within an agri-care-system. The total amount of milk production is radically reduced and milk can be viewed as an expensive by-product of a cattle-husbandry without slaughtering while the cattle live their lives also in illness, age and with handicaps.

Keywords: slaughterfree, milk, vegetarian, care-ethics, animal rights

Introduction

Farmers occasionally keep single cattle on the farm until they die a natural death, if there is a special story connected with the animal (cf. Bock *et al.*, 2007). Shared times of illness (Zeitelhofer, 2009: 121), an outstanding breeding performance, or the status as 'outdoor pets' (Wilkie, 2005) on hobby-farms can secure a long live without slaughtering for single animals. Besides this, some farmers prefer not to slaughter all of their cattle (Baur, 2008; Lodrick, 1981: 160). Which ethical understanding motivates farmers to secure the sustenance of the cattle and give them the possibility of a natural death – independent from their productivity performance? Which ethical approach does emerge? Albert Schweitzer (1990: 330) pointed out to the basic dilemma of living on the life of other living beings which also want to live. Ethical vegetarians and vegans try to avoid this dilemma at least partially. Thus the question arises: Is there any kind of milk-production for lacto-vegetarians, that does not (indirectly) cause the fattening of milk-calves and the slaughtering of unproductive cattle? Under which conditions?

Method

Qualitative-interpretative methods are especially suited for emerging topics and phenomena about which little is known (Rosenthal, 2005: 18). Interviews within 5 Case studies – four in Europe and one in India – were conducted, using narrative and semi-structured interview techniques. The cases were selected by theoretical sampling (Strauss and Corbin, 1996: 148-165) during the ongoing research, drawing on emerging results and opportunities during field research, while looking for minimal and maximal internal contrasts. The Data of the 5 Case Studies were coded and categorized (Charmaz, 2006) separately using the coding system of grounded theory with open coding, axial coding, selective coding and the software of Atlas.ti. Additionally, text-segments were selected according to their content, change of speaker, description, narrative, or unintelligibility and analysed by sequential micro analysis (Rosenthal, 2005). Beginning with the interviews and their transcription, memos were constantly generated and further developed. This led to the emerging of categories and subcategories with their respective properties and dimensions. After analysing the single cases, the anonymized cases were compared along the categories within the three given interests of research: the biography of the farmers, the ethics of action (focus of this paper) and the management of an agri-care 'system that works' (cf. Meyer-Glitza and Baars, 2012).

Results: cross comparison of the care-system

Together with the agrarian-system (Table 1) the care-system emerged as the main constituent of the analysis. Besides animal welfare, care is motivated by aspects of Care-ethics (Donovan and Adams, 2007), non-violence (ahimsa) (Lodrick, 1981) and animal rights (Regan, 2004; Hall, 2010). Along with the daily chores of cattle husbandry it practically includes caring for ill and handicapped animals, as well as ageing and dying animals. There are 5 core rules of a shared ethical understanding and practice of the farmers (cf. Meyer-Glitza and Baars, 2012): (1) universal; (2) unconditional; (3) lifetime of care; (4) familising; and (5) prevention and the public sphere.

Table 1. Farm stiles, farms and their properties in the year 2012.

	Sanctuary	Agricultural sanctuary		Vegetarian milk farm	
	Farm 1 (EU)	Farm 2 (EU)	Farm 3 (India)	Farm 4 (EU)	Farm 5 (EU)
Height above sea level in m	1	150	610	250-500	
Number of cattle	35	42	ca. 300[a]	35	45
LU/ha	1.1	3.5	(21)[b]	0.8	1.4
Milking				X	X
Breeding				X	X
Ox draft			X		X
Dung products		X	X		
Horticulture, agricultural products		X	X	X	X
Sanctuary	X	X	X	X	

[a] Cattle and buffalo.

[b] The figure is not really informative as the animals graze in the forest and other common grazing areas that are not included in the figure.

Universal

Male, old, unproductive and maimed cattle which otherwise would be slaughtered are kept or taken on the farms and cared for. In principle and additionally, all kinds of animals (from all kinds of species) are looked for. The widening circles of care (e.g. Donovan and Adams, 2007) expand from man and his family to the pets, the cows, bulls, buffalo and other farm animals, they include the wild animals and finally the whole bio- and ecosphere. This inclusion was not anticipated at the begin of the research as the focus was on milked cattle. Instead of a non-slaughtering cattle husbandry – or even a cowcentrism as a form of specisism (Ingensiep and Baranzke, 2008: 125) on one farm – the husbandry could be called a non-slaughtering animal husbandry.

Within the sanctuary function farmers 'rescue' animals from slaughtering and bring them as 'refugees' (Hall, 2010: 49) into the save haven of their farm where the cattle are cared for. There emerges a threefold course of rescue-protection-care. All three of them are positive duties compared to a prohibition of killing (cf. Wolf, 2004: 42). To rescue connotes a heroic act, compared to the daily chores of caring for the animals. Protection encompasses the land of the farm, a legal form (e.g. a trust), a community which ensures continuity and (financial) support, and the promise of no slaughtering. This promise is in contrast to Wilkie (2005), who reports that the status of cattle as pets has a 'transient nature' and that 'decommodified' animals can be 'recommodified' (at any time).

Unconditional

Beginning with the theoretical sampling, the function as a sanctuary became a crucial focal point – contrary to the initial plan to interview farmers that milk their cows. To emphasize their function as a sanctuary some of the farms intentionally take cattle which are obviously sorted out as non-productive and show signs of maltreatment. Although by most farms (some) cattle-products are wanted or welcomed, productivity is no precondition to be cared for, it is not a category of life and death within the function as an asylum (Hall, 2010: 35). For Donaldson and Kymlicka (2011: 139) animal rights can include a 'cautious commercialization of the use of cow's milk (that) could lead to it becoming a luxury good, resulting in a limited but stable cow community' if cows were allowed to mate along their own inclinations and to raise their calves.

The unconditional caring for the cattle has a strong impact on the productivity, financial situation, organisation and the working of the agri-system. Besides public relations or a holiday flat, products of the farms are e.g.: different types of (fermented) fertilizers, produce, milk, dung-products (like massage oil, insent sticks, soap, distilled urin for traditional Indian ayurvedic products), draft power. The two milking farms milk their cows after calving for years. Farm 5 milks up to 10 years, with an average of a 3 years long lactation. On farm 4 there have been spontaneous and partially recurrent lactations in spring by 5 dried off cows after 2-3 years without being in calf again. Three of them have been milked, but milk production has rather become a by-product of cattle keeping and farming. For the Hare Krishna farm (farm 5) which provides the adjacent Hare Krishna Hindu-temple and community with milk, the milk production remains focal. Other vegetarian milk brands like 'Ahimsa milk' and 'Cow nation' (both respectively supplied by one farm) sold in 2011 milk in London for 2.40 £ per litre or 1.60 £ per pint (568 ml) respectively (Meyer-Glitza and Baars, 2012).

The agri-system of the 5 farms can be differentiated into 3 farm stiles, which include: (1) the pure sanctuary without any animal products being sold and the manure solely being used for the pastures; (2) the agricultural sanctuary where the above mentioned dung-products are sold (oxen may be used for draft); and (3) the vegetarian milk farm that produces milk, breeds and works the oxen on farm 5.

Lifetime of care

The cattle coming onto the farm and the cattle born on the farm are protected to be able to live their 'whole life' (Sezgin, 2014: 178). The lifetime of care includes caring for the cattle from birth or time they arrive until their death. This includes the (time-) intensity of care while living together for a long time and the 'intimacy' during illness and dying. 'We are not killing our old people, and one way or the other we have found a solution because we have old age homes or the old people are with the family, whatever it is, but it is not being questioned, so in the same way it should be for cattle' (farm 2). In this way the functional role of the cattle keeper resembles modern institutional roles like that of old age homes, asylum homes, a hospital, workshops for the handicapped, a kindergarten, a hospice and psychological support for 'traumatized' animals. This is pioneering work as there is hardly any literature and knowledge available about cattle with arthritis, obesity, cancer or their geriatric and palliative treatment, as is for the geriatric treatment of dogs and cats (cf. Kraft, 2003). Farm 1 and 4 euthanize most of their dying cattle as a 'preference respecting euthanasia' (Regan, 2004: 113f), while the Hare Krishna community explicitly rules out euthanasia.

The integrity of the animals as their ability to express species-specific behaviour and to integrate influences of the environment (Schmidt, 2008) is cut off by slaughtering. Additionally, slaughtering is the ultimate destruction of the bodily (Schmidt, 2008) integrity of 'a living whole' (Verhoog, 2005). Taking this into account, a lifetime of care and integrity as a lifetime-progression-integrity (Lebensverlaufsintegrität) adds a dimension of ontogenesis and individual psychological development to the properties of animal integrity.

Familising

The cattle are mostly treated as individualities and taken into a moral sphere akin to that of pets and family members. The time frame, intensity of care, emotional involvement and the developing bond can be compared to that found in the above mentioned institutions and within the human family. Still, there is 'an overlapping but different space from humans in the family. Even people who think of their pets as children know this is not literally true' (Cohen, 2002). Three aspects of familising are identified:

1. Familising as a rhetorical metaphor to claim support for the animal (rights) issues and for the farm in public debate and in contact with visitors (see rule 5). This includes the option of a godparenthood for an animal to financially support their sustenance.
2. In the process of familising, functional family roles are allotted, alike the cow as a mother that nourishes man. The functional roles are not static and can change from animal to animal and situation to situation, when the animal e.g. becomes brother or sister, father, child or aunt and man becomes father or child. An additional understanding of the family as a clan, a farming or religious community (farm 5) includes many families into a culturally protected space of a broader 'family'. On farm 5 cattle keeping is a religious way of life and of worshipping God while the Hindu scriptures and the Guru are the sources of authority. Analysing the scriptures this farm arrived at five basic principles of cattle keeping (no killing, handmilking, calves suckle from their mothers, meaningful work for the ox and natural vegetation feed). Within a god-given symbiotic human-cattle-bond their cowcentrism includes the castration of the bulls, the drinking of milk and its use for the temple rituals.

   In an attempt to grasp the human-cattle relation cattle are also called employee, fellow worker (Wilkie, 2005) and friend on farms. At the same time some farmers say that their cows basically prefer to relate to their conspecies. Along a continuum within the given 'local human animal culture' (Segerdahl, 2007) the individual cattle stay and move – depending on the actual situation and the history of the human animal interaction – more towards the conspecies-pole or the human-pole of the continuum.

3. Familising as a bond within the framework of attachment theory resembles our common understanding of the core-family. Relating to the human-animal-bond attachment is usually looked at within pet-human bonds (Zilcha-Mano *et al.*, 2011). In regard to cattle, the quality of the bond depends on the time the cattle have been on the farm, the number of cattle, their use/function and history of illness, the daily physical contact (Bock *et al.*, 2007) and the amount of work by hand.

Prevention and the public sphere

The farmers promote a common right of life for cattle and other animals by campaigning, public relations and direct action like rescuing cattle (from illegal slaughter in India). For this they sometimes collaborate with official bodies, networks and other farmers. The selling of the products and the exemplary work on the farms (that can be visited) supports their work in the public sphere. In line with this, the main product of farm 1 is the raising of awareness for animal rights by being an example (for visitors), repeated contacts with the media, their internet presence and products like calenders, t-shirts, etc. In India farm 2 runs an ox-adoption programme which hands out rescued draft oxen to local farmers as long as they are strong enough. The oxen are taken back and cared for if they become ill or to old. Farm 2 is part of a NGO which includes also a wildlife rescue centre, an animal hospital (including Animal Birth Control for male and female cattle and Buffalo), agriculture, a vegan shop, a cattle hospital and a cattlehospice. Being a vegetarian the head of the NGO became vegan after a post mortem operation of a cow taken up from the streets disclosed 40 kg of plastic in the rumen. As these (street) cows starve with a stomach full of plastic, the NGO continues with lubetomic operations to take out the plastic from their rumen. The 'plastic-cow' project works in collaboration with other NGOs to ban plastic along the roads and reached up to the Indian Supreme Court.

Conclusion

The individual life of the cattle and other animals is the main product of the farms. A limited number of the combined care- and agri-systems has milk production. This farms have the potential to supply lacto-vegetarians with milk products by a radically reduced and expensive milk production in comparison to other dairies of a similar herd size. Besides the central role of milk especially for the Hare Krishna farm, milk can be a by-product of cattle-husbandry (and dung production) while the cattle are allowed to live their lives also in illness, age and with or without handicaps. The building up of a cattle husbandry without slaughtering requires new systems of husbandry that adjust e.g. to the housing of handicapped and dying animals, to extensive feeding, to new (dung) products, to draft power or techniques for the daily turning over of downers and to medical treatment of old and dying animals – including palliative treatment. This form of husbandry usually has and requires the financial support by some kind of community and is usually to be found within organic farming practices. Compared to a vegan agriculture and from a perspective of organic farming, the cattle still secure the benefits of cow manure (Oltmanns, 2013), they convert grass(land) into milk, dung and additionally draft-power. For the farms that only want to rescue cattle their actual aim is the ending of breeding and commercial animal use, including dairying (farm 1 and 2). They see sanctuaries as a temporary solution.

The cattle have a right of life (a lifetime integrity, or more exactly a lifetime-progression-integrity), a right of bodily integrity (which is violated by the castration of the bulls as a compromise to be able to keep them) and of freedom (especially regarding the mother-calve bond) to the extent possible (Regan, 2003; Sezgin, 2014: 192). And they have a right of care if kept on the farm. The moral status of the familised cattle resembles that of people with severe intellectual disabilities. Referring to new theories about disability and citizenship, Donaldson and Kymlicka (2011: 106f) draw attention to the 'subjective good' of people with severe intellectual disabilities and to their 'individualized capacities' in negotiating challenges and by developing 'trusting relations' over time. Dependence is not perceived as a shortcoming, but as a fact of life. The main idea is that of a 'trust based 'dependent agency' ... that is exercised in and through relations' that 'assist the expression of agency' (Donaldson and Kymlicka, 2011: 104). And this applies also to cattle that depend on human care for their species-specific needs, individual preferences and right of life. Ethological research about behaviour and herd-structure of the cattle at different stages of life and productivity, corresponding to the health status would help adjust to the needs of the animals. The extended time frame of the human-cattle-relation as well as the medical treatment of the animals and further adjustments to a cattle husbandry without slaughtering also warrant further research.

References

Baur, G. (2008). Farm sanctuary. Touchstone, New York, USA, 286 pp.

Bock, B., Van Huik, M.M. and Prutzer, M. (2007). Farmers relationship with different animals: the importance of getting close to the animals. International Journal of Sociology of Agriculture and Food 15(3): 109-125.

Charmaz, K. (2006). Constructing grounded theory: a practical guide through qualitative analysis. Sage, Thousand Oaks, USA, 208 pp.

Cohen, S.P. (2002). Can pets function as family members? Western Journal of Nursing Research 24(6): 621-638.

Donaldson, S. and Kymlicka, W. (2013). Zoopolis. A political theory of animal rights. Oxford University Press, Oxford, UK, 352 pp.

Donovan, J. and Adams, C.J. (2007). Introduction. In: Donovan J. and Adams C.J. (eds.) The feminist care tradition in animal ethics. Columbia University Press, New York, USA, pp. 1-15.

Hall, L. (2010). On their own terms. Bringing animal-rights philosophy down to earth. Nectar Bat Press, Darien, USA, 330 pp.

Kraft, W. (ed.) (2003). Geriatrie bei Hunden und Katzen. 2te Auflage. Parey Verlag in MVS Medizinverlage, Stuttgart, Germany, 340 pp.

Lodrick, D.O. (1981). Sacred cows, sacred places. Origins and survivals of animal homes in India. University of California Press, Berkeley, USA. 307 pp.

Meyer-Glitza, P. and Baars, T. (2012). Non-killing cattle husbandry. In: Rahmann, G. and Godinho, D. (ed.) Tackling the future challenges of organic animal husbandry. 2nd Organic Animal Husbandry Conference, Hamburg, Trenthorst, Germany, 12-14 September 2012. Landbauforschung 362(vTI): 184-187.

Oltmanns, M. (2013). Vieharme Landwirtschaft. Brauchen wir Tiere für eine Bodenfruchtbarkeit? Forschungsring-Materialien Nr. 27. Forschungsring für biologisch-dynamische Wirtschaftsweise e.V., Darmstadt, Germany, 16 pp.

Regan, T. (2003). Sentience and rights. In: Turner, J. and D' Silva, J. (eds.) Animals, ethics and trade. The challenge of animal science. Earthscan, London, USA, pp. 79-86.

Regan, T. (2004). The case for animal rights. Second edition. University of California Press, Berkeley, USA, 425 pp.

Rosenthal, G. (2005). Interpretative Sozialforschung. Eine Einführung. Juventa Verlag, Weinheim, Germany, 241 pp.

Schmidt, K. (2008). Würde oder Integrität – Verlangt die gentechnische Veränderung von Tieren neue tierethische Konzepte? ALTEX 25(4): 313-320.

Schweitzer, A. (1990). Kultur und Ethik. C.H. Beck Verlag, München, germany, 372 pp.

Segerdahl, P. (2007). Can natural behaviour be cultivated? The farm as local human animal culture. Journal of Agricultural and Environmental Ethics 20: 167-193.

Sezgin, H. (2014). Artgerecht ist nur die Freiheit. C.H. Beck, München, Germany, 301 pp.

Strauss, A. and Corbin, J. (1996):. grounded theory: Grundlagen Qualitativer Sozialforschung. Beltz, Psychologie Verlags Union, Weinheim, Germany, 227 pp.

Verhoog, H. (2005). Animal Integrity. In: Marie, M., Edwards, S., Gandini, G., Reiss, M., Von Borell E. (ed.) Animal bioethics. principles and teaching methods. Wageningen Academic Publishers, Wageningen, the Netherlands, pp. 97-112.

Wilkie, R.M. (2005). Sentient commodities and productive paradoxes: the ambiguous nature of human-livestock relations in Northeast Scotland. Journal of Rural Studies 21(2): 213-230.

Zeitelhofer, S. (2009). 'Vom Umgang mit dem Vieh' – Eine qualitative Untersuchung zur Mensch-Nutztier-Beziehung in Niederösterreich. Dissertation. Universität Wien, Vienna, Austria, 236 pp.
Zilcha-Mano, S., Mikulincer, M. and Shaver, P.R. (2011). An attachment perspective on human – pet relationships: conceptualization and assessment of pet attachment orientations. Journal of Research in Personality 45(4): 345-357.

Studies about characteristics and treatments of cans industry wastewater

M. Botis
University 'Politehnica' of Timisoara, Mechanical Faculty, B-dul Mihai Viteazul nr. 1, Romania; botis.mihaela@gmail.com

Abstract

Wastewaters from vegetables and fruits cans industry results from washing of the raw products, peeling and cutting, boiling, sterilization, cooling process and equipments washing. Effluent from cans processing contains soluble organics, suspended solids, peels, seeds, trace organics which releases gases, causes taste and odor and a specifically turbidity and color. This paper analysis wastewater parameters from cherries, bean and pears and presents conclusions about the most polluted wastewater from cans industry. By comparing the obtained values graphically represented, especially for bio-chemical oxygen demand and chemical oxygen demand results the most polluted wastewaters from pears, cherries and bean cans processing industry.

Keywords: bio-chemical oxygen demand, chemical oxygen demand, effluent

Introduction

The economics of the wastewater lie in the amount of product loss from the processing operation and the cost of treating this wastewater material with a high impact on the societal life. The cost for product loss is self evident, however, but the cost of wastewaters treatments depend on its characteristics. The environmental consequences if the pollutants agent are not removed can seriously affect the ecological ramifications: if inadequately wastewater were to be discharged to a stream or river, an eutrophic condition would develop within the aquatic environment due to the discharge of biodegradable, oxygen consuming compounds (high values of bio-chemical oxygen demand (BOD) and chemical oxygen demand (COD)) so the ecological balance of the receiving stream, river or lake would be upset.

Industrial activities in food sector have one of the highest consumptions of water and are one of the biggest producers of effluent per unit of production. Wastewaters from vegetable and cans industries results from different processes as peeling, cutting, boiling as well as from cleaning with caustic solutions (sodium hydroxide) acid solutions, rinsing, disinfection (Botis Nistoran, 2014). Large amount of water used for cleaning can contain detergent, sanitizers, base, salts and organic matter depending on the process from which they result. The soil quality is also affected by wastewater infiltration into the ground, especially when they are discharged without any treatment.

Disposal of wastewater into the rivers, land and fields without a proper treatment will cause a serious problem to health and hygiene. Wastewater volume fluctuated widely because of differences in production, therefore water concentration and weight of waste generated per unit of product processed (g waste/kg product) also changes (Soderquist *et al.*, 1971).

Effluents (wastewater) resulted from vegetable and fruit cans industries contains soluble organics which released gases, causes tastes and odor and modified the color of water (Botis Nistoran, 2014). In this study were analyzed wastewater from cherries, bean and pears industry, the parameters that are influenced by the pollution and a comparative study between them.

There are many investigations for finding solutions for cheaper treatment, easy disposal and utilization of wastewater from vegetable and fruits cans industry as well as for the factory sludge used as a fertilizer for acid soil (Soderquist *et al.*, 1971).

Materials and methods

Wastewater (effluents) quality was determined by estimating physical, chemical and biological characteristics of waste water. The collected samples were analyzed for temperature, pH, total solids, total suspended solids, bio-chemical oxygen demand (BOD), and chemical oxygen demand (COD), total nitrogen and total phosphorus values.

At the processing plant representative samples from each major unit operation have been collected. The processes monitored included the canning of cherries, bean and pears where a sufficient number of samples were collected to permit statistical analysis of the analytical results.

The green bean canning process involved an initial washing followed by grading, snipping of ends, cutting, blanching and canning. The major contribution to the total pollution load was from washing and grading at the blanching process (Soderquist *et al.*, 1971).

At the cherries, process involved washing, grading, pitting, canning, cooking. The greatest quantity of wastewater results from pitting followed by stemming. The solid in wastewater were largely soluble or colloidal.

At the pear canning process, after initial grading, the pears were peeled and then flumed in sodium chloride brine to fresh water rinsing tank to remove the excess of brine solution. After rinsing the pears were trimmed and then conveyed to the chopping process to be processed for canning.

Results and discussion

Wastewater quality from cherries, pears and beans processing was analyzed establishing the physical-chemical characteristics.

Physical characteristics of wastewater

Temperature

Temperature influences the chemical and biological reactions taking place into the water (Tikariha and Sahu, 2014; Masu and Morariu, 2012).The water discharged from industries has generally a high temperature and affects the land. Temperature of wastewater is between 13-35 °C depending on the weather.

Hydrogen ion concentration (pH)

PH is an important factor expressed as the negative logarithm of the hydrogen ions concentration. The presence or absence of various ions can have direct relation with pH of the effluent. The pH values of wastewater was varied from 3.9-7.1. The pH of wastewater indicates the acidic nature of effluent (Soderquist *et al.*, 1971).

Total dissolved solids

Total dissolved solids varying from 0.71-2.06 kg/t in waste water for the analyzed cases. Increase in concentration of dissolved solids is because of a greater concentration of solid particles from waste.

Chemical characteristics

Biochemical oxygen demand (BOD)

BOD is defined as the amount of oxygen required by microorganism while stabilizing biological decomposable organic matter in waster aerobic conditions (Tikariha and Sahu, 2014; Soderquist *et al.*, 1971). During oxidation organic pollutants are oxidized by certain microorganism into carbon dioxide and water using to dissolve oxygen. BOD is an important parameter that indicates the level of water pollution by the oxidizable matter and the oxygen used to oxidize inorganic materials such as sulphides and ferrous ions.

Chemical oxygen demand (COD)

COD is the oxygen required for chemical oxidation of the organic matter without the help of strong chemical oxidant and it is useful in indicating the toxic condition and presence of biological resistance substances (Tikariha and Sahu, 2014). COD values were higher during the winter and lower in rainy months because of input of higher concentration of organic matter during winter.

Nitrate nitrogen (NO_3^-)

Nitrate nitrogen ranged from 0.0009-0.0049 kg/t in waste water resulted from pears, beans and cherries cans industry. The higher values were obtained due to greater rate of oxidation, while lower value was obtained due to lowering in oxygen, which reduced the rate of oxidation.

Total ortho phosphate (PO_4^{-2})

Total ortho phosphate were between 0.0027-0.03135 kg/t in waste water. Higher and lower value of ortho phosphate was dependent on the amount of discharge and the temperature.

These effluent characteristics for pears, beans and cherries are presented in Table 1 and the comparative results for BOD and COD. which represents the degree of pollution for the waste water from pears, cherries and beans cans industries are presented in Figure 1 and 2.

Table 1. Effluent characteristic for pears, cherries and pears.

Effluent's type	Temperature (°C)	pH	Total dissolved solids (kg/t)	Biochemical oxygen demand (kg/t)	Chemical oxygen demand (COD) (kg/t)	nitrate nitrogen (NO_3^-) (kg/t)	Total ortho phosphate (PO_4^{-2}) (kg/t)
Pears	13.5-17.5	3.9-4.5	0.71-1.29	7.73-10.39	11-19.25	0.0009-0.0018	0.0027-0.0086
Cherries	22-28.5	5-6	0.276-0.811	8-12.48	13.56-21.15	0.0032-0.013	0.0233-0.0305
Bean	25.8-31.3	5.1-7.1	0.71-2.06	1.18-2.49	2.23-4.69	0.0014-0.0049	0.0061-0.0135

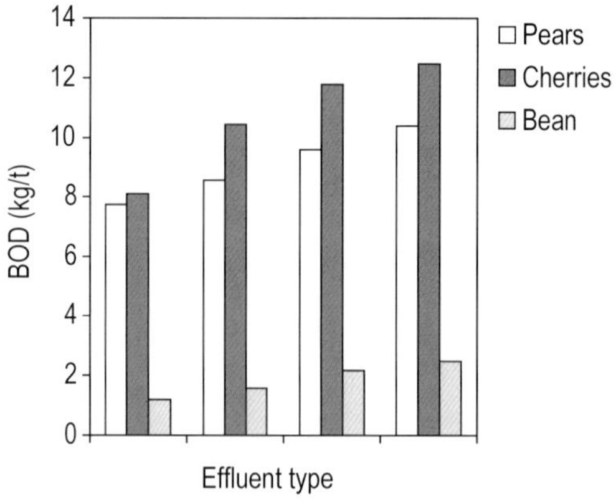

Figure 1. Biochemical oxygen demand (BOD) for wastewater from pears, cherries and bean cans industry.

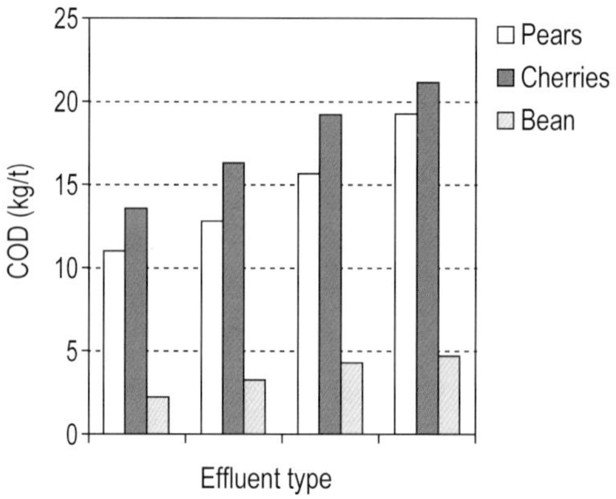

Figure 2. Chemical oxygen demand (COD) for wastewater from pears, cherries and bean cans industry.

Conclusions

BOD and COD values were higher in the waste water from pears and cherries cans industries indicate the polluted nature (Masu and Morariu, 2014). The greatest portion of the pollutant load is generated at the cherries processing, especially from pitting and this stream should be incorporated into recovery system. A high value of BOD and COD destroy the ecological balance from the river and also has a negative influence from societal and ethical point of view. Cherries and pears wastewaters have a larger amount of organic materials, soluble and colloidal, that requires a greater quantity of oxygen to oxidize and decompose them into carbon dioxide and water. The most acid wastewater results at the pears processing and in the case of bean processing, the total dissolved solids was generated during washing-grading process.

By analyzing the two graphics it can be concluded that the higher values for organic substances are at the cherries and pears processing that requires more oxygen for theirs decomposition also indicating a more polluted nature of the resulted wastewaters.

Although the organic substances from these wastewaters are degradable, the low concentration of nitrogen and phosphorus requires adding of some nutritive substances to stimulate the biological purification. If these wastewaters are not cleaned appears an acid fermentation that has a negative influence on the river quality, as a receptor, as well as a negative influence on the aerobe microorganisms that acts in wastewaters treatment (Soderquist *et al.*, 1971). Wastewaters treatment consists, in this case, from sand and sludge removing in a sedimentation tank and biological treatments in bio-filters.

References

Botis Nistoran, M. (2014). Ecotechnologies and recycling in food industry. University De Vest, Timisoara, Romania, 130 pp.

Masu, S. and Morariu, F. (2012). Recycling biotechnologies of the residual products. University Politehnica, Timisoara, Romania.

Soderquist, M.R., Blanton, G.I. and Taylor, D.W. (1971). Characterization of fruit and vegetable processing wastewaters, ASAE Paper No. 71-418, American Society of Agricultural Engineers, St. Joseph, USA.

Tikariha, A. and Sahu O. (2014). Study of characteristics and treatments of dairy industry wastewater. Journal of Applied & Environmental Microbiology 2(1): 16-22.

Urban Elika, towards a better use of food resources in EU urban environments: focusing on decision making processes

L. Escajedo San-Epifanio[1], M. De Renobales[2], P. Dapena Montes[3] and I. Glz-Baskaran[3]*
[1]University of the Basque Country/EHU, Constitutional Law and History of Political Thought, Faculty of Social and Communication Sciences, P.O. Box 644, 48080 Bilbao, Spain; [2]University of the Basque Country/EHU, Biochemistry and Molecular Biology, Faculty of Pharmacy, 01006 Vitoria-Gasteiz, Spain; [3]Sortarazi Association (NGO), P.O. Box 644, 48080 Bilbao, Spain; leire.escajedo@ehu.es

Abstract

Food insecurity and hunger are growing problems present also in the wealthy developed world. Financial crisis, market instability and increasing food prices contribute to aggravate the problem. According to Caritas-Europe, 79 million people are food insecure in the EU. That happens while the current food wastage in EU27 is 179 kg per capita (89 million ton per annum). Reducing food waste through donations or redistribution of surplus food to people in need or food banks is a frequent option. In fact, a kind of food security net has rapidly emerged in the EU, collecting surplus food and distributing it to target population. Urban Elika, a collaborative and participatory forum funded by the 'University of the Basque country', aims to contribute to a better use of food resources in EU urban environments. Among other research areas, the group analyzes the ethical dimensions involved in the range of recovery practices that are currently collecting food surpluses and offer them to food insecure people. It is crucial to realize that many ethical issues over the short, medium and long-term arise from this kind of redistribution of surplus food. In this paper we describe some of these issues. In the short term, successful emergency relief has to be rooted in thoughtful decision making despite the compelling urge to give help under the pressure of extreme circumstances. There are legal and safety concerns to be addressed and there is a lot to learn from those NGOs and charities that provide hunger relief and at the same time develop programs to promote human rights while respecting and dignifying each individual. In the medium and long term, the focus has to be on the ways to promote nutritional inclusion. Hunger is a political issue. Hunger is a failure of our current food system, probably one of its main failures, and both food waste and food poverty need the full attention of the public and governments. It is not only a question of managing surplus food. Unless our efforts tend to reorganize the primary food system, the food insecure people will continue to be excluded.

Keywords: food insecurity, food waste, nutritional inclusion, public participation

Giving away or redistributing surplus food is not a simple matter

Food insecurity and hunger are growing problems also in the wealthy developed world (Haddad *et al.*, 1999; Lang, 1999; McGill, 2009; Warshawsky, 2001). The combined consequences of the financial crisis, market instability and increasing food prices severely aggravate the situation. According to Caritas-Europe, 79 million people are food insecure in the EU (Caritas report, 2014). Simultaneously, current food wastage in EU27 is 179 kg per capita (89 million ton per annum), as several EU institutional publications, reports and statistics stated in the last two years. 'The food banking solution to food waste and hunger' (FEBA, 2013) proposes to reduce food waste by donating or otherwise redistributing surplus food to people in need or to food banks. However, in our opinion, this is not as straightforward as it might appear at first sight. It is crucial to realize that this kind of redistribution of surplus food raises many ethical issues over the short, medium and long-term.

Why is the matter not as simple as it seems? Referring to another contemporary food issue – obesity-, bioethicist Ben Mepham said that 'our rapidly changing world is now revealing a succession of developments which, although when viewed in retrospect seem to have been emerging gradually over a substantial period of time, now threaten to assume critical status within a few years' (Mepham, 2010). Undoubtedly, food waste and current food insecurity rates are two phenomena that have emerged gradually over the last decades and have now reached a scandalous level. Using food surplus to feed people in need should force us to pay attention to two very different underlying situations: food waste and increasing food insecurity.

It is a well-known fact that for several decades our food system produces more food than we need and simultaneously an increasing number of people do not have sufficient access to that food, mostly because they are priced out of the food system Both facts are the negative consequences of a number of actions and decisions at various levels which we need to identify if we do want to correct them to move towards a more sustainable and equitable society. If we focus only on how to manage the surplus food, we mask food insecurity and we renounce to make political decisions to correct the situation. Besides, we have to take into account that there are very different kinds of food banking strategies, from local to international levels, and their impact on the food system has positive and negative effects that are to be assessed. We take the risk of depoliticizing those issues, as if no decisions or human actions were behind those facts. In the long term, such attitude can take us further away from an equitable distribution of food resources. No one becomes included by receiving free handouts (Busatto, 2007); social integration involves promotion of human rights, equality of opportunity and participation for all people, including disadvantaged and vulnerable groups and individuals.

In this project we identify and analyze how decisions regarding immediate food attention are made in EU urban environments and their consequences in the short, medium and long term to reduce food poverty. Thus, we carefully take into account the considerations of entities involved in urban food aid. First of all we will refer to those elements contributing to a reasoned decision-making process regarding immediate food attention based on a human rights approach. There are legal and safety concerns to be addressed (see below) and there is a lot to learn from those NGOs and charities that give hunger relief and at the same time develop programs to promote human rights while respecting and dignifying each individual.

Hunger is a political issue. It is one of the main failures of our current food (and economic) system and it needs the full attention of both the public and governments. The dramatic scale of food insecurity is helping to make it more visible, despite some governments – like the Spanish one- maintaining that poverty rates are just statistics that may be read and interpreted in many different ways (press conference on the Caritas Spain report, Spanish Minister of Economy, 28 March -2014). Reducing food insecurity is not only a question of what to do with surplus food. From a human rights approach, the question should not be whether we may reduce food poverty by using surplus food, but rather why we accept a food system that excludes millions of people when it has been shown that it produces much more food than we need to feed everyone (Cotugna and Beebe, 2002; Lambie-Mumford, 2013; Riches, 2002). No matter how creative we are to manage the surplus food we produce, unless our efforts tend to reorganize the primary food system so that food insecurity will disappear, the food insecure people will continue to be excluded from our food system.

Hunger relief is needed, but not just in any way. Identifying the best practices

The goal to help people in need sounds straightforward, but professionals and volunteers that are immersed in everyday practice find many obstacles. An outsider who has never been actively involved in that kind of activities can easily forget about the economic, political and social context in which social problems are defined. Shortcomings in human and financial resources are also to be taken into account.

In the last five years the amount of formal or informal charities that have emerged in Spain to attend food insecure people is impressive. Sympathy for the people suffering the crisis has promoted solidarity and altruism in many ways, and food donation is probably one of the most outstanding activities (Lledó, Simón and Bilbao-Urquidi, and Caritas-Spain, personal communications).

The Urban Elika research group strives to identify good and not-so-good practices to improve emergency hunger relief programs, and to help those immersed in its daily practice. Taking food surpluses and giving them away is a complex process. The characteristics of each type of food (i.e. packaged and non-packaged, perishable or not, refrigerated, cooked) need to be carefully considered to manage them in a way that guarantees their hygienic safety. It is crucial to develop adequate, safe protocols. The NGO 'Nutrition without borders' in collaboration with the government of Catalunya and some other entities have developed a clear and easily understandable guide to address some of those issues (Generalitat de Catalunya, 2013).

Besides the food safety, there are other issues involved in emergency hunger relief. According to D. Andrew, our food choices shape the world and the way we provide food access to the food insecure involves many food choices as well (Andrew, undated). Successful hunger relief has to be based on a thoughtful decision-making aimed, as its final goal, to the social inclusion of the food insecure people. In this sense, we are identifying the good practices of some charities and NGOs that not only give out food but also develop programs to create conditions for the beneficiaries' food inclusion. These programs offer empowerment, nutritional counseling and assistance to face the overall and complex social situations of food insecure individuals and families. The work carried out by nonprofit solidary grocery shops is worth mentioning. Managers of these shops reach agreements with (mostly) local providers to buy basic food items at a substantial discount. In some instances, concerned supporting individuals purchase those food items at regular market prices. As a result, the solidary grocery shop can sell those food items to food insecure people at a price considerably below market values. It is important to realize that food items are sold, not given out free. Food insecure people pay according to their small incomes, thus dignifying them. This model hardly deviates from the primary market. Thanks to the joint efforts of providers, concerned supporting individuals, and volunteers, and to the nonprofit character of these shops, food insecure people can access the food items they need in a dignified manner.

How can we contribute to a decision-making framework that takes into account medium- and long-term issues?

In our opinion, medium- and long-term answers to food waste and food poverty must also regard the ethical issues arising from our local, national and global food systems. Otherwise, it is likely that the decisions taken may result in even more, rather than less, food insecurity. As we have already said, food waste and food poverty are consequences of a wide range of decisions, actions and also lack of actions occurring in the last decades. Although food waste and food poverty may seem to be somehow linkable, they are not and both need to be separately studied in depth if we truly want to reduce their incidence in the next decades. Food insecurity occurs when some people do not have economic access to sufficient, safe, nutritious food (FAO, 1996), regardless of how much food is wasted just around the corner. If in the long-term actions directed to contain food surplus do not result in social inclusion of food insecure people, we run the risk of keeping them excluded from our primary food system.

Riches and Silvasi shared this view after analyzing the outcomes of thirty years of food banking in Canada (Riches and Silvasi, 2014). Some actions that are efficient as emergency hunger relief may be counterproductive if we keep carrying them on too many years. In their paper, these authors compare experiences in different countries. Rather than a temporary response to the effects of the recession, some food aid tools have been corporatized – that is, those tools are now dictated by large food multinationals- and are somehow institutionalizing a kind of secondary food market.

To adopt policies for an intermediate future, a human rights approach requires identifying the causes of the current food insecurity in our urban societies. Nutritional inclusion will not happen if food insecurity is only seen as a way to cope with food surplus, as some institutions are doing nowadays. For example, the EU parliament called in 2012 for urgent measures to halve food wastage in the EU, because up to 50% of edible and healthy food is discarded in EU households, supermarkets, and restaurants and along the food supply chain each year. Likewise, the Directorate-General for Internal Policies of the Union published in 2013 a paper titled 'Technology options for feeding 10 billion people. Options for cutting food waste' (Lowe *et al.*, 2008; Priefer *et al.*, 2013). From the first line of the text, the reduction of food waste is described as an important lever for achieving global food security. The remainder of the proposed measures are mainly directed to improve the food supply chain (i.e. revising the regulations on food date labeling, awareness campaigns, and economic incentives). Is it hunger just a problem of food logistics? We think a deeper analysis to see the lights and shades of those proposals that promote the donation of food surplus is needed, including food production parameters (amounts, localization), distribution and final consumer prices. Our decisions about food waste should bring us to envision food systems that are secure and fair for all human beings.

Acknowledgements

Urban Elika is funded by the University of the Basque Country (University & Society Research Grant Call for 2014-2016). Leire Escajedo is also member of the Research Team Grant of the Directorate on Scientific Policy of the Basque Country IT 743-13 'Multilevel constitutionalism and integration of diversity (political, legal-institutional and cultural-religious)'.

References

Andrew, D.(undated). Eating is a moral act. NCRLL reports and presentations.

Busatto, C. (2007). Solidary governance for creating inclusive societies. A contribution to the expert group meeting on 'creating inclusive society: practical strategies to promote social integration'. Expert group meeting, 10-13 September 2007, Paris, France.

Caritas report (2014). The European crisis and its human costs – a call for fair alternatives and solutions. Available at: http://tinyurl.com/mnc375e.

Cotugna, N. and Beebe, P.D. (2002). Food banking in the 21st century: much more than a canned handout. Journal of American Dietetics Association 102 (10): 1386.

European Federation of Food Banks (FEBA) (2013). Every crumb counts, food wastage declaration. Available at: everycrumbcounts.eu.

FAO (1996). Rome declaration on world food security. World food summit, 13-17 November 1996, Rome, Italy. Available at: http://tinyurl.com/q4e77hb.

Generalitat de Catalunya (2013). Guia de prácticas correctas de higiene para el aprovechamiento seguro de la comida. Available at: http://tinyurl.com/mwly4br.

Haddad, L., Ruel, M.T., Garrett, J.L. (1999). Are urban poverty and undernutrition growing? Some newly assembled evidence. World Development Journal 27(11): 1891-1904.

Lambie-Mumford, H. (2013). 'Every town should have one': emergency food banking in the UK. Journal of Social Policy 42: 73-89.

Lang, T. (1999). Food policy for the 21st century: can it be both radical and reasonable? In: Koc, M. (ed.) For hunger-proof cities: sustainable urban food systems. IDRC, Orrowa, Canada.

Lowe, P., Phillipson, J. and Lee, R.P. (2008). Socio-technical innovation for sustainable food chains: roles for social science. Trends in Food Science & Technology 19(5): 226-233.

McGill, A. (2009). Food security: local and individual solutions to a global problem. In: Barbosa-Canovas, G., Mortimer, A., Lineback, D., Spiess, W., Buckle, K. and Colonna, P. (eds.) Global issues in food science and technology. Elsevier Inc., New York. USA, pp. 43-64.
Mepham, B. (2010). The ethical matrix as a tool in policy interventions: the obesity crisis. In: Gottwald, F.T., Ingensiep, H.W. and Meinhardt, M. (eds.) Food Ethics, Springer, New York, USA, pp. 17-29.
Priefer, C., Jörissen, J. and Bräutigam, K.L. (2013). Technology options for feeding 10 billion people. Options for cutting food waste. EU parliament. Directorate-General for Internal Policies of the Union.
Riches, G. (2002). Food banks and food security: welfare reform, human rights and social policy. Lessons from Canada? Social Policy and Administration Journal 36(6): 648-663.
Riches, G. and Silvasti, T. (2014). First world hunger revisited: food charity or the right to food? Palgrave MacMillan, London, UK.
Warshawsky, D.N. (2001). Urban food insecurity and the advent of food banking in Southern Africa. AfricanFood Security Urban Network (AFSUN). Available at: http://tinyurl.com/l3rl5w9.

Author index